D1346142

STÅLFELT'S
LANT ECOLOGY

Plants, the soil and man

A Halsted Press Book

STÅLFELT'S PLANT ECOLOGY

STÅLFELT'S PLANT ECOLOGY

Plants, the Soil and Man

Translated by Dr Margaret S. Jarvis
and Dr Paul G. Jarvis

A Halsted Press Book

John Wiley & Sons
New York

LONGMAN GROUP LIMITED
London
Associated companies, branches and representatives
throughout the world

First published in Swedish under the title *Växtekologi*.
This English translation was first published in 1972.

Published in the U.S.A. by Halsted Press, a
Division of John Wiley & Sons, Inc. New York

Stålfelt, Martin Gottrfid, 1891–1968
  Stålfelt's plant ecology.

"A Halsted Press book."
Translation of Växtekologi.
  Bibliography: p.
  1. Botany-Ecology.   2. Plant-soil relationships.

I. Title.
QK901.S7813   581.5'26
ISBN 0 470 81970 7   72-8715

Printed in Great Britain by
William Clowes & Sons, Limited
London, Beccles and Colchester

Translators' note

Professor Stålfelt wrote the following preface to the first Swedish edition in 1960, thus clearly emphasising his considerable perception regarding future trends in the development of ecology. Well in advance of the current widespread concern about the destructive impact of man on other organisms and on the environment, he wrote a fundamental text dealing with the interactions between plants and the environment, the main emphasis of which is on 'those aspects of plant ecology which have been or can be disturbed by man'.

Professor Stålfelt treats in detail, in the classical manner, inter-actions between plants and environment, especially the soil environment, with particular regard to the stability of systems and their re-actions to disturbance. It might be thought that in the 1970s his original text would be too dated. However, no comparable text has appeared in the English language. In presenting this detailed treatment of biological processes, Stålfelt has drawn upon a huge body of literature, especially literature of Scandinavian and German origin. This synthesis of information is as valuable now as it was then, especially to those English-speaking ecologists who are attempting to understand the functioning of ecosystems and to the University student of ecological processes, for whom this book was primarily written.

In this translation, a certain amount of repetition has been removed, a few additional references added, and the statistical data brought up to date, but no comprehensive revision has been attempted. During the translation of the first chapters, Stålfelt saw and approved the English text and added material that he was proposing to include in a revised Swedish edition. Unfortunately he died in 1968, before the revision had progressed very far, and so the text closely follows that of the second edition.

Preface to the First Edition

It has long been known that living things are dependent on their environment and on the changes which take place in it. Plants and animals are dependent on the properties of the soil, the water, the topography of the area, the climate of the region, neighbouring organisms, and the other features of the environment. Different plants and communities may live their lives in the same geographical region but not necessarily in the same environment. In a more or less stable plant community, a system of equilibria has developed between the plant and the environment, i.e. between the biological-chemical-physical processes in the environment and the plant's reaction towards these processes. This equilibrium system changes in response to changes in the environment. For example, a qualitative or quantitative change in the composition of the vegetation results in changes in the processes taking place in the soil, and consequently changed conditions for the plants and other organisms in the area.

The aim of ecology is to study the relation between the environment on the one hand, and the existence and development of individual organisms and communities on the other. Plant ecology, like animal ecology, seeks to analyse this relation and to record the changes which are taking place and which have taken place in the past.

Animal life depends on plant life. Plants use solar energy for manufacture of organic compounds from which animals and men are constructed and from which they derive energy. Thus man and animals obtain the necessities of life from plants. As animals, particularly man, have increased in numbers, their demands have increased to such an extent that the consumption of plant material now equals or exceeds production. Man alone is generally responsible for this over-consumption. Damage caused by animals has always been local and temporary; only man has been capable of organised exploitation over ever-increasing areas causing permanent damage to an extent significant for production as a whole.

Man has made use of most natural resources but none as extensively as plant and animal communities. Where man has settled he has consumed or cleared the natural vegetation, ploughed up the soil, sowed and harvested, planted and felled. These external and visible changes have set in motion other processes which cannot easily be observed and measured but which in the long run have altered the productive capacity of the land, e.g. changes in the amount and structure of the soil, its content of air, of water and of nutrients, its crumb structure and content of living organisms, i.e. the animals, fungi and bacteria by whose activity the condition of the land is built up and maintained. At the same time, the interference has upset the equilibria, built up over centuries, between the plants and the environment and the resulting disturbance has meant the decimation or total disappearance of many species, and possible establishment of a number of species undesirable to man.

For a long time the changes have escaped notice, partly because they have been relatively slow, partly because they have been masked by changes in detail. Changes observed by one person in a lifetime may not often be spectacular enough to seem significant. Only by putting together biological, geological, pedological, historical and ethnographical observations can the changes which have taken place over several centuries be assessed. Research in this century has brought a degree of order to the details and allowed the main features of the changes to become apparent. It has become clear that many of the procedures by which man has obtained the necessities of life from plants are destructive in effect. However, it has also become apparent that much of the damage can be mitigated or avoided if its nature and origin is understood; and that some of the damage already done can be repaired.

As far as possible, the presentation in this book has taken account of historical changes, especially when they affect the biological-chemical-climatic balance system between plants and their environment. Thus a relatively large amount of space has been devoted to those factors which have changed most in historical time, as a result of man's activity, and those concerned with the soil environment, particularly the microbiology of the soil.

The role of organisms and plant organs living beneath the soil surface, bacteria, fungi, root systems and underground stems, is often forgotten, although this group of plants is very rich in species and in numbers of individuals and has an all-important effect on soil processes, on soil productivity and on the plant life above the soil. A study of the ecology of plants becomes, to a large extent, a study of the ecology of the edaphic flora, i.e. of the flora living below the soil surface. Factual information about this is mostly scattered in recent specialist literature.

The title of the book *Plant Ecology* is misleading, in so far as not everything covered by the term plant ecology is included. Some of the best-known aspects, discussed in most botany textbooks, have been omitted, e.g. pollination and seed dispersal, and, to some extent, the role of light and temperature. Instead, most weight has been laid on those aspects of plant ecology which have been or can be affected by man. It is of special significance for man to be aware of these.

The aims of this book are therefore as follows:

to describe how plants live in a state of dependence on one another and on the soil, water, animals and, in general, on those features of the environment which can be changed by man;

to describe how the processes of decomposition by the soil micro-flora are in equilibrium with the processes of synthesis by green plants;

to describe how changes in this equilibrium can lead to increase in soil fertility, or vice versa;

to describe how the plant world builds up its reserves of nutrients and soil, and how in the course of time it has assembled its own capital of nitrogen and of salts and to a large extent built the warehouse in which the capital reserves are stored;

to show how the mobile equilibrium systems between the plants and the environment are built and how these systems work together in the accumulation of the capital;

to show which systems are stable and resistant to change and which are labile and sensitive;

to show how various kinds of outside interference can shift the position of equilibrium and favour or hinder accumulation;

to show how the plant world, if possible, makes good the damage and renews its productivity, and how the healing process can be helped or hindered by outside interference;

to describe the interaction between the plants themselves, and between the plants and the environment, and how this can result in a production system which with moderate taxation functions with a net profit so that there is a positive yield;

to describe how this production system is exploited by man and animals;

to describe how man, knowingly or unknowingly, changes the environmental conditions and equilibria so that the nutrient capital sometimes increases, sometimes is dispersed;

to describe how man, when he exploits the plant world, is not always satisfied with the annual production but sometimes harvests part of the accumulated capital; and how man often fails to distinguish between capital and interest;

to give an account of the historical change in productivity from ancient times up to the present day, of the food potential of the plant world for man, of the part which has been exploited already, and partly consumed, and the part which remains still unused.

At the moment these questions are mainly economic, but in the long term they have sociological implications, e.g. in the design of built-up areas. They are important to nature conservation. The nature conservation movement, which developed as a result of over-exploitation, must analyse environmental situations in order to find suitable methods for conservation and restoration. Interest in conservation may be considered by many people as an interest in curiosities or an expression of enthusiasm for nature, combined with lack of realism. But its basic aim is nothing less than the careful husbandry of the most important capital resource of man, namely the living natural environment, with what it has built up in the past, which man has partly exploited in recent centuries and which can now be exploited even more rapidly and severely because of increasing technological knowledge.

Contents

The plant environment

Habitat and the plant community

Like other organisms, plants are dependent on their surroundings, on the living and physical components of the environment, generally called 'the habitat'. The term 'habitat' refers to the place where the plants occur. It may apply to the specific areas within a district in which the species is found; it may mean the whole region in which a species or group is distributed; or it may simply be that piece of ground occupied by a single plant. The habitat may be a tussock, a mountainside or a lake. Habitat and geographical locality are not the same: one geographical locality may offer several different plant habitats side by side. For example, if one species grows in the shade cast by another then the environmental conditions and, thus, the habitat are different, although the geographical locality is the same; or two herbs growing beside one another, one surface-rooted and the other deep-rooted, have different sources of supply of water and nutrients and are, in a sense, growing in different habitats. The biological term 'habitat' therefore embraces all the conditions—edaphic (from the Greek *edaphos*—soil), climatic, and biotic—which influence the organisms growing there. Each plant in its habitat is affected by microclimatic factors, such as temperature, light, humidity, wind, etc.; by edaphic factors, such as soil water content, nutrients, acidity, etc.; by the proximity and behaviour of other organisms; and by the height of the locality above sea level, the slope of the ground, etc. All these are part of the biological habitat which they characterise in a broad sense.

The characteristics of the habitat are constantly changing. In the soil, a large number of physical, chemical and biotic processes are going on, and these affect the plants. In the air, there are changes in the temperature, light and wind. The habitat is dynamic, it is like a kaleidoscope, where the combinations and proportions of the parts are continuously changing. The plant may exploit the variations as long as they are advantageous, but if the effects are unfavourable it may avoid them by some kind of adaptation. It may modify its characteristics, for example, by increased frost or drought resistance, or genotypic variation and selection may lead to the development of more resistant variants. At the same time, the plants themselves influence the habitat, for example, by acting as windbreaks, by shading or by altering the properties of the soil. Hence, there is another element in the interplay of forces: competition or collaboration between individuals and between species.

Whether a species can establish itself in a new area and spread there, or can retain its position in an area already colonised, depends to a large extent on its ability to exploit the environmental factors. Success is achieved only to the extent that the characteristics of the habitat fit the characteristics of the species, and to the extent that the species has a greater ability than its competitors to make use of the resources offered by the habitat and to avoid or to adapt itself to those which are unfavourable. There are therefore certain inevitable links between the conditions of the habitat and the requirements of the species in it.

The plant community

Each habitat attracts the species with requirements that it can satisfy. There is selection among the plants whose fruits, seeds or other dispersal units are distributed within the habitat. Habitats of the same type therefore come to support plant communities with a broadly similar composition. This recurring combination of species gives the impression of a community, and gives a particular character to the vegetation. It is mainly these selection processes which are the basis for the development of the associations of species which are called 'plant communities'. These are often so easily distinguished that they have long had special names, such as 'heath', 'forest', 'bog', 'fen', etc. They can usually be recognised on sight. Within these units other smaller ones can be distinguished. There are, for example, *Calluna* heaths, *Erica* heaths, lichen heaths, etc., and the term 'heath' thus really covers a group of plant communities. The study of the composition, association and distribution of plant communities is called 'plant sociology' and is a part of plant geography (for further reading see Sjörs, 1956; Hanson, 1958; Greig-Smith, 1964).

The aim and methods of plant ecology

Every detail of the composition and development of vegetation is dependent on the environmental factors. On the other hand, several of these factors are themselves dependent on the vegetation. A tree needs light for production of its food materials, but at the same time it shades other plants in the vicinity; its water balance is affected by the wind, but at the same time the tree reduces the wind velocity; it uses up some of the soil constituents and provides humus in exchange, etc. In this way, a reciprocal relationship is built up between plants and their environment, and this can lead to change in the conditions of the habitat and, therefore, to change in the conditions experienced by the plants. The exploration of this reciprocal relationship is the aim of plant ecology.[1]

The word 'ecology' is derived from the Greek word *oikos*, which means house, home or dwelling. Plant ecology is the study of the environment affecting the plant, the way in which the plant utilises natural resources and the result of this utilisation. Thus, it is the study of the significance of the various environmental factors to a plant species or community and of the ability of the species or the community to exploit or to control them. It is the study of the means by which plants succeed in their habitats, including adaptation, competition and collaboration, and, finally, of the result of the reciprocal relationship between plant and habitat reflected in the spread or the decline of the species or the group. Plant ecology is therefore the study of the relation between the plants and the environment and its significance for the occurrence, development and distribution of the plant in a particular habitat or in a wider geographical area.

When ecological research is restricted to the investigation of single species and their habitats, it is concerned with a relatively small number of environmental factors and plant characteristics, but when it is concerned with the study of plant communities, there are numerous environmental factors, plant characteristics and possible combinations between them. In theory, innumerable combinations might be expected, bringing insurmountable difficulties in unravelling the situation. In fact, the analysis is simplified because some of the characters of the habitat, or some of the common features of the components of the vegetation, are so pronounced that they dominate the ecological relationships, so that the same relationships are often encountered again and again. The study of plant communities therefore seeks to reveal these common features

[1] Haeckel, *Generelle Morphologie der Organisme*, Berlin, 1866, defined 'ecology' as the science of the relationship between organisms and the environment, using the term 'environment' in a wide sense to include all the factors affecting the existence of the organisms.

and their significance to the formation and development of the community.

Variation, selection and the formation of new races all play an important part in the reaction of the plants to the environmental factors, both in adaptation and in competition. Progress in the struggle for survival often depends on whether the species can develop suitable races.

The problems of plant ecology are the same as those of the biological aspects of crop production in agriculture, horticulture and forestry. In plant ecology and in crop ecology the problems are concentrated around the magnitudes, proportions, combinations and changes of the environmental factors and the effects they have on the plants, i.e. on yield, on the balance between vegetative growth and flowering and fruiting, on regeneration, on spread or decline, on resistance to unfavourable conditions, etc. It is only possible to compensate for deficiencies in the natural properties of the locality by cultivation, manuring or crop rotation when the properties of the habitat and the requirements of the cultivated plants are known. The possibility of growing plants in an area where they do not occur naturally depends on the successful elucidation of these conditions and the development of methods of compensation for the deficiencies in the habitat. Some of the oldest botanical work, and some of the oldest of all scientific work, comprised investigations of this kind. Hitherto, it has been studies made in the context of agricultural or forestry research which have made the greatest contribution to the development of plant ecology. Plant ecology is therefore one of the foundation stones of economic crop production.

In addition, there is another practical aim of plant ecology and that is to provide a scientific basis for nature conservation. Knowledge of the laws governing the natural equilibria between species and between vegetation and its environment is essential for the assessment of the consequences of any interference with nature. In many cases, interference by man has led to undesirable changes, for example, the harvest of economically valuable species has resulted in the development of vegetation which is less desirable. In other cases, plant communities have disappeared which ought to have been retained, both from economic and aesthetic points of view, and vegetation may even have been completely or partly destroyed. Large areas of the earth's surface have been converted to infertile land because man's interference in the balance of nature has been so drastic and so long-lasting that the vegetation has not only been displaced, but there is no longer any possibility of natural recolonisation: plant ecologists must then try to find ways and means of repairing the damage.

Historical aspects

Since written records began, man's existence has always been dependent on the utilisation of plants. Only plants are able to build up organic material from inorganic units, and to produce the substances from which the bodies of man and animals are built up and maintained. Even if the food itself is animal, then it is derived from plants. A rich animal life requires species-rich and luxuriant vegetation, which together provide the conditions for man's existence. The oldest cultures probably arose in areas so richly endowed by nature that there were sufficient plants and animals to support a large population for a long time. It is in just this sort of area that the remains of ancient civilisations have been found: the Babylonians and Egyptians, on fertile flood deposits; and the Aztecs and Mayas in rich and fertile areas of South and Central America. The fragmentary remains of these civilisations bear witness to the close connection between the lives of the people and the plant world: grain has been discovered in the graves of Egyptian kings; various decorative plants were used for beautifying the graves and the places of religious significance or used as motives for decoration in architecture or handcraft; and plants were used medicinally, or for making clothing and household goods, showing that the people used the products of the plant world for similar purposes as they are used today (Täckholm, 1941; Jessen, 1948).

Old documents often include accounts of the plant world. The first occupations mentioned in the Bible are those of the farmer and the herdsman. Orchards and fertile fields are praised by Homer and Hesiodos (seventh century BC). Hippocrates (460–377 BC) named more than 200 herbs used in medicine and in his writings are found the first attempts at a detailed description of the physiological characteristics of plants.

Aristotle (384–322 BC) was the first to carry out research in the way which has developed into presentday scientific method. Like the other Greek philosophers of the time, he sought to achieve understanding of the world of material objects by speculation, but he also used his insight and his experience of the objects in a way which was new and which eventually became the keystone of science. He made observations based on the senses of touch, taste, smell and sight, and on the results of tests such as boiling or burning. By these simple means he was able to suggest several important principles—for example, that all living things obtain their food from their surroundings, and that their food consists of several things, even if it seems to be only one thing, as in the case of plants and water, because soil is mixed with the water.

Aristotle used both his philosophy and his ideas on how man and animals feed and grow in interpreting his observations. He thought that all living organisms, plants included, had a soul, and that the soul was the basis and the reason for existence in the material world. He thought that 'plants find their food in the soil and since they stay still and do not run about they do not need to carry food with them in their bodies like animals do. They have no stomach, and the roots take up food from the soil in the same way as the vessels in animals take up the nutrient juices from the stomach' (after Jessen, 1948, pages 17–18). The food consisted of soil and water and it was prepared and converted in the soil by a heat force; from this food the soul built up the material from which the plant is constructed. The plant itself was thought to play no part in this process, although the plant governed the disposal of the material to fruits, seeds or other organs. These ideas imply that plants depend on their environment for soil, water and heat, which are termed 'environmental factors' in presentday terminology.

The connection between plants and their environment was given further consideration by Theophrastus (371–286 BC), some of whose writings are concerned with further development of Aristotle's ideas (see Hort, 1949). In one of his works, Theophrastus gave a detailed description of the dependence of plants on their environment:

The habitat and the air [climate] also have a large effect. Many trees —palms, pomegranates, etc.—which in Greece are infertile or have only poor fruit, are fertile with good fruit in Babylon and Egypt, whether they are taken there as seeds or as cuttings. On the other hand, there are others from colder regions which do not flower there. Soil and treatment are also important. However, the idea that wheat can grow from barley and barley from wheat, even in the same field, must be regarded as a myth. . . . Every plant prefers a particular kind of soil, and does best on it. This is why every region has its own characteristic plants. . . . Position and climate have an important influence on shoot growth. Where there is mild, clear air, as in Egypt, shoots develop in a short time, but in places with a raw, colder climate they take longer. In places like Egypt, many plants retain their leaves [for example, figs and vines], whereas they lose them in the winter in our area. . . . A winter with a lot of snow or rain, but no ice, is the best. Copious rain is good at the time of sowing, but is best for trees in the winter, just before bud break, and after flowering. Cold winds are favourable to most plants, except in spring. Soil which is too rich can be harmful. Very salty water is harmful, but many plants like some salt in the water—date palms, cabbage, turnips, etc. Clean drinking water is not as nutritious for plants as that containing some decomposing material. Each tree seeks the locality which suits it best. It is thought that the fruits become good to eat partly through heat and partly through cold, but there is no need to assume that the same phenomenon has different causes. The effect of cold in making fruits good to eat is probably some sort of rotting effect. Trees that grow in shady places or in dense stands are tall and slender. Their wood is looser, moister, and weaker, because it is not made dense by the sun and wind, or by cold. Because of this moistness and weakness, figs, dates and almonds drop their unripe fruits too readily. In figs this is prevented to some extent by small flies which suck up excess moisture. (After Jessen, 1948, pages 23–35.)

Theophrastus discussed many environmental factors—precipitation, temperature, light, soil, water, salts, insects, etc., and many of his conclusions are still valid today. His main aim was to clarify the problems which were important in agriculture or horticulture.

For nearly two thousand years the ideas of Aristotle and Theophrastus were accepted by learned men. Linnaeus (1751) in his *Philosophia Botanica* followed Aristotle, and wrote that the soil is the stomach of plants ('*plantarum ventriculus est terra*'); and (1749) he also suggested that soil is taken up by tree roots. It was a long time before Aristotle's approach—basing conclusions on the results of observation and experiment—was used again. Instead, during the Middle Ages, natural phenomena were explained on the basis of speculation and intuition; and science was invested in magic and mystery.

During the seventeenth century, there were attempts to extend and develop the ancient hypotheses by observation and experiment, mainly by a series of French workers. By that time, a procedure for distillation had been learned from the Arabs, and this provided a new method for investigating the composition of plants. The results were in general agreement with Aristotle's conclusions. Among the products of distillation were 'air' (the escaping gases); 'the soul' (invisible like the 'air', but with a penetrating smell, mainly organic acids which distilled over with the water); 'oil' or sulphur (the oily, inflammable products); 'salt' (the soluble, crystalline part of the residue); and 'soil' (the insoluble part of the residue). The Irishman Robert Boyle (d. 1691) was one of those who carried out distillation experiments. These observations and others like them affected eighteenth century plant ecology in that oil was recommended as a nutrient for plants.

In this way, important conclusions about plant composition were reached, but it was more difficult to get to grips with the major problem of how the constituents were formed. Knowledge about growth of the human body, and of human food requirements, proved to be more of a hindrance than a help, since nutrient uptake by plants was assumed to be analogous with that by man, as it had been by Aristotle. The problem was thought to be to find in plants the organs and processes which were known in man. Plant water uptake, and water movement in the plant were thought to be

equivalent to the circulation of the blood. By the end of the sixteenth century, Cesalpino (d. 1603) had worked out a detailed description of plant nutrition and nutrient uptake along these lines.
The postulated analogies between plants and animals seemed so
natural and so obvious that later research indicating that other
principles were involved was not readily accepted.

Water, soil, humus and salts

Water. The first experiments to find out the importance of water
as a plant nutrient were carried out by the Belgian J. B. van Helmont
(d. 1644), according to Sachs (1875). He planted a *Salix* shoot in a
pot and watered it for five years, during which time it increased in
weight from 2·3 kg to 74·4 kg (5 to 164 lb) while the soil in the pot
decreased only 57 g (2 oz). Since only water had been added, van
Helmont concluded that all the plant material, both the com-
bustible constituents and the ash, was built up from water by the
action of the 'life force'. This interpretation was in accordance
with contemporary ideas of alchemy, but even so it met with some
opposition. The French physicist Mariotte (d. 1684) was among its
opponents; he sought to demonstrate that it was the substances
taken up with the water which were converted in the plant and
which were its source of food.

J. Woodward (1699), an Englishman, demonstrated an important
principle by showing that in three months a *Mentha* plant took up
and lost about forty-six times more water than its own water
content, while another Englishman, S. Hales (1727), confirmed
this by extensive experiments, and so it was established that most of
the water taken up from the soil is lost in transpiration. However,
this was not generally accepted since it contradicted the current
idea of the plant sap circulating through the organs along a definite
pathway, analogous to blood circulating in animals.

In 1804 Theodore de Saussure, a Swiss, first showed that water
was a component in the synthesis of plant material; he studied
carbon dioxide assimilation using quantitative methods and found
that the increase in plant weight was greater than that equivalent to
the amount of carbon dioxide taken up and could only be attributable to water taking part. This was subsequently confirmed by two
agricultural chemists, a Frenchman, J. B. Boussingault (1841) and
a German, J. von Liebig (1840).

Soil. While water was thought to be the real food of plants, soil was
considered significant only as a rooting medium and a store of
water, a protection from excess heat or cold. The incompleteness of
this hypothesis was shown up by the results of quantitative analysis
of agricultural experiments. For example, Woodward's experiment

with *Mentha* showed that the increase in plant weight varied, depending on the 'impurities' in the water; the more impurities, the better the growth of the plants. He concluded that a soil constituent was the plant food: this might be of a fossil nature, or one of the products of the decomposition of plants or animals. As mentioned above, Theophrastus had made the same suggestion 2000 years earlier.

A Scottish lawyer, J. Tull, who devoted himself to agricultural studies, realised the importance of soil structure, and by introducing improved methods of tilling the soil (1731) he tried to build up the crumb structure which he thought important for nutrient uptake by the roots. He considered that this increased the soil surface from which nutrients could be taken up. Tull may therefore be regarded as the discoverer of soil structure as an environmental factor, although its significance to plants was not generally recognised until much later.

Humus. Further study of the soil led to the discovery of other food materials: in 1757, a Scotsman, Francis Home, made the observation that fertile soil contained oil, and so oil was included among them, and certain salts were also shown to have favourable effects. However, interest became centred more and more upon the plant and animal remains contained in the soil, which were called the 'humus'. It was thought that the humus, which was derived from living organisms, was the material from which living plants were built up, the conversion being made under the influence of the 'life force'.

This *humus theory* won general acceptance, and continued to be accepted even during the first half of the nineteenth century. The Swedish chemist Wallerius (1761) was one of those who developed this idea. Linnaeus (1747) describes it in the following way:

> When plants and animals decay, they become humus and the humus is then used as food by the plants which germinate and grow in that locality, so that the tallest oak and the most unpleasant nettle are constructed from the same materials, the fine, black humus particles, by Nature or a *Lapis philosophicus* (the philosopher's stone) which the Maker set in each seed for the conversion of the humus into its own species. When animals die they are changed to humus, humus to plants, and the plants are eaten by animals, and are changed into the organs of the animals. The soil which is changed into grain then goes as grain into the human body where it is changed into flesh, bones, nerves, etc. When man dies and decays the substance goes back into the soil from whence it came. When a plant germinates in the resulting humus it grows well, and the humus is converted to plant material. Hence the loveliest young girl's cheek may be changed to the ugliest henbane and the strong arm of a youth to the flimsiest pondweed.

The henbane may be eaten by insects, the insects by birds, the birds by man.

Linnaeus followed this train of thought at a time when he had cause to wonder if humus from a graveyard should be used to enrich the soil in fields and gardens, and he concluded that it should not. 'Nature teaches us that we should not eat the bodies of our fathers or our children, and I do not know who might wish to, unless it were an inhuman cannibal.'

Definite evidence against the humus theory came first from de Saussure (1804) who showed in his work on plant nutrition that the soil provided only a relatively small proportion of the plant material, and that plants obtain their carbon from the air and not the humus in the soil. Notwithstanding, the humus theory persisted for several decades and was accepted by many leading scientists, including J. J. Berzelius (d. 1848). Discussion centred mainly upon the origin of the carbon bound in organic material. Supporters of the humus theory thought that oil and coal were particularly valuable sources of plant food because of their high carbon content. However, criticism of the theory became more and more severe and in a paper published in 1840 Liebig rejected it and showed that the amount of humus in the soil did not decrease as a result of nutrient uptake by plants and that, anyway, its content of carbon was not enough to meet plant requirements. At about the same time, in 1834, J. B. Boussingault succeeded in growing plants in humus-free sand, so that the humus theory in its original form was discarded.

However, later study has shown that humus does fulfil a function of the kind indicated by the humus theory, since humus has been found to be the food of a large number of soil microorganisms, including some heterotrophs, which assimilate organic molecules or parts of molecules which they obtain from the humus. Furthermore, carbon dioxide released from the soil by the respiration of the microorganisms is reassimilated by the canopy of vegetation above.

Salts. Before their composition was known, it had been found that certain salts or saltlike compounds had a favourable effect on plant growth. Glauber, in 1650, found that additions of saltpetre (potassium nitrate) to the soil resulted in increased yields: he could detect the substance in soil saturated by urine and manure and so he concluded that it was present in animal excrement and was derived from plants, and, hence, the favourable effect of stable manure was attributed to saltpetre. According to Russell (1936), in the second half of the eighteenth century Francis Home found that potassium sulphate and Epsom salt (magnesium sulphate) also increased plant growth, and a similar effect was attributed to alkaline phosphates by the Earl of Dundonald in 1795. Alkali was obtained from plant ash. The idea that the substances in the ash were nutrients which circulated between the soil and the plant had been expressed as early as 1653, by a Frenchman, Palissy. However, in the seventeenth and eighteenth centuries it was generally assumed that these substances were formed in the plant itself. From his experiment with a willow cutting, van Helmont drew this conclusion, and Boyle (1661) from his distillation experiments. At the end of the eighteenth century, it was still thought that the alkalis and salts were a product of the life force.

There were various ideas about the function of salts, but Tull thought that they had a favourable effect because they brought about a more open soil structure. As long as the humus theory was accepted, salts were thought to increase the solubility of the humus, enabling it to be taken up more easily by the roots.

In 1804 de Saussure's investigations of conditions for growth, and his quantitative analyses of humus and plant ash yielded results which were vital to the understanding of the role of salts. By comparing the ash content of seeds with that of the resulting plants, which had only received distilled water, he showed that the plants themselves formed no ash and that growth was limited by the ash content of the seed; that the salts present in the ash were also found in the humus; that normally the composition of plant ash changed with the age of the plant or the plant organ; that young plants were rich in potassium and phosphate whereas old plants contained relatively more calcium and silica. However, these conclusions were not generally recognised by his contemporaries, and the humus theory remained firmly established since it offered an explanation which seemed probable and which fitted in with current ideas about the life force.

The new ideas first gained ground towards the middle of the nineteenth century, when Liebig (1840) published his work on the dependence of plants on the salts in the soil and the release of salts from minerals as a result of the process of weathering, a process which was a newly recognised environmental factor to be considered. The relation between crop yield and the content of nutrient salts in the soil was clear to Liebig: he realised that a part of the mineral content of the soil was removed in the harvest and that this must be replaced if loss of soil fertility was to be avoided. He wrote, 'The harvest from a field increases or decreases to the same extent that the supply of minerals in fertilisers is increased or decreased.' Thus, he thought of adding salts as such to the soil. He recommended a particular mixture of salts for this purpose and so initiated the use of mineral fertilisers.

At first, Liebig's 'mineral theory' was subject to lively discussion and criticism. At the same time, many workers were testing new techniques for cultivating plants. Plants were grown on artificial substrates, the composition and nature of which could be varied. Sand, sugar, coal, and water, with the addition of salts, were tried. In this way, Liebig's theory was gradually proved correct. Towards the end of the nineteenth century, a list of the minerals essential to plants could be drawn up, including salts containing potassium, calcium, iron, nitrogen, sulphur and phosphorus. Later work has enabled the list to be lengthened by the addition of boron, manganese, zinc, copper, and molybdenum, essential in at least some cases. These nutrient salts are present in varying quantities and proportions in the soil and together they make up an important group among the various environmental factors.

Air and light

The first mention of air as a food for plants was made by Nehemiah Grew, an English teacher, in a work published in 1682. Grew had made studies of plant anatomy, and, as a result, he suggested that certain plant parts were analogous to the lungs of animals, and that plants lived partly on air.

In 1727, Stephen Hales put forward a more firmly based hypothesis. He suggested that the 'air' produced in fermentation and distillation processes is actually the same air which has been taken in and bound by the plants during their lifetime, and, hence, that air is a plant nutrient.

Joseph Priestley in the 1770s made the discovery that air which has been 'spoiled' by animal respiration, or by combustion or decomposition, can be purified by green plants. Priestley (1776) enclosed a shoot of *Mentha* in air which was so 'spoiled' that a candle would not remain alight in it and he found that after ten days the air would again allow the candle to burn. Interpreted according to the phlogiston theory, this showed that the plant had replaced the phlogiston of the air. However, at that time, two of the constituents of air were discovered: oxygen by Scheele in 1772 and by Priestley in 1774; and carbon dioxide by Lavoisier in 1781.

Scheele set up experiments to prove Priestley's assertion that plants purify the air, but obtained results in opposition to Priestley's; and when Priestley repeated his own experiments he also obtained a different result. The explanation of these contradictory results came from the Dutchman, Johan Ingen-Housz (1779), who discovered that plants could either purify or 'spoil' the air, depending upon whether their green parts were illuminated or not. In sunlight the air was purified, but 'during the night or in dark places during the day, a harmful air was breathed out'. In this way, Ingen-Housz had expressed the principle of the carbon dioxide assimilation and respiration of plants. In subsequent investigations he showed that air was the source of carbon for green plants.

The Swiss Jean Senebier was working on the same problem at the same time. In 1788 he put forward the theory that green plants split carbon dioxide with the help of sunlight, releasing the oxygen and retaining the carbon; and that this carbon is the phlogiston of the plants. The importance of carbon dioxide as a nutrient was investigated further by de Saussure (1804), who found that the increase in weight of plants which were taking up carbon dioxide was greater than that equivalent to the carbon taken up. He attributed the excess to the binding of water, too, and concluded that water was a nutrient, and not only a solvent, as had previously been thought. De Saussure also succeeded in demonstrating that plants can produce carbon dioxide, like animals, and that this process consumes the oxygen in the air; and that oxygen is necessary for various life processes.

Again, it was several decades before these experimentally obtained results gained general acceptance. The current philosophy governed prevailing ideas, and the idea that plant carbon came from the air seemed unlikely as compared with the more easily understood humus theory, and, once again, Liebig contributed to the breakthrough of the new ideas. He showed that the carbon dioxide reserves in the air were sufficient to supply the carbon requirements of the vegetation; and that the loss of oxygen from the air in burning, decay and animal respiration was balanced by the gain of oxygen from the process of carbon dioxide assimilation by plants.

This series of fundamental discoveries, which began in the last half of the eighteenth century and continued in the first half of the nineteenth, provided the foundations for development of modern ideas about the greatest of all synthetic processes, carbon dioxide assimilation by plants, and established that carbon dioxide, oxygen, water, and light were vital environmental factors.

Organisms

In the early history of botany there is very little reference to the dependence of plants on one another, or on other organisms. Certain phenomena, such as competition for space and light, had certainly been observed very early and considered to be important because of their significance for crop production, but they were thought to be obvious and aroused no particular interest. Floral biology was considered much more interesting. It was observed very early (see the quotation from Theophrastus on page 7 that the development of figs, for example, depended on insects visiting the trees, although the reason for this was unknown. Linnaeus (1747) was the first to suggest the correct explanation. It was clear to him that seed development required the transference of pollen (*farina seminalis*) to the pistil (*tuba*). Later, J. G. Koelreuter's (1761–1766) and C. R. Sprengel's (1793) demonstration of the importance of insect visits to flowers for cross-pollination initiated research on this subject. However, it was only towards the end of the nineteenth century that the pioneer nature of their work was recognised.

The importance of animals in seed and fruit dispersal was also realized by Linnaeus. In his *Philosophia Botanica* (1751, p. 86) he gives examples of seed and fruit dispersal by birds, rodents and other animals.

In 1837, the Frenchman, Cagniard-Latour, and the Germans, Schwann and Kützing, showed that alcohol fermentation was caused by fungi; and in the late 1850s the French chemist, Louis Pasteur, showed that some fermentation processes were caused by bacteria. The discovery of yeasts and of bacteria had been made by a Dutchman, Antoni van Leeuwenhoek, the 'father of microscopy' during the years 1673–1683. This was the beginning of the investigation of the part played by various organisms in the many processes which bring about the humification of dead organic matter. Humification has various ecological effects. The fungi, bacteria and other organisms (protozoa, earthworms, etc.) which take part in fermentation, decay and the other processes of humification change the environment for all the vegetation and are therefore important environment factors. On the other hand, their own existence depends on the plants and animals living in the locality, whose dead parts are their source of food. Hence, the plants and animals living largely above the soil surface are, in turn, environmental factors for the humifying organisms. Study of the interplay between the two groups and the processes of humification is one of the most important aspects of plant ecology.

Symbiosis between leguminous plants and bacteria was first discovered by Hellriegel and Wilfarth (1883). Thus, one of the sources of the nitrogen supply for the world of living organisms had been found: that is, the assimilation of nitrogen from the air by these bacteria. Another pioneering contribution to knowledge of the interrelationships between organisms was the discovery of substances which are secreted by various organisms and which initiate or stimulate various metabolic processes. E. Wildiers in 1901, or possibly even Liebig, made the first observations of this kind.

There are many other aspects of interrelationship between organisms, studied and described under the names of parasitism, epiphytism, mycorrhiza, competition (for light, space or nutrients), etc. Research into these interrelationships mostly began during the last hundred years.

Around the beginning of the twentieth century, important contributions to modern plant ecology were made by Eug. Warming, a Dane, in a series of publications, the first of which, *Plantesamfund*, came out in 1895; and by Schimper in his *Pflanzengeographie auf physiologischer Grundlage* (1898).

Practical applications of plant ecology

As the environmental factors and their modes of action were discovered, knowledge about them has been applied to crop production. The use of mineral fertilisers, based on the discovery of the salt requirements of plants, is perhaps the most important application hitherto. There are also the various methods for fixing atmospheric nitrogen; methods for bringing about good conditions of soil moisture, aeration and populations of soil organisms, and for retaining the reserves of fine material; the adaptation of agricultural methods according to the requirements of the crop plants for light, heat, water and space; and the effects of plant species on one another as they appear in various systems of crop rotation.

2

The relation between the plant and the environment

Environmental factors: Delimitation and interactions

Development of the plant is primarily determined by its genetic make-up or genotype. How this genotype can influence development depends, in turn, on the environmental factors, i.e. the forces and objects in the environment which stimulate or inhibit the constitutionally determined development processes. As knowledge of the plant's environment has increased, more and more factors have been identified which affect plant performance and development. A number are soil factors, others are climatic or related to the living organisms in the environment. Hence the environmental factors are usually classified as edaphic (Greek *edaphos*—soil), climatic and biotic. In addition there are certain others, such as fire and topography of the ground.

Edaphic factors, or soil factors, comprise the chemical, physical and biotic factors which originate in the soil or are part of the character of the soil and which in one way or another have an effect on plants. Soil nutrient content, acidity, water and nutrient storage capacity, osmotic properties, particle size, crumb structure, density, aeration, water conductivity, anaerobic respiration, litter decomposition and other processes initiated by soil organisms, etc., are included.

Climatic factors include light (e.g. illuminance and day length), any other forms of radiant energy, air temperature, carbon dioxide, wind, precipitation and humidity of the air. All these change with the habitat and are usually different from the climate as a whole. The climate important to a particular plant is not primarily that characteristic of a region as a whole (the *macroclimate*), but that prevailing in the habitat itself, with dimensions of a few square metres or less (the *microclimate*). Even within one habitat the plants may experience different microclimates, trees and the herbs growing in their shade, for example. The herbs receive less light and precipitation than the trees, and are less affected by wind, but in general experience a higher relative humidity. To describe the microclimate is an important object of ecology.

The *biotic* factors include the living organisms which affect the plants in the habitat. Above ground, the plants compete for space, shade one another and produce litter which may be advantageous or harmful to neighbouring plants; animals graze and in other ways behave so that some plant species are favoured and others inhibited. Underground, plants compete for space, water and nutrients and in addition are influenced by fungi, bacteria and other plants, and animals of various kinds. These may take part in the soil-forming processes and hence influence the environment of the habitat, or they may be parasites, symbionts or antibionts.

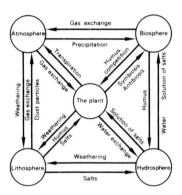

FIG 1. Examples of direct and indirect relationships between plants and the environment. (Partly from Mattson, 1938.)

There is a reciprocal relationship between the plants and the environment such that the environmental factors not only affect the plants, but in many cases are also affected by them. The effects are partly indirect in that one factor may influence another which, in turn, affects the plant (Fig. 1). Such a causal chain may, for example, be as follows: the climate affects the fungi which contribute to the decomposition of plant remains, thus affecting soil formation, which alters the edaphic conditions experienced by the plants.

In their effects on plants, the environmental factors are never completely separate from one another, but always interact to some extent. The right combination of edaphic, climatic and biotic factors is necessary for the occurrence of a suitable environment. On their own, the edaphic factors, like any of the other groups, are unable to provide this.

Mattson (1938) has tried to show the relationships schematically (Fig. 2). He divides the environmental factors into four groups or

spheres; the lithosphere (L) or the mineral factors (Greek *lithos*—
stone); the atmosphere (A) or the climatic factors; the hydrosphere (H) or the water factor; and the biosphere (B) or the
organisms. Only in certain combinations can the spheres create

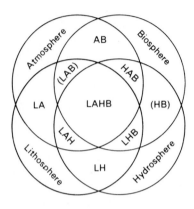

FIG 2. The theoretically possible
combinations between the atmosphere (A), lithosphere (L), hydrosphere (H) and biosphere (B).
(After Mattson, 1938.)

conditions which will support life. The combinations which are
theoretically possible are the following:

1 **Combinations which will not support life.** LH (lithosphere +
   hydrosphere). Example: stone, sand and clay at the bottom
   of a body of water with a depth so great, or with such a composition, that oxygen from the air does not reach the lithosphere.
AB (lithosphere + biosphere), e.g. dry litter.
LA (lithosphere + atmosphere), e.g. dry sand (dry desert regions).
LAB (lithosphere + atmosphere + biosphere), a derivative of LA, if
   organic material is added.
2 **Combinations which will apparently support life.** B (biosphere) forms a habitat, for example, for endogenous parasites. But since the parasites' requirement for minerals,
   water and oxygen is met by the host plant, which has itself
   obtained them from the other spheres, these must also be
   included as habitat factors, even though indirect. The inde-
   pendence of the parasite of the other spheres is therefore
   only apparent, not real.
AHB (atmosphere + hydrosphere + biosphere). Exogenous parasites, saprophytes and plants growing on peat, etc., have no
   direct contact with the lithosphere, but they take up mineral
   substances which are derived from it. The organic substrate
   on which they grow has been built up with contributions
   from the lithosphere. Another group of plants, the epiphytes, has no direct connection with the lithosphere, but

obtains mineral nutrients from the dust, carried in wind or water, the salt constituents of which are derived from the lithosphere.

HB (hydrosphere + biosphere), e.g. dy,[1] and

LHB (lithosphere + hydrosphere + biosphere), e.g. gyttja,[2] occur on the bottom of bodies of water, where organic material is deposited. In such habitats live a collection of anaerobic microorganisms which do not take up oxygen from the surrounding water nor, for HB, minerals from the lithosphere, but obtain these substances from the organic material on which they live. However, since this is formed with contributions from the atmosphere and the lithosphere the organisms are only apparently independent of these spheres.

3 Combinations which will support life. LAH (lithosphere + atmosphere + hydrosphere). Example: newly exposed surfaces deep in the ground, e.g. in a quarry, or the fresh surface of a newly split rock provide, with water, conditions which will support vegetation. The surface of the earth was in this state before the development of organic life, as was the ground exposed by the retreat of the ice at the end of the Ice Age. In nature, this combination is also present in sterile form, as in sterile salt soils.

LAHB (lithosphere + atmosphere + hydrosphere + biosphere) has developed from LAH as a result of the development of organisms and now comprises, to all intents and purposes, all plant habitats. On a surface in Glacier Bay, Alaska, which was covered by a glacier forty to fifty years ago, and which is now covered by plants, including alder, Crocker and Major (1955) found an accumulation of organic material in the uppermost 45 cm of the soil which was equivalent to 4 kg/m^2 of carbon and 0.3 kg/m^2 of nitrogen. The amount of litter covering the ground was as much as 5 kg/m^2. In this way conditions suitable for plant life are derived from a combination of the various physical spheres. Where a sphere is missing, life cannot be supported for any length of time.

Every one of the four spheres comprises environmental factors and each factor varies quantitatively so that an infinite number of combinations between the factors is possible. It is these numerous

---

[1] Muddy material formed of plant residues and deposited from nutrient-poor water. (FAO, *Multilingual Vocabulary of Soil Science*, 1966.)

[2] Sedimentary peat consisting mainly of plant and animal residues precipitated from standing water. (FAO, *Multilingual Vocabulary of Soil Science*, 1966.)

combinations and the interrelationships between the groups of factors which make clear delimitation between the groups, and their effects, so difficult or impossible. In spite of this, division into edaphic, climatic and biotic factors is justified, since it allows the situation to be surveyed more easily.

Quantitative variation in the environmental factors

Even adjacent places may differ in soil water content, soil temperature, salt content, etc., and in one and the same place each of these

FIG 3. Gravel pit colonised by pine, birch, aspen, heather, raspberries and other deep-rooting species. Ekerö, Uppland, Sweden.

properties will change with time. Temperature, light and, to a certain extent, soil water content, have daily and annual cycles of change. The species and number of individuals growing in any locality change from one instant to the next, resulting in changes in the competition for nutrients, water, etc. The plants react to such differences and changes in the environment with variable sensitivity. As a rule, each species thrives only within a particular part of the total range of each factor and departures from this part of the range have unfavourable consequences.

Minimum, optimum, maximum. Hardening

Conceptions of the relation between the performance of a species and the magnitude of the various environmental factors are similar to corresponding ideas about the relation between a physiological process and the magnitude of a factor which affects the process, e.g. the relation between temperature and growth (Fig. 4). Growth is most rapid at a particular temperature, and both lower and higher temperatures result in decreased growth. Growth ceases altogether at a particular minimum temperature and at a particular maximum temperature. Sachs (1860) called the cardinal points the optimum, minimum and maximum temperatures, respectively, for the particular process. Temperatures above the maximum or below

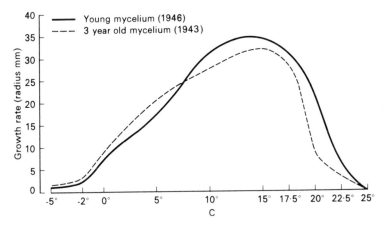

FIG. 4. The effect of temperature on the growth rate of fungal mycelium. Duration of growth, 20 days. Material, young (4 months) and older (3 years) mycelium of *Phacidium infestans*. Optimum temperature *c*. 15°C. (After Björkman, 1948.)

the minimum have harmful effects which at their most severe may be lethal. However, this relation between plant and temperature is not a constant one. If the temperature regime changes, so does the relationship (Fig. 5). This relation also changes with time; for example, growth is most rapid at the beginning of a period of optimum temperature and then falls off, depending on a so-called time factor. Neither are the maximum and minimum temperatures constant, but change in the same way as the optimum. Because of this the plant may become *hardened* to high or low temperatures if exposed to increasing or decreasing temperature, respectively. The plant gradually becomes conditioned to the prevailing temperature, and acquires a new temperature dependence. In natural conditions

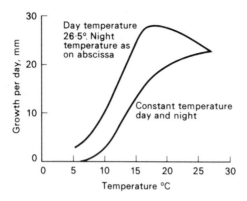

FIG 5. The relation between temperature and shoot growth of tomato plants. The optimum temperature is *c.* 30°C at constant temperature, but is lower at fluctuating temperatures. (After Went, 1944.)

this sort of shift takes place so that the optimum adjusts to changes in the environment, such as those consequent upon the seasons (Fig. 6).

FIG 6. The relation between temperature and net daily carbon dioxide assimilation in a moss *(Hylocomium proliferum)*. The numbers on the curves refer to the period of assimilation (in hours) per day. In natural conditions this will vary from 3–6 hours in late autumn in northern latitudes, to 18–24 hours in midsummer. The dotted line represents the seasonal shift of the optimum temperature towards a lower value during the autumn, and towards a higher value during the spring. (After Stålfelt, 1937.)

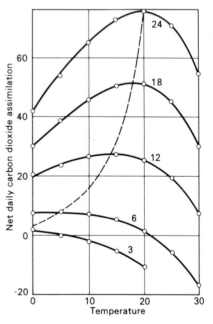

Limiting and inhibiting effects

In spite of their variation, Sachs' three cardinal points are characteristic expressions of the relations between a physiological process and external factors. Since the effects of environmental factors on the development of a plant are in principle the same, and follow the optimum curve, the same terms have been applied in plant ecology. If the relation between plant and environment is to be characterised briefly, then it is said that a particular factor is optimal, suboptimal, etc. Unfortunately, the definition of this concept has altered to some extent as the terminology has developed, so that the terms 'minimum' and 'maximum' have come to include values just over the minimum or just under the maximum, and optimum has come to mean not only the particular value which allows the maximum effect, but also the values immediately on

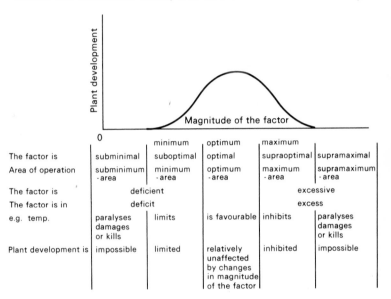

		minimum	optimum	maximum	
The factor is	subminimal	suboptimal	optimal	supraoptimal	supramaximal
Area of operation	subminimum -area	minimum -area	optimum -area	maximum -area	supramaximum -area
The factor is	deficient			excessive	
The factor is in	deficit			excess	
e.g. temp.	paralyses damages or kills	limits	is favourable	inhibits	paralyses damages or kills
Plant development is	impossible	limited	relatively unaffected by changes in magnitude of the factor	inhibited	impossible

FIG 7. **Diagrammatic representation of the relation between the development of a plant and the magnitude value of a particular environmental factor, e.g. temperature. The diagram illustrates the meaning of some of the terms used in the ecological literature, and in this book, to refer to the mode of operation or magnitude of a factor under consideration.**

either side (Fig. 7). A factor may be called a 'minimum factor' immediately its value falls outside the lowest limit of the optimum range, i.e., the factor's value is no longer sufficient to allow optimum development of the plant; the lower the factor, the more the development is *limited*. In the same way, a factor is said to be 'maximum' when it exceeds the optimum range and *inhibits* development. The development of the plant depends on the interplay of limiting and inhibiting factors.

In the literature of plant ecology there are certain other terms which have their origin in the concept of the optimum curve.

Some of them, and particularly those used in this book, are summarised in Fig. 7, in which the effect of temperature on the development of the plant has been chosen as an example. By the 'development' of the plant is meant the summation of all the processes which determine success in the struggle for existence, not only normal growth and formation of organs, but also hardiness to unfavourable conditions, the ability to compete with other plants, to remain established in the habitat and to spread.

Development is the result of all the physiological processes, such as assimilation, respiration, organ formation, growth, etc., and its

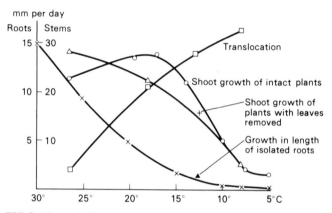

FIG 8. The relation between night temperature and various physiological processes in tomato plants. (After Went, 1944.)
☐ translocation with guttation sap.
○ shoot growth of intact plants.
△ shoot growth of plants with the leaves removed.
× growth in length of isolated roots.

cardinal points are dependent on that of each of these separate processes, and also on their duration (Fig. 8). For instance, at least in some instances, a relatively long season of growth can compensate for a relatively low rate of growth: the amount of material which a plant needs for its development may be produced at suboptimal temperatures, provided that the time available for production is sufficient. The time available for some processes is shorter than for others and the need for optimum conditions is therefore greater. Seeds often ripen in a shorter period than is available for assimilation, and ripening is consequently more dependent on its optimum temperature, since the plant survives a suboptimal temperature for assimilation more easily than a suboptimal temperature for seed-ripening. Thus, the temperature optima for development and for the single separate physiological processes cannot be equated. Analysis of the significance of the temperature factor requires the determination of the temperature optima for various separate physiological processes and also the optimum for development as a whole. The latter can be found by growing the plants at different temperatures and comparing the results. Together with the maximum and minimum, the optimum temperature reflects the plant's requirement for heat. The same kind of relation obtains between the development of the plant and other external factors, even if there is not always a maximum.

Knowledge of the position of these cardinal points is of basic significance to ecology, since species and varieties differ from one another in their habitat preferences, i.e. in the position of the optima, and have different abilities to utilise or to tolerate the various environmental factors. For example, photosynthesis of alpine and arctic plants is characterised by low temperature optima. For lichens, the optimum is about $+10°C$ and there is positive net photosynthesis at $-5°C$, or even lower in some species (Lange, 1965). A clear understanding of the relation between the environmental factors and the development of the plant therefore requires knowledge of the cardinal points. Only when these are known is it possible to discern the special needs of the species in respect of its environment and its ability to maintain its position in the habitat. Determination of the optimum curve is a method which has long been used in crop production. The suitability of the soil and other environmental factors for various sorts of plants has been assessed with the help of the optimum curve and the management needed to provide optimum conditions determined.

In ecology, these questions are investigated in both field experiments and those in the laboratory; field experiments alone do not allow the mode of operation of the factors to be analysed and clarified. On the other hand, there are risks involved in extrapolating from laboratory experiments to corresponding conditions in the field. A process may go on in a different way in the field, because the environment is different. In laboratory experiments, usually only one or a few of the factors operating in the field are the subject of study; most are excluded or kept constant. Because of this simplification and standardisation of the conditions, plants in laboratory experiments react more uniformly and regularly than they do in the field. For example, in the culture of microorganisms on an artificial medium, growth is sometimes slow, although the most common nutrients have been added, but can be stimulated by the addition of soil water or a soil extract (e.g., see Nilsson *et al.*, 1938, 1939). In similar experiments, Rodhe (1948) found an alga which required relatively large amounts of phosphorus for its development in an artificial medium, but only small amounts if sea water was added to the culture vessel.

Several of the methods used in crop production or in food conservation are based on the relations expressed by the optimum curve. When plants suffering from drought are watered, the water supply is increased from a suboptimal value to a value nearer the optimum; temperature and illuminance are altered by heating the glasshouse during the winter and by supplementary lighting; deficient nutrition is improved by fertilisers; in conservation by drying, the water content is reduced to a value which is subliminal for microorganisms; conservation by deep-freezing uses subliminal temperatures; and conservation by salting uses supramaximal salt contents.

Sensitivity, tolerance, competitive ability, selection, ecological amplitude

The optimum curve reflects the relative *sensitivity* or *tolerance* of a particular species to an ecological factor. A relatively long interval between the minimum and the maximum corresponds to a high degree of tolerance, i.e. the species is able to exist when the factor

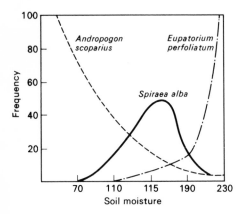

FIG 9. An example of the selective effect of soil moisture. 'Frequency'=no. of occurrences as per cent of total no. of localities investigated. 'Soil moisture'=water-holding capacity of the soil in vol. per cent. (After Curtis, 1955.)

has a relatively low value or a relatively high one. The species then has a large ecological amplitude for the factor and has a relatively low sensitivity to changes in it. When a species is extremely insensitive to changes in a factor, the optimum curve has a broad and level top.

The more sensitive a species is to a factor, the narrower the optimum and the shorter the interval between the minimum and maximum; the broader the optimum, the greater the tolerance, so that the optimum curve is an expression of the *competitive ability* of a species. However, the greater the tolerance, the more localities the species can invade and the greater the possibility of the species maintaining a position where it has become established. A process of *selection* is involved. The prevailing environment in a habitat *selects* among those species which seek to invade it; the result of this selection depends partly on the tolerance of the individual species towards the various environmental factors (Fig. 9). A species with a wide tolerance has a greater chance of success: aspen and pine have a wide tolerance of variations in soil water and nutrient content and they are therefore found in both dry and wet localities and on fertile or poor soil; while alder has a preference for damp places, beech for dry and ash for nutrient-rich soils. For these last three species, the *ecological amplitude* for moisture or nutrients is narrower than for pine or aspen, and they become established less often.

These qualities are of fundamental significance to the distribution of a species. Species with broad amplitudes are relatively insensitive to variations in the environment and have a good chance of becoming widespread. However, if a species is very sensitive to changes in just one factor then this factor will have a dominating influence on the distribution of the species.

The explanation for the strikingly large number of cosmopolitan and widely distributed species of water plants is probably that the water supply is always optimum, in particular for submersed forms; they avoid the limiting or inhibiting effects of water experienced by land plants, effects which are often one of the most important reasons for their more restricted distribution.

The interplay of factors

A plant encounters the best conditions in the habitat where all the environmental factors are optimum. In such a place, development of the plant is really only limited by the number of individuals of the same species present: development is as luxuriant as the genotype allows. However, if one of the factors is not at the optimum, development is limited or inhibited, and goes only so far as is allowed by that factor; there is, of course, a similar limitation if two or more environmental factors are suboptimal or supraoptimal. If for example factors A, B, C and D are suboptimal in the way shown diagrammatically in Fig. 10, development of the plant is limited by the size of the deficits and the interplay of the deficient factors.

How this interplay occurs, and how a deficit or an excess operates, have been the subject of many investigations and interpretations. This question is of great significance, because plants in natural conditions are generally subject to the simultaneous influence of several suboptimal and supraoptimal factors. It also has important practical consequences because many methods used in

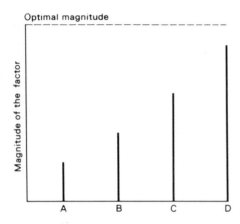

FIG 10. **Four suboptimal environmental factors are deficient, A most and D least. See text for discussion of their limiting effect on plant development. Dotted line = optimum magnitude of the factors. (After Meyer and Anderson, 1939.)**

the commercial production of plants depend on the way the question is answered. Agricultural research has long been concerned with this question and the first experiments done to clarify the problem were with cultivated plants.

The law of limiting factors

In the middle of the nineteenth century, the German agricultural chemist Justus von Liebig studied the relation between yield at harvest and the magnitude and interplay of environmental factors, mineral salts in particular. He concluded that it was only the most deficient factor, i.e. factor A in Fig. 10, which determined the yield; he called this the 'minimum' factor. Liebig's interpretation of the effect of deficient factors is known in the agricultural and ecological literature as *the law of limiting factors* or *the law of the minimum*. The term 'minimum' applies to the most deficient factor among several which are suboptimal. However, when this term was later introduced to the physiological literature by Sachs it was used to denote a particular point in the range of variation of a factor (see page 21). In the ecological literature the meaning has changed and 'minimum' now denotes a whole section of the factor's range, namely, all the values which are limiting (see Fig. 7). According to the original interpretation of the law, an increase in the suboptimal factors B, C and D has no effect on the plant as long as A is the most deficient factor (Fig. 10). But if A becomes less deficient than B then A ceases to be limiting and B takes its place. After that, continued increase in A has no effect, as long as B remains the most deficient. Liebig seems to have thought that the development of a plant, or the harvest from a field, increased proportionally to increase in the most deficient factor. This interpretation of the law has been represented by diagrams like that in Fig. 11. For several decades, Liebig's law formed the basis for understanding the relation between plant production and environmental conditions.

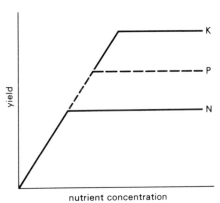

FIG 11. The interpretation of the law of the minimum according to Liebig. Yield is shown to be limited by three plant nutrients, one at a time, in the order in which they are deficient: in this case, first nitrogen, which is most deficient, then phosphorus, then potassium. (After Liebig, 1840, 1862.)

Mitscherlich's law

At the beginning of the twentieth century certain experimental results, particularly in agricultural research, did not fit in with Liebig's ideas. Mitscherlich (1909) found that the relation between increase in yield and increase in a suboptimal factor was not expressed by a straight line, as had hitherto been expected, but was logarithmic (Fig. 12), and that the yield depended on all the suboptimal factors and not only on the most deficient one. Other similar experiments gave the same result and led to a new formulation of the relation between the plant and its environment, known as *Mitscherlich's law*. This states that the relation between development of the plant (for example, the yield) and environmental factors (nutrients, light, water, etc.), or production factors as they were also called, is expressed as follows:

1 The yield is affected by changes (increases or decreases) in any one of the factors, not only by the one which is most deficient or most in excess, but the effect is greatest when the change is in one of these, or, in general, in the factor which is furthest from its optimum.

2 The smaller the interval between the actual value of a factor and its optimum, the smaller the effect of a change in that factor.

The yield is thus relatively insensitive to changes in a factor which is at or near the optimum. The further the factor is from the optimum, the greater the plant's sensitivity to the deficit or excess.

3 To a certain extent, environmental factors can substitute for one another, a favourable level of one compensating for a deficit in another. For example, a high temperature may partly compensate for a lack of nutrient-rich minerals in the soil because it causes increased weathering (for other examples, see pages 326, 411).

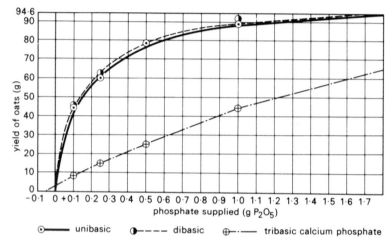

FIG 12. **The relation between yield of oats and phosphate supply.**

Mitscherlich's law has reformulated Liebig's law; *the law of limiting factors* is therefore still discussed, although now with the meaning attributed by Mitscherlich.

Analysis of the law of limiting factors

Experiments of the kind used by Mitscherlich are hardly suitable for showing the effects of single factors, because the yield, which Mitscherlich used as a measure of the effects of the environmental factors, is such a complex characteristic.

Yield is the result of many physiological processes, several of which may be dependent on the same environmental factor. For example, many physiological processes are dependent on temperature, the total effect of which is therefore equal to the sum of all the separate effects. A particular temperature may be optimal for one process and suboptimal for another. Analysis of the law therefore requires investigation of the separate physiological processes.

The first experiment to show the limiting effect of single factors in detail was by Blackman (1905). He attempted to show the relationship between the amount of carbon assimilation and the supply of light and carbon dioxide. At first, the result seemed to be in accordance with Liebig's law, which at the time was still generally applied. According to Blackman's interpretation of his results, assimilation was limited only by carbon dioxide concentration as long as the deficit of carbon dioxide was greater than that of light (Fig. 13, A–B). Only when the carbon dioxide concentration increased so that the deficit of light was greater than that of carbon dioxide was there any effect of light. Thereafter, the assimilation rate was determined only by the illuminance, and increased in proportion to the illuminance until another factor became limiting (Fig. 13, B–D). At any instant, the rate of the process was determined only by one factor, the one most deficient, or in shortest

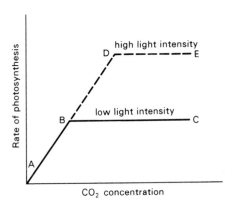

FIG 13. According to Blackman's principle of limiting factors, the rate of photosynthesis at low CO_2 concentration and low light intensity is determined by CO_2 concentration alone, if that is the most deficient factor (A–B in the diagram). At higher CO_2 concentrations the rate of photosynthesis is determined by light intensity alone (B–C). At higher light intensities, CO_2 concentration continues to be limiting (B–D). (After Blackman, 1905.)

supply, and the rate of the process increased in proportion to that particular factor. Blackman's interpretation agreed in principle with the contemporary idea of the minimum law (Figs 10 and 11). In fact his results were so variable that the diagrams upon which he based his interpretation could also have been drawn with curved lines and no sudden changes of direction, but Blackman thought it was an obvious, axiomatic interpretation.

Later investigations, however, failed to confirm Blackman's conclusions. Harder's (1921) experiments with the aquatic moss *Fontinalis* may be quoted as an example (Fig. 14). Increase in assimilation rate was not proportional to increase in carbon dioxide concentration, but was relatively less as the carbon dioxide concentration neared its optimum. The curve was logarithmic, and had no sudden bend. The experiments also showed that the process was limited by two factors at the same time, not by one at a time, as

Blackman had thought. For example, Fig. 14 shows that at a carbon dioxide concentration of 0·05 per cent, assimilation increases as a result of an increase in either carbon dioxide concentration or illuminance. Harder's experiments therefore support Mitscherlich's ideas and not Liebig's. They have also been regarded as supporting the minimum law as enunciated by Mitscherlich. At the same time, Lundegårdh (1921) carried out experiments of a similar kind and came to the same conclusions.

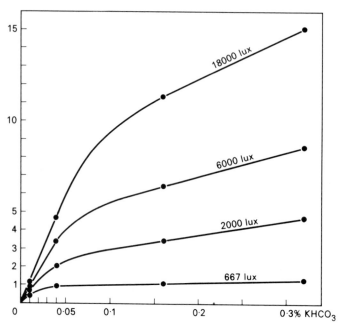

FIG 14. The relation between carbon dioxide concentration, light intensity and photosynthesis of the aquatic moss *Fontinalis*. (After Harder, 1921.)

However, if a unicellular organism in a single layer arrangement is used in the same kind of experiment, the relation between the physiological process and the environmental factor is of a different kind, in closer agreement with Liebig and Blackman than with Mitscherlich. Van den Honert (1930) used a unicellular alga of the genus *Hormidium* (Fig. 15). In contrast with Harder's and Lundegårdh's results, the relationship between assimilation and carbon dioxide was linear both at low and at high carbon dioxide concentrations and the transitional region between the two takes the form of a sharp bend not unlike the corresponding portion in Liebig's or Blackman's diagrams. Van den Honert considered that the agreement with Blackman's diagram would have been even closer if the cells and the chloroplasts had been small enough to allow a uniform supply of carbon dioxide to the different parts of the chloroplasts: it must be assumed that the supply is not uniform, even though the *Hormidium* cells are small. Determination of the carbon dioxide concentration at which carbon dioxide ceases to be limiting, i.e. the point at which the curve in the diagram changes direction abruptly, is difficult because the different parts of the chloroplasts are not exposed to the same carbon dioxide concentration at the same time. As carbon dioxide diffuses through the chloroplasts, the limiting concentration is reached in the various parts in relation to their distance from the carbon dioxide source. When an optimum carbon dioxide concentration is reached at a particular point in a chloroplast and, hence, assimilation at that point is not limited, the concentrations are still suboptimal at points further away from the

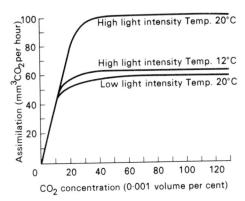

FIG 15. The relation between carbon dioxide concentration, light intensity and photosynthesis of a unicellular alga *Hormidium*. (After van den Honert, 1930.)

source and perhaps supraoptimal at points nearer to it. Measurements of assimilation at any carbon dioxide concentration are therefore an expression of the sum of all these effects. As a consequence, the transition from limiting to non-limiting values is gradual for the plant as a whole and not abrupt at a particular carbon dioxide concentration. The curve is therefore logarithmic, and not angular.

The relationship which Blackman expected between environmental factor and reaction (Fig. 13) is therefore only attained in material which is so thin that all parts of it are affected simultaneously by changes in the factor under investigation. Deviation from the Liebig–Blackman formulation of the law of limiting factors, so that the form of the diagram is logarithmic rather than linear, depends on the extent to which the material differs from this theoretical ideal. In *Hormidium*, there is an insignificant deviation and the relationship between factor and reaction differs little from the Blackman curve (*cf.* Figs. 13 and 15). However, a plant, with as simple a construction as *Fontinalis*, is already so elaborate that there are differences in carbon dioxide supply not only between the different parts of the chloroplasts but also between the chloroplasts in different cells, resulting in the logarithmic relation shown in Fig. 14. The more cell layers light has to penetrate, the less steep is the curve: mutual shading of leaves and plants has the same effect. Increase in illuminance causes a more gradual increase in assimilation rate in a closed plant stand than in a single leaf (Fig. 16).

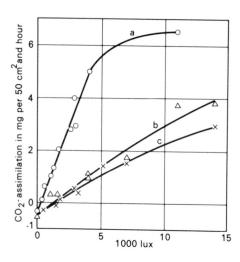

FIG 16. The relation between light and carbon dioxide assimilation in oats. (a)=single horizontal leaves; (b) and (c)=stands of oats, (c) having approx. twice as many leaves as (b). The optimum light intensity was relatively low for (a) (10 000–15 000 lux), but higher (> 20 000– 30 000 lux) for (b) and (c). (After Boysen-Jensen, 1932.)

What has been said about the lack of uniformity of carbon dioxide concentration within an organism also applies to other environmental factors. The temperature of a directly illuminated organ decreases from the outside inwards; a nutrient taken up by the roots is transported to different cells or organs in different amounts. If the relationship between such factors and a particular physiological process is determined, it is usually found to be of the same sort, in principle, as that between assimilation and carbon dioxide.

Liebig's formulation of the minimum law is therefore true for plants as a whole only if the magnitude of the environmental factor in question remains the same throughout the plant. This requirement is generally not fulfilled, different values of the factor occurring in the cells, in the organs and in the vegetation. The relationship between a physiological process and an environmental factor is therefore a summation effect which, in principle, can be interpreted in the way that Mitscherlich described.

Liebig's formulation is now only of theoretical interest. However, in agriculture, and in other branches of commercial plant production, Mitscherlich's formulation is of fundamental significance.

Ecogenous equilibria

Plant communities, like plant species or individuals, have become part of a biotic-edaphic-climatic complex in the course of their development, according to the environmental conditions and the genetic, physiological and morphological character of the individual species. This has occurred through a process of selection of the most suitable of the species and varieties available. When a species has become established in a community in this way it seems as though it were 'adapted' to the complex in which it occurs. Only those species which can tolerate the particular environmental complex and which have sufficient competitive ability can survive within the community.

Species have been selected for a community on the basis of their particular requirements for light, heat and water and their resistance to extremes of these environmental factors, and also their relationship with other organisms in the community, e.g. symbiosis, antibiosis, and their reaction to harvest by animals and man.

In a community which has developed without being disturbed by too much outside interference, there is an equilibrium among the species themselves and between the species and their environment. The environment meets the needs of each species without overstepping the limits of its resistance. These equilibria which came into being during the development of a plant community—ecogenesis—are a special group of natural equilibria and may be called *ecogenous*, as distinct, for example, from the physiological equilibria among the various component reactions in living cells or tissues.

As long as equilibrium is maintained the species live in harmony with the environment. However, this harmony can be destroyed in many ways. A new species with high competitive ability may invade the community or there may be a sudden change in one of the environmental factors. Fire, or a temporary large increase in numbers of an animal species, or a temporary period of unfavourable climate, come into the second category. Man also comes into this category. Man was formerly just one of many species, with no particularly dominating influence, but now has a special position and can exploit the whole of nature as he wishes. His interference cuts right across the balance of the ecological complexes, modifying and reshaping their maintenance and equilibria. Of all the environmental factors most important to plants, it is really only light and temperature which have up to now escaped large-scale change; other conditions have been changed and exploited by man on such a scale that plant communities, landscapes and the flora and fauna of whole regions have been radically altered. The components of the environment most readily available to man have been hardest hit, in particular water, plant nutrients, fine earth,[1] and plant and animal species which can be eaten or used in other ways. The immediate changes which result from man's interference have had secondary effects. These may have affected other ecological complexes, or plant or animal species, which are of no immediate interest to man but which may form an indispensable part of the production system on which his existence depends.

[1] Fine earth: soil passing through 2 mm sieve without grinding primary particles. (FAO, *Multilingual Dictionary of Soil Science*, 1966.)

3

Water

The components of the water balance

Retention and control systems

Under natural conditions most environmental factors are subject to such variation that many plants or plant communities sometimes suffer from a deficit, sometimes from an excess. So it is with light, temperature or nutrients, and also with water. Through their limiting or inhibiting action such factors decisively influence the selection of species by the environment. It is primarily temperature and water that determine the broad pattern of distribution of plants.

The water supply to plants varies between the extremes where water is the surrounding medium, as for aquatic submerged species, or where there is an insignificant amount of water, for example, in deserts where there may be so little that vegetation can develop only exceptionally. The water balance of submerged species presents only secondary problems, for example, in connection with the oxygen requirement. For plants growing in air and soil, a water deficit or excess has more direct effects: such plants are in the boundary region between the more or less moist soil on the one side, in which water is held with a force of a few bars or parts of an bar, and air on the other side, in which the suction force may reach several hundred bars, depending on the relative humidity (see Fig. 17). At a relative humidity of 99 per cent (at 20°C) the suction force of the air is 14 bars, at 80 per cent it is *c.* 290 bars (Fig. 18). In northern Europe in spring and early summer the relative humidity may be 30 or 40 per cent or even less. On such occasions drought may be decisive for the distribution or existence of many species. Evergreen species in early spring experience particularly extreme conditions, when the ground is still frozen and the suction force of the air reaches several hundred bars.

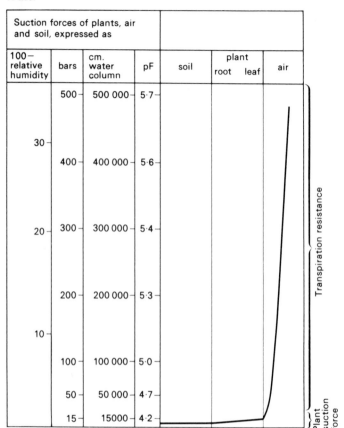

FIG 17. Schematic diagram of the position of land plants at the boundary between the soil and the air. The suction forces of the plants, the air and the soil are given in bars, in the corresponding height of a water column in cm, and as the log (pF) of the height of the water column. The suction force of the plant changes as the cell water content changes. The resistance to transpiration, associated with the outer layer of the plant (cuticle, epidermis and stomata), applies at the position of the large difference in suction force between the plant and the air. (Mainly after Gradman, 1928 and Laatsch, 1954.)

The ecological effects of water change with the wetness of the medium. They are fundamentally different for submerged species and for species which grow wholly or partly in air. For the first land plants, thought to have developed from aquatic forms, the transition to life on dry land must have been beset with various difficulties including the complicated problem of water supply. If the solution to this problem can be deduced from study of the

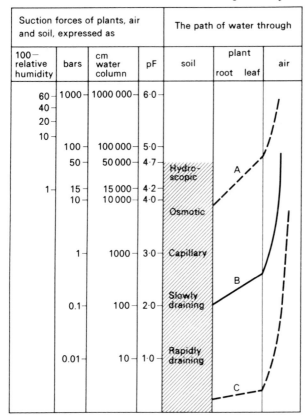

Suction forces of plants, air and soil, expressed as				The path of water through			
100 − relative humidity	bars	cm water column	pF	soil	plant root	leaf	air
60	1000	1000 000	6·0				
40							
20							
10							
	100	100 000	5·0				
	50	50 000	4·7	Hydro-scopic	A		
1	15	15 000	4·2				
	10	10 000	4·0	Osmotic			
	1	1000	3·0	Capillary			
					B		
	0.1	100	2·0	Slowly draining			
	0.01	10	1·0	Rapidly draining			
					C		

FIG 18. Same schematic diagram as Fig 17, except that the ordinate is a log scale of pF values. The hatched area represents soil water with varying availability to plants. B: Plant suction force and the path of water in fairly good soil moisture conditions. A: Lower soil water content and higher plant suction force. C: After rain, higher soil water content and lower plant suction force.

presentday flora, it seems that the ways in which land plants have satisfied their water requirements and protected themselves against drought have developed along two main lines. The first is exemplified by mosses and lichens, the water balance of which fluctuates widely according to the weather. These plants grow only when it is damp and their water balance is determined directly by precipitation and evaporation, they are mainly passive; water is taken up and evaporated more or less as by blotting paper, and evaporation proceeds until the cells and organs are air-dry. Such a

water economy demands high desiccation resistance of the protoplasm, which is one of their characteristics.

The second line was taken by the phanerogams and pterido-
phytes. Their water turnover is determined not only by physical
factors, but also by the physiological activity of the plants themselves. By the development of the cuticle and periderm, uncontrolled evaporation of water has been reduced, and by the development of stomata it has been brought under control by the plant.
In addition, these plants have special organs for water uptake, the
roots, which have made it possible to use the soil as a store of
water and so to ensure a more even supply. This stabilisation of the
water supply enables metabolism to go on more or less uninter-
rupted during the vegetation period and not in random stops and
starts as in the bryophytes.

The cuticle-periderm-stomata-root system would be more or
less perfect for its purpose, if the cuticle and periderm were completely impermeable to water and if the stomata could close her-
metically. However, the cuticle and periderm are not completely
impermeable, and the stomatal mechanism, at least in its simpler
forms, cannot completely prevent evaporation. The system therefore is imperfect, water loss is not completely controlled and there
is an uncontrolled cuticular transpiration as well as that under stomatal control. The amount of water lost through the cuticle and
through apparently closed stomata is certainly small in comparison
with stomatal transpiration, but it is large enough to endanger the
survival of the plant at times of water shortage and persistent
drought. A species' ability to survive a drought mainly depends on
the extent to which it has succeeded in minimising the uncontrolled part of the water loss. This is different in different species.

In plants which grow in dry places and are particularly liable to
water shortage, there are many ways in which the cuticle and the
guard cells are specialised, as are other morphological, anatomical
and physiological features. It is these specialisations which give
the plants the characteristic appearance which has long been
known as 'xeric', the basis of the terms 'xeromorph' and 'xerophyte' (see Warming, 1918, page 198).

Physiological features

1 The osmotic characters of the cells. The osmotic value and
the wall pressure together determine the suction force of plant
cells. As a water deficit develops, the wall pressure decreases and
the concentration of the cell sap increases, so that there is an increase in suction force. The suction force may also increase as a
result of an increase in the osmotically active contents of the cell.

Thus, suction force and water uptake can to some extent be regulated by the changes brought about by a change in water deficit. This regulation is incomplete, so that water deficits may often increase to several per cent or even 20 or 30 per cent of the maximum plant water content.

At sunrise plants have their highest water content, in the early afternoon their lowest: during the evening and night they make up the deficit. The osmotic value of the cells changes in accordance with these changes in water content; it increases during the morning and is maximum when the water content is lowest, when the demand for water is greatest. Pisek and Cartellieri (1931) compared values at 6.00 h and 14.00 h and found that in *Stachys recta* the osmotic value rose from 14 bars at 6.00 h to 19 bars at 14.00 h; in *Viburnum lychnitis* from 10 to 13; in *Peucedanum oreoselinum* from 18 to 20; in *Sedum acre* from 7 to 7·5; and in *Sedum maximum* from 5·8 to 6·7 bars.

There is an annual cycle of osmotic values in addition to this diurnal cycle. At a certain time of year a maximum is reached, and at another time there is a minimum. These times, and the amplitude, vary with species and climate. In herbs and deciduous broad-leaved trees the osmotic value is lowest in spring and highest in summer: in evergreens it is highest in winter (see Walter, 1951).

The osmotic value increases with illumination of the plant organ, and with height up the stem. It is higher on the sunny side of a tree than on the shaded side. For example, in spruce needles it may be 21 bars on the sunny side of the crown and 17 on the shaded side; 15 in the lower part of the crown and 17 in the upper (Pisek and Tranquillini, 1951). It is lower in roots than in stems or leaves, e.g. in *Alisma plantago*, 8 in the roots and 15 in the leaves; in *Hippuris vulgaris*, 9 in the roots and 11 in the leaves (Hannig, 1912).

By such adjustments of the osmotic value, the plant may regulate its water uptake within certain limits, but the extent of such adjustments is limited to a particular range for a particular species. The position of this range varies—in some species it may comprise high values, in others low, but for any one species it is strikingly constant. Some values measured for the same species in central Europe and North America illustrate this: in *Polygonum bistorta* in Europe 6 to 12 bars, in North America 8 to 11; in *Dryas octopetala* 13–15 and 11–14, respectively; in *Silene acaulis* 10–14 and 9–11; in *Campanula rotundifolia* 11–13 in both regions (Walter, 1951, page 278).

The wide differences in osmotic characteristics among species are best illustrated by comparing groups of plants from different habitats. According to a comparison made by Walter (1951, page

292) plants in moist, shady places have values which are generally between 6 and 8 bars. In response to the changes in water loss, or the time of year, these figures may rise to double, or fall to half. In fen plants and reeds the osmotic values are generally higher and vary between 5 and 25 bars.

The flora of dry habitats is characterised by high osmotic values, which again vary according to the type of plant. In soft-leaved species values are about 8 bars; in sclerophyllous species 20–40; in trees and shrubs 10–30; alpine species 8–16; steppe, prairie and heath plants 10–30; in the Mediterranean sclerophylls 10–35 bars, and so on. These figures, which refer to normal climatic conditions, may increase to double. Geophytes and succulents present in dry localities have low osmotic values, generally about the same as those measured among the vegetation of damp localities. The values are also surprisingly low in desert species, e.g. 10–20 bars, but in extreme cases up to 50 or more. In tropical rain forests values of 10–15 bars have been measured for trees and shrubs; 5–10 in small shrubs and herbs; 6–10 in lianes.

2 Protoplasmic desiccation resistance. The water content of plants changes with the balance between water uptake and water loss. The condition of *saturation* or *full turgidity* is attained when the plant has taken up as much water as the elastic properties of the cells and tissue allow and reached a certain stiffness or firmness. However, this water content is not constant, nor is it specific for the leaf or for the species; it changes according to the external conditions and the function and development of the organ, yet it does remains within a relatively small range. During wet weather, plants may remain fully turgid or almost turgid throughout the day, but in dry weather turgidity is only reached during the night. A transpiring plant may suffer some net loss of water and acquire a deficit in its water balance. The term *water deficit* refers to the difference between the actual water content and the content at full turgidity. Lower deficits, which can be tolerated by a plant, cause certain physiological changes, but only reversible ones; higher deficits cause irreversible changes, and the plant is damaged. The highest deficit which can be tolerated by a plant marks the limit above which there are *lethal deficits*. This *critical deficit* is different in different species and may be regarded as a measure of the desiccation resistance of the protoplasm. Lichens and mosses have high desiccation resistance, but plants with stomata are much more sensitive and some are damaged even by relatively small water deficits. Pisek and Berger (1938, page 153) determined the following critical values for herbs, broad-leaved and coniferous trees, Ericaceae and succulents, expressed as a percentage of the saturated weight of the leaves (see Biebl, 1962).

Picea abies	25
Pinus silvestris	44
Fagus silvatica (sun leaves)	21
Fagus silvatica (shade leaves)	29
Quercus robur	28
Betula pendula	34
Caltha palustris	49
Stellaria nemorum	50
Stachys recta	55
Veronica beccabunga	69
Arctostaphylos uva-ursi	46
Sedum maximum	46

These examples show that higher plants can also lose a large part of their water content without damage. Because of this, some species can, within limits, exist in dry places and survive periods of drought. The range of reversible deficits changes with age and development of the plant and with external conditions so that it is adaptable to some extent. For example, it is higher during dry summers and lower during rainy ones (Arvidsson, 1951).

Whether there is a relationship between the magnitude of the lethal deficit and the occurrence of the species in particular habitats is not clear. It might be supposed that species in which the lethal deficit, and hence the desiccation resistance, could change, have an advantage in competition for space in the habitat. In an investigation of deficits in plants of the dry heath on the Baltic island of Öland, Arvidsson (1951) found that species which occurred both on the heath and in forest had more variable desiccation resistance than those found only on the heath.

3 Periodic leaf development. A number of plants avoid the most severe climatic conditions by periodic leaf development. Bud break and shoot extension occur, for example, at the beginning of the summer, or of the rainy season: when winter or the dry season begins the leaves or shoots fall. Trees and bushes green in summer or in the rainy season, geophytes, and annuals, which survive the unfavourable period in the form of seeds, are of this kind.

Xeromorphic features

1 Thickening of guard cells and cuticle. In xerophytes the stomata are often sunken in the epidermis so that the pathway through the guard cells is lengthened. In addition, lips and projections from the surface of the guard cells serve to make it even longer. The diffusion of gases is decreased by the increased path length. The guard cells have their greatest, and most rapid, regulatory effect when the opening is small, e.g. less than 0·002 mm. A change of 0·001 mm in stomatal aperture has a relatively large effect on transpiration rate if stomatal apertures are small, but only an insignificant effect if the stomata are wide open. Large numbers of stomata with small apertures are therefore more effective and more rapid regulators of transpiration than smaller numbers with wider aperture, even though the total pore area is the same. A characteristic feature of xerophytes is the high frequency of stomata in the leaf epidermis. The same is true of sun forms (species or leaves) in comparison with shade forms. For example, *Impatiens noli-tangere* has about 100 stomata per mm^2 in sun-grown seedlings, but only about 50 in shade-grown seedlings.

The impermeability of the cuticle is increased in some xerophytes by cutinisation of the guard cell walls bordering the stomatal pores or by the deposition of a thick covering of waxes on the surface of the cuticle.

These features reduce cuticular transpiration and residual stomatal transpiration. There is a certain amount of water loss even through 'closed' stomata, at any rate in large-celled leaves. The more a species has been able to reduce its uncontrolled water loss the greater the chance it has of cutting down harmful water loss in times of water shortage. Xerophytes have progressed furthest in this respect and are therefore better equipped than other plants to control their water balance. The following examples from an investigation made in Austria by Pisek and Berger (1938) show how greatly species differ. Cuticular transpiration (transpiration after the stomata have attained maximum closure, in mg water, per gram leaf or shoot, per hour) was measured in the same evaporation conditions.

Impatiens noli-tangere	110–160
Caltha palustris	45–75
Stachys recta	25–45
Betula pendula	20–35
Quercus robur	15–35
Oxalis acetosella	15–30
Arctostaphylosos uva-ursi	3–7
Hedera helix	2–4
Picea abies (leaves of the current year)	1·6
Pinus silvestris ,, ,, ,, ,, ,,	1·5
Opuntia camanchica	0·1

2 Surface area development. The more a plant spreads out its leaves and branches the greater the absorption of light and heat and the more effective the wind in replacing the air around the leaves and shoots. Plants with a large surface area per unit weight of leaf or shoot are therefore less suited to open habitats exposed to strong sun than those with a relatively smaller surface. The importance of surface area development to plant water balance is apparent from its variation among plant species and from the distribution of the species in habitats differing in insolation and dryness. A large surface area is a normal feature of plants in shady places and in certain moist habitats, whereas a small surface area is characteristic of plants in dry, very sunny habitats. The following examples from Pisek and Berger (1938) show the range of variation of surface area development. The figures give the surface area (dm^2) per unit saturated weight (g).

Impatiens noli-tangere	2·20
Oxalis acetosella	1·76
Fagus silv. (shade leaves)	1·62
Fagus silv. (sun leaves)	1·38
Pulmonaria officinalis	1·16
Betula pendula	1·05
Convolvulus arvensis	0·83
Veronica beccabunga	0·64
Arctostaphylosos uva-ursi	0·46
Picea abies (leaves of the current year)	0·42
Sedum maximum	0·12
Opuntia camanchica	0·03

Branch arrangement, leaf position and leaf movements may also affect the exposure of the plant to light. Short dense shoots and leaves, particularly those grouped into cylindrical, club-shaped or spherical formations, give the plant as a whole a small surface area and therefore give better protection against insolation and wind. The orientation of the leaf and also temporary or permanent inrolling of the leaf edges or the whole leaf may have the same effect.

3 Sclerenchyma: relative vessel lumen. A normal feature of the construction of a xerophyte is the large number of sclerenchymatous elements. The leaf veins are closely packed, the vascular bundles and groups of fibres are well developed, the epidermal cells are sometimes thick-walled and the leaves are firm and leathery. One of the favourable consequences is that the leaves are protected against wilting if water loss is so great that the cells lose turgidity. Wilting is harmful: it deforms the tissues and the vascular system, so that the passage of water is hindered, in extreme cases the stems and petioles may even break. This mechanical strengthening is accompanied by change in the conductivity of the xylem elements. Conductivity can be related to the dimensions of the conducting channels (L), i.e. the sum of the cross-sectional areas of the lumina of the xylem elements in an organ, e.g. a stem. If the weight of those tissues which obtain their water by conduction through L is termed G, the ratio L to G is an expression for the *relative vessel lumen* available to the organ. This ratio changes with degree of xeromorphy and with habitat conditions. For example, it is largest in heath plants, smaller in meadow plants, and smallest in aquatic plants; in trees it increases with height above the ground and is therefore larger for the topmost branches than for the lower ones.

4 Water storage. The water economy of succulents is based on the principle of building up a store of water in the plant when there is rain and utilising it economically during a subsequent dry period. Because of the massive water storage tissue in the leaves (leaf succulents) or stems (stem succulents), succulent species can store many times as much water as other plants. Their ability to retain water is also great, because of sunken stomata and wax-covered cuticle (see 1) and small surface area (2). But the osmotic values of the cells are strikingly low compared with other plants in the same habitat. The root system is shallow but extensive and therefore comes rapidly into contact with water even after only a small amount of rain.

Hydromorphic features

Aerenchyma. In places where water is abundant, gas exchange in the soil is hindered by the water, so that the oxygen supply to the roots, or other living material in the soil, is poor. The gas contained in the intercellular space system of the tissues, and conduction through this system, probably allows the aquatic plants (hydrophytes) and amphibious plants of fens and bogs (helophytes) to tolerate the wet soil and water. There are large intercellular spaces because of the presence of aerenchyma, which is a type of parenchyma including wide pores and channels. Plants growing on well-aerated soil lack aerenchyma and are usually sensitive to flooding; they need a lower soil water content than the helophytes and are therefore displaced if the water table rises. For the same reason, many cultivated plants are damaged if the pore space in the soil fills with water.

Combinations of adaptations

These physiological and morphological features which affect the water balance enable plants to live in different conditions of dryness. Aerenchyma counteracts the disadvantages of an excess of water; other special features promote conservation of water. They are complementary to the cuticle-periderm-stomata-root system which, because of certain imperfections, can only maintain the water balance in the least demanding habitats, such as damp, shady places. A single species usually has only one or a few of the anatomical, morphological or physiological characters which have been mentioned, in one of several possible combinations. A particular combination will be most suitable for invasion of a certain locality, and for survival there. Hence, these combinations of adaptations will be one of the factors which determine whether a species occurs in a habitat and the formation of communities.

Soil water

Different kinds of soil water

In the soil, water exists in the free and in the bound state.

1 Free water. Free water, that is water which can move in response to gravity, occurs as *surface water, gravitational water* and *ground water*.

SURFACE WATER forms bodies of water or streams above the ground, moves downwards as gravitational water, or evaporates.

GRAVITATIONAL WATER moves in response to gravity below the soil surface. It forms underground rivers and ground water.

GROUND WATER is water which collects in the soil, in cracks, holes, and in all spaces bigger than the capillaries. It moves easily in a vertical direction, but more slowly horizontally (Tables 1 and 2).

TABLE 1 Rate of horizontal water movement in different soils, calculated for a slope of 1:54 (Rindell, 1919, according to Tamm, 1940).

SOIL	PARTICLE SIZE (mm)	RATE OF WATER MOVEMENT (m/year)
Fine sand	0·01–0·2	150–500
Medium sand	0·2 –0·5	750
Coarse sand	>0·5	3 500
Gravel	1·0 –5·0	10 000

The rate of movement depends on the porosity and thereby on the soil structure. In sandy, compressed boulder clay, Hesselman (according to Tamm, 1931) found a rate of sideways flow of 5·5 m/year when the slope of the ground was 1 in 30 and 8·7 m/y when it was 1 in 15: the coarser the soil, the more rapid the water movement (Table 1). In holes in the ground, wells, for example, water collects from the surrounding soil until the height of the surface of the water (the water table) corresponds to the pressure in the pore system from which the water comes. The depth of the water table increases and decreases in response to precipitation and evaporation and is affected by drainage.

TABLE 2 The permeability in horizontal and vertical directions of peat humified to different degrees. Rate of water movement through sample slice 0·1 m² surface area and 5 cm thick, pressure 2 cm water. (After Malmström, 1928.)

TYPE OF PEAT	DEGREE OF HUMI- FICATION (ON A 10 POINT SCALE)	ORIGIN POSITION OF THE SAMPLE IN THE SOIL	VOLUME OF WATER PASSING THROUGH (litre/h)
Scirpus caespitosus —			
Sphagnum	2	horizontal	5·5
		vertical	29·4
Sphagnum fuscum	3	h	12·3
		v	59·4
,, ,,	4–5	h	2·5
		v	7·5
,, ,,	7	h	0·24
		v	0·24
Dy		h	0·016
		v	0·036

In soils rich in colloids or other fine material, for example, in some non-cracking clays and in highly humified peat, there is only an insignificant amount of free water. Since the movement of water is also insignificant there is usually no water table (Table 2). In peat, water is often at different heights in holes lying close to one another.

2 Bound water. Soil materials are generally hydrophilic, and the particle surfaces are wet unless drying has proceeded so far that air has completely replaced the water. Some of the water adheres firmly to the surfaces of stones, gravel, sand, etc., forming a film; this is called *film water*. The water nearest the surface is bound most firmly, further from the surface less so, and furthest out it is held only by surface tension. At a certain film thickness the effect of the force of gravity is greater than the forces maintaining the film and the water runs away as free water.

The water films of two wet surfaces close to one another join together and form a concave meniscus as a result of the attraction between the water molecules and between the water and the surfaces. Together, the two surfaces can support, and lift, a layer of water, *capillary water*, greater than that which either can support separately, an effect which is most pronounced when the water is surrounded on all sides by wet surfaces, as it is in a capillary tube. The narrower the capillary, the greater the height to which the water rises (this height is inversely proportional to the radius of the capillary: the constant of proportionality is the coefficient of surface tension). In soil, the particles lie at varying distances from one another and form irregular pores, but when the surfaces of the particles lie close together, capillary rise can occur. Two particles which touch one another form an angular space in which water is held by capillarity. The water furthest in is held most firmly. The conditions for capillary movement depend on the extent to which these capillaries are connected with one another. The finer the soil, the smaller the capillaries and the greater the force required to move the water.

If the particle surface is very dry, adhesion of water is hindered by adsorbed gases, and movement of water from wet to dry soil is therefore insignificant at first. For the same reason, rain water may run off dry soil without being absorbed.

In organic soil particles, and in soil colloids, water is taken up into or between the micelles and colloidal elements, which are therefore pushed apart so that there is an increase in particle volume (see page 271). This water, called *imbibitional water*, is held more firmly than that in the capillaries. Water is also bound firmly to salts. The soil water is a dilute salt solution, which becomes more concentrated as the soil dries out and therefore binds the remaining water in dry soil very firmly (tens or hundreds of bars); such water is termed *osmotically bound water*. When the soil is dry, and there is an equilibrium between the water vapour pressure in the soil and in the air, the soil may take up or give off water as the humidity of the air changes, and any water taken up is called *hygroscopic*. Lebedeff (1927) found that the water content of the uppermost 1 cm layer of chernozem soil increased from 1·99 per cent to 5·26 per cent in twelve hours at night, while the relative humidity of the air varied between 50 and 92 per cent; some of the additional water came from the air, some as vapour from deeper soil. The latter is particularly important as an example of one of the ways in which water moves in the soil.

The forces binding these categories of water include electrostatic attraction between water molecules and electrically charged particles or particle surfaces. Because of their dipolar character, water molecules are attracted to these particles or surfaces and form a sheath around them, a phenomenon called *hydration*. The water nearest the surface is bound most firmly, that further out is bound more loosely, and still further out there are free water molecules (see Fig. 60). Bound soil water consists partly of such *hydration water*.

The force by which water is bound can be measured in various ways, e.g. by salt solutions which take up water from the soil (solution hygrometer, Hansen, 1926; Gradmann, 1928), or by water-filled porous pots, embedded in the soil. The water in the pots is in liquid film continuity with that in the soil and the pots are connected to manometers, on which the suction force can be read (tensiometers). (For further reading see, e.g., Baver, 1970; Walter, 1951; Slatyer, 1967; Kramer, 1969.)

Bound water cannot run-off and therefore cannot be removed by drainage. Neither can it be pressed out. However, both bound and free water can move by diffusion or because of vapour pressure differences and is called *diffusion water* or *vapour*. Usually, vapour movement is outwards from the soil surface to the atmosphere, but it can also be in the opposite direction. Temperature changes may accelerate or inhibit the process and therefore affect the movement of water: when the soil surface cools, during the night or during winter, the vapour pressure is reduced and water vapour may move to the surface from the air or from deeper soil and condense there, and when the surface warms up, some of the water evaporates, some goes back into the soil. Litter or a plant cover reduce these changes in temperature which are therefore of most significance in bare soil or soil with little vegetation.

During and immediately after rain there is free water (surface water or gravitational water), which can be taken up relatively easily by plants, and also bound water (capillary water, hydration water, etc.), which is more or less easily available, depending on the way it is bound. Imbibitional water and hydration water are firmly bound and can be utilised by plants only if the roots develop a suction force of several bars or tens of bars.

3 The suction force of the plant, the air and the soil. The most easily available water, the surface water and gravitational water, is lost first, through uptake by plants and by evaporation, then capillary water is lost, and then the other fractions in turn, depending on how firmly they are bound. The water remaining becomes more and more firmly bound and needs larger and larger suction forces to remove it (Fig. 18). As this occurs, the water deficit in the root cells, the vascular tissue and the leaf cells increases, the cell sap becomes more concentrated, and the cell suction force increases, so that even relatively firmly bound water can be utilised (Fig. 18). However, as the deficit increases, the cells lose turgor, the turgidity of the tissues is reduced and gradually the plant wilts. At first the cells regain turgidity during the night, when water uptake but little or no transpiration occurs. However, if the drought continues, the deficit gets larger every day and sooner or later reaches a lethal point at which permanent wilting begins.

If evaporation were the same every day, first wilting and permanent wilting would begin when the suction force of the remaining soil water reached particular values corresponding to these changes in cell turgor, and could be used as indicators of the condition of the soil water, and attempts have been made to use the turgor status of the plant in this way. In the literature, the terms 'permanent wilting percentage' (German *Welkungspunkt*) have been used to denote a particular stage of drying out of the soil. Of the two stages, first wilting and permanent wilting (the onset of a lethal deficit), the first is more easily determined: it is reversible, and begins fairly rapidly, whereas permanent wilting is irreversible and begins more slowly, and earlier in some sorts of cells than in others.

Soil permeability to water

The penetration of water downwards in soil depends on the amount of non-capillary pores and on the wettability of the particles, wettability depending partly on the dryness of particle surfaces. In addition, infiltration can be hindered by trapped air. Coarse soils, such as sands or gravels, are more permeable than boulder clay, for example, in which the permeability depends on the proportions of fine and coarse material. The higher the clay content, the lower the permeability, since the clay particles fill the pores between the coarser particles.

In the same way, the pores between the coarser particles in silt and clay soils are filled by colloidal clay. The permeability of clay soils is increased by freezing and by drying out to the extent that aggregation is increased and cracks are formed. Aggregation influences the permeability of mull soils and the humification of peat. Highly humified peat has low permeability, while in peat which is little humified the permeability is relatively high.

Permeability varies with the changes which affect the porosity. In soils which are rich in colloids, the permeability decreases as water is taken up, since swelling of the colloids reduces the pore volume; calcium-saturated soils change less in this way than do sodium-saturated soils because of the weaker hydration of calcium ions. Swelling also closes some of the cracks in clay (see Chapter 13 for further discussion).

Permeability influences water uptake by the soil after rain, snow thaw or from other sources. It therefore also affects surface run-off; the less permeable the soil, the more water runs off the surface. On clay soils and dense humus soils, where there is little infiltration, there is more risk of run-off and erosion by water than on other soils. Litter and a well-developed plant cover reduce erosion.

Soil water capacity

Water which filters down through the soil moves along three paths: in wide, non-capillary pores, in capillary pores, and along diffusion paths in airspaces or colloids. When these pores are full then the soil is water-saturated, but the amount needed for saturation depends on the water-holding capacity of the soil. This can be determined as the maximum water-holding capacity of the soil, which is the amount of water per unit weight or volume of water-saturated soil. Soil which has been under standing water for some time is water-saturated, as is the soil below the water table. Soils at the bottom of lakes, and, periodically, soils of marshes and lake and sea-shores, are in this condition, and, as a result, vegetation on them has several special hydrophytic characters. Various difficulties arise in determining the water-holding capacity of soils in natural conditions because the movement of gravitational water takes time, often two or three days or more, during which water is lost by evaporation and by transpiration, if plants are growing there. In practice, the term *field capacity* is used to denote the water content measured in the soil after the gravitational water has drained away (although perhaps not completely) for two or three days, during which time there has been evaporation and transpiration.

The field capacity of the soil is of primary importance in determining the water supply of plants: the finer the soil, and the more colloids and capillaries, the larger the field capacity. It is largest in clay, humus and peat soils. The following values have been measured: 270 per cent by weight for inorganic colloids and 440 per cent for organic colloids (Lutz and Chandler, 1946); 75–1488 per cent for *Sphagnum* peat (Malmström, 1928) and 13 per cent by weight for sandy boulder clay (corresponding to *c*. 34 per cent by volume, Tamm, 1940). The larger the field capacity, the more water is stored in the surface layers of the soil, and the more even and long-lasting the water supply to the roots. If field capacity is small, water sinks deeper into the soil and becomes less easily available to plants, although roots which can penetrate down to deep-lying stores of water have a relatively long-lasting supply, which is protected against evaporation.

Near the water table the soil is usually water-saturated, but in the layers above it is usually less moist. Only exceptionally is the water-holding capacity of soils on dry land fully used. Soil moisture is at a maximum after the snow has melted or after a long period of rain; after shorter periods of rain the soil is not generally brought to field capacity completely.

Very dry soil can only slowly be rewetted. The air in the capillaries prevents the penetration of water, and swelling of the colloids also takes a long time. When rain starts, penetration into the uppermost layer of the soil starts at once, but since saturation of the capillaries and the colloids takes some time, the storage of the top layer cannot immediately be filled, and water also passes through to deeper layers. Thus, water is taken up in several horizontal zones simultaneously. Saturation of the litter cover takes a long time. A lysimeter experiment in a coniferous wood with a moss and litter cover showed that three consecutive rainy days were needed to saturate these two layers, and that the water taken up was equivalent to 3·9, 3·6 and 3·0 mm of rain, respectively, and that 2·6, 22·4 and 18·2 mm, respectively, filtered through to deeper layers (Stålfelt, 1944).

A measure of the actual soil moisture content can be obtained by calculating the *relative soil moisture content* (R) as $R = (f . 100)/F$, where f is actual water content of the soil and F is the field capacity (Halden, 1926). This relative soil water content can be used as an expression for the actual water supply of the plants, provided that comparison is limited to a particular species on a particular type of soil; it cannot be used to compare the water supply on different soils or for different species, because water uptake by plants does not depend only on the amount of water in the soil, but also on the forces which bind it and on the suction force which can develop in the plant. The plant and the soil compete for water, both using those forces (of osmosis, or adhesion) which they can develop.

Plants themselves have a direct effect on the water-holding capacity of the soil through their production of humus. When sands or gravels, or other soils which retain only a small amount of water, are mixed with humus, some of the non-capillary pores become capillary and the colloidal component of the water capacity is increased. When colloids swell, other pores may become smaller and therefore of capillary dimensions. Hence during a rainy period, when water is continuously absorbed into the soil, the field capacity automatically increases.

Various agricultural practices aim at altering the water-holding capacity of the soil. Rolling arable land compresses some of the non-capillary pores to capillaries, or the addition of humus or of clay to dry sandy and gravelly soil fills the non-capillary pores with capillary-forming colloids, but not all of the water retained in the soil by such means need necessarily be available to plants, because of the large forces binding it in the colloids.

The actual water content of the soil depends on several factors: the water-holding capacity and permeability of the soil and the distance from the water table; the precipitation, run-off and evaporation (temperature, wind, relative humidity) and the consumption of water by the vegetation.

Water use by plants

Methods of measurement

The study of the water balance of plants is an old one, but in spite of this there is no known method of measuring directly the amount of water used by a plant under natural conditions. However, there are several methods for indirect measurement.

1 Plants are planted in pots or other containers and the water loss is determined by weighing. The disadvantage of this procedure is that generally it is suitable only for small plants, the rooting space is abnormally small, and roots are damaged by transplanting.

If the experiment is done on a large scale and the plants are grown in large containers sunk into the ground (lysimeters), larger plants, areas of vegetation and even small trees can be studied. However, the difficulty of distinguishing between water evaporating directly from the soil surface and water lost by transpiration is increased.

2 Another method is to pass air over leaves, shoots or whole plants and to determine the increase in the humidity. This can be done gravimetrically, by passing the air through a water-absorbing medium and determining the increase in weight, or, more commonly now, electrically, by passing the air over a psychrometer or an electronic humidity sensor. However, the leaves or plants must be enclosed in a chamber for this method and the rate of water loss then depends on the environmental conditions in the chamber (e.g. temperature, humidity, air movement) which may be quite unlike those outside in the natural environment.

3 Attempts have been made to calculate water loss from whole stands of plants, e.g. a piece of forest, from measured values of some of the following: temperature, vapour pressure, wind speed, turbulent transfer and the radiation balance. Variation in these factors is very large, both between points and between occasions, but with appropriate sampling techniques, electronic sensors and recording equipment, these are probably the most promising methods of the future.

4 The most common method is the 'cut-leaf' method. A leaf or shoot is cut off the plant and its water loss during the subsequent minutes is determined by weighing on a rapid-weighing torsion balance. One of the disadvantages of this method is that the weighed part may be small in relation to the whole plant, e.g. to a tree whose water loss is to be estimated. Another is that the transpiration from the weighed shoot may be altered by cutting it off and by changing its environment.

In spite of the disadvantages, measurements made using two or more methods give broadly similar results.

A comparison between the water use of different species or communities requires the choice of a basis for comparison: possible bases are the surface area, weight or volume of the plant or of the ground covered by the plant or permeated by its roots. The nature of the investigation determines which will be best. If the water balance of a species or a community is to be studied, knowledge of water use and of water supply is required, i.e. of transpiration and of precipitation, or, more exactly, of the water supply to the ground occupied by the roots of the plants. Hence, in this case, the water balance should be expressed per unit of ground surface. Since precipitation is usually expressed in millimetres it is convenient to express water use in the same units. The problem is then to determine how many millimetres of precipitation are available to the plant and how many it consumes in transpiration.

The dependence of transpiration on water supply

If transpiration of a tree is measured for several consecutive days, using the cut-leaf method, it is found that the rate of water loss changes not only in relation to changes in the factors affecting evaporation (temperature, relative humidity and wind speed), but also in relation to the water supply in the soil. Immediately after the soil round the roots has been wetted transpiration is high, but after only two or three days it is less and it continues to decrease as long as the weather remains dry. A spruce tree (in South Sweden) transpired 1·9 kg water per kg leaves per day, after a lot of rain, but only 0·8 kg after three weeks of dry weather. Pisek and Tranquillini (1951) measured transpiration of spruce trees in forests in Austria and under similar conditions observed decreases of about the same amount. The stomatal regulation system is responsible for this change; it cuts down water loss when the roots have used up the easily available water in their immediate vicinity. As the soil dries out, the water deficit in the cells increases and the remaining water is used less rapidly. Water consumption by the plant is therefore dependent on the amount of water in the

TABLE 3 Water use by pine and spruce, expressed as mm precipitation (litre water per m² soil surface). (From measurements in Austria, Germany and Sweden.)

	DAILY TRANSPIRATION (mm)	ANNUAL TRANSPIRATION (mm)	EXPERIMENTAL PERIOD	PLACE	METHOD	AUTHOR
37 year spruce	2·1–4·9		8 days July–Sept.	Tharandt	Cut-leaf	Schubert, 1939
100 year spruce	—	320	—	,,	,,	,,
4–8 year pine	—	300	veg. period	Eberswalde	Lysimeter	,,
Spruce 3–4 metre in height	1·5	250	4 days June–Aug.	Innsbruck	Cut-leaf	Pisek & Cartellieri, 1939, 1941
40 year spruce	1·9	211	May–Aug.	North Skåne (dry place)	Cut-leaf	Stålfelt, 1944
,,	3·4	378	,,	North Skåne (wet place)	,,	,,
Stand of spruce	4·3					Polster, 1950

soil; it depends also on the type of plant and on the density of the vegetation. Dense vegetation uses the most easily available water more rapidly so that there is a greater lack of water during drought.

Table 3 is a summary of information about the transpiration of conifers, from Sweden, Germany and Austria. The trees are of different sizes and the methods are different but nevertheless the results are broadly similar. Spruces and pines of various ages use about 200–300 mm precipitation per year. Russian investigations gave similar results (Molchanov, 1963). Depending on the climatic conditions, the soil and the character of the forest, water use was between 140 and 430 mm per year.

TABLE 4 Transpiration of various plant communities, expressed as mm precipitation. (From measurements in Austria and Germany.)

COMMUNITY	DAILY TRANSPIRA- TION (SUMMER) (mm)	MONTHLY TRANSPIRA- TION (SUMMER) (mm)	ANNUAL TRANSPIRA- TION (mm)	PLACE	AUTHOR
Alpine dwarf shrub heath		30	90	Innsbruck	Pisek and Cartellieri, 1941, page 278
Arctostaphylos— Calluna ass.		47	189	,,	,,
Dry meadow (Brometum)		60	195	,,	,,
Damp meadow (Arrhenateretum)		101	323	,,	,,
Wet meadow (*Caltha-Cirsium olerac.* ass.)		364	1125	,,	,,
Birch stand		78	349	,,	,,
Hazel stand		92	414	,,	,,
Iris pseudacorus stand	20–58			Berlin	Kiendl, 1954
Glyceria maxima stand	18–45		< 1600	,,	,,
Phragmites stand			500–1300	,,	,,

In general, transpiration is highest on days following rain and then rapidly falls as soil moisture decreases. Similarly, water use in a moist place with a high water table is greater than in a dryer place (for example, the figures of 378 and 211 mm in Table 3). The dependence of water use on water supply is shown more clearly by Table 4, in which communities with different water supplies are compared, and it can be seen that consumption in hydrophytic communities and moist meadows is many times greater than on dry heaths.

Other experimental results point in the same direction and show that the highest transpiration occurs not only in response to conditions causing high evaporation, but also, and to a greater degree, in response to good water supply. According to investigations made by Henrici (1946, page 13) in Pretoria, South Africa, *Pinus insignis* used *c.* 760 mm per year, *Eucalyptus* 1200 mm, and *Acacia molissima* 2500 mm. *Eucalyptus* and *A. molissima* were growing in marshy areas; precipitation was 780 mm per year. This shows that *Eucalyptus* and *A. molissima* used a large amount of the ground water, having a draining effect on the marshy land.

FIG 19. Because some of the snow has been intercepted by the trees, the forest floor may be free of
snow when open fields are still snow-covered, especially
in early spring.

Thus, the water turnover of plants is determined primarily by water supply and evaporation. If there is an abundant water supply, there is no reduction in its use and transpiration reaches maximum values. If there is a lack of water, plants can reduce their consumption, by stomatal regulation cutting down transpiration, to a fraction of the maximum.

The amount of precipitation which reaches the roots

Only part of the precipitation which falls reaches the roots, the rest evaporates from the soil surface or the surface of the vegetation or is lost as surface run-off.

An analysis of the plant water balance therefore requires knowledge of the precipitation reaching the rooting volume in the soil. If this is to be compared with the precipitation elsewhere in the locality, the proportions of precipitation reaching the various other layers of the vegetation must also be known.

Stålfelt (1944) made an analysis of precipitation in a forty-year-old spruce stand in south Sweden. Measurements were made for sixty-seven days (June to August), on seventeen of which rain fell. The closure of the canopy was such that about half the ground was covered by the tree crowns, the remainder was open. Precipitation was measured near the tree trunks, under the periphery of the crowns, under the gaps in the canopy, and in the open. The following values of precipitation were obtained:

in the open, 124 mm;
beneath gaps in the canopy, 92 mm, i.e. 74 per cent of that in the open;
on the crowns of the trees, 132 mm (106 per cent);
under the periphery of the crowns, 113 mm (91 per cent);
under the inner part of the crowns, 29 mm (23 per cent).

It is apparent from observations of this kind that precipitation on trees is greater than in the open; the reason is that raindrops usually fall more or less obliquely, so that less rain falls in the lee of the tree and hence the crowns receive relatively more water and the gaps in the canopy less.

Spruce trees also act as a roof to some extent. The effect is greatest near the trunk, because the water passing through the crown is led outwards. More drops therefore fall from the tips of the branches than near the trunk.

The retention of water in the crowns of trees and in the vegetation as a whole is termed *interception*, so that one may speak of the interception of a forest, the interception by mosses, etc.

Analysis of precipitation on vegetation should thus include determination of the interception by the various layers, the tree layer, field layer, etc. Such an analysis was made for the same forty-year-old spruce stand, on a moraine. Rain gauges and lysimeters were used (Stålfelt, 1944).

The results, summarised in Table 5, show the following:

1 In the stand of well-grown, forty-year-old spruce with the crowns covering half the ground, the rain shadow effect of the trees was such that the canopy received an extra 36–42 per cent of the precipitation in the open because in the gaps in the canopy precipitation was only 58–64 per cent of that in the open. Lukkala (1942) found similar values (20–50 per cent) for the rain shadow effect of trees in woods in Finland.

2. Interception by the crowns of the trees was up to rather more than half of the water falling on them. The value varied according to the height and density of the crowns. Because the stand was fairly open, the crowns were relatively dense. In addition, interception depends on the intensity and duration of the rain. A light and short fall of rain, *c.* 1 to 2 mm, generally fails to penetrate the tree layer. However, this is not without significance for the water balance of the trees, since it reduces transpiration, and, in

TABLE 5 Water balance in a forty-year-old spruce forest growing on moraine in North Skåne. The values are means for five spruce trees, four areas between trees, and four years. Mean precipitation for the four years, in the open field was 298 mm for May–Aug.; 495 mm Sept.–April; 793 mm total. The rain shadow effect of the trees is shown by the difference in precipitation between the open field and the areas between trees. (After Stålfelt, 1944.)

TREES RAIN—SHADOW EFFECT	PRECIPITATION (IN % OF PRECIPITATION IN THE OPEN FIELD)	
	MAY–AUG.	SEPT. TO APRIL (in mm)
Interception by trees	36	42
	57	58
Precipitation between trees	64	58
Interception by mosses, litter, etc. on soil surface		
under tree crowns	18	9
between trees	31	15
Precipitation reaching tree roots		
under crowns	24⎫ (90 mm)	(202 mm) 38⎫
between the trees	34⎭	44⎭
Trunk flow	0·1–2·0	
Transpiration by the trees		
dry place	211	
damp place	378	

addition, a certain amount may be taken up by the leaves. In general, the more intense the rain, and the longer it goes on, the higher the proportion that gets through the crowns.

3. The ground layer of mosses and litter intercepted significant amounts, which were larger beneath the gaps in the canopy (13–31 per cent of precipitation in the open) than under the trees (9–18 per cent). Interception by the ground layer was larger in summer than in winter. Other examples of interception are given by Ovington (1965). Rutter (1963) found that there was inter-

ception of 32 per cent of the annual precipitation of 700 mm in a Scots pine plantation in England. Law (1958) measured interception in a stand of *Picea sitchensis* (*c.* 10 m in height) and found it to be 43 per cent in summer and 36 per cent in winter. Annual precipitation was about 1000 mm.

4 The proportion of the precipitation which reached the tree roots, that which passed through the vegetation, was 24 per cent under the crowns in summer and 34 per cent in the gaps in the canopy; in winter it was 38 and 44 per cent, respectively.

5 The total water supply to the roots was 90 mm during May to August and 202 mm in the rest of the year, i.e. a total of 292 mm. In the same summer period 211 mm were used in transpiration by the trees, in a dry locality where the roots were not in contact with the ground water. This comparison shows that during the summer the forest generally used up the water received as precipitation during the summer months and, in addition, some of that stored in the soil in the autumn and winter months. If there had been more water available the trees would have used more, as shown by the fact that transpiration of trees in a wetter locality (where the roots were in contact with the ground water) was 378 mm, which was more than the precipitation reaching the root zone during the year. Trees in a wet place therefore have a draining effect. The ground water which they used did not come from the afforested boulder clay but from fens, bogs, lakes and water-courses, which in that locality cover an area about as great as the dry land.

The selective effect of the water balance

The soil moisture content is determined primarily by the ratio of precipitation to evaporation, i.e. by the humidity or aridity of the climate. Climates in which there is more precipitation than evaporation, and which therefore give rise to water-courses, are termed *humid*: climates in which all the precipitation evaporates, or which give rise only to local rivers which rapidly dry up, are termed *arid*.

The distribution of plant species in regions of varying aridity or humidity and in regions with different water tables is a result of selection which has brought together species of a similar hydromorphic or xeromorphic character, and with similar physiological attributes of drought resistance, or ability to survive flooding. Through selection and competition these plant communities come to occupy a particular position in relation to water. Vegetational zonations have developed, for example, from a lake to dry land, from the foot of a mountain upwards, and on a larger scale, from humid regions in the north to arid ones in the south.

From a lake to dry land, successive zones of hydrophytes, helophytes, mesophytes and xerophytes can be distinguished:

1 The open water surface supports plankton communities (*Cyanophyceae*, diatoms, *Chlorophyceae*, etc.) with or without floating or submerged phanerogams (*Lemna, Hydrocharis, Utricularia, Ceratophyllum*, etc.) and cryptogams (*Sphagnum* sp., *Hypnum* sp. etc.).

2 The floating-leaf zone (species of *Nymphaea, Nuphar, Potamogeton, Ranunculus*, etc.).

3 The reed zone (species of *Scirpus, Phragmites, Typha, Butomus, Sparganium, Acorus*, etc.).

4 The carr zone (closed stands of sedges, with occasional herbs, trees and bushes, for example, species of *Alisma, Iris, Rumex, Caltha, Ranunculus, Lythrum, Oenanthe, Cicuta*; alder, pine, birch, aspen, juniper, *Vaccinium* spp. etc.). Since prehistoric times the carr zone has been exploited by man and used for mowing and for grazing, which has partially changed it to meadow.

5 The dry land zone. From the water's edge inwards over the land there is often a series of communities which have replaced the original forest. Where the forest grew on fertile, flat ground, in inhabited regions it has been felled and replaced first by meadow (for mowing and for grazing), then by arable land. On ground which is rocky or infertile the forest remains, with scrub nearer to the water. As one moves away and the ground rises there is often a zonation, for example, with the following series:

(a) Scrub (*Rosa, Rubus, Prunus*, etc.)

(b) Deciduous wood (birch, aspen, oak and, on fertile soil, lime, ash, elm and other more demanding trees, also whitebeam, guelder rose, honeysuckle, *Ribes* species, herbs including many hydrophytes, grasses, ferns)

(c) Spruce forest (spruce, occasional deciduous trees, mosses)

(d) Pine heath (pine, lichens, mosses, *Ericaceous* species)

(e) Lichen heath (occasional pines, lichens, mosses)

On a large scale, such a zonation occurs over whole countries and continents. Between the tundra communities of Arctic regions and the deserts of the subequatorial belt, the species and communities are mainly distributed according to their demands for, and their hardiness to, water and temperature. They form plant belts which succeed one another in a particular sequence, for example, heaths, coniferous forests, deciduous forests, maquis scrub (sclerophyllous species) steppes, semideserts, deserts. In the same way there is a zonation across the equatorial and subequatorial latitudes, where rain forest, forest, savanna, semideserts and deserts follow one another, and grade into one another (for further information see Sjörs, 1956; Walter, 1951).

Within these zones and belts, the occurrence of individual species and the composition of the community is determined mainly by the balance between the availability of water to the plants and the responses of the plants to a shortage of water. On the other hand, the availability of water in the soil is dependent upon the vegetation. The equilibrium between water consumption by the vegetation and the soil moisture content settles at a certain level, so that there is a balance between availability and use. This equilibrium is disturbed by a change in the composition of the plant community or in the water supply, for example, by cultivation of the land, during which ground is broken up, the natural plant community is replaced by a community of cultivated plants, and sometimes the ground must be drained. One of the many factors affecting the results of cultivation is the water balance of the cultivated species. Is it optimal or is there a deficit or an excess of water? In wet areas an excess of water is common; drainage may counteract this. In arid regions it is usually the lack of water which limits cultivation. Water deficits may also arise in humid or fairly humid regions where the soil holds relatively little water (in elevated sand or gravel soils, moraines) but supports forest or another plant community with a large demand for water.

The balance between water supply and water use

In Table 5 it can be seen that a tree in a dry locality transpires the rain which falls in the summer and also a part of the winter precipitation, so that there is probably very little or no water contributed to the ground water. Measurements of the amount of gravitational water and other soil water in the experimental area showed that the precipitation which penetrated the vegetation was mostly retained in the uppermost 20 cm layer of the soil. Only 35–45 mm per year reached a depth of 75–150 cm, and no gravitational water penetrated deeper than this. However, some water from deeper layers, i.e. from the ground water, was supplied to the soil below the 150 cm limit since during the winter the water moved upwards at this depth (Stålfelt, 1944).

The lysimeter measurements in Eberswalde (Germany) (Table 3) showed that even pine trees only four to eight years old can use up so much of the precipitation that there is no addition to the ground water during the summer months (Schubert, 1939), thus a water deficit probably developed in these young pine trees as well as in the forty-year-old spruces. Water use might have been larger if more water had been available. It was not possible to find out whether a deficit developed which was sufficient to inhibit the growth of the trees. However, growth is inhibited in places where the climate is arid or semiarid. The soil moisture content in such places is still more dependent on the plant cover. According to Walter (1939, page 813), *Acacia* plantations in South Africa have caused neighbouring springs to dry up; *Robinia* planted on steppe in south Russia grew normally at first, while the trees were small, but after several years, when the trees had grown large, the water available was insufficient. Thornthwaite and Mather (1951) showed that summer rainfall in an experimental area near Birmingham, England, was insufficient for the needs of the vegetation and a deficit of about 100 mm developed. In autumn and winter a small excess built up and streams were formed. If the ground water does not compensate for the low precipitation, a soil water deficit builds up. In a pinewood in Berkshire, England, Rutter (1967) found that the soil water deficit increased from April to July, the deficit disappeared gradually from August to November, and from December to March there was some run-off and percolation through the soil.

This sort of information shows that vegetation with a large water requirement can drain wet ground. This mostly occurs in wooded areas since forest, with its many layers of vegetation and large area of leaves, is one of the largest water consumers among the plant communities of dry land. Direct observations of this draining action have been described in the literature, but are relatively rare as they are in their infancy. Tamm (1955) cites examples from south and central Sweden where stands of alder and ash have drained the ground so that high productivity has become possible. Marshy areas in Italy have been drained with the help of *Eucalyptus* plantations (Walter, 1932, p. 504). A similar effect of *Eucalyptus* and *Acacia* in South Africa has already been mentioned. Thurmann-Moe (1941) gives further examples.

Where ground has been drained by a forest, a rising water table and increased run-off would be expected if the forest were felled. Such an effect has been recorded in several instances. Thurmann-Moe (1941, page 73) found that the water table rose 22–27 cm as a result of felling a mixed spruce and birch wood in Norway (Hedmark); and 14–24 cm as a result of felling spruce. Lukkala's (1942) investigations in Finland showed that the water table was higher on clear-felled ground than on forested ground. Similar changes in the soil water balance may come about without any intervention by man. Love (1955) cites an interesting example: stream-flow increased after beetle infestation of the forest of the White River plateau in western Colorado had killed the trees over more than

700 square miles. Similar instances in East Prussia have been described by Breitenfeld and Mothes (1940) and in Russia by Molchanov (1963). In a New Hampshire Forest, cutting the vegetation in a 15 hectare watershed resulted in increased water output. The run-off for 1966 exceeded the expected amount by about 40 per cent (Bormann and Likens, 1970).

Soil erosion is another possible consequence of disturbing the equilibrium between water consumption by the vegetation and water content of the soil. If the vegetation is removed, no water is intercepted; all the rain reaches the ground; surface water run-off increases and channels may form and erode the wet soil, e.g. in arable areas and on road embankments (for further discussion see Chapter 17).

What has been said so far about the relation between precipitation and vegetation has referred to the precipitation which reaches the roots or is intercepted by the aerial parts of the plants. Precipitation retained in the surface layers of the soil evaporates more slowly in the forest than in the open, so that after rain the surface remains wet longer. The common misconception that forest protects the soil from drying-out is probably based on this easily made observation.

4

Litter and litter production

Organic soil types are derived from the waste products of the populations of living organisms, consisting of dead plants or plant parts, dead animals and the excretory products of animals.

Plant material is quantitatively most important. Abcission of various parts is a normal occurrence for higher plants, and goes on throughout the life of the plants, and when the plant as a whole dies, all the material it has produced goes towards soil formation. Wherever plants grow, the ground is covered by a layer of dead plant material which may be up to several centimetres thick, for example, in a deciduous wood. This forest material was formerly used as a floor covering in cowsheds and barns, with the result that organic material was transferred to arable land, the fertility of which was increased at the expense of the forest. Mineral fertilisers were first used in agriculture in the second half of the nineteenth century: before that any form of nutrient supply to arable land was of the greatest importance since productivity was limited by nutrient deficiency to a much greater extent than it is nowadays. Man's interest in the dead plant material is shown by the fact that there is often a special word for it: *litter* in English: *Streu* in German; *förna* or *strö* in Swedish. Litter consists of dead plants or parts of plants in which the processes of decomposition have not yet gone far enough to destroy the morphological character of the organs. The division between litter and its products is not a sharp one. The term is also applied to dead animals or parts of animals and excrement. *Plant litter* and *animal litter* can be distinguished. Litter deposited in water is broken down in a different way from that on dry land.

Different kinds of plant litter

Sernander (1918) recognises several categories of litter, dividing them according to which layer of the plant community they come from.

1 Fallen litter. This comprises material from the tree layers and the field layers, made up of many different organs and parts of organs. It falls more or less seasonally.

LEAF FALL is quantitatively the most important, and is of particular significance to the soil because the leaves consist of relatively young tissues with a lot of parenchyma, which is easily humified, and nutrients, important to the animals, fungi and bacteria in the soil.

SHOOT FALL. This is greatest during autumn and winter. Some trees, such as aspen, poplar, willow and oak, normally drop

FIG 20. Beechwood, Skåne, South Sweden. The soil is covered by litter made up of last year's leaves, beechnuts, cupules and twigs.

branches by the development of an abcission layer at the base of the shoot. In many other deciduous trees and conifers only the dead shoots fall. Shoot fall of this sort includes *crown-thinning* and *branch-pruning*. Crown-thinning is the process by which the shoot system is thinned out, so that a certain illuminance is maintained within the crown. In branch-pruning, the lowermost branches on the trunk abciss leaving that part of the trunk free of branches. Shoot fall occurs for other reasons, too, for example, as a result of wind or weight of snow, disease, insect pests, bites by squirrels or other animals. During winter, the ground under spruce trees may be covered by shoots bitten off by squirrels.

Whole trees and bushes which die and fall down also belong in this category.

BARK FALL. Around trunks of old trees such as pine and *Eucalyptus* there is often a raised area, formed mainly of the bark which has been shed.

BUD SCALE FALL. This makes an annual contribution to the litter, but is quantitatively and probably also qualitatively only of minor significance.

FRUIT AND SEED FALL. Cupules of beech and oak, for example, and the cones of coniferous trees are particularly prominent in this group. The cones are often pulled apart by squirrels or crossbills. Significant amounts of readily available nutrients are supplied to the soil organisms with the seeds.

POLLEN-RAIN. This is quantitatively more significant than it might seem, particularly the enormous quantities produced by pine and spruce. Because of the high protein content this sort of litter is of high quality.

FALL OF PARASITES OR EPIPHYTES. Fungi, mosses and lichens which live parasitically or saprophytically on trunks or branches fall down to the ground. In wind or rain, significant amounts of lichens may fall.

FIELD LAYER LITTER. Dwarf shrubs have the same sort of leaf and shoot fall as trees. The shoot system of *Vaccinium myrtillus* is thinned in the same way as the crowns of trees, by the abcission of some of the current year's vegetative shoots. Shoot fall in heather is particularly copious.

The above-ground shoots of grasses and herbaceous plants usually fall as a whole forming thick litter layers which are edaphically important.

2 Ground layer litter. In the ground layer of the vegetation, consisting mainly of mosses and lichens together with the basal parts of the plants in the upper layers, an often considerable amount of litter is produced. Because of the mode of growth of the plants, the dead parts retain their connection with the living for a relatively long time, so that the same shoot or shoot system can be part of both the litter and the living vegetation. Mosses do this, and thus they may grow on top of their own litter: the same is true of a number of lichens. The decomposition of the litter may be fairly rapid, as it is for *Hylocomium* species, and so their litter layer is relatively shallow. In other cases, particularly for *Sphagnum*, the litter forms deep layers. Because of their bulk and other properties these layers have long been used for various practical purposes and so have been given a special name, *peat*. This term has a considerably wider meaning than the term *litter*. It includes not only dead moss remains which retain their morphological character and can therefore be identified, but also their decomposition products, in which the features of the organs are indistinguishable. Peat can be of high or low humification. There are thus different terms for different types of dead plant remains: peat, when the plant remains form deep, compact layers; and litter, when they form a loose and comparatively thin layer on the surface of the ground. Besides *Sphagnum*, other genera of mosses, such as *Drepanocladus*, *Campylium*, etc. and species of *Carex* and other sedges are important peat-formers.

3 Underground litter. Under the surface of the soil are a large number of plants and plant organs which by their death add litter to the soil. The root systems under a closed cover of vegetation are usually so dense that the soil is thoroughly exploited. The finest roots and root hairs have a relatively short life and are rapidly replaced. Kalela (1957) found that the amount of roots per unit volume of soil in a pine wood of the *Vaccinium* type increased from early spring to late July and then decreased, until at the end of October it was about the same as in early spring. The maximum amount, in the summer, was about double that in the winter; this cycle was independent of the age of the trees, but was affected, for example, by soil moisture. The roots were killed by drought and renewed after rain. In a watering experiment by Kalela the weight of roots increased by *c*. 40 per cent and the number of root tips increased fifty-fold. Heikurainen (1957) showed the same periodicity in bogs with dwarf shrubs and pines. He calculated that roots with a diameter of 1 mm or less were renewed about every third year and those with diameter 2–5 mm about every tenth year. The material contributed to the underground litter by this cycle of root replacement is probably quantitatively important, and nutrient-rich because of the large number of meristematic cells in the root tips.

Another contribution to the underground litter is made by underground stems. Underground parts of higher plants are often as large or larger than the above-ground parts. Mork (1946) found that about half the mass of the vegetation was underground in old stands of *Vaccinium myrtillus*, *V. vitis-idaea* and *Calluna vulgaris*. Other plants contributing to underground litter are the saprophytes (fungi and bacteria) which live on litter and after their death become part of it.

Because of their origin, the fallen litter, ground layer litter, and underground litter form layers on top of one another, but there is no sharp demarcation line between them. Where the layers meet, the different sorts of litter are mixed. The ground layer of vegetation grows through the fallen litter which is therefore covered by ground layer litter. The soil animals move from one layer to another and transfer litter from the uppermost layers down into the soil where it mixes with the underground litter, and so on.

Litter can also be classified according to the organs from which it originates; as leaf litter, bark litter, twig litter, wood litter, etc. (see Julin, 1948).

4 Underwater litter. On the bottoms of lakes and other bodies of fresh water and in bays of the sea, litter derived from plankton and higher plants accumulates. This may form a thick layer, particularly in places to which material is carried by wave action, and may fill shallow bays or lie in lines along the water's edge.

Quantity of litter

There are relatively large amounts of leaves, short shoots, parts of fruits and root hairs in litter, because all these organs have a short life. Root hairs function only for a few days or weeks; flowers, fruits, and the shoots on which they are borne, often last only one vegetation period, and so do the leaves of the majority of herbaceous plants, grasses and deciduous trees. However, the leaves of many conifers often have a life of several years, sometimes ten or more in spruce, but even so, the life of the organ is only a fraction of that of the tree as a whole. In most plants the foliage is partially or completely replaced annually; it is primarily the nutrient reserves of the plant which are used for this. Trees and shrubs growing in environments where the climate or soil is poor, so that the production of organic material is only just sufficient for the renewal of the foliage and vegetative short shoots, may therefore make very little growth in trunk or branches. The leaves and flowers which develop are soon converted to litter. If the nutrient supply is poor, the development of the crown may be stunted as in the dwarfed spruces found on high mountains or bogs. But even these make a significant contribution to litter formation.

The above-ground parts of herbs and grasses are often shed, partly or wholly, as early as the end of the vegetation period. The resulting litter is particularly valuable as it is mainly leaves and fruit material.

In pedology and ecology it is important to analyse, quantitatively and qualitatively, the different sorts of litter—leaves, short shoots, branches, fruit parts, bark, roots, etc. Such studies have primarily been made in forestry or agricultural research (see Bray and Gorham, 1964). The amount of litter formation in other vegetation is little known and requires further study.

In most cases, the principle that all living material sooner or later becomes litter can be applied; generally, the only exception is that of material destroyed by fire. A balance between photosynthetic production and litter production is therefore to be expected, although litter is often moved from one place to another by animals, man, wind or water, for instance, harvested crops are removed from arable land; timber, seeds and fruits, litter for cowsheds, etc., are removed from forests. Felling timber affects litter

TABLE 6 Dry matter production by arable crops.

	DRY MATTER PRODUCTION (kg/ha/year)	
Autumn-sown wheat (grain + straw)		
normal yield	500– 9 000	Åkerman, 1951
maximum yield	11 000	Müller, 1948
Autumn-sown rye (grain + straw)		
normal yield	4 000– 6 000	Åkerman, 1951
Barley (grain + straw) normal yield	5 500– 7 000	,,
,, ,, maximum yield	11 100	Müller, 1948
Oats, normal yield	7 000– 9 000	Åkerman, 1951
Oats, maximum yield	11 700	Müller, 1948
Turnips, maximum yield (roots)	14 800	,,
Sugarbeet, maximum yield (roots)	16 800	,,
Grassland (hay)	7 500	Ovington, 1957
Potato (tubers) maximum yield (Rothamsted)	50 000	Watson, 1971
Wheat (grain) maximum yield (Rothamsted)	7 500	,,

production because part of the vegetation becomes litter prematurely in the form of waste from the felling operations. Lutz and Chandler (1946) say that the calculated amount of waste from clear-felling Douglas fir on the Pacific Coast is about 800 metric tons per hectare.

Yields in forestry or agriculture can be used in assessing litter production, and some examples are listed in Tables 6 and 7. However, figures for harvests refer only to the economically important parts of the plant, fruits and straw, for example, for agricultural crops, and wood in the trunk for forest trees. Stumps, roots, twigs, etc., have not usually been included, nor the material produced by the weeds and other plants growing with the crop species. The tables therefore include only a part of the real production of the area in question. The magnitude of the omitted amount can be assessed to some extent from the examples in Table 7 in which the yield of leaf, stem and underground parts in *Pinus sylvestris* (Ovington, 1957), *Fagus sylvatica* (Möller, Müller and Nielsen, 1954) and *Fraxinus excelsior* (Möller, 1945) are shown separately. Figure 21 is an example of a detailed analysis.

Ovington and Pearsall (1956) suggest that the greater production of the evergreens may be associated with their ability to photosynthesise during suitable periods in the winter months. Other studies showed that the dry weight yields from agricultural crops, arable and pasture, were about 4300 to 4800 kg/ha.

TABLE 7 Dry matter production by forest trees.

	DRY MATTER PRODUCTION BY AERIAL PARTS (kg/ha/year)	
Spruce trunks. Norrland	640	Romell, 1939
,, Skåne	8 000	,,
,, Denmark	6 600	Möller, 1945
Pine trunks ,,	4 600	,,
Oak ,, ,,	4 000	,,
Ash ,, ,,	4 600	,,
Beech ,, ,,	6 200	,,
Ash, 12 years ,,	7 450	Boysen-Jensen, 1932
Beech, 22 years, all above-ground parts	13 700	Boysen-Jensen, 1932
Deciduous hardwood plantations, 20–47 years. England	4 200	Ovington and Pearsall, 1956
Pine plantations, 20–47 years. England	8 000	Ovington and Pearsall, 1956

	LEAF	STEM	UNDERGROUND	TOTAL	
Pine. England	3 300	7 200	1 900	12 400	Ovington, 1957
Beech	2 500	7 400	1 500	11 400	Möller *et al.*, 1954
Ash	2 700	6 700	1 500	10 900	Möller, 1945

Examples of production by wild plants are provided by other studies made in England. The annual dry weight production measured as standing crop of leaves and shoots of grasses, sedges and rushes, as well as leaves of *Pteridium aquilinum*, usually lies between 4000 and 14 000 kg/ha (Pearsall and Gorham, 1956). *Phragmites communis* has especially high dry-matter production: the shoots alone are responsible for producing between 5300 and 15 800 kg/ha dry matter annually. The authors suggest that this variation is dependent mainly on the quantity and type of inorganic nutrients available.

The amounts of fallen litter from forest trees and from some of the important plants in the understorey vegetation in the forest have been measured in various investigations (Tables 8 and 9). These figures are again incomplete because root and other underground litter is not included. Also, because of varying methods employed, the figures are not fully comparable since some of the weights are of oven-dried material, and others of air-dried. Some refer to fresh litter and others to litter collected several months after it had fallen, when some loss of material would have occurred. Nevertheless, as Romell (1939) pointed out, it is interesting that litter production by forest trees is about the same throughout the

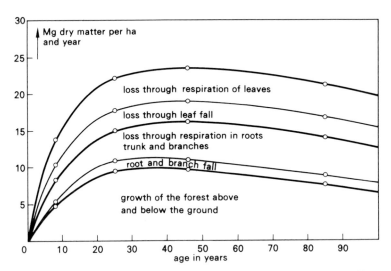

FIG 21. Dry matter production in a Danish beechwood. (After Möller, Müller and Nielsen, 1954.)

temperate zones (Bray and Gorham, 1964). It is between 2 and 4 Mg/ha/year, except in two Norwegian localities which are not comparable with the others because of their altitude. The differences are probably mainly attributable to differences in soil conditions.

The field layer and ground layer also produce a large amount of litter, often as much or more than the tree layer (Table 9). The moss layer alone may sometimes produce an amount of material of the same order as that of the wood in the trunks of forest trees (Romell, 1939). The bushy lichens, particularly those like *Cladonia* and *Cetraria*, may also make a significant contribution. However, the role of lichens as producers of soil material is generally overestimated, especially for encrusting species (Cooper and Rudolph, 1953).

It is as yet hardly possible to draw any definite conclusions about the total amount of litter formed by all the layers in different types of forest. However, it is apparent from the tables that the amounts are large, at any rate on the better sites, and sometimes not much less than yields on arable land. Since the forest is usually left only

TABLE 8 Production of fallen litter by forest trees.

	LITTER PRODUC-TION BY THE ABOVE-GROUND PARTS (DRY WT) (kg/ha/year)	
Spruce, 40–140 years, 800 m altitude. S. Norway	3300–1465	Mork, 1942
Spruce. Central Sweden	3364	Lindberg & Norming, 1943
Spruce, 30–90 years. Bayern	5258–3273	Ebermayer, 1876
Pine, 25–100 years. Germany	3397–4229	,,
Conifers. New England, Minnesota, Florida	1900–4000	Lutz & Chandler, 1946
Conifers and oak. N. Carolina	2900–3400	,,
Deciduous forest. New York	2700–3350	,,
Birch, 50 years, 100–150 m altitude. S. Norway	1876– 800	Mork, 1942
Birch, 40 years. Central Sweden	1865	Knudsen & Hansson, 1939
Aspen, 42 years. Central Sweden	1958–2120	Andersson & Enander, 1948
Beech, 5–200 years. Denmark	2100–3000	Romell, 1939
Beech. South Sweden	2000–3000	Lindquist, 1938
Beech, 30> 90 years. Germany	4182–4044	Ebermayer, 1876
Oak, birch, alder, aspen, hazel, rowan. Central Sweden	4000	Julin, 1948
Oak, elm, ash, beech (leaf litter). Central Sweden	2100	Lindquist, 1938
Birch, rowan, with some other deciduous trees (leaf litter). Dalarna	2046	Sjörs, 1954

on the poorer soils, the values are probably lower than the potential production by forest in optimum conditions. In other plant communities, such as meadows or marshes, production is probably also high. Investigations of these, and others, for example, heaths and bogs, would yield interesting results.

This applies to litter production above ground. The underground organs, roots, rhizomes, bulbs, corms, etc., also produce significant amounts, and there is also the contribution from soil organisms, mainly bacteria, fungi and algae. Very little is known of the annual production by aquatic plants, often accumulated in thick layers of underwater litter. Reed communities, and seaweeds, are particularly productive. Seaweeds may form beds and banks along shallow sea coasts. The most important contribution to the underwater litter and gyttja (see page 222) which accumulate at

TABLE 9 Production of ground layer litter and field layer litter in different kinds of forest.

TYPE OF LITTER	LITTER PRODUC-TION (DRY WT) (kg/ha/year)	
Homalothecium sericeum. Denmark	500	Romose, 1940
Hylocomium spp. (mainly). Central Sweden	700	Romell, 1939
Vaccinium myrtillus (93%) and other Ericaceae (7%). S. Norway	800	Mork, 1946
Field and ground layers in a north Swedish *Vaccinium* forest	900	Romell, 1939
Vaccinium vitis-idaea, V. myrtillus, Hylocomium spp. with some heather and grasses	1455	André, 1947
V. vitis-idaea (63%) and *V. myrtillus.* S. Norway	1400	Mork, 1946
V. uliginosum (78%) and other Ericaceae (22%). S. Norway	1500	,,
Empetrum hermaphroditum (69%) and other Ericaceae (22%). S. Norway	2100	,,
Calluna vulgaris (91%) and other Ericaceae (9%). S. Norway	2600	,,
Calamagrostis, Rubus, Holcus, Lathyrus, Galium, etc.,		
in a beechwood	4367	Wittich, 1933
in a pine wood	2533–4060	,,

the bottom of lakes, eventually causing a succession, is probably from reed communities.

Climate has a predominant influence upon litter production. Total annual litter production in arctic–alpine forests averages 1 Mg/ha, in cool temperate forests 3·5 Mg/ha, in warm temperate forests 5·5 Mg/ha and in equatorial forests 11 Mg/ha (Bray and Gorham, 1964).

5

The constituents of litter

Cells, tissues or organs which die and become litter are made up of the same components as living cells. They contain mineral salts, organic compounds such as carbohydrates, lipoids, proteins and protein derivatives, organic acids, lignin, suberin, cutin, waxes, glycosides, alkaloids, tannins, ethereal oils, resins, together with enzymes, pigments, etc. The bound chemical energy in the litter is very important in the process of humification and the role of the humus as a soil constituent.

The mineral salts in the litter

Except for the ammonium ion, all the cations taken up by plants from their environment, and some of the anions, are left in the ash remaining after plant material has been burned. The composition of the ash depends on the mineral salts present where the plants were growing, since plants take up all salts with which they come into contact, whether these are beneficial, harmful or without physiological effect. Although the plants are unable to exclude completely any of the ions in the soil solution, the ions are not taken up in the same proportions in which they occur. Some ions are taken up in relatively larger amounts, some in smaller amounts. Hence, the quantitative composition of the ash may be different in some respects from that of the external medium. For example, the ratio K to Ca is different, and usually higher, in the plant than in the soil (Iljin, 1933). The cause of this may be physico-chemical, or physiological. The ions differ in size, hydration and charge and, therefore, in their tendency to be adsorbed, which affects uptake. The readiness with which they pass through the epidermal cell walls and the cytoplasmic membranes is also affected by the differing concentrations of the ions in the external medium (Table 10).

TABLE 10 The relation between mineral soil and leaf
calcium content. (Müller, 1934, page 83.)

MINERAL	CaO CONTENT AS % OF LEAF DRY WEIGHT	
	SPRUCE	BEECH
Basalt	1·52	1·04
Chalk	3·57	3·19
Granite	0·42	0·99
Gneiss (Zöblitz)	0·42	0·79
Gneiss (Hirschberg)	0·51	0·93

If the concentration of a salt in a locality changes, there is a
double effect on salt uptake. The conditions for the uptake of the
salt itself are changed and so is the uptake of other ions. For example, increased potassium concentration results in increased uptake of potassium and decreased uptake of other cations (Fig. 22).

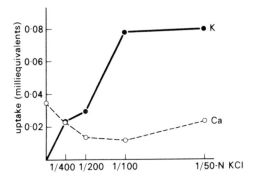

FIG 22. Uptake of potassium and calcium
by oat seedlings from solutions with a
constant amount of calcium chloride and
varying amounts of potassium chloride.
(After Burström, 1934.)

The uptake of any ion is dependent on the species and quantity of
other ions present, i.e. on the ionic balance of the solution. The
acidity of the medium also affects salt uptake (see Chapter 16).

Physiological factors are involved in salt uptake, and salt uptake
is affected by respiration (Lundegårdh and Burström, 1933);
energy is required and work is done by the protoplasm. In this
again, certain ions are favoured in relation to others. To some extent, the protoplasm can make a choice.

Because of the inherent differences between species, the results of
this choice are not the same; some species take up more calcium,

others more potassium, and so on. Hence, the composition of the ash from different species may be different, even if they grew in the same environment (Table 11).

TABLE 11 Composition of the ash of some semiaquatic plants growing in the same place. (After Chodat, 1920.)

	K_2O %	Na_2O %	CaO %	SiO_2 %
Stratiotes aloides	30·8	2·7	10·7	1·8
Nymphaea alba	14·4	29·7	18·9	0·5
Chara sp.	0·2	0·1	54·8	5·9
Phragmites communis	8·6	0·4	5·9	71·5

Among bog plants, *Menyanthes trifoliata* accumulates a relatively large amount of sodium, but little silicon, whereas the *Cyperaceae* accumulate silicon, especially if the peat is not particularly acid (Malmer and Sjörs 1955). Johnson and Butler (1957) found that some species of pasture grasses contained ten times more iodine than others growing in the same soil.

TABLE 12 Ash content of different organs of the same species. (After Ebermayer, 1876.)

	ASH (% OF DRY WT.)		
	BIRCH	SPRUCE	BEECH
Trunk wood	0·40	0·17–0·46	0·44–0·88
Branches	0·84	0·32–1·87	1·80
Bark	5·38	0·94–2·02	3·0 – 3·90
Leaves	6·39	4·52	5·57

The highest ash contents are found in deciduous broad-leaved trees and in herbaceous plants, the lowest in mosses and lichens (Table 13). There are also large differences in composition of the ash. Unfortunately, the data in Table 13 are incomplete, in that they include only those elements present in the largest amounts. The ash also contains elements present in such small amounts that they cannot be measured by the same techniques; these are known as the *trace elements* and considerable attention has recently been given to them (see Chapter 15).

The causes of differences in the amount and composition of the ash may be environmental or protoplasmic. Protoplasmic differences are particularly important and are the cause of the particular demands on the mineral content and composition of the environment. Hence, a species can establish itself only in an environment which meets these demands sufficiently; its protoplasmic characteristics have led to a choice of environment.

TABLE 13 Nitrogen and ash content of various types of litter; percentage composition of the ash. (After Andersson & Enander, 1948 (1); Aarnio, 1935 (2); Aaltonen, 1948 (3–6.)

	1 ASPEN LEAVES	2 SPRUCE NEEDLES	3 *Vaccinium* *myrtillus*	4 *Calluna* *vulgaris*	5 *Hylocomium* *proliferum*	6 *Cladonia* *rangiferina*
Ash (% of dry wt.)	9·53	7·18	4·99	4·23	2·73	1·87
N (% of dry wt.)	0·58	1·13	1·75	1·21	1·21	0·51
Ash % composition						
SiO_2	16·4	63·94	5·6	54·8	12·5	67·0
MnO	0·40	—				
Al_2O_3	0·46	3·16				
Fe_2O_3	0·16	0·36				
CaO	35·0	22·0	28·2	13·0	12·5	4·8
MgO	4·6	3·09	13·6	8·1	6·6	3·2
Na_2O	0·13	1·27				
K_2O	10·3	2·2	20·1	9·0	31·2	10·2
CuO	0·026	—				
P_2O_5	1·35	2·98	8·0	5·2	18·5	7·0
SO_3	2·6	0·95				
Cl	1·7	0·07				
B	0·03	—				

Protoplasmic differences also have an effect on the distribution of mineral ions within the plant so that some organs contain larger amounts of salts than others, particularly the peripheral parts of the plant such as the leaves and the bark (Table 12). This is significant to the role of litter in soil formation processes, because it is the peripheral parts, and the leaves especially, which make up a relatively large proportion of the litter. The dominant species of forest or scrub—the trees and shrubs—produce large amounts of leaf, bark and twig litter every year.

Some characteristics of ash composition are apparent in Table 14, which shows the change in amounts of various elements as the leaves age. The contents of potassium, phosphorus and nitrogen are largest in young leaves and decrease as the leaves get older, the reverse is true of calcium and magnesium, and also of silica. However, a decrease of potassium content, for example, expressed as a percentage of leaf dry weight, as in Table 14, does not necessarily mean that the absolute potassium content of the leaves decreases. As the leaves get older, various substances accumulate in them, and their dry weight therefore increases. During and after the yellowing of the leaves in autumn, the content of some elements, such as potassium, phosphorus and nitrogen, decreases. Compared with living leaves, especially young leaves, litter is rich in calcium, magnesium and silica, but relatively poor in the physiologically important potassium, phosphorus and nitrogen compounds (see Fig.

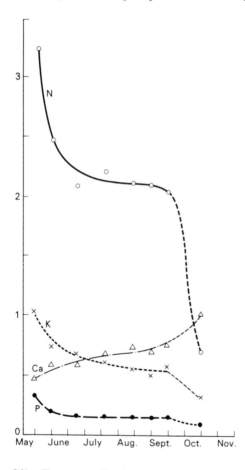

FIG 23. The contents of K, Ca, P and N in birch leaves during the vegetation period. (After C. O. Tamm, 1951b.)

23). Factors affecting transpiration rate, especially light, also affect salt uptake.

Thus, the amounts and proportions of the constituents of the ash are determined by the interplay of several factors: the type and age of the plant and the organ, the supply of minerals in the locality, the acidity of the soil and the climate. Using information about the amount and composition of litter of the type in Tables 13 and 15, it is possible to make some assessment of the approxi-

TABLE 14 Composition of leaf ash at different times during the vegetation period; mean for nine species of deciduous broad-leaved trees. The figures are % leaf dry weight. (Alway *et al.*, 1934.)

	ASH	Ca	Mg	K	P	S	N
1 June	6·59	1·20	0·37	1·49	0·29	0·20	2·99
1 July	6·89	1·49	0·47	1·16	0·25	0·18	2·36
1 August	8·22	1·92	0·49	1·09	0·22	0·17	2·24
1 September	10·24	2·23	0·54	0·99	0·22	0·18	1·85
11–16 October (directly after leaf fall)	11·32	2·85	0·57	0·41	0·19	0·14	0·80

TABLE 15 Examples of the amounts of mineral salts (kg/ha) carried to the ground in the litter annually. (After Andersson & Enander, 1948 (1); Romell, 1939 (2–3); Lindberg & Norming, 1943 (4); and Knudsen & Hansson, 1939 (5).)

TYPE OF LITTER	1 ASPEN LEAVES	2 MOSSES (mainly *Hylocomium* spp.)	3 STEMS AND LEAVES OF *Vaccinium myrtillus*	4 SPRUCE NEEDLES	5 40-YEAR-OLD BIRCH WOOD
Dry wt. of litter kg/ha/year	1958–2120	690	440	3089	1865
Ash	172				93·1
N	10·5	6·7	7·0	29·7	22·0
SiO_2	28·1			101·5	6·13
Mn_3O_4	0·74				1·28
Al_2O_3	0·79			1·42	0·34
Fe_2O_3	0·27			0·92	0·28
CaO	60·3	3·9	5·3	63·7	33·25
MgO	7·9	1·7	1·9	6·7	9·33
Na_2O	0·22			0·84	0·22
K_2O	17·7	5·5	4·3	14·0	10·25
CuO	0·045			0·027	—
P_2O_5	2·3	3·4	1·7	4·88	2·49
SO_3	4·5			13·0	3·92
Cl	2·9			2·2	0·19
B	0·052				

mate amounts of mineral salts returned to the soil in the litter. However, there is insufficient information to build up a general picture. Not enough is known about the amounts and proportions of ash constituents in different species and in different plant communities.

The supply of litter to the soil does not, of course, bring about any net increase in mineral nutrient reserves; the salts which are returned in the litter came from the soil in the first place. The figures are therefore a measure only of the fraction of the mineral nutrients cycling between plant and soil contained in the litter at the time. Other salts in the cycle are in the vegetation, or in the humus constituents of the soil; and others, in solution or in a bound form, are on their way to or from one of these fractions. A complete picture of the cycle of one element requires a measure of the amount of the element in each of the fractions. The vegetation's requirement for a particular element cannot therefore be assessed from the quantity contained in the litter or in the living plants; the amounts in the other fractions, and the time the element is retained in each fraction, must also be known (see pages 300, 411, 450).

The organic constituents of the litter

Litter consists mainly of organic substances (Tables 12–15). These are particularly important in soil formation, as they are sources of easily available energy. In contrast to the salts, they are the product of synthesis by the plant and mainly comprise material which originates from water or air. The more litter is produced, the higher the production of soil-forming, energy-rich substances.

TABLE 16 The composition of rye plants at different stages of development. (Waksman & Tenney, 1927.)

COMPOSITION (% OF DRY WT.)	PLANT HEIGHT (25–35 cm)	JUST BEFORE HEAD FORMATION	JUST BEFORE FLOWERING (stems and leaves)	JUST BEFORE GRAIN RIPENING (stems and leaves)
Pentosan	16·6	21·2	22·7	22·9
Cellulose	18·1	27·0	30·6	36·3
Lignin	9·9	11·8	18·0	19·8
Water-soluble substances	34·2	22·7	18·2	9·9
Fats and waxes	2·6	2·6	1·7	1·3
Total nitrogen	2·5	1·8	1·0	0·2
Ash	7·7	5·9	4·9	3·9

A summary of the most important substances in litter is given in Table 16. This also shows the change in composition with age of plant. The ratio of protoplasmic material to cell wall material decreases as the cell develops from the meristematic stage to maturity; the ratio of the constituents of the protoplasm and the cell wall does the same. Some substances decrease in amount, others increase, and new substances are synthesised. Young cells contain relatively large amounts of nitrogenous compounds, expecially amino acids, water-soluble substances, such as sugars, and ether-soluble fats and lecithin, etc. As the cell matures, the amounts of these compounds decrease and others, especially cellulose, hemicellulose and lignin, increase. The older the parts of the plant, the higher the content of wood constituents and cellulose and the lower the content of water-soluble carbohydrates, nitrogenous compounds and mineral salts.

According to Waksman (1938, page 95), litter generally has the following percentage composition:

Cellulose	20–50
Hemicelluloses (pentosans, hexosans, polyuronides)	10–28
Lignin	10–30
Tannins, pigments, cutin, suberin, fat, wax	1–8
Protein	1–15

Acidic and basic compounds in the litter

The contents of acidic and basic constituents in the litter, and the ratio of one to the other, are of particular significance to the biological and the physico-chemical decomposition of the litter. The readiness with which the litter is attacked by microbes and soil animals is probably partly determined by these factors. The acidic constituents, or, rather, the excess of acidic over basic constituents, are also very important in weathering and leaching.

Important information about these properties can be obtained by measuring the acidity and buffer capacity of the litter. Hesselman (1926) carried out an investigation on various types of litter. He mixed the pulverised material in a dilute solution of KCl and measured the pH value, then added increasing amounts of a dilute solution of an acid or a base, recording the pH at each stage. The more acidic substances contained in the litter, the more base is added in the titration. The amount of base added in the titration is therefore a measure of the content of acidic substances in the litter. When acid is added, there is an alkaline buffer effect. Hence, the litter contains both acidic and basic substances. Figure 24, from Hesselman (1926), illustrates the similarity between the buffer effects of fresh moss and conifer needle litter; and the difference between the litter of conifer needles and leaves of broad-leaved deciduous trees.

Hesselman found that the pH of litter of coniferous needles, ericaceous dwarf shrubs and mosses was relatively low, and that of litter of leaves of broad-leaved deciduous trees and of herbs was relatively high. However, the values varied according to when and where the litter was collected. The pH value of the litter was higher in calcareous areas than in areas poor in lime.

There are changes with time in the pH values of aqueous extracts of different kinds of leaf litter. The pH increased in extracts

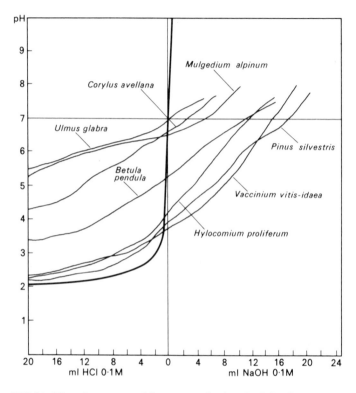

FIG 24. Titration curves (electrometric titration) for fresh leaf litter of forest plants. An example showing the content of acid and basic buffers in fresh leaf litter, and the acidity of the leaf extract. The buffer content was determined by titrating a suspension of 5 g powdered leaf material, dried at 98°C in 200 ml 0·1 N KCl. The thick line is the titration curve for the KCl solution alone. The more acid buffers there are in the litter, the further to the right are the titration values, as more NaOH must be added to increase the pH to neutral. (After Hesselman, 1926.)

in contact with air but decreased in anaerobic conditions (Nykvist, 1963). Sjörs (1959) studied the pH changes in leaf litter lying on the soil. During thirty-four days in late autumn, the pH of seven kinds of leaf litter rose by an average of 0·9 (maximum 1·6, minimum 0·3).

Further analysis of these questions has been made by Mattson and his fellow workers (1941 onwards). They measured the following properties in litter and in living plant material:

pH in a suspension of the material in water.

Bases (B), measured by back titration of the ash.

Acidity (H), the amount of acid buffer substances, measured by electrometric titration, to pH 7, of the original material.

Acidoids (A), colloidal acids, measured by titration, to pH 7, of the electro-dialysed material.

In electro-dialysis, the soil is saturated with H^+ and OH^- ions and therefore becomes unsaturated with other cations and anions.

Organic acids (a), calculated from the other values.

The base saturation, or base status, of the material can be expressed as $(B.100)/(A+a)$, i.e. the amount of base-saturated substances in terms of the total amount of acid substances.

Tables 17 and 18 list these values, in order of decreasing base saturation, for litter of a number of herbs, grasses, trees, dwarf

TABLE 17 Base saturation of litter (leaf litter for trees). Acids and acidoids determined at pH 7 (electrometric titration). (Mattson and Koutler-Andersson, 1944.) B, H, A and a in me per 100 g.

	pH	BASES B	ACIDITY H	ACID-OIDS A	ACIDS a	$A+a$	BASE STATUS $\dfrac{B.100}{A+a}$	TOTAL NITRO-GEN %
Fraxinus excelsior	5·5	217	31	95	151	246	88	1·5
Ulmus glabra	5·3	281	49	136	194	330	85	0·8
Fagus silvatica	4·8	106	34	55	84	139	77	1·0
Alnus incana	4·6	99	59	85	71	156	64	2·5
Quercus robur	4·4	105	67	70	101	171	62	1·2
Betula pubescens	4·6	115	82	82	115	197	58	0·7
Pinus silvestris	4·0	28	32	34	27	61	46	0·5
Sphagnum cuspidatum	3·1	18	103	88	34	122	15	1·1
S. fuscum	2·8	18	123	121	19	140	13	0·6
Cladonia rangiferina	3·8	4	30	23	11	34	13	0·8

shrubs, mosses and lichens. Comparison of the various species leads to the important conclusion that base saturation is high in litter from herbs and leaves of broad-leaved deciduous trees (particularly ash and elm), less for conifers and dwarf shrubs, and least for mosses and lichens. In general, the values for bases, ash content, acids and pH change in the same way as base status, whereas acidoid content varies more irregularly. There are naturally differences depending on locality, season, age, and other factors, but these do not affect the general sequence. The two examples for *Calluna vulgaris* from different places, in Table 18, illustrate the local variation. Mattson and Koutler-Andersson (1945) found that ash content, acids, acidoids, pH and base saturation increase with age of spruce needles, whereas acidity and nitrogen content decrease.

Tables 17 and 18 also show that litter in general contains an excess of acidic substances, and the extract has an acid reaction.

TABLE 18 Base saturation of litter from herbs, trees (leaves), and ericaceous dwarf shrubs (leaves and young shoots). (After Mattson and Karlsson, 1944.)

	ASH	pH	BASES B (me/100 g)	ACIDITY H (me/100 g)	BASE STATUS	TOTAL N %
Urtica dioica	23·6	7·4	286	−17	106	3·9
Myosotis arvensis	16·5	7·1	137	2	99	2·0
Caltha palustris	15·1	6·0	217	23	90	2·4
Angelica sylvestris	13·4	5·6	168	26	87	2·3
Dryopteris filix-mas	10·8	6·0	108	28	79	3·0
Juncus effusus	6·8	6·1	48	14	77	1·8
Alnus glutinosa (suckers)	9	5·3	136	47	74	4·0
Alnus glutinosa (trees)	5·3	5·1	73	48	61	3·3
Eriophorum vaginatum	3·2	5·3	31	22	59	1·2
Betula pendula	5·1	4·9	75	59	56	2·3
Epilobium angustifolium	4·9	5·3	68	63	52	1·3
Calluna vulgaris	4·0	4·5	32	31	51	1·0
Vaccinium myrtillus	3·8	4·1	53	56	49	1·1
V. vitis-idaea	3·9	4·4	39	45	47	0·9
Pinus sylvestris	2·5	4·5	28	39	41	1·1
Hylocomium sp.	?	4·5	28	52	35	0·7
Picea abies	3·8	4·0	27	53	33	0·9
Sphagnum acutifolium	?	4·4	34	68	33	0·6
Polytrichum commune	5·4	4·5	22	58	28	0·9
Cladonia	2·6	4·1	2	45	5	0·5

Urtica dioica is the only exception among these examples. In the main, the nitrogen content (in percentage of the dry weight of the material) follows the base status. The leaves of herbs and broadleaved deciduous trees contain most nitrogen; coniferous needles and dwarf shrubs come next; mosses and lichens contain least (see page 185).

Other substances of ecological significance

Among other substances present in cells and organs when they die, i.e. as litter is formed, and among those which are the products of decomposition of the litter, are many which have an inhibitory or stimulatory effect on soil organisms. Some are of considerable significance to the growth of bacteria, fungi or algae. Nilsson *et al.* (1938), for example, found a substance in oat straw which promotes growth of *Bacteria radicicola*. Lindberg (1944) and Melin (1946) showed that various kinds of tree leaf litter contained substances which affect the growth of mycorrhizal fungi: some promote growth, and have effects like vitamins; some have an inhibitory effect and can be classified as antibiotics (see Chapter 15). Fifty-two out of a hundred lichens investigated by Burkholder and Evans (1945) had an antibiotic effect on bacteria. There is a large number of substances found in litter, or formed there, including many substances specific to various plants. Because of this, and because of the special requirements and sensitivity to soil organisms, it seems that these inhibitory and stimulatory substances are important in shaping the conditions in which the organisms in the soil grow, and therefore for plant life as a whole. Very little research has yet been done towards sorting out this complex of problems.

The energy content of the litter

The great majority of soil organisms are saprophytes, which generally obtain both material and energy heterotrophically. Some of the chemical constituents of the litter are assimilated directly, others after they have undergone some chemical change. Some of the litter is used as an energy source. The energy is made available through various chemical processes, for example, splitting of the molecules or oxidation. These processes usually take place in small steps, and a particular saprophytic species utilises only part of the material, and part of the energy, in the litter. One group of organisms uses the sugar contained in the cells, another the cellulose, a third the lignin, etc. What is left by one species is used by another, which in its turn only uses a part. The details of this system are only fragmentarily known and the processes which have received most attention are those with products such as alcohol and acetic acid, of economic importance. The formation of these compounds can serve as an example of the successive steps in decomposition:

The yeast fungi use the sugar constituents of the litter, which are broken down:

$$C_6H_{12}O_6 \rightarrow 2\ CO_2 + 2\ CH_3CH_2OH + 100\ kJ.$$

The acetic acid bacteria carry out the next stage, oxidation of the alcohol to acetic acid:

$$CH_3CH_2OH + O_2 \rightarrow CH_3COOH + H_2O + 490\ kJ.$$

The result of the activity of the yeast fungi is thus that only one unit is split off the molecule, and a comparatively small proportion of the energy contained is released. The oxidation carried out by the acetic acid bacteria releases a considerably larger amount of energy, but not all that is contained in the molecule. This is released only when all the hydrogen and carbon atoms are oxidised:

$$C_6H_{12}O_6 + 6\ O_2 \rightarrow 6\ CO_2 + 6\ H_2O + 2830\ kJ.$$

No exact measure is available of the amount of energy supplied to the soil annually in various sorts of litter, but various estimates of its magnitude have been attempted. The litter comprises mainly energy-rich organic compounds (see Table 16), the heat of combustion of which can be measured; van Suchtelen (1923) has made this sort of estimate and calculations and arrived at an average value of 19·3–21·0 kJ/g dry weight of organic material. Taking the figure of 19·3, and assuming that annual litter production is 2000 kg/ha organic material (according to Tables 8 and 9 considerably more may be produced) the energy supply is as follows: 2000 kg/ha/year litter gives 38 600 000 kJ, which is 10 500 kJ per day. This is equivalent to about 3 kg anthracite per day, or 1100 kg per year.

6

Litter-decomposing organisms

Litter is decomposed mainly by plants and animals which use some of the energy released and assimilate some of the products. These organisms are called heterotrophs. Their energy and nutrient supply is not derived by assimilation of mineral raw materials like that of autotrophs, but from organic materials present in plant and animal litter. All animals are heterotrophs, so are bacteria, fungi, and some other plants found above or below the soil surface.

The plants living in the soil, called *soil microphytes*, are most active in litter decomposition. They include a small group of partial heterotrophs, which fulfil their energy requirements from the litter, but are otherwise autotrophic. By far the largest group of soil microphytes obtain both energy and material from the litter and so are completely heterotrophic.

The organic substances and easily available energy in the litter are essential to the existence of the heterotrophs and it is through the activity of the heterotrophs that litter and other waste material is removed from the soil surface.

Large amounts of material are consumed by the heterotrophs, which generally use only part of the nutrient and energy content. Hence, their excretory products are usually rich in organic substances and in energy and they provide a substrate for other heterotrophs in turn.

Heterotrophs living mainly above the soil surface

Plants

Some bacteria, actinomycetes, and fungi parasitise living plants and animals and, consequently, the organic constituents of the hosts are subject to decomposition. Sometimes only the products of metabolism are consumed, or the protoplasm and cell walls may be attacked, leading to the death of the cells or organs. Extensive damage will kill the host organism. Decomposition by parasites is quantitatively relatively unimportant. It is more important if the organism which has been attacked is killed, or parts of it are killed, so that it is prematurely converted to litter.

Animals

There are parasitic animals whose role in decomposition processes is comparable with that of plant parasites, but these are insignificant in comparison with non-parasitic animals.

Animals are traditionally classified as herbivores or carnivores. Herbivores may eat a certain amount of litter, but generally prefer living plant material which is thus prematurely converted to litter. Farm animals, a significant number of wild animals, and also some insects have a large annual turnover of plant material, the remains of which are incorporated with the soil humus after passage through the animal's alimentary canal. On pasture, animal faeces can be an important constituent of the soil humus complex. The traditional agricultural practice of putting farmyard manure on arable land means that this form of humus is utilised for crop production.

The activities of herbivores in humification and mineralisation of litter are not limited to grazing and subsequent decomposition. The litter is affected in other ways. Trampling may crush or break up some of the material in the litter, and living organs, roots, rhizomes or fungal hyphae, may be destroyed and converted to litter; these effects are particularly marked along paths. Some animals scratch in the ground, many birds, for example; others make their home there, pigs root in the soil, foxes and badgers make holes and passages, voles, moles, mice and other rodents dig tunnels.

Insects are some of the litter-decomposers active above the ground. Their ravages may be so extensive that large quantities of living plant material are converted to humus. In some years, all the foliage of many trees and other plants is eaten by early summer. Some plants put out new leaves subsequently. Lindquist (1938) lists trees such as oak, elm, ash and alder as prone to attack by insect larvae, and beech, lime, sycamore and willow as more resistant.

Carnivores generally eat herbivorous animals, so their food is made up of material which originates from autotrophs, but has also been through one cycle of decomposition and resynthesis. Only a small proportion of the animal victim is converted to new living material and the rest eventually becomes humus.

Heterotrophs in the soil and in water

Soil organisms, particularly bacteria, actinomycetes and fungi, protozoa, worms and insects, are responsible for most of the litter decomposition. Together with representatives of many other plant and animal groups they form a whole living community in the soil (an *edaphon*) which includes more individuals and probably more species and varieties than the communities above ground. The simplest types have a short life cycle and reproduction may be so rapid that there are many generations in a few weeks or even a few hours. This is particularly true of the bacteria. It allows rapid adaptation to the environment, and consequent high resistance to unfavourable external conditions and capacity to break down substances which are poisonous to other organisms (see Wright and Grove, 1957). In part, this resistance depends on the primitive nature of these organisms.

A remarkably high proportion of soil microorganisms (bacteria, actinomycetes, lower fungi, algae, protozoa) are cosmopolitan. For example, the most important species of soil bacteria occur everywhere. This is probably because temperature, humidity and light are much less variable underground than above ground, so that they are less likely to have limiting and inhibiting effects.

The distribution of soil microorganisms, particularly soil microphytes, is not limited by dispersal methods to the same extent as the distribution of higher plants. Because of their small size and the copious production of spores and other dispersal units, fungi and bacteria are spread very easily by wind, water and animals, or in soil or plant material which is moved in some way from one place to another. Dispersal by animals, especially insects, is probably the most effective.

Some soil organisms are free-moving, particularly the bigger ones; others are not, and remain mainly in the colloid films around the soil particles. These colloids are dispersed by large quantities of water in the soil and then the microorganisms may be released from the absorbing forces which held them fast. Protozoa, ciliate bacteria and other motile forms may therefore be temporarily free-moving.

Table 19 gives a summary of the differences between soil microphytes and higher plants. The most fundamental is the difference in their mode of nutrition: the metabolism of the autotrophs is mainly concerned with synthesis whereas that of the heterotrophs results in decomposition. The study of litter decomposition is closely correlated with the study of the metabolism and ecology of soil organisms.

TABLE 19 A comparison of the characteristics of soil microphytes and higher plants.

HIGHER PLANTS (AUTOTROPHS)	SOIL MICROPHYTES (HETEROTROPHS)
Live above soil surface	Live below soil surface
Microclimate variable	Microclimate more constant
Distribution depending on geography	Largely cosmopolitan
Large plants	Small plants
Small numbers of individuals	Large numbers of individuals
Spread slowly	Spread rapidly
Little metabolism in relation to body weight	Much more metabolism in relation to body weight
Metabolism mostly concerned with synthesis	Metabolism mostly concerned with breakdown
Requiring energy	Releasing energy
Complicated construction	Simple construction
(Large-scale synthesis requires considerable organisation)	(Breakdown has relatively simple requirements)
Sensitive to environmental variation	Hardy
Long lifespan	Short lifespan
Slow adaptation	Rapid adaptation

Bacteria

A large part of the organic material is broken down by bacteria, which occur in large numbers and increase rapidly. In favourable conditions, they may divide every 20 minutes. At this rate, a single individual increases to 1600 billion (1 600 000 000 000 000) after 24 hours. If a bacterium is represented by a cube with sides 0·001 mm long, then one milliard bacteria (one thousand million) have a volume equivalent to 1 mm^3 and weigh approximately 1 mg. The volume of the descendants of one bacterium in 24 hours would be 1·6 litre. This is one of the theoretical extrapolations which can be made for many biological phenomena, but which are usually unreal. The potential increase in numbers of plants which produce large numbers of seeds (birch or dandelion, for example), or of animals with a large capacity for reproduction (rodents, flies, etc.) can be calculated in the same way. Extrapolation of figures for human population increase at a particular time involves the same risk of arriving at results which are biologically wrong. A biological activity which potentially goes on at an accelerating rate usually, directly or indirectly, causes the onset of some external or internal resistance which inhibits the activity so that it stops or reverses.

Bacterial metabolism is rapid and extensive, and uses a relatively large range of organic materials, partly because of extreme specialisation: any one species is involved in only a small part of the metabolic activity involved in litter decomposition. Specialisation implies limitation of the range of activity of the species, but on the other hand, it has a certain capacity to change this range because of its ability to adapt.

The magnitude of bacterial metabolism is best illustrated by a comparison with body weight. The material metabolised daily may be more than ten times the body weight. In comparison, brewer's yeast metabolises three times its body weight, *Aspergillus niger* one-sixth of its body weight, and man one-fiftieth.

Calculation of the rate of increase of bacteria has been based on laboratory experiments in which the conditions were generally more favourable than natural conditions; the results may therefore be taken to indicate maximum rates. In natural conditions, the processes are slower and the rate of increase is only a fraction of the maximum possible. In soil, there are generally 2–200 million bacteria per gram of soil; in favourable conditions there may be many times more. Such wide differences in numbers are attributable to variation in the prevailing environmental conditions, of which the following are probably of special significance.

1 The type and amount of litter and the carbon/nitrogen ratio. Litter is comparatively rich in carbon and poor in nitrogen, since it contains about 50 per cent carbon and rather more than 1 per cent nitrogen. Carbon can be assimilated by the microorganisms only to the extent that nitrogen is available; it does not matter whether the nitrogen is present in organic form or as ammonium or nitrate, as bacteria can assimilate inorganic nitrogen. However, there are only small quantities of ammonium or nitrate in the soil and they are taken up preferentially by other organisms, so there is competition for them. The nitrogen content of the litter is therefore a factor which limits bacterial growth and when the nitrogen is used up bacterial assimilation and reproduction stops.

Bacterial growth is also dependent on various growth substances, including vitamins, which are present in the litter in variable amounts, e.g., aneurin, nicotinic acid, biotin, riboflavin, pantothenic acid, etc. Carbon dioxide may be included in this group, since on some bacteria it has effects like those of a growth substance. The requirement for growth substances is linked with the specialisation characteristic of bacterial nutrition. Bacteria lack the characteristic ability of green plants to synthesise growth substances in sufficient quantity to meet their own requirements. They may never have had this ability, or they may have lost it as a consequence of their heterotrophy. The absence of this ability is a factor limiting bacterial growth.

Bacterial activity is most intensive during the first stage of litter decomposition, and it subsequently increases again only when there is a supply of fresh material, for example, when an arable field is manured. If there are one million bacteria per gram of soil in unfertilised arable soil, then there may be an increase to 1·7 million as a result of adding mineral fertiliser and to 3·7 million after adding farmyard manure (Russell, 1952). There is a state of equilibrium between the micropopulation and the available food supply.

TABLE 20 Numbers of microorganisms in forest soil (under Douglas fir) in Oregon. (From Powers and Bollen, 1935, page 324.)

	MILLIONS PER g DRY SOIL	
	BACTERIA	ACTINOMYCETES
Litter	66·8	0·88
Upper part of humus layer	47·5	11·6
Lower part of humus layer	34·5	9·2
Transition zone	7·6	4·7
Upper part of bleached layer	2·0	0·64
Lower part of bleached layer	4·3	0·73
Accumulation layer	0·60	0·08
Subsoil	0·18	0·08

The number of bacteria varies with soil depth (Table 20), probably because of differences in food supply. The figures in Table 20, and other similar ones, must be regarded as approximate. The results of the measurements and calculations depend to a large extent on the methods of isolation which are used.

There are particularly large numbers of bacteria in the rhizospheres of higher plants, probably because of the litter produced as a result of root replacement (see Chapter 4) and sloughing-off of epidermal cells and root hairs, and because of the secretory products of the roots which may also act as nutrient sources for the bacteria (for references see Grümmer, 1955; Gyllenberg, 1956).

2 Oxygen. Aerobic and anaerobic bacteria. Some bacteria—the *obligate aerobes*—require a good supply of oxygen; others utilise the oxygen contained in the litter and therefore need little or no access to atmospheric oxygen. They are tolerant of large variations in oxygen supply and are called *facultative anaerobes*. A third group—the *obligate anaerobes*—are specialised for life in media with no supply of atmospheric oxygen, for example, at the bottom of lakes and other bodies of water; if there were no anaerobionts, the organic material deposited in such places would remain undecomposed. The activity of these organisms is more limited than that of the aerobic species, partly because they are dependent on the oxygen content of the material, and decomposition is not as complete or extensive. Material lying under water (including gyttja) is therefore broken down more slowly than material lying on land.

3 Acidity (pH). In general, bacteria do best in more or less neutral media, though there are exceptions to this. The optimum pH depends on other properties of the media, such as temperature, nutrient content and osmotic value. There are some anaerobionts which tolerate very acid conditions, and they may radically change the acidity of the substrate as a result of their own metabolic activity. The acetic acid bacteria and lactic acid bacteria do this, and they grow well even on very acid substrates.

Silage making is a method of preserving animal fodder which makes use of the sensitivity of bacteria to acidity. The fodder is stored in such a way that its acid content increases sufficiently to inhibit the growth of undesirable bacteria. The green plant material is packed tightly into containers, known as *silos*, so that air is excluded and respiration of the living cells and growth of aerobic bacteria is inhibited. Lactic acid bacteria, which are faculative anaerobes, are able to grow and produce lactic acid, the content of which increases gradually up to about 2 per cent by weight until the activity of the lactic acid bacteria is inhibited. The development of other bacteria is prevented by the acidity.

Another method, known as the AIV method because it was developed by A. I. Virtanen, involves acidifying the material by addition of a mixture of hydrochloric and sulphuric acids. Use of the right amounts of acids, and packing and isolating the material, results in a hydrogen ion concentration which brings respiration and bacterial activity to a minimum.

4 Humidity. The optimum amount of water for bacteria is about 40–60 per cent of the soil water-holding capacity, when about half the pore volume of the soil is water-filled. Like other plants, bacteria are killed by desiccation; the spores are more resistant than the bacteria themselves. Because bacteria are sensitive to lack of water, food and other organic material can be preserved by drying. Archaeological finds of very old organic materials have mostly been preserved either because of lack of water or because they have been in water. In arid regions, objects from past cultures have been preserved until the present day because of the low humidity in the places where they lay. In wetter places, objects have been preserved in peat, where an excess of water has hindered humification.

5 Salts. The salt content and composition in the environment is an important factor affecting bacteria. In general, bacteria require the same nutrient salts as higher plants. The salts act not only as nutrients, taking part in cell growth and maintenance through metabolic processes, but have also osmotic and antagonistic effects. The supply of bases is particularly important, probably because of their effect in neutralising hydrogen ions. Like other environmental factors, the salt content may have a selective effect between species, for example, the marine bacteria tolerate higher concentrations of salts than freshwater species. The centuries-old method of preserving food by salting, to prevent fermentation and rotting, is based on the sensitivity of bacteria to salt concentration.

6 Light. All colourless bacteria, which are the majority, are damaged by ultraviolet light, and large doses may be lethal. Again, the spores are more resistant than the vegetative cells.

Ultraviolet light is used to sterilise air or equipment.

A small group of bacteria, the sulphur bacteria, are dependent on light for their growth. They contain a pigment, bacterial purpurin, and are able to assimilate carbon dioxide photosynthetically, using light as the source of energy (see page 167).

7 Temperature. The spores are particularly resistant to extremes of temperature: they survive heating to 100°C or more and cooling to absolute zero, −273°C. The vegetative cells are more sensitive, but even these may tolerate large temperature changes.

In cold regions and cold habitats, some bacteria may grow at temperatures of 0° to 30°C, with an optimum of 15° to 20°C. They are said to be *psychrophilic* or *cryophilic* (cold-preferring) (see Biebl, 1962).

In warm places, the bacteria mostly prefer high temperatures. Some, called *thermophilic*, have a high optimum temperature of about 50° to 55°C and tolerate temperatures of 60° to 90°C. Their minimum temperature is also high, between 25° and 45°C. Some of the anaerobic bacteria in soil, manure and compost belong to this group.

The *mesophilic* bacteria are a third type between these two extremes. They live at temperatures between 5° and 43°C, with an optimum of about 37°C. This group includes the species which are most common.

The temperature characteristics of bacteria are not constant. Sensitivity to extreme temperatures, and the position of the optimum, may be affected by other environmental factors, such as humidity, acidity, nutrient status, or by the age of the bacterial cells or culture. For example, sensitivity to high temperature is often increased by low pH and high humidity, and partly desiccated bacteria are therefore more resistant than those in moist conditions. On the other hand, sensitivity may be reduced if there is a good protein supply.

Food conservation by deep-freezing and heat-sterilisation is based on the sensitivity of bacteria to temperature.

Bacteriophage

It is of ecological interest that bacteria are attacked, or eaten, by many other organisms. Among these are the bacteriophages, which are viruslike agents found universally in soil, sewerage, animal excrement, etc. Their size is about 10–60 nm (1 nm $= 10^{-6}$ mm). Some bacteriophages are parasitic and they live and reproduce within the host bacterial cells. Others are more comparable with herbivores or carnivores and kill the bacteria by dissolving their cell membranes. Bacteriophages are specific to some extent, like the bacteria themselves, and each type of phage seems to attack only one species, or group of closely related species, of bacterium. Like bacteria, they are damaged or killed by ultraviolet light. The nature of bacteriophages is still incompletely understood. They probably cannot be regarded as organisms in the usual sense of the word, but rather as some sort of agent, perhaps as enzymatically active proteins.

Actinomycetes

The actinomycetes show some similarities to bacteria, some to fungi. They are very sensitive to acids and show a preference for habitats which are weakly alkaline. This may explain their rarity in

TABLE 21 Numbers of microorganisms in an arable soil. (After Barthel, 1949, page 113.)

	MILLIONS PER g SOIL
Bacteria	1000–3000
Actinomycetes	1–10
Mould fungi	0·03–1
Green and blue-green algae	0·1
Protozoa	0·01–1

raw humus, which is generally very acid. They are common in the slightly acid or more or less neutral humus which is formed in fairly dry conditions in deciduous woods or herb communities.

They do best in warm places, even if the temperature rises to 60–65°C, and are more common in dry soils than in wet ones, possibly because of their high oxygen requirement. They are typically aerobic.

Counts of actinomycetes have varied between a few thousand and several million per gram of soil.

The actinomycetes decompose many carbon- and nitrogen-containing compounds, such as cellulose, hemicellulose, protein and possibly lignin. The characteristic smell of soil is thought to be due to the volatile substances released mainly as a result of decomposition processes carried out by these organisms.

Fungi

Determination of the numbers of fungi in the soil is particularly difficult because of the way in which the fungi grow. The 'individuals' consist of mycelia intertwined with each other and impossible to separate.

Information about numbers of fungi is therefore indefinite. It is perhaps more feasible to measure the total weight of the fungi in a volume of soil, but even this can yield no more than approximate estimates. The number of viable spores and mycelial fragments has been found to be between a few thousand and a million per gram of soil.

Garrett (1951) divided the soil fungi into various ecological groups: saprophytic sugar fungi, cellulose-decomposers, lignin-decomposers, coprophilous fungi, root-infecting fungi, and predaceous fungi which attack nematodes or other soil animals.

According to Garrett, it is the sugar fungi which are the first of these groups to exploit the nutrient supply in the litter. They colonise the new substrate rapidly and consume the most easily available carbohydrates, such as the sugars and starch. However, there are only small quantities of these substances in the litter, because they have previously been transferred to storage tissues or used up in the production of reproductive structures.

Next come the species which attack protein, lignin, cellulose and other structural components. Many of the fungi which form toadstools are effective decomposers of this kind (see, for example, Lindberg, 1948, a and b; Lindberg and Korjus, 1955).

Fungal mycelium is most abundant in the litter layer and the uppermost humus layer. Hence, the fungi appear to be most active during the earliest stages of litter decomposition. They are subsequently less active, become litter themselves, and are attacked by bacteria.

The fungi require constantly moist soil and a good oxygen supply. They are more tolerant of acid substrates than the bacteria and they are therefore found in acid soil types, such as raw humus. They may alter the soil reaction by their metabolic activity, primarily by production of ammonia or organic acids.

Proteins and amino acids, released by decomposition, are the main nitrogen source for the fungi, but inorganic compounds may also be used. Fungi use ammonia in preference to nitrate. Lindeberg (1944) tested fourteen *Marasmius* species and found that thirteen could assimilate ammonia and only one took up nitrate.

Fungi use relatively less nitrogen than bacteria. For every part by weight of nitrogen, thirty to fifty parts of cellulose are consumed. This corresponds to thirteen to twenty-two units of carbon per unit of nitrogen. If litter rich in nitrogen is decomposed by fungi, there is a residue which is still rich in nitrogen, later released as ammonia.

Fungi use relatively more carbon than bacteria: the fungi consume 20–60 per cent of the metabolised carbon, whereas the bacteria usually use only 5–10 per cent, 30 per cent at the most. Thus, the fungi utilise the material in the litter more economically. This same characteristic is shown by the ratio of respiration to growth: the fungi produce a relatively large amount of mycelium per unit of material respired.

Various growth substances (vitamins) are necessary for fungal growth, and these may be present in the litter or may be produced by other soil organisms. Some of the fungi, like higher plants, are able to synthesise their own growth substances, but the majority are heterotrophic in this respect and thus they take up such substances from litter and humus, or obtain them from other fungi in the vicinity (see Fries, 1948, for example).

The large hymenomycetes and gasteromycetes produce fruit bodies only when the soil is moist and the air is relatively humid, and when there is a good supply of growth substances and of nitrogen. In Scandinavia, these conditions are best fulfilled in late summer and autumn.

Algae

Green and blue-green algae are found both on the soil surface and below it. The blue-green algae are more common in warm regions whereas the green algae predominate in cooler places. Like other soil organisms, many of the algae are cosmopolitan. Fehér (1948) examined the distribution of 685 species from various parts of the world and found that most of the common ones were more or less ubiquitous. He observed that there were no characteristic associations of species of soil algae in the various regions, probably because the climate in the soil is relatively uniform and therefore does not have the selective effect necessary for the development of such groups.

Green and blue-green algae differ also in sensitivity to acidity: the green algae are the more tolerant and are found on acid soils, whereas blue-green algae prefer more neutral substrates.

The soil algae are facultatively heterotrophic. In light, on the soil surface, they develop chlorophyll, assimilate carbon dioxide and live autotrophically like other green plants, but in the dark they live like saprophytes, heterotrophically.

Nitrate is the most important source of nitrogen, but they can also use ammonia.

The largest number of soil algae is found 2 to 3 cm below the soil surface, where there may be 100 000 to 3 000 000 per gram of soil. They have the same moisture requirement as the bacteria and actinomycetes, i.e. a soil water content corresponding to 40 to 60 per cent of the water-holding capacity.

The importance of the algae in litter decomposition and other soil processes is as yet unknown. They certainly may play a part in the production of organic material on young soils and bare soils in general, for example, on volcanic ash.

Protozoa

Large numbers of protozoa of various types are present in soils containing organic material which is in the process of humification, i.e., in humus, raw humus, underwater litter and gyttja, and in water in contact with such soils. Waksman (1932) counted 2500 to 10 000 per gram of moist soil (see Table 23, page 106). Fehér and Varga (1929) counted similar numbers and observed that there were more in arable soil than in woodland soil and that there were two annual maxima. Short-term variation in numbers may be considerable. In twenty-four hours there may be an increase from a few hundred to several hundred thousand per gram of soil, but little is known of the possible reason for this.

Such variations in numbers of individuals make ecological studies rather difficult, and little is known of protozoan requirements. According to Fehér and Varga (1929), they are more sensitive to changes in soil moisture content than to differences in temperature, humus content, and soil aeration. They are tolerant of a wide range of pH.

The protozoa absorb various dissolved substances from the soil water. They probably have a more important effect on soil processes generally, because they consume large numbers of bacteria and other soil microbes such as algae, yeast cells, and other protozoa.

Protozoa vary in size from about a hundredth to a quarter of a millimetre. Since an amoeba may be a thousand times as big as a bacterium, the number it consumes is considerable. Cutler and Crump (1935) calculated that *Naegleria gruberi*, a typical soil protozoan, consumed about 130 000 bacteria before it divided. Hence, the protozoa may limit the increase of the bacterial population in water, gyttja, dy, and waterlogged soils. In these substrates, the protozoan population fluctuates as a result of changes in bacterial numbers, and there is a relationship between numbers of individuals in the two groups. Their requirements are broadly similar, with the exception of the amount of water they need. There are few protozoa in dry soil, where the film of water around the soil particles is thin or non-existent so that the movement of the animals would be restricted. In dry periods the animals encyst, and resume full activity only when the soil has been sufficiently wetted again by rain. Their effect in inhibiting bacterial development is probably insignificant in dry soil (Forsslund, 1948).

Invertebrates

In the uppermost layers of the soil, particularly the top 3 to 5 cm, there are large numbers of invertebrates, which live on various kinds of food. Those which feed on plant or animal litter or the products of litter decomposition are called *saprophagous*. Others live on faeces, and are *coprophagous*. Some eat living plant material; others are animal predators.

The invertebrates, particularly the larger species, also influence soil processes in other ways. Some make passages, which break up the soil and aid aeration and drainage; others bite up the litter, carry it about, and mix it with soil materials. In the alimentary canal of the animals, the material is changed in such a way that it can more easily be decomposed by bacteria and coprophagous animals. The activity of earthworms and millipedes is of particular importance. In addition, the soil animals contribute to the dispersal of fungi, bacteria and other microbes.

1 Annelid worms. Earthworms, which are oligochaete annelids, have a large effect on the turnover of organic material in the soil. Their significance was first described by Darwin (1881), but they had previously attracted the attention of soil scientists. The ecology of earthworms has been the subject of numerous investigations, and, after bacteria, they are probably the soil organisms whose significance is best understood.

In general, earthworms require a good calcium supply and prefer neutral or fairly neutral soils. In mull soils (brown earths) they are common, but are infrequent or absent from peat or raw humus (with the exception of certain species). They are usually encountered in the soil in deciduous woodland, meadows and cultivated land, but are less frequent in coniferous woodland. A high soil moisture content is essential, but earthworms also require good aeration and therefore avoid waterlogged soil and heavy clays. They have a high resistance to desiccation and can survive the loss of about 75 per cent of their normal body water content, which makes up about 80 per cent of their body weight (Roots, 1956). Nevertheless, the earthworm population is rapidly reduced by drought, and there are few earthworms in dry soils (Julin, 1948). Earthworms are also frost sensitive. Hence, earthworm activity is greatest in autumn and spring, when the soil is usually moist, and less in winter and summer, because of cold and dryness, respectively.

In various types of vegetation in south Sweden, Lindquist (1938) counted 630 000 to 2 590 000 earthworms per hectare. There were 30 000 to 6 230 000 per hectare in deciduous woodland soil in central Sweden (Södermanland) and 40 000 per hectare in raw humus (Julin, 1948). Similar numbers have been estimated in American investigations, for example, 2 800 000 (Oregon) and 5 900 000 to 11 500 000 per hectare (New York) (Lutz and Chandler, 1946; similar data in Thorp, 1949; and see Svendsen, 1955). The total weight of earthworms in the soil may be 50 to 75 per cent of the total weight of all the soil animals (see Table 23, page 106).

There are several known species of earthworms. Length varies between 2·5 and 25 cm and weight 0·05 and 7 g. Earthworms probably live longest of all soil organisms: in the laboratory, they live for six years, becoming sexually mature at anything after six or eighteen months.

There are several ways in which earthworms may affect soil processes. Their food consists of litter, which is consumed in large quantities, and is mainly collected from the surface of the soil. Deciduous leaves are preferred, but fungal hyphae are also eaten. Some leaves are taken in preference to others: oak and beech leaves may come last in order of preference, pine and spruce needles are avoided (Julin, 1948).

Earthworms make burrows in the soil, by eating the soil particles in front of them. Each worm inhabits a system of burrows with an opening at the soil surface and burrows going in all directions. Most species live in the soil just below the surface, but there are others which go down to 15 or 20 cm; a few species go very deep: *Allolobophora longa* and *A. nocturna* are found as much as 2 m below the surface and *Lumbricus terrestris* has been found even deeper. The permeability of the soil to air and water is increased by their burrows, and penetration of roots or other underground plant organs is facilitated, and their activity may increase soil porosity by up to 27 per cent.

Earthworms make an effective contribution to the development of mixed layers in the soil, i.e. to mull formation (see Chapter 10), by finely dividing and mixing litter material with humus and mineral soil. Thus, the presence of earthworms is a prerequisite for the development of mull soils, and a large earthworm population is often a characteristic feature of such soils (Romell, 1939; Forss-lund, 1948).

Humus and litter is changed mechanically and chemically during its passage through the alimentary canal of the earthworms, and this is particularly true of *Allolobophora longa* and *A. nocturna*. The organic matter is crushed and mixed with mineral particles, and mixed with chalk which has been consumed by the worm, passed through the vascular system and secreted by glands in the alimentary canal. This process is of particular significance for the continued decomposition of the material by bacteria, and is connected with the earthworms' selection of food. Deciduous leaves, with a high content of basic minerals, are preferred. The faecal material deposited by the earthworms is enriched chemically and has a high content of nitrogen, phosphorus, potassium, calcium and magnesium. It has high base-saturation, high base exchange capacity, and high pH value. It also has a higher content of bacteria than normal soil (Kollmannsperger, 1951/52).

Earthworm activity thus promotes subsequent decomposition of the litter by microorganisms since earthworm faecal material is more suitable than other humus, and because of this, and because of the large quantities which are produced, it is thought that it is the most important constituent of mull.

Other annelid worms which play a part in soil processes belong to the group Enchytraeidae. They are small in comparison to earthworms, up to only 1 to 2 cm long, but their numbers are correspondingly larger. Forsslund (1948) found up to 600 per litre of soil, which is equivalent to 600 million per hectare in a layer 10 cm thick. He considered them to be of considerable significance in the soil. They are saprophytes, but otherwise little is known of their activities.

2 Eelworms (Nematoda), Nematodes are only about 0·5 to 1·5 mm long, but very large numbers are present in the soil. There may be up to 37 000 million per hectare in mull soils and 7000 to 15 000 million per hectare in raw humus (Forsslund, 1948). Investigations in England and Denmark have shown numbers between 1 and 20 million per square metre in cultivated soil

(equivalent to 10 000–200 000 million per hectare). This corresponds to a weight of 1 to 2 g (see Table 22).

The roles played by the nematodes in the soil are various. Some
live on partly decomposed organic material; some on protozoa,
fungi, algae or bacteria, including cellulose-decomposing bacteria;
some are predatory and may eat other nematodes; some are plant
parasites. By eating smaller organisms, the nematodes have a regulatory effect on the balance of the microbial populations in the soil,
in the same way as the protozoa and other organisms which live on
fresh food (see page 100).

TABLE 22 Numbers of small animals in soils at Rothamsted,
England, in millions per hectare to a depth of *c*. 22 cm. (From
Russell, 1952, page 173.)

| | ARABLE SOIL | | |
	MANURED	UNMANURED	GRASSLAND
Springtails (Collembola)	100	70	135
Beetle larvae (Coleoptera)	14·6	2·1	5·7
Fly larvae (Diptera)	48·0	9·4	27·5
Other insects	8·4	12·3	27·0
Centipedes and millipedes (Myriapoda)	11·1	4·4	4·4
Mites (Acarina)	13·6	4·7	5·4
Spiders (Araneae)	0·42	0·17	3·0
Woodlice (Isopoda)	0·1	0·12	—
Slugs and snails (Gasteropoda)	0·1	—	0·12
Oligochaetes	6·4	1·5	21·0
Nematodes	3·7	0·5	18·3
Time of investigation	Feb. 1936 to Jan. 1937		April 1936 to March 1937

3 Arthropods. Like the soil fauna in general, little is known about
the significance of the arthropods in soil processes. Numbers of
species, and of individuals, are large, and it is known that the
arthropods are responsible for a significant proportion of the turnover of animal material in the soil. Forsslund (1948) counted up to
60 000 arthropods per litre of soil in raw humus in coniferous
forest in north Sweden, equivalent to 3 million per square metre
in a 5 cm thick layer of raw humus.

They occur to some extent in the litter, but are most numerous
in the humus layer, where they feed on fungal hyphae, algae and
other living plant material, partially decomposed litter and faeces.
Some are predatory.

Strikingly large numbers of arthropods live in raw humus and
probably have an important effect on the development of this form
of humus and on its properties. The main difference between the

fauna of podsols and brown earths (for an explanation of these terms see Chapter 14) is that podsols have large numbers of small animals, but few larger ones, whereas brown earths have large numbers of both. Hence, brown earths contain a considerably larger weight of animals, often as much as five times more than podsols.

Since the turnover of organic matter is more or less dependent on body size, turnover by animals in mull soils is probably also many times greater than in podsols. Larger animals are more effective in mixing soil materials, an essential process for mull formation (see Chapter 14).

The activity of the arthropods alone is sufficient to cause the organic material in the litter to enter into the cycle of conversions which leads directly or indirectly to the final stage of mineralisation. At every stage of the cycle the material is divided into three main fractions, which may unite again at a later stage. The first part comprises the structural components, which may be transferred from plant to animal, from animal to plant, from one plant to another or from one animal to another. The second is faecal material, or animal litter, which is further exploited by animals or plant microorganisms and hence is in part incorporated into the structural component part again. The third is the material which reaches the final stage of mineralisation: substances which are broken down in respiration to carbon dioxide and water, or are converted by metabolism to urine and uric acid, which decompose to ammonia after release from the organism. The substances entering this phase are no longer organic, and this represents the end of the process of litter decomposition. The material in the other two fractions reenters the cycle, part of it moving into the final stage each time.

Knowledge of the distribution of the various groups of arthropods in different soils and plant communities is still scanty. However, there have been important investigations of a few soil types. SPRINGTAILS (Collembola) are an important group of soil arthropods. In north Sweden, Forsslund (1948) found 3·7 million per square metre of raw humus, in a layer 10 cm thick. Only about a third as many were found in mull. Springtails are about 1 to 3 mm long. Their food consists of fungal hyphae and rotting organic matter.

MITES (Acarina) are particularly frequent in raw humus where there may be three times as many as in mull (Lutz and Chandler, 1946). According to Russell (1952), there may be as many as 100 million per hectare. They are 0·1 to 1 mm long. They are mainly predatory, but fungal hyphae are the main source of food for one large group of mites. Others live on spores, rotting leaves and faecal material, and some live on cellulose-decomposing bacteria, or live symbiotically with these bacteria. Fungal mycelium is the main food of the Orabatidae.

Mites are probably the dominant groups of animals in some soils (see Tables 22 and 23).

MOSQUITO LARVAE may occur in large numbers when they may have a marked effect cn humus structure mainly as a result of their faeces production (Forsslund, 1948). The larvae of some species

TABLE 23 Approximate numbers of microorganisms and small animals in the uppermost 15 cm of a moderately fertile cultivated soil. (After Stöckli, 1950, page 278.)

	NUMBERS (per g soil)	WEIGHT (kg/ha)	TOTAL WEIGHT (kg/ha)
A Microphytes			
Bacteria	600 000 000	10 000	
Fungi	400 000	10 000	
Algae	100 000	140	
Total	600 500 000	20 140	20 140
	(per litre)		
B Protozoa	1 500 000 000	370	370
C Metazoa	(per litre)		
Nematodes	50 000	50	
Springtails	220	7	
Mites	150	4	
Enchytraeids	20	15	
Woodlice	14	50	
Insects, beetles, spiders	6	17	
Molluscs	5	40	
Earthworms	2	4 000	
Total	50 417	4 183	4 183
Total A, B, C			24 694

live underground for some time and then assemble in large groups, which creep out into the soil surface. Mosquito larvae are mainly saprophagous, but they also eat fungal hyphae.

BEETLES (Coleoptera) and their larvae are mostly predatory, but some are saprophagous or coprophagous. Their burrowing in the soil is also important.

APHIDS AND FLY LARVAE are sometimes present in soil in significant numbers. Forsslund (1948) found aphids in raw humus, forming colonies on spruce roots. There may be up to 700 per litre of soil.

ANTS AND TERMITES have more localised effects because they build nests, in and on the soil or in rotting wood. They chew up plant material, dig passages and eat animal litter, but they are also predators and plant parasites. They transport material both in horizontal and vertical directions, but do not mix organic and inorganic materials. In some places the effect of ants and termites on the soil may be considerable, and comparable with that of earthworms.

CRUSTACEANS are most frequent in the sea, but some occur in fresh water and some small species live on land. Their size varies from microscopic planktonic species to fairly large ones. The crustacea are of considerable importance in the conversion of organic material partly because of their consumption of algae and other organisms, partly because of the mechanical effect which they exert. In this respect, the species which dig passages under stones, roots, etc., on the sea or lake bottom are particularly important, as organic material is mechanically broken up and mixed with the mineral soil as a result of their activity. The crustaceans thereby contribute to the development of gyttja.

CENTIPEDES AND MILLIPEDES (Myriapoda) are found in the same types of soil as earthworms. They live on dead plant material and possibly hyphae, but may also eat other living plants. Some are predatory and may attack earthworms and other large soil inhabitants. Their numbers may be large (Table 22), and since they are large in size, their total body weight per unit of soil may be considerable. Investigations in England (Russell, 1952) showed that Myriapoda and earthworms alone might account for 90 per cent of the dry weight of the animal population of the soil.

WOODLICE (Isopoda) are thought to be significant because of their transport of materials, because they make burrows in the soil down to 50 to 100 cm, and because they live in soil which is much too dry for earthworms.

4 Slugs and snails. SLUGS AND SNAILS live mainly in the litter layer, in moist shady places. They eat fungal hyphae and fresh litter (Forsslund, 1948). Some eat living plants if no suitable litter is available.

Summary

Stöckli (1950) attempted to calculate the weight of the various soil organisms in a fairly fertile, cultivated soil (Table 23). The calculations are necessarily based partly on approximate estimates;

and the result depends on the choice of the soil. Thus Table 23 is intended only as an example of the orders of magnitude and the proportions which can occur. The soil may contain 25 tons of living organisms per hectare in a layer 15 cm thick. Taking the amount of organic material in the soil to be 3 per cent of the dry weight, then living organisms make up 8 per cent of this amount, or 0·24 per cent of the total weight of the soil. Of this, about 80 per cent is made up of plant microorganisms and the rest is mainly earthworms (Table 23).

7

Humification and mineralisation of the litter

Litter is broken down by chemical and biological processes into the constituents from which it was built up, carbon dioxide, water and salts (Table 24), and the energy which was bound when the organic material was built up from these constituents is released. These processes can be termed *mineralisation* of the litter.

Mineralisation may sometimes be very rapid, particularly in the oxidation of litter taken up by soil organisms, when compounds

TABLE 24 Average elemental composition of the dry material of five mature maize plants. (Miller, 1938, page 284.)

ELEMENT	PERCENTAGE OF TOTAL	ELEMENT	PERCENTAGE OF TOTAL
Oxygen	44·43	Sulphur	0·17
Carbon	43·57	Iron	0·08
Hydrogen	6·24	Silicon	0·17
Nitrogen	1·46	Aluminium	0·11
Phosphorus	0·20	Chlorine	0·14
Potassium	0·92	Manganese	0·03
Calcium	0·23	Undetermined	
Magnesium	0·18	elements	0·93

such as carbohydrates are oxidised directly to carbon dioxide and water. However, only very small proportions of the litter produced annually return in this way to the air, water and soil. Mineralisation of most of the litter takes a long time, either because the reactions themselves proceed slowly or because decomposition processes are at times replaced by syntheses. In the course of such mineralisation, a number of new substances, which are intermediates between the litter and its inorganic end products, are formed. These transitional substances usually occur in mixtures. Such mixtures, brown or black in colour, are called *humus*; and the processes leading to humus formation are called *humification* of the litter.

Litter is rich in energy: it contains the solar energy which was bound when the inorganic constituents were linked together in the process of carbon dioxide assimilation by autotrophic plants. Like other energy-rich compounds, the substances in litter are chemically unstable and can be split fairly easily, releasing some of the bound energy. The decomposition of litter is a consequence of its energy-rich character and the instability which this brings about.

To some extent, the decomposition processes are initiated by the enzymes and the reactive substances in the cells themselves (autolysis), but in the main they result from the activity of animals and plants which make use of the products and the energy released.

Autolysis

The first phase of decomposition is the result of reactions which take place within the protoplasm and which are initiated by the cells themselves. When a cell dies, plasmatic control of the structure of the protoplasm and of the distribution and transport of the various substances in the cells and tissues ceases. As a result, the chemical and physical potentials which existed in the living cells tend to equalise, the system of equilibria which was part of the organisation of the cells is destroyed, and the unstable, energy-rich compounds are transformed to more stable ones less rich in energy. Enzymes take part in these reactions. While the cells are still living, the structure of the protoplasm, in particular the lipoid films, ensures that the enzymes are localised and separate from one another. The death of the cells and loss of plasmatic control of the structure results in the release of enzymes and other chemically active substances, which diffuse freely around the cells and intermix, so that reactions not previously possible can take place. The enzymes act on the constituents of the cells themselves, and the cells undergo autolysis.

In part, autolysis consists of enzymatic oxidations and reductions. In living cells there is a balance between these reactions, but when the cells are dead oxidations predominate. Autolysis can be prevented by high temperatures, which destroy the enzymes.

Autolysis has long been used by man in the production of tea and tobacco. During the so-called *fermentation*[1] of the leaves, their

[1] From Latin *fermentum*, which has the same meaning as the Greek-derived word *enzyme*.

chemical composition is changed by autolytic processes. A certain amount of heat is produced as a result of the decomposition and oxidation of the energy-rich compounds. When the protoplasmic structure has broken down, the soluble contents of the cells can diffuse. Dead leaves or other organs lying in water lose substances such as carbohydrates, acids and mineral salts, and a number of brown substances which precipitate when in contact with air, and sink to the bottom. Some of the humus in soil consists of similar substances Litter which has lain in water for some time colours it brown. Loss of soluble substances is most rapid at first. Romell (1939, page 375) found that 50 per cent of the potassium and 20 to 30 per cent of the phosphorus was lost from fallen litter after leaching for only one day.

Biotic decomposition

Animals break up litter mechanically by biting and chewing, and by trampling, digging and other activities. The more finely divided the litter, the larger the surface which can be affected chemically. The material is prepared by such mechanical processes for the subsequent chemical decomposition to which it is exposed, first in the alimentary canal and then in the animal cells. Some of the litter reaches the final stages of decomposition to inorganic substances, carbon dioxide and water, within the animal body.

The action of the soil microphytes is exclusively chemical. Through production of appropriate enzymes, these organisms bring about oxidations, reductions and hydrolysis, with some resynthesis at the same time. Some of the products of decomposition are taken up by the cells of the microphytes and used partly as a respiratory substrate and partly as a raw material for synthetic activity, in which case they again come to form part of a living organism.

Of all the processes involved, oxidation is the one most characteristic of the course of decomposition. The uptake of oxygen and production of carbon dioxide (soil respiration) can therefore serve as a measure of the activity of the soil organisms.

Decomposition is accompanied by various easily observed phenomena which have long been known and which have been given various names. For example, *putrefaction* is the term traditionally applied to decomposition involving production of unpleasant-smelling compounds; and *fermentation* implies that gases are produced. Slow decomposition in the soil, with no noticeable production of gases or smell, is called *humification*.

Decomposition of material containing little or no nitrogen

Litter usually contains small quantities of sugar and starch, but large quantities of cellulose and lignin (see Table 16, page 82).

Many enzymes are involved in the decomposition of carbohydrates and similar compounds. For example, an extract pressed out of fungi may contain several enzymes or enzyme complexes such as cellobiase, lichenase, xylanase, proteases and phosphatases (see Janke, 1949).

The extent, rate and course of decomposition are primarily determined by the enzyme systems of the various microphyte species, but the final result also depends on environmental conditions such as moisture, temperature and the availability of nitrogen and oxygen (Table 25). There may be hydrolyses,

TABLE 25 The effect of additions of nitrogen on the decomposition of rye straw. 5 g samples. Units in mg. (After Waksman, 1952, Table 39.)

ORGANISMS	NH_4^+ ADDED	CO_2 LIBERATED	TOTAL PLANT MATERIAL DECOMPOSED	HEMICELLULOSES DECOMPOSED	CELLULOSE DECOMPOSED	NH_3—N ASSIMILATED
Control	—	37	—	—	—	—
Control	+	37	—	—	—	—
Trichoderma	—	333	251	190	47	—
Trichoderma	+	648	504	280	327	25
Humicola	—	588	602	199	125	—
Humicola	+	1106	908	339	461	31

syntheses, oxidations or reductions, or splitting and rearrangement of the molecules; and the products may include carbon dioxide, water, organic acids, various alcohols, methane, hydrogen, etc.

The microphytes benefit from the less complex organic compounds which they can take up and use, and from the energy released, the amount of which depends on the course of the decomposition processes.

The following reactions are examples of these processes.
Complete oxidation:

$$C_6H_{12}O_6 + 6\ O_2 \longrightarrow 6\ CO_2 + 6\ H_2O + 2820\ kJ.$$

Partial oxidation:

$$C_6H_{12}O_6 + 4\tfrac{1}{2}\ O_2 \longrightarrow 3(COOH)_2 + 3\ H_2O + 2060\ kJ.$$
oxalic acid

Anaerobic splitting:

$$C_6H_{12}O_6 \longrightarrow 2\ C_2H_5OH + 2\ CO_2 + 92\ kJ.$$
ethyl alcohol

$$C_6H_{12}O_6 \longrightarrow 3\ CH_3COOH + 63\ kJ.$$
acetic acid

The more complete the decomposition and oxidation, the more the energy released. Complete oxidation of one gram molecule of glucose results in the release of 2820 kJ, which is about forty-five times as much as that released in the formation of acetic acid. Hence, in this case, anaerobes must have a turnover of forty-five times as much material as aerobes if they are to obtain the same amount of energy.

Many of these processes have long been exploited by man in the production of substances such as acetic acid and alcohol, in baking, and in preparation of tobacco or fibres such as linen.

Sugars and the simpler carbohydrates

Decomposition. Monosaccharides, disaccharides and some poly-saccharides are broken down by bacteria and fungi (sugar fungi), and also by algae, which use hexoses, for example. The simpler carbohydrates are also used by soil animals, particularly by wood-lice, centipedes and millipedes. Water-soluble carbohydrates are leached out of litter, and decomposition is rapid at first. Leaching usually begins during the first rain following deposition of the litter, and subsequent microbial activity results in the formation of large amounts of organic acids within a few weeks (Mattson and Koutler-Andersson, 1941).

Resynthesis. After hydrolysis of the various carbohydrates by the soil microphytes, some of the products are assimilated. Hydro-carbons, alcohols, acids, proteins, amines and amides may be used as respiratory substrates. Some of the simpler products of hydroly-sis are used as raw materials for synthesis, by condensation, of complex carbohydrates which become part of the microbial body. Hydrolysis and condensation proceed as follows, for example:

$$C_{12}H_{22}O_{11} + H_2O \underset{\text{condensation}}{\overset{\text{hydrolysis}}{\rightleftharpoons}} 2\ C_6H_{12}O_6.$$

Resynthesis by some colourless bacteria involves the assimilation of CO_2. *Escherischia coli* changes pyruvic acid to oxaloacetic acid in this way:

$$CH_3.CO.COOH + CO_2 \longrightarrow COOH.CH_2.CO.COOH.$$

Some of the products of decomposition may be used as storage compounds. Fungi and bacteria, which have no plastids, store carbohydrate not as starch but as glycogen, a water-soluble substance related to amylopectin, as animals do. When required, the glycogen is converted to glucose by the enzyme glycogenase.

Another carbohydrate commonly present in fungi is the disaccharide, trehalose, which is split to form glucose by trehalase.

Complicated compounds may sometimes be formed, for example, by *Lentinus lepideus*, which synthesises an aromatic compound, p methoxyl-cinnamic acid-methyl ester, from xylose and glucose. This compound is of particular interest because of its close resemblance to the units in the lignin molecule (see Nord and Vitucci, 1948).

Cellulose and hemicellulose

Microbial degradation of cellulose is a series of hydrolyses brought about by enzymes (cellulase and cellobiase) which are present in the majority of soil microphytes.

The final product is glucose and the processes are generally:

cellulose $\xrightarrow{\text{cellulase}}$ cellulobiose $\xrightarrow{\text{cellobiase}}$ glucose
$n\ C_6H_{10}O_5$ (disaccharide $C_6H_{12}O_6$
            $C_{12}H_{22}O_{11}$)

The way in which the glucose is used depends on the requirements and enzymatic make-up of the organism, and on the prevailing environmental conditions. When there is a good oxygen supply, glucose is mostly used in the following way:

glucose $\begin{cases} \xrightarrow{\text{oxidation}} & CO_2 + H_2O \\ \xrightarrow{\text{storage}} & \text{glycogen} \\ \xrightarrow{\text{synthesis}} & \text{cell materials} \end{cases}$

In anaerobic conditions, various intermediates may be formed. These are present only temporarily and when oxygen is available again they are oxidised by other organisms to carbon dioxide and water.

glucose $\begin{cases} \xrightarrow{\text{fermentation}} & \text{intermediates} \\ \xrightarrow{\text{synthesis}} & \text{cell materials} \end{cases}$

Hemicellulose is broken down in a similar way, but the end

products are hexoses, such as galactose and mannose, or pentoses, such as xylose and arabinose.

All these processes require intimate contact between the microbes and the material, because the enzymes are secreted by the microbial cells and the products are partly utilised by them. How this contact is achieved is not fully known. It has been shown that bacteria of *Clostridium* species arrange themselves parallel to the long cellulose micelles so that they are in contact over a relatively large area. It is also known that bacteria which attack cotton fibres, for example, begin to act at the periphery of the fibre, so that the fibre surface becomes corroded; and fungi tend to attack from the lumen of the fibre, from which the hyphae grow into the tertiary cell wall and thence into the secondary wall (Baker and Martin, 1939).

Nor is the mode of decomposition known in detail. Up to now, detailed study has been restricted to only a few types of litter, the straw of cereal crops, for example (see Table 25). Decomposition begins as a result of vigorous attack by fungi on the cellulose and hemicellulose, accompanied by the production of a considerable amount of heat and carbon dioxide. Within a few days a large part of the material has been used, sometimes as much as half. However, there is then a marked slowing down and further decomposition seems to be concentrated more on the products formed by the microorganisms.

A large number of species of bacteria, actinomycetes, fungi and other organisms take part in the decomposition of cellulose and hemicellulose. (Review by Janke, 1949.) Two main directions can be distinguished, according to the organisms involved, viz. microorganisms or larger animals.

1 Decomposition by microorganisms
(a) BACTERIA
(i) *Anaerobionts*. In anaerobic conditions, such as may prevail in water or mud, in thick layers of litter, or in the alimentary canal of ruminants, cellulose is broken down mainly by bacteria of the genera *Clostridium*, *Vibrio*, *Ruminococcus* and *Ruminobactes*. Some *Clostridium* species live symbiotically with other microorganisms, such as *Escherischia coli*, *Bacillus aerogenes* and *Bacillus putrificum*. These release nitrogenous compounds which are used by the *Clostridium* (Janke, 1949, page 400).

This sort of interaction between organisms is probably the reason why species in pure culture cannot achieve as high a rate of decomposition as occurs in natural conditions (Janke, 1949, page 400).

In the alimentary canal of ruminants, especially in the rumen and the caecum, anaerobic bacteria break down cellulose into substances which can be digested by the animal. While they are active, these bacteria take up cobalt, among other things, which is considered to be an essential requirement for the production of vitamin B_{12} (anti-pernicious anaemia factor). The digestibility of the cellulose depends on the lignin content of the material, because lignin has an inhibitory effect on the activity of the bacteria (Lindeberg, 1944; Janke, 1949, page 430).

Anaerobic decomposition is incomplete. Carbon dioxide and water are formed, but only in small amounts, and this carbon

TABLE 26 Microbial protein synthesis and the quantity and composition of gas evolved during the anaerobic rotting of rice straw. Temperature of decomposition 30°C. Material 100 g straw, rotted for six months. (After Acharya, 1935; Russell, 1952, page 253.)

DEGREE OF ANAEROBICITY	MILD	MODERATE	STRICT
Carbon dioxide (litres)	32·5	17·4	10·8
Methane (litres)	2·1	6·9	10·6
Hydrogen (litres)	0·93	5·6	0·1
Nitrogen (litres)	14·2	12·2	0·09
Protein synthesis (relative)	0·33	0·27	0·07
Loss of dry matter (g)	34·0	37·0	35·0

dioxide is mainly derived from calcium carbonate. The end products are mostly intermediates such as volatile acids (e.g. formic, acetic, propionic and butyric acids), non-volatile acids (e.g. lactic acid and other oxy-acids), alcohols (e.g. ethanol), hydrogen and hydrogen sulphide (Wikén, 1942; Fåhraeus, 1944; Janke, 1949). In aerobic conditions these intermediates are relatively permanent, since the so-called methane bacteria are about the only organisms which attack them to any extent (Wikén, 1942), and they tend to accumulate. Bubbles of gases, such as methane, may rise to the surface of the fermenting material. The methane results from decomposition of the intermediate products by methane bacteria. If large quantities are formed, the process is called *methane fermentation*. The relation between oxygen supply and formation of methane, hydrogen and carbon dioxide is shown in Table 26.

Lack of oxygen results in increased production of methane and decreased production of nitrogen and carbon dioxide. Hydrogen production first increases and then decreases, protein synthesis decreases, and loss of dry matter is more or less unchanged.

Even after the breakdown of cellulose has gone so far that carbon and hydrogen have been converted partly to carbon dioxide and water, partly to methane and gaseous hydrogen, microbial activity still continues. Methane and gaseous hydrogen are attacked by other bacteria and used in the following ways:

First, in hydrogen production and oxidation. The term butyric acid bacteria (*Bacillus amylobacter*, *Clostridium pasteurianum* or *Cl. butyricum*) covers a number of similar, strictly anaerobic bacteria which break down sugars and other carbohydrates and certain alcohols (glycerin, mannitol) with the production of butyric acid, hydrogen and other compounds (butyl alcohol, acetic acid, acetone, etc.). The reaction may be represented as:

$$C_6H_{12}O_6 \longrightarrow CH_3.(CH_2)_2.COOH + 2 H_2 + 2 CO_2 + 75 \text{ kJ}.$$
butyric acid

The process was first described by Louis Pasteur in 1861.

Like other gases formed in the soil, at least some of the hydrogen diffuses out into the air; and as it diffuses into media which contain more oxygen, it may be used by certain aerobic bacteria which are autotrophic for carbon, and which obtain the energy required for chemosynthetic carbon dioxide assimilation from the oxidation of hydrogen:

$$2 H_2 + O_2 \longrightarrow 2 H_2O + 469 \text{ kJ}$$

$$CO_2 + H_2O + 473 \text{ kJ} \longrightarrow CH_2O + O_2.$$

In this chemosynthesis, the role of the oxidation of hydrogen is the same as that of sunlight in the process of photosynthesis by green plants. Carbon dioxide assimilation is an endothermic reaction and proceeds only when there is a supply of energy.

The oxidation of hydrogen in connection with chemosynthesis is possibly quite common in soil bacteria. However, nothing is yet known of its biological significance, nor of the external conditions necessary for the activity of the organisms. Some of these bacteria are very tolerant of oxygen and may even live in a mixture of oxygen and hydrogen (cracking gas bacteria, e.g. *Bacillus pantotrophus*). They are probably faculatively autotrophic, and can also use organic material as a respiratory substrate. They can therefore obtain their energy in two ways, from the oxidation of hydrogen, or from normal respiration, and thus they differ from other autotrophic bacteria.

Second, in methane production and oxidation. Methane bacteria (e.g. *Methanobacterium omelianski*), which are strictly anaerobes, attack the alcohols and fatty acids released in cellulose fermentation and oxidise them by transferring a part of their hydrogen to carbon dioxide, which is thereby converted to methane (marsh gas, CH_4). The alcohols and fatty acids are the hydrogen donors and the carbon dioxide is the hydrogen acceptor (Wikén, 1942; Janke, 1949):

$$2\ C_2H_5OH + CO_2 \longrightarrow CH_4 + 2\ CH_3COOH + 88\ kJ.$$

The methane bacteria can also reduce carbon dioxide to methane, in the presence of molecular hydrogen:

$$CO_2 + 4\ H_2 \longrightarrow CH_4 + 2\ H_2O + 260\ kJ.$$

Methane bacteria occur in gyttja in marshy ground, lakes and other bodies of water, in bogs and fens, and in wells (Wikén, 1942). In such places, methane production manifests itself as bubbles of gas in the water, particularly if the bottom is compressed or otherwise disturbed.

In sewage works, where breakdown of organic material is mainly anaerobic, there is so much methane produced that it is of economic use.

In places where methane is produced, other processes of reduction are going on at the same time, e.g. conversion of nitrate to nitrite and ammonia, and sulphate to hydrogen sulphide. Such places have a characteristically unpleasant smell.

Some of the methane produced in soil is used as a source of energy by aerobic bacteria, such as *Bacillus metanicus*, which are autotrophic for carbon and which obtain energy for chemosynthetic carbon dioxide assimilation from the oxidation of methane:

$$\tfrac{1}{2}\ CH_4 + O_2 \longrightarrow \tfrac{1}{2}\ CO_2 + H_2O + 444\ kJ.$$

These bacteria, too, are often called *methane bacteria*.

In similar ways, other hydrocarbons, such as hexane, octane, paraffin, and the aromatics benzol, xylol, toluol, naphthalene, etc., are utilised by microorganisms. The end products are again carbon dioxide and water. Decomposition, including oxidations and reductions, gives rise to a number of intermediates, including alcohols and organic acids.

(ii) *Aerobionts.* Considerably more cellulose is decomposed aerobically than anaerobically. Several species of bacteria are involved, particularly of the genera *Clostridium*, *Cellvibrio* and *Cytophaga*, together with a number of actinomycetes (review by Janke, 1949).

It is probable that at least some of these organisms use other carbohydrates in preference to cellulose, but can use cellulose if nothing else is available. Fåhraeus (1944, 1947) found that a form of *Cytophaga*, which apparently specialised in using cellulose, preferred glucose if it were available.

The end products of aerobic decomposition are carbon dioxide and water. Intermediates are formed but do not usually accumulate since they are immediately used by other organisms, as long as their growth is not limited by lack of nitrogen. The most important role of the actinomycetes in cellulose decomposition seems to be the further conversion of intermediates as they arise (Janke, 1949).

Some intermediates are also used as carbon sources by nitrogen-autotrophic bacteria, which assimilate atmospheric nitrogen. The genera *Clostridium*, *Cellvibrio* and *Azotobacter* include some nitrogen-fixing bacteria which live symbiotically with cellulose-decomposing species, obtaining carbon-containing compounds from them and providing nitrogenous compounds in return. Since these nitrogen autotrophs are partially anaerobic, the symbiosis requires some limitation of oxygen supply and hence is confined to thick deposits of organic material (Janke, 1949). (See Chapter 8 for further discussion.)

(b) FUNGI. It is possible that more cellulose may be decomposed by fungi than by bacteria. A large number of species are involved, including most Hymenomycetes and numerous moulds, including species of *Aspergillus*, *Penicillium*, *Trichoderma* and *Botrytis*. Many are regarded as harmful to man, since they attack timber, paper and so on, or are plant parasites.

Plant parasites attack organic material before it has become litter. When the leaves fall and the shoots die down in autumn, they are often already more or less completely destroyed by parasites. Old tree trunks infected by wood-rotting fungi have rotten centres which eventually become hollow. Like autolysis, decomposition by microorganisms has already begun at the time when litter is formed. Some components have already been broken down and the way to bacterial infection is open.

Fungi are more economical than bacteria in the use of nitrogen (see page 98), and this is apparent in cellulose decomposition. Use of a certain amount of cellulose is matched by use of a corresponding amount of nitrogen. There seems to be no luxury consumption, so that there is a quantitative relation between cellulose decomposition and use of nitrogen by fungi.

Fungi take up nitrogen in its inorganic form and prefer ammonium to nitrate. In competition for nitrogen, fungi have an advantage over other organisms. Because of this, the addition of large quantities of cellulose (e.g. straw or sawdust) to soil may result in nitrogen deficiency for higher plants, since cellulose-decomposing fungi develop so profusely that they use up all the available nitrogen. After death of the fungi, the nitrogen is returned to the soil, but in an unavailable form, so that it must again be

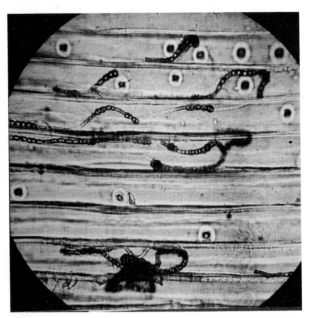

FIG 25. *Pullularia pullulans* **in the wood of pine. The photograph shows how the hyphae have penetrated the tracheids. (From photographic collection of the Royal College of Forestry, Stockholm.)**

converted by microorganisms before it can be taken up by higher plants.

The detailed way in which cellulose is broken down is still unknown. Fungi produce various enzymes, such as cellulase and cellobiase, which split the cellulose molecules into cellobiose and glucose respectively (Fåhraeus, 1944).

2 Decomposition by animals. Apart from bacterial decomposition of cellulose in the alimentary canal of some mammals and birds, little is known of the ability of animals to make use of carbohydrates of high molecular weight. Some invertebrates, such as snails, termites, mites and nematodes, can live on cellulose and hemicellulose, but it is not known whether digestion is by the enzymes of the animals themselves or by bacterial enzymes. In termites it seems to be bacteria which are responsible, since there are cellulose-decomposing bacteria in the alimentary canal, living symbiotically with their hosts. They are also dependent on a symbiotic relationship with various protozoa which are present, but how this operates is not clear. The activity of the bacteria is affected in some way by the protozoa because if the protozoa are killed then the destruction of cellulose ceases (Janke, 1949). This process is of considerable importance in the tropics, where termites are largely responsible for cellulose decomposition.

Several species of worms, including earthworms, and also amoebae, produce cellulase as well as chitase and are thus potentially able to attack both cellulose and chitin (Tracey, 1955).

Cellulose eaten by animals is important in humification of litter because the chemically resistant epidermal tissues of plants are broken up mechanically, thus exposing the central tissues rich in cellulose, and increasing the surface area available to attack.

TABLE 27 Percentage composition of lignin and cellulose. (After Russell, 1952, page 231.)

	CARBON	HYDROGEN	OXYGEN
Lignin	61–64	5–6	30
Cellulose	44·5	6·2	49·3

Some invertebrates use only the parts of the litter which are mostly hemicelluloses, such as leaf parenchyma, and leave parts with more cellulose, such as the networks of leaf veins, untouched.

Lignin

The cellulose in the vascular bundles and in most tissues with a mechanical function is impregnated with lignin. Its Pauschal formula may be written $C_{10}H_{15}O_5$ or $C_6H_{6-8}(OCH_3)(OH)(O)$. There are various conflicting opinions about its chemical structure, but it is generally accepted that the main part is made up of phenol derivatives, probably phenolic alcohols, which are condensed into long chains containing various groups. Lignin contains more carbon and less oxygen than cellulose (Table 27), and also contains small quantities of nitrogen. In the lignin of the Papilionaceae there may be as much as 3 per cent, in grasses only 1·5 to 2·0 per cent and in wood and cereal straw still less.

Lignification of the cell walls begins in young cells and proceeds so far that in the mature cell wall the lignin content may be 25 to 35 per cent. However, the cellulose content is 50 per cent or more.

Lignin alone does not form cell membranes; it occurs only as an impregnant of cellulose. It is not known whether the two components are bound closely together in any way. According to Nord and Vitucci (1948), the binding is merely of the absorptive type. In any case, lignin impregnation always makes the cellulose more resistant to decomposition: a lignin content as low as 15 per cent causes a decrease in microbial activity, and 20 to 30 per cent of lignin decreases activity so much that such material is considered unsuitable as a source of humus in agriculture. If there is even more lignin, for example 40 per cent in coconut fibres, it acts like a preservative (Russell, 1952, page 234).

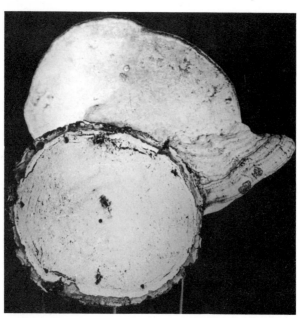

FIG 26. *Polyporus fomentarius* growing on pine. The fungus uses lignin more than cellulose, so the wood becomes white, porous and full of air (white rot). When dry, this wood can be made to ignite by a heavy blow alone, so it was formerly used as tinder. Tinder was also prepared from the inner parts of the fungus itself, which ignited as easily as the rotted wood. The fungus forms thick, tough blankets of hyphae between the annual rings and between the wood and the bark. (From photographic collection of the Royal College of Forestry, Stockholm.)

This protective effect probably results from the relative positions of the two main components in the cell wall. The cell wall is built up of cellulose micelles, which, in turn, are made up of a large number of long chain molecules (Frey Wyssling, 1935). The lignin is packed in around these micelles as a hard, amorphous matrix, isolating them from the immediate surroundings, to that microorganisms must first penetrate the lignin sheath before reaching the cellulose (Frey Wyssling, 1935).

Lignin is not broken down chemically by worms or arthropods (Forsslund, 1948) nor probably by any animal. The relatively small amount of lignin ingested as food is broken up mechanically, but this is insignificant compared with the total amount of lignin.

Bacteria seem to be of no great significance in lignin decomposition. Only lignin in a soluble form can be attacked by bacteria. However, this occurs in the presence of ammonia, for example, and since ammonia may be present in soil in natural conditions it is not improbable that the lignin in the soil is changed so that it becomes available to some bacteria (Fischer, 1953).

Lignin is attacked by many fungi, particularly by Basidiomycetes. Bavendamm (1928) showed that wood-destroying fungi produce enzymes (polyphenol oxidases) which catalyse the oxidation of the polyphenol molecules present in lignin. Lindeberg (1948a) found enzymes of this type in species of several genera of Hymenomycetes (*Clavaria, Collybia, Flammula, Lepiota, Marasmius, Mycaena*). He demonstrated the extra-cellular existence of polyphenoloxidase in *Marasmius foetidus* (Lindeberg, 1948a; Lindeberg and Korjus, 1949, page 103). Lindeberg's work (1944, 1948a) showed that most toadstools decompose lignin to a greater or lesser extent: of forty-six species tested, no less than forty-four could decompose lignin, and they could all decompose cellulose. Some species of the genera *Clitocybe, Marasmius*, and *Clavaria* were found to decompose twice as much lignin as cellulose. *Collybia butyracea*, grown on beech leaf litter, used 77 per cent of the lignin and 16 per cent of the cellulose. *Clavaria gracilis* is also a lignin specialist and used 39 to 53 per cent of the lignin in beech leaves, but none of the cellulose.

Fungi which use more lignin than cellulose, and also those which use both to more or less the same extent, leave a white residue of cellulose. They are called *white rot fungi*, and the humification process is called *white rot*. The colour of the residue is brown if only the cellulose is used. This so-called *brown rot* is caused by a number of *Coniphora* species (for example, Lindeberg, 1946; Falck, 1930 cited by Lindeberg, 1948a). Decomposition of lignin involves loss of methoxyl groups, oxygen uptake, and a darkening in colour of the material, together with chemical binding and absorption of proteins, and increasing solubility in alkalis (literature in Fischer, 1953).

Pectin

Pectin is the collective name for a group of compounds built up of several components, mainly esters and metal salts of pectic acid. Other constituents are arabinose and galactose residues. Pectic acid itself is an oxidation product of galactose, closely related to the hexoses. The salts include Ca—Mg-pectate. Some pectins are water-soluble, others are insoluble or merely swell in water. The cell middle lamella consists mainly of calcium pectate, together with other insoluble or slightly soluble forms. The middle lamella is like mortar, holding adjacent cells together. Other pectins lie between the cellulose and hemicellulose micelles in the primary layer of the cell wall, or form layers between hemicellulose lamellae, in

FIG 27. The stump of a spruce with the trunk attacked by rot while the tree was alive. The peripheral annual rings, which were still living when the tree was felled, were not affected by fungi until after felling. The inside of the stump is filled with humus from the wood, in which *Vaccinium vitis-idaea* is growing.

collenchyma, for example. There are soluble pectins and pectins which swell in water in the cell sap of fleshy fruits.

Pectin is attacked by several fungi, including *Mucor* species, and bacteria, e.g. *Clostridium* species such as *C. butyricum*. Decomposition gives rise to various substances, including acetic, butyric and propionic acids.

Organisms which break down pectin attack only a small proportion of the cell material—the amount of pectin is relatively small—but their activity is very important in promoting the further humification of the litter. The middle lamellae are dissolved so that the cells separate from one another and become more easily accessible to soil microorganisms. The internal surface area of the material is increased so that there is room for more bacteria and fungi.

In the preparation of fibres from flax, hemp and other plants with long fibres, retting processes have long been used, and the most important stage is the breakdown of pectin. The material is left to rot in water or in a damp atmosphere: this process is stopped when the bundles of fibres, or the individual fibres, are sufficiently separated from one another, but before there is any significant breakdown of the cellulose in the fibres. Pectin breakdown is not the only process involved: sugars and the simpler carbohydrates, and probably also hemicellulose, are broken down, so that the parenchyma which surrounded the bundles of fibres is destroyed.

Fats and lecithins

Plant litter generally contains only small amounts of fats and lecithins, in contrast to animal litter and sewerage, in which there may be large quantities.

Fats are broken down by many aerobic and facultatively anaerobic bacteria, including *Bacterium fluorescens*, *B. punctatum*, *B. pyocyaneum*, *B. prodigiosum*, all of which also live on other substances and are not restricted to fat. There are also numerous fungi which consume fat, particularly within the genera *Aspergillus* and *Penicillium*.

Fatty acids accumulate as a result of microbial fat consumption. The process probably involves hydrolysis of the fat molecules to glycerol and fatty acids followed by relatively rapid resorption of the glycerol. Little is known about the decomposition of the fatty acids.

Cutin, sporopollenin, suberin, chitin and wax

The outer layers of plants and animals often contain a number of substances which are chemically relatively inert, and which are therefore resistant to litter-decomposing microorganisms. Cutin is particularly resistant, especially that in spores and pollen, known as *sporopollenin*. Because of this high resistance, spores and pollen have sometimes been preserved intact for thousands of years, as in peat bogs, for example.

Wax is found as a secretion on the outer surface of the epidermis of plants. It helps to keep the surface dry, thereby minimising attack by microorganisms, the spores of which need water in which to germinate. A number of soil organisms grow best in the film of water surrounding wet litter.

Chitin ($C_{18}H_{30}N_2O_{12}$) is a constituent of the cuticle of insects, and also occurs in fungal hyphae. It is rapidly broken down by certain bacteria (Veldkamp, 1952).

These substances are rarely important quantitatively. However, suberin sometimes occurs in relatively large amounts. Bark shed from the trunks of forest trees may accumulate around the base of

FIG 28. Pine stump. The cut surface of the sap wood
(the peripheral annual rings which are not impregnated with wood tar substances) overgrown by lichen
(*Evernia prunastri*). There are only scattered, poorly-
grown lichens on the cut surface of the heart wood.
This is resistant to rotting because it is impregnated
with antibiotic substances and it remains after the sap
wood has undergone humification. Pine stumps at this
stage were formerly used for wood tar extraction. It
is relatively easy to pull up such stumps after the sap
wood and the peripheral root system have rotted away.

the trunks of older trees, forming deep layers of litter in which suberin is one of the main components.

Notwithstanding, these compounds are often qualitatively important in hindering humification of the organs, because they form a protective coat around less resistant substances. The great majority of the microorganisms would not be able to gain access to this food source unless the way was opened by other organisms or means. It is not known whether there are any species of fungi or bacteria which can alone break through the whole system of outer layers. There are bacteria which can break down wax and utilise the carbon it contains, but wax forms only part of the protective covering.

The microorganisms often get in through the gaps in the protective layers; some fungi, particularly parasitic species, can penetrate the stomata and other pores and attack the organ from within. The epidermis is also destroyed, so that the way is opened for other microorganisms, while the outer layers may also be broken up by the gnawing and chewing of herbivorous animals. Destruction by the direct effect of the atmosphere is another possibility. The cuticle of the older parts of plants eventually wrinkles and splits, showing that some sort of weathering has taken place, probably caused by the action of water and gases in the air. Similar processes probably go on in litter and eventually bring about the mineralisation of the resistant substances. Other pathways by which infection may enter are through wounds.

Other substances containing little or no nitrogen

There are also various specialised substances in the litter, such as tannins (condensation products of phenol alcohols), phenols, glycosides, saponins (a group of glycosides), terpenes, pigments, vitamins and other growth factors. Some of these plant products, especially ethereal oils, balsams, resins and rubber, are used commercially.

Little is known of the way in which they are broken down in the soil. Some, like the tannins, are resistant, and some are even poisonous to microorganisms, e.g. phenol. However, because of the high degree of adaptability of the microorganisms and their well-developed specialisation, there are some species which are able to attack these substances.

In litter with large amounts of tannic acids, phenols, etc., for example, that derived from oaks or conifers, humification is slow because of their conserving effect. For this reason, the heart wood of these trees humifies more slowly than the wood from the periphery of the trunk; the cells of the heart wood are dead, and so may be regarded as litter even while they are still in the living tree. They are impregnated with substances which are antiseptic and which ooze out of wounds in the form of resin or rubbery latex, from which wood tar is produced by dry distillation. The substances are a mixture of several constituents, including phenols (the active constituents of creosote, which is made from wood tar), tannins, acids, alcohols and hydrocarbons. Trees with wood lacking large amounts of tannin- and tar-forming compounds, for example, spruce and many common broad-leaved deciduous trees, eventually become hollow if they are infected by rot fungi, humification begins in the dead annual rings of such trees even while the trees themselves are still alive.

As the litter decomposes, several growth factors synthesised by living plants are released. Some of these are probably taken up directly by soil microphytes, many of which are heterotrophic for vitamins in one or several respects (see page 99).

The decomposition and resynthesis of nitrogenous compounds

The nitrogenous compounds in the litter are mainly the same substances as occur in the living organisms. Proteins are most common, and there are also amino acids, amides, nucleic acids, alkaloids, purine bases and phosphatides.

Decomposition by microorganisms

Proteins. The simpler proteins are built up like peptides, i.e. of linked amino acids, which can be separated from one another by hydrolysis.

With relatively few exceptions (the nitrogen-autotrophic forms), bacteria, actinomycetes and fungi fulfil their nitrogen requirements from the nitrogen compounds in the litter, or from their breakdown products. The assimilation of nitrogen from the litter therefore determines the growth and rate of reproduction of the microorganisms, and, in turn, their growth and rate of reproduction determines the rate of decomposition of the litter. The nitrogen content of the litter is therefore of the greatest significance for the processes of humification and mineralisation. The proteins are among the first components of the litter to be utilised by microorganisms.

Humification probably proceeds in the same way as digestion in the mammalian alimentary canal, i.e. by hydrolysis brought about by enzymes (proteases). This assumption is supported by the following observations: enzymes of the same type as those in the alimentary canal have been found in plants; a similar pepsinlike enzyme, papaine, is present in the latex of *Carica papaya* (papaw), and there are other proteolytic enzymes in carnivorous plants; *Penicillium* species produce proteolytic enzymes, which are important in cheesemaking. So-called maturity in cheese is due to qualitative changes resulting from protein hydrolysis brought about by certain *Penicillium* species, e.g. *P. roqueforti* and *P. camemberti.*

Many bacteria secrete nitrogen-containing enzymelike substances. Some of these are proteases; others belong to the group of bacterial products which are called *toxins,* and attack plant and animal tissues, causing various diseases. In general, it seems that the proteolytic bacteria produce extra-cellular enzymes which attack proteins of large molecular weight and which are active from the very beginning of litter decomposition. In addition, there are

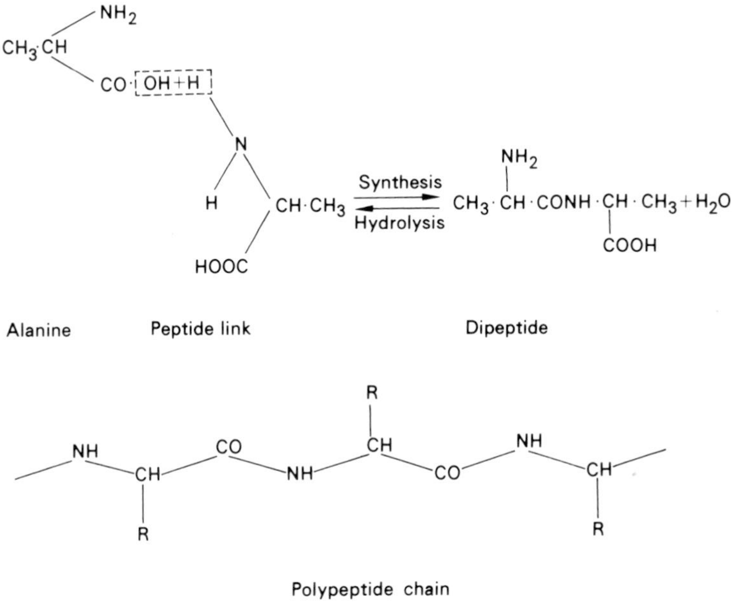

Alanine   Peptide link   Dipeptide

Polypeptide chain

FIG 29. Diagram showing peptide formation.

intracellular enzymes which act within the cells and mainly attack proteins of smaller molecular weight.

The decomposition of nitrogen-containing litter releases products of the same type as those resulting from protein breakdown in the mammalian alimentary canal. Decomposition by bacteria produces peptones, polypeptides and amino acids, i.e. the same compounds produced as a result of hydrolysis of proteins by boiling with dilute acids and alkalis; aromatic and evil-smelling compounds such as indole, skatole and tryptophan, which are characteristic of the type of bacterial decomposition known as putrefaction; fatty acids and other carboxylic acids such as acetic acid, butyric acid, valeric acid, etc.; inorganic end products such as carbon dioxide, water, salts, hydrogen sulphide, ammonia, methane, gaseous hydrogen and nitrogen; and phosphoric acid, if the material contained nucleoproteins.

Urea, uric acid, hippuric acid. Another nitrogenous product which occurs in large quantities is urea ($NH_2.CO.NH_2$). Urea production by man and animals has been estimated to be about 150 000 Mg per day. Urea probably serves as a nitrogen reserve in fungi, which contain significant amounts (Table 28), and is used in spore maturation.

TABLE 28 Urea content of fungi. (From Kostytschev, 1926, page 379.)

	UREA CONTENT ($\%$ dry weight)
Lycoperdon saccatum (puff-ball)	2·9
L. marginatum	5·8
Bovista nigrescens	11·2
Psalliota campestris (field mushroom)	6·2
Pholiota spectabilis	2·5
Cortinarius violaceus	0·5

Urea is broken down by a large number of common species of soil bacteria, aerobes and anaerobes. The appropriate enzyme, urease, is also found in some mould fungi and even in legumes. The process requires a high pH. Ammonium carbonate, which is unstable and decomposes easily, is formed first:

$$NH_2.CO.NH_2 + 2\,H_2O = (NH_4)_2\,CO_3$$
$$(NH_4)_2\,CO_3 = 2\,NH_3 + CO_2 + H_2O.$$

Hippuric acid, which is present in urine, is broken down by some of the same bacteria as urea, to glycine and benzoic acid. When the glycine is broken down, ammonia is released.

$$\underset{\text{hippuric acid}}{C_6H_5\,CO.NH.CH_2COOH} + H_2O = \underset{\text{benzoic acid}}{C_6H_5COOH}$$
$$+ \underset{\text{glycine}}{CH_2.NH_2.COOH.}$$

Many other substances rich in nitrogen occur commonly in natural conditions. These are broken down in similar ways. Uric acid ($C_5H_4N_4O_3$) is present in bird and reptile excrement and is contained in large quantities in guano.

Deamination of amino acids. In subsequent stages, the amino acids are deaminated, i.e. the amino group is split off to yield ammonia. This process represents an important source of nitrogen for bacteria and fungi. Stephenson (1949) has described the various ways in which deamination may take place. Only a few examples will be given here.

Aerobic bacteria may carry out the process by oxidation:

$$CH_3 . CHNH_2 . COOH + \tfrac{1}{2} O_2 = CH_3 . CO . COOH + NH_3.$$
d-alanine   pyruvic acid

Proteolytic anaerobes and faculative anaerobes obtain energy in other ways, the most important of which are probably the following:

1 REDUCTION. In some cases the amino acids undergo reduction to fatty acids at the same time as deamination. The enzyme,

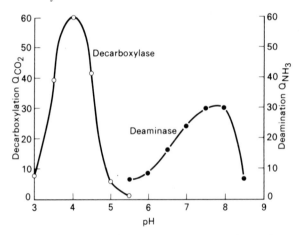

FIG 30. The relation between enzyme activity and acidity. (After Gale, 1940, from Stephenson, 1949.)

aspartase, found in some bacteria (*Bacillus coli*), splits aspartic acid in this way to form fumaric acid and ammonia. The reaction is reversible, and the enzyme is found in higher plants, for example, in germinating peas, young leaves of red clover and in other seedlings, so synthesis of aspartic acid is probably by the same or a similar process.

$$COOH . CH_2 . CHNH_2 . COOH \underset{\substack{\text{germinating peas} \\ \text{young leaves}}}{\overset{B.\ coli}{\rightleftarrows}}$$
aspartic acid

$$COOH . CH{=}CH . COOH + NH_3.$$
fumaric acid

The acid product is unsaturated, because the reduction takes place without hydrogen. In other cases, hydrogen is used and saturated fatty acids are formed.

$$R.CH_2CHNH_2.COOH + 2H \longrightarrow R CH_2.CH_2.COOH + NH_3.$$

The hydrogen comes from a hydrogen donor, or if none is present, from endogenous material which has been reduced in the cells of the organism.

Cystine, an amino acid which contains sulphur, is reduced in this way in the presence of hydrogen or reducing substances such as glucose, with the formation of hydrogen sulphide.

$$COOH.CHNH_2.CH_2S.S.CH_2.CHNH_2.COOH + H_2 + 4H_2O$$

cystine

$$= 2H_2S + 2NH_3 + 2CH_3.COOH + 2H.COOH$$

formic acid

In anaerobic oxidation, energy can also be released in the absence of a hydrogen acceptor, e.g.:

$$5HOOC.(CH_2)_2.CHNH_2.COOH + 6H_2O \longrightarrow$$

glutamic acid

$$6CH_3 COOH + 2CH_3(CH_2)_2COOH + 5CO_2 + H_2 + 5NH_3.$$

acetic acid   butyric acid

2 RECIPROCAL OXIDATION AND REDUCTION OF AMINO ACIDS. There are microorganisms which use some amino acids as hydrogen donors and others as hydrogen acceptors. By simultaneously attacking one acid of each type, they can bring about the oxidation of one acid and the reduction of another.

The same acid may act in both ways. For example, one molecule of alamine can be oxidised while two are reduced.

$$3CH_3.CHNH_2.COOH + 2H_2O = 2CH_3.CH_2.COOH$$

alanine   Propionic acid

$$+ CH_3COOH + CO_2 + 3NH_3.$$

acetic acid

3 HYDROLYTIC DEAMINATION. Some amino acids, such as histidine, phenylanaline, tryptophan and tyrosine, are attacked by yeast and deaminated, forming CO_2 and alcohols.

$$R.CH.NH_2.COOH + H_2O = NH_3 + CO_2 + R.CH_2OH.$$

Decarboxylation of amino acids. Several amino acids are decarboxylated by a number of bacteria, with the formation of the corresponding amines. However, decarboxylases are not as common in bacteria as are deaminases. The two types of enzymes are active in different pH ranges.

Resynthesis of nitrogen compounds by microorganisms

Several of the nitrogenous intermediates and end products are re-assimilated by the proteolytic microorganisms. Some bacteria assimilate amino acids as such; others synthesise their amino acids from inorganic nitrogen compounds, ammonia in particular; other microorganisms do both. For example, yeast assimilates ammonium salts, amino acids and peptones. The resulting nitrogenous products are built into protein molecules, so that they again become cell constituents. In other words, the nitrogen from the litter becomes bacterial nitrogen. The resynthesis probably occurs immediately after the release of the nitrogenous products of humification, so that synthesis and decomposition proceed simultaneously.

TABLE 29 An example of increase in protein content during humification. Changes in the composition of rye straw after humification for two months. (After Waksman, 1938.)

	CONTENT, AS % OF ASH-FREE DRY MATTER	
	FRESH LITTER	AFTER 2 MONTHS AEROBIC HUMIFICATION
Total material	100	58·0
Cellulose	41·5	18·3
Pentosan	26·0	10·3
Lignin	22·5	20·0
Protein	1·2	3·4

Since microorganisms are able to utilise inorganic nitrogen as well as nitrogen from the litter, their growth is to some extent independent of the nitrogen content of the material in the litter. If the material contains less than 1·2 to 1·3 per cent nitrogen (on a dry weight basis), the microphytes also take up any available ammonium salts. On the other hand, if the nitrogen content is high, more than 1·8 per cent, some of this nitrogen is converted to ammonia. In this way, ammonia acts as a buffer in microphyte nitrogen economy. Because of this, humification of material poor in nitrogen may lead to a temporary decrease in the forms of nitrogen available to higher plants, particularly ammonium salts and nitrate, and there may be an increase if the material is rich in nitrogen. Tables 29 and 30 show experimental results which illustrate this. During humification there is a relative increase in the amount of nitrogenous substances.

Decomposition of material poor in nitrogen can therefore be stimulated by the addition of nitrogen, for example, in the form of ammonium salts, or urine, faeces or other animal waste, litter from legumes or young plant material. Microbial activity is promoted by this material, more of the carboxylic acids formed as a result of decomposition are used up in the formation of amino acids, and synthesis of microbial protoplasm increases.

In contrast, adding large quantities of material such as straw, which is poor in nitrogen, to an arable field may cause nitrogen deficiency in the crop, because the microorganisms use up the ammonium and nitrate in the soil to make up for the lack of nitrogen in the straw. Only when the microorganisms die and are humified does the nitrogen go back into the soil.

TABLE 30 An example of increase in protein content during humification. The composition of peat and of some peat-forming species. (After Waksman, 1938, page 157.)

| | CONTENT AS PER CENT OF ASH-FREE DRY MATTER | | | |
	ETHER-SOLUBLE FRACTION	CARBO-HYDRATE	PROTEIN	LIGNIN AND LIGNIN COMPLEX
Cladium, green plants	1·2	51·3	7·5	30·3
Carex, green plants	2·6	48·0	7·3	21·7
Sphagnum, green plants	1·5	53·6	6·1	7·2
Lower moor peat	1·2	9·9	20·8	55·9
High moor peat	4·0	36·7	6·7	38·8
Woody peat	3·3	8·4	14·7	62·6
Gyttja (sedimentary peat)	0·9	19·9	27·6	31·6

Yeasts contain large amounts of proteins and fats. *Rhodotorula*, *Nectaromyces*, *Torulopsis*, *Oidium* and other microorganisms synthesise large quantities of protein as they grow. Their protein content may be up to 60 to 80 per cent of the dry weight. If nitrogen is deficient, fats are synthesised instead of proteins (Mothes, 1954). This is the basis of industrial production of proteins and fats from yeast. The raw materials are industrial by-products, such as molasses and sulphite lye (which contains pentoses and hexoses), to which ammonium sulphate and superphosphate are added. A suitable variety of yeast is inoculated into the mixture, preferably *Torula utilis*, *Mycotorula japonica*, *Oidium lactis* or *Saccharomyces cerevisiae*. About 100 kg of yeast is obtained from 343 kg of molasses (containing *c.* 50 per cent sugar), 34 kg ammonium sulphate and 54 kg superphosphate. Thus, the yield is equivalent to

50 to 60 per cent of the sugar. The yeast multiplies rapidly. *Torula utilis*, growing in a container holding 2270 litres, can produce 35 kg pure yeast protein after only eight hours. The dried, compressed yeast can be used as animal fodder; and also in the production of human food. Experiments have shown that animals can use as much as 90 per cent of the yeast protein; and yeast is also rich in fat. Industrial production of protein began in Germany during the First World War, using *Torula utilis* (Mothes, 1954; Gilbert and Robinson, 1957), when there were several factories in operation, with a total annual production of 15 000 Mg.

Mycorrhiza

A special type of nitrogen decomposition is characteristic of mycorrhiza, which in their symbiotic relationship with the roots of forest trees and other plants bring about, or assist in, nitrogen

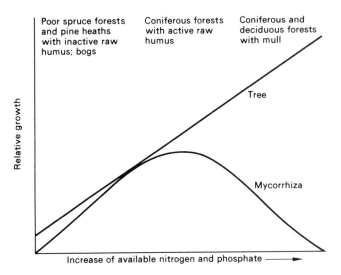

FIG 31. **The relation between the development of mycorrhiza and tree growth, at varying soil nitrogen and phosphorus contents. (After Björkman, 1944–46.)**

uptake by the hosts. Melin and Nilsson (1953, 1954) have shown that mycorrhizal fungi (*Boletus variegatus*, *Cortinarius glaucopus*) take up nitrogen compounds (e.g. ammonium nitrate, glutamic acid) and other nutrients (phosphoric acid), which then pass into the cells of the host (pine).

The development of mycorrhiza depends on the nutrient status of the host (Björkman, 1942, 1956) and therefore occurs on some soil types, but is absent, or insignificant, on others. In coniferous woods with active raw humus, in which microorganisms effect rapid decomposition, an excess of carbohydrates builds up in the tree roots, since the plants growing there are photosynthesising rapidly, but experience a deficit of available nitrogen and phosphorus. This means that not all the carbohydrates produced can be converted into proteins. The excess carbohydrate is used by the mycorrhizal fungi, which develop luxuriantly in such conditions (Fig. 31). In the more fertile soils typical of deciduous woods there is usually so much nitrogen and phosphorus that all the carbohydrates are used within the host cells. The amount left over for the fungi is insignificant and mycorrhiza are only sparsely developed. In the very poor soil of pine heaths or bogs, mycorrhiza are also sparse, probably because the environmental conditions allow only low productivity and little growth. (See Björkman (1956) for discussion of other symbiotic relationships.)

Breakdown of nitrogenous compounds by animals

Since soil animals eat bacteria and fungi, the nitrogenous compounds go through a series of processes of synthesis and decomposition. The nitrogen obtained by other animals from plant material, litter or living plants, is converted in the same way.

Some of the nitrogen is a constituent of the animal cells—the nitrogen changes from microbial nitrogen or litter nitrogen to animal nitrogen; some is lost in the excrement, either because it is indigestible or after it has taken part in metabolism and been converted into urea, uric acid or ammonia. Ammonia is especially common in earthworms.

8

Nutrient cycling and its dependence on soil microphytes

The carbon cycle

World reserves of carbon

Carbon is present in the air, in water, in living organisms, litter, humus, peat, gyttja (see page 222), dy (see page 220), and in fossils and carbonates.

The carbon dioxide content in the inner layers of the atmosphere, surrounding the aerial parts of plants, is about 0·03 per cent by volume (or 300 cm³ CO_2 per m³ air), equivalent to about 0·57 g CO_2 per m³ of air. This concentration decreases with height above the ground, because of the relatively high density of carbon dioxide (1·81 kg/m^{-3} at 20°C). The CO_2 content close to the soil surface is much higher than that a few decimetres above the ground, because of the diffusion of CO_2 out of the soil. The air in the spaces between the soil particles is rich in CO_2, released as a result of biological processes going on in the soil. The concentration of CO_2 there may be up to one hundred times as great as that in the atmosphere (Lundegårdh, 1924).

The total amount of CO_2 in the atmosphere has been calculated from analyses of air to be 2350×10^9 Mg,[1] which is equivalent to 641×10^9 Mg of pure carbon (Table 31).

In the course of time, carbon has been accumulated in living organisms. The amount has been estimated to be equivalent to over 300×10^9 Mg of CO_2, or about one-seventh of the present total amount in the atmosphere (Table 31). Earlier estimates (Boresch, 1931) of the carbon content of living organisms were much higher than this, equivalent to about half of that contained in the atmosphere. Most of the carbon bound in this way is in plants, and there is only a relatively small amount in animals and man.

[1] 1 Mg $= 10^6$ g $= 1$ metric ton.

Litter, humus, raw humus, peat, gyttja and dy contain carbon, but little is known about the total amounts involved. In warm regions these types of soil are rare, at least in plant communities in well-drained areas. The amount of carbon in the soil is therefore insignificant in comparison with that in living organisms. The same is probably true of woods and forests in warm and temperate regions.

However, in cold regions there is a relatively large amount of carbon in the soil, since the plant communities there form soils

TABLE 31 World carbon reserves calculated as CO_2 in sedimentary minerals, in the hydrosphere, atmosphere and biosphere, etc. (Revelle and Suess, 1957; Boresch, 1931; Steeman Nielsen, 1957.)

	AMOUNT OF CO_2	
	$Mg \times 10^9$	IN RELATION TO ATMOSPHERIC CO_2
Carbonate in sediments	67 000 000	28 500
Organic carbon in sediments	25 000 000	10 600
Atmospheric CO_2	2 350	1
Living organisms on land	300	0·13
Material from dead organisms on land	2 600	1·1
Inorganic carbon in the oceans	129 700	55·0
Living organisms in the oceans	30	0·01
Dead organisms in the oceans	10 000	4·3
Total carbon in the oceans	140 000	59·4
Total photosynthesis per year	110	0·04
Marine plant photosynthesis per year	44–55	
Animal respiration per year	5–10	

which often include relatively thick layers of organic matter, particularly raw humus and peat. In arctic plant communities, especially tundra, the ground is covered by a layer of peat and raw humus, with a carbon content probably many times higher than that of the vegetation. Peat bogs and fens have the highest carbon content of all. Thick layers of peat from these communities are utilised as fuel with a high carbon content.

Other carbon-rich soils are arable soils with a high humus content and some prairie soils.

In large parts of the water-covered areas of the world, there is probably at least as much carbon in the gyttja as there is in the vegetation. In borings several metres deep (maximum 19 m) taken from the sea bottom, carbon was found in the form of plant and animal remains (Arrhenius, 1950).

World stocks of fossil carbon were derived from the atmosphere and from the lakes and oceans, and were bound in previous epochs by photosynthesis by plants.

An old estimate (Schroeder, 1919) assesses Europe's reserves of fixed carbon as 700×10^9 Mg. If 75 per cent of this is taken to be pure carbon, the equivalent amount of CO_2 is 1928×10^9 Mg, which is somewhat less than the figure for CO_2 given in Table 31. Again, according to Schroeder (1919), the total world reserves of fixed carbon are four to ten times larger than Europe's and therefore equivalent to $7712–19\,280 \times 10^9$ Mg CO_2. Bolin (1970) estimates world coal and oil as equivalent to $36\,000 \times 10^9$ Mg CO_2.

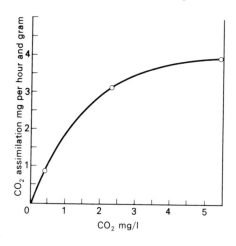

FIG 32. The relation between rate of assimilation and carbon dioxide concentration. (After Stålfelt, 1924.)

The world's known oil reserves are reckoned to be 12×10^9 Mg, but this is probably only a small part of the total oil resources present in the earth's crust (*Petroleum*, 1951, page 17). Unfortunately, all these calculations are based on uncertain estimates; the figures in Table 31 suggest that the amounts of organic carbon in sediments are considerably greater than those quoted above.

In any case, the supply of carbon dioxide available to the organic world was greater in previous epochs than it is at the present day. The decrease is less than the equivalent of presentday amounts of fossil carbon, because some of this carbon was derived from CO_2 dissolved in the oceans and lakes. Nevertheless, the higher concentration of CO_2 would have been significant, because photosynthesis increases in response to small increases in CO_2 supply above the presentday natural level (Fig. 32). For example, if the ambient CO_2 concentration increases by 10 per cent, then the assimilation rate also increases by about 10 per cent. These measurements refer to periods of only a few minutes or, at most, a few hours. Figures for longer periods, for example for some days or for the whole of the annual vegetation period, provide more satisfactory evidence on which to base an estimate of the ecological effect of change in the CO_2 content in the atmosphere (Table 32). It seems probable that the rate of assimilation in previous epochs may have been higher than it is at the present day.

TABLE 32 The effect of increases in atmospheric carbon dioxide concentration on plant growth rate. (From Hurd, 1968; Hughes and Cockshull, 1969.)

CO_2 CON-CENTRATION ppm	RELATIVE LEAF AREA INCREASE		RELATIVE DRY WEIGHT INCREASE	
	Callistephus chinensis	TOMATO	*Callistephus chinensis*	TOMATO
325	100		100	
350		100		100
600	106		132	
900	120		173	
1000		124		141

A considerable amount of CO_2 is stored in water in both free and bound form, as bicarbonate or carbonate. The proportions of ions present change with pH of the water (Fig. 33). Water in contact with air takes up or releases carbon dioxide until there is an equilibrium between the partial pressure of carbon dioxide in the air and in the water. The position of equilibrium is affected by

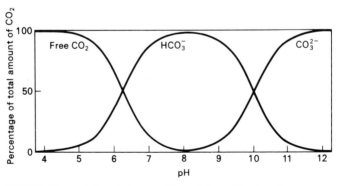

FIG 33. The relation between dissociation of carbonic acid and the acidity of the solution. (After Emerson and Green, 1938.)

many factors, such as temperature, amount of vegetation, humification and weathering processes. Because of this, and even more because of variation in carbonate content of the water, there are large differences in the carbon dioxide content of the water in rivers, lakes, etc. For example, in a river flowing through relatively calcareous soils (Fyris river, Uppsala) there is about 130 times as much CO_2 per unit volume as in air, whereas in another river in Sweden (Ljusnan), with water poor in lime, there is only about 30 times as much (Romell, 1926).

The carbon dioxide in the water contained in well-drained soils is mainly bound as carbonate, and is transported in that form from the rivers to the sea, where it accumulates. Sea water contains about 5 per cent by volume of CO_2, about 167 times more than the air of the inner layers of the atmosphere. The CO_2 is mainly present as carbonate and bicarbonate, with only a small proportion in the form of free acid.

Attempts to calculate or estimate total amounts of CO_2 in the sea differ from one another, but they demonstrate that there is considerably more CO_2 in the sea than in the air (Table 31) (Bolin, 1970).

The carbonate–carbon cycle

In the chemical weathering of carbonates, neutral carbonate is converted to bicarbonate, so that some of the free carbon dioxide in air or water becomes bound.

$$CaCO_3 + CO_2 + H_2O \longrightarrow Ca(HCO_3)_2.$$

Weathering of silicate minerals also uses free carbon dioxide, which becomes bound by the alkali ions which have been released, e.g.

$$KAlSi_3O_8 + H_2O + CO_2 \longrightarrow H_4Al_2Si_2O_9 + K_2CO_3 + SiO_2.$$

felspar   kaolinite or crystallised aluminium silicic acid

The calcium and potassium bicarbonates are carried in streams and rivers into lakes and to the sea. During this passage, or later, some of the carbon dioxide is released, either through assimilation by aquatic plants in which bicarbonate is converted to neutral carbonate, or because only cations are taken up by the plants, or because the acid bicarbonate enters a medium with a low carbon dioxide content and becomes converted to neutral carbonate.

There is no net gain in total reserves of carbon dioxide in air and water from the carbonate-carbon dioxide cycle. However, it does provide carbon dioxide to aquatic plants in higher concentrations than from dissolved atmospheric carbon dioxide alone, and as a result of the progression of the cycle many alkali and alkaline earth metals important to plant and animal life are transported, distributed and made available to organisms.

Carbon dioxide bound in minerals is released as a result of lime-burning. However, the quicklime later takes up carbon dioxide again so there is no net gain of carbon dioxide to the atmosphere. However, there is a net gain if carbon dioxide is released from carbonates by strong acids. Mineral weathering releases sulphates and phosphates which are taken up by plants and eventually used in the formation of calcium sulphate and calcium phosphate, for example. As a result, an equivalent amount of carbonate ions may be released, but probably not enough to bring about any significant increase in the total amounts of carbon dioxide in air and in water.

It appears that the only large-scale release of carbon dioxide from minerals takes place in volcanoes and hot springs. The amount of carbon involved is unknown. It is probable that in previous epochs, especially during the cooling of the earth's crust, so much carbon dioxide was produced that the carbon reserves in the air, in water and in organic materials, could all have been derived from this source (Boresch, 1931).

The organic carbon cycle

Land plants use in photosynthesis an amount of carbon dioxide estimated to be 60×10^9 Mg annually, which is about one-fiftieth of the total amount of carbon dioxide in the air (Schroeder, 1919; Boresch, 1931; Bolin, 1970).

In addition, large amounts of carbon dioxide are used by plants in the sea. Steeman Nielsen (1954; 1957, page 89) measured the carbon dioxide assimilation of plankton in the sea at different depths and in different regions and calculated that the maximum daily production was equivalent to 0.15 g carbon per m^2. This corresponds to a total net annual production of *c.* 15×10^9 Mg of carbon (or 55×10^9 Mg of carbon dioxide) for the water-covered part of the earth's surface.

In the soil, small amounts of carbon dioxide are assimilated by sulphate bacteria, nitrate bacteria, methane bacteria and other carbon-autotrophic microorganisms.

In photosynthesis about 10 656 000 kJ of energy is bound per Mg of carbon dioxide assimilated. This process therefore involves one of the largest turnovers of energy in nature, comparable with evaporation of water or the expansion of the gases in the air brought about by increase in temperature. In the same way that these physical processes are the source of wind energy and of the energy responsible for the movement of precipitated water, of which the energy in waterfalls and rivers is but a minor part, carbon dioxide assimilation provides the energy by which the metabolism of

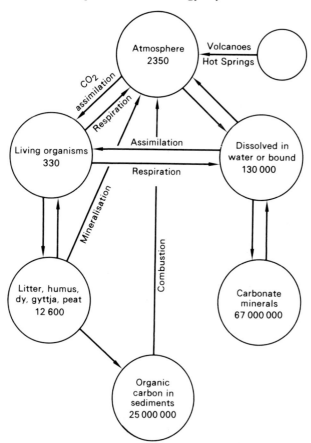

FIG 34. The organic and inorganic carbon cycle, and world distribution of carbon. The figures represent equivalent amounts of CO_2 in Mg $\times 10^9$.

living organisms is maintained, and a large proportion of the sources of energy which have been exploited by man, such as wood, coal and oil.

The assimilated carbon returns to the atmosphere or to water by various pathways, in the form of carbon dioxide (Fig. 34). One pathway is via plant or animal respiration. Carbon dioxide respired by animals has been estimated at $5-10 \times 10^9$ Mg per year, which is equivalent to 5 to 9 per cent of the carbon dioxide assimilated by plants (Boresch, 1931). Plants are responsible for the other 91 to 95 per cent, through respiration, and various processes of decomposition. Some respiratory carbon dioxide is returned to the air as a result of soil respiration, i.e. the uptake of oxygen and release of carbon dioxide by soil. Carbon dioxide is produced in the soil as a result of humification and mineralisation processes, and respiration by soil organisms, roots and other underground parts of plants. Walter and Haber (1957) found that the amount of carbon dioxide produced in the soil in thirty-six different plant communities varied between 1·5 and 5·5 kg carbon dioxide per hectare and hour. Similar values (2 to 7 kg per hectare and hour) had previously been found for forest soils by several investigators (Romell, 1932). Bolin (1970) estimates soil respiration as 25×10^9 Mg carbon per year (equivalent to *c.* 90×10^9 Mg CO_2).

In an undisturbed plant community, there will be gradual development of an equilibrium between soil respiration and litter production or, more precisely, between the amount of carbon moving from soil to air and that moving from air to soil.

The two directions of flow can be represented diagrammatically:

$$\text{CO}_2 \text{ turnover of organisms} \begin{cases} \begin{array}{c} \text{respiratory CO}_2 \\ \longrightarrow \\ \longleftarrow \\ \text{CO}_2 \text{ uptake} \end{array} \text{air} \\ \begin{array}{c} \text{litter-production} \\ \longleftarrow \\ \longrightarrow \\ \text{soil respiration} \end{array} \text{soil} \end{cases}$$

When a plant community has attained a certain stability, so that litter production and soil respiration are balanced, there is an equal movement of carbon dioxide in each direction. As a rule, no plant community remains undisturbed long enough for such an equilibrium to develop. Romell (1928) cites one example: he compared soil respiration with carbon dioxide assimilation in a pine forest, and found that the amounts of carbon dioxide involved in each process were more or less the same. Since processes other than soil respiration must be supported by carbon dioxide assimilation (for example, cell respiration, loss of material from the site), soil respiration must utilise material in addition to the amount equivalent to the supply of new litter, so that the reserves of organic material in the soil must have been encroached upon and depleted to some extent. In time, this would lead to a decrease in soil respiration so that equilibrium would again be established. In this particular forest, the opposite condition must have obtained earlier, with litter production greater than soil respiration, so that the reserves of organic carbon in the soil were built up. During the transitional period of change from net carbon accumulation to net loss, equilibrium would have existed for a time.

The combustion of wood, carbon, oils, and their products, releases significant quantities of CO_2. In the early 1950s, 1.5×10^9 Mg coal per year were mined. If the carbon content is taken to be 75 per cent of the weight of the mined coal, this is equivalent to 4.1×10^9 Mg CO_2 per year.

In the same period, world production of mineral oils, crude oil and heavy oils averaged 0.93×10^9 Mg per year (*Petroleum*, 1951, page 15), equivalent to *c.* 3.4×10^9 Mg of CO_2.

American calculations (Revelle and Suess, 1957) of CO_2 production from fossil fuels give a figure of 6.4×10^9 Mg per year for the 1940s, equivalent to 0.27 per cent of the CO_2 in the air; 4.7×10^9 Mg per year for the 1920s; and 0.54×10^9 Mg for the 1860s. Bolin (1970) calculates that current consumption of fossil fuels is sufficient to increase the CO_2 in the air by 0.7 per cent per year. Most of this CO_2 dissolves in the oceans or contributes to the slow rise in atmospheric CO_2 concentration which has been occurring.

Comparatively little CO_2 is produced by burning wood. Besides trees felled for use as fuel, estimates must include wood used in building, furniture manufacture, paper-making, etc., which sooner or later is generally burned or allowed to rot.

World production of sawn timber was an average of 1366 million m^3 for 1951–54. Assuming a specific gravity of 0.7 and a carbon content of 0.5, this is equivalent to about 0.5×10^9 Mg of carbon. If wood for household use is included, then the annual total is equivalent to about 1×10^9 Mg of carbon. This use of wood is equivalent to premature mineralisation of litter, in terms of its position in the carbon cycle.

In previous epochs, the amount of carbon dioxide released as a result of animal respiration, humification, and combustion was less than that used in photosynthesis. Hence, there was a net accumulation of carbon which is now found in living organisms, in soil organic matter, and in a fossil form (coal and mineral oils). All these forms of carbon still take part in the organic carbon cycle, although to very varied extents. Living organisms, litter, humus and gyttja are responsible for by far the largest part of the turnover; the part played by peat is insignificant, and that of coal and oil is practically non-existent.

If utilisation and production of carbon dioxide were equal at the present time, there would be no net increase in fossil carbon nor in any other form of organically bound carbon. Whether this is in fact so is not known. It can be shown that carbon is being laid down at the present time, e.g. in some peat bogs, in lakes which are being gradually encroached by vegetation, and in the gyttja of the sea, at least where the sea bottom is relatively recent. As these layers increase in thickness, carbon dioxide production per unit of surface area is increased until an equilibrium is reached between breakdown and deposition. Deposition of carbon would therefore be expected only in relatively young sites, such as those laid bare by the passage of inland ice, or in places which have recently experienced a change in climate. This deposition of carbon from the atmosphere is balanced, or partly balanced, by the release of bound carbon in volcanoes and hot springs and in the combustion of fossil fuels by man.

To determine any changes in the carbon dioxide content of the air, on a large scale, requires extensive and continuous study. Some such studies have been made (Egnér and Eriksson, 1958; Bolin, 1970). The changes involved would be so slow that even over several years or several decades the total change might be obscured by analytical errors or natural variations. Besides, any changes would only partially be expressed as a change in the carbon dioxide content of the air, since this is in equilibrium with sea water, which has a large content of CO_2 acting as a buffer (Revelle and Suess, 1957; Bolin, 1970). Callendar (1958) has collected together the results of measurements made in various places and at various times during the twentieth century and he suggests that the average carbon dioxide content of the air has risen from about 290 parts per million in 1909 to about 350 ppm in 1956. Measurements in Hawaii show a change from 313 ppm in 1959 to 321 ppm in 1969 (Bolin, 1970), an annual increase of c. 0·7 ppm.

The nitrogen cycle

Over the centuries, plants themselves have bound nitrogen from the air, by converting it into an organic form, thereby accumulating the nitrogen reserves which are part of all organic material, living or non-living. Nitrogen released from litter in an inorganic form as a result of humification processes is taken up by plants and again becomes a constituent of organic material. In this way the bound nitrogen is not lost, but retained by a series of reactions which together make up a closed cycle. Within the cycle, and in constant movement, are the nitrogen reserves from which the presentday organic world is built up and maintained.

The description of the humification processes shows that ammonia is the most frequent end product of the humification of nitrogen compounds. However, this is usually short-lived, since it is the raw material in the metabolism of nitrifying bacteria. The oxidation processes of these bacteria convert ammonia nitrogen to nitrate nitrogen. This is of outstanding importance, since it means that practically none of the nitrogen from the litter is lost, the nitrogen cycle is closed, and the nitrogen reserves are retained by the biological world. Besides the nitrification of ammonia, there are a number of other processes which form part of the nitrogen cycle. Those that play a part in the decomposition of litter have already been described, and some of the others are described below.

Nitrification

Some bacterial processes are oxidations, with products such as organic and inorganic acids. Acetic acid fermentation, by which alcohol is converted to acetic acid, is an example. Nitric acid is formed in soil as a result of bacterial oxidation of ammonia. This process is called *nitrification*, and has long been known, although it was not until the last half of the nineteenth century that the part played by bacteria was understood.

In 1877, Schloesing and Müntz showed that the oxidation of ammonia was a biological process. Later, Winogradsky (1890) succeeded in isolating the agents, the bacteria subsequently known as *nitrifying bacteria*.

Nitrification takes place in two stages: first, the oxidation of ammonia to nitrite, then of nitrite to nitrate. The bacteria concerned in each reaction are rigidly specialised.

Winogradsky (1890) found several species of bacteria which were nitrite-producers. He placed them in the genera *Nitrosomonas*, *Nitrocystis* and *Nitrospira*. Ammonia is oxidised to nitrite in the cells of these bacteria:

$$NH_3 + 1\tfrac{1}{2}\,O_2 = HNO_2 + H_2O + 331 \text{ kJ.}$$

Nitrate-producers include several related forms of bacteria, which Winogradsky called *Nitrobacter*. They convert nitrite to nitrate:

$$KNO_2 + \tfrac{1}{2}\,O_2 = KNO_3 + 90 \cdot 4 \text{ kJ.}$$

The energy released in the oxidation is used by the nitrifying bacteria for the chemosynthetic assimilation of carbon dioxide, which they obtain partly from the air and partly from carbonates in the soil. About 6 to 7 per cent of the energy released in the oxidation is bound in assimilation. For every gramme of carbon assimilated, 35 g nitrogen are oxidised by nitrite bacteria, and 100 to 135 g nitrogen are oxidised by nitrate bacteria.

The oxidation of ammonia and nitrous acid is the only source of energy available to the nitrifying bacteria. No organic material is oxidised. These bacteria assimilate ammonia for their own nitrogen economy, so they are able to grow in inorganic media. They are markedly aerobic: nitrification is promoted by good soil aeration (see Table 51, page 276).

The nitrifying bacteria are widely distributed in both acid and alkaline soils. However, nitrification is most rapid at a pH of 8 to 9, and is therefore promoted by the addition of carbonate, which also increases the supply of CO_2 necessary to the bacteria for chemosynthesis.

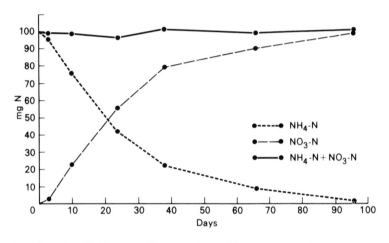

FIG 35. The nitrification of ammonium nitrogen in a medium clay (100 mg $NH_4 - N$ added). (After Nömmik, 1954.)

In acid soils, the pH optimum for the bacteria is lower, but even so there is usually little nitrification in these soils, e.g. in peat soils, mainly because of lack of oxygen. Soil drainage and intermixing of mineral soil promotes nitrification, because aeration is improved, mineral content is increased, and acidity is decreased. One of the many ecological effects of cultivating soil is the change of structure which allows increased gaseous exchange and, consequently, more rapid nitrification.

Raw humus and podsolised soils are usually also characterised by slow nitrification, probably because of the high acidity. After clear-felling, the pH of a forest soil usually rises, nitrification increases and new plant species become established. After a few years, the pH falls again and nitrification decreases (Olsen, 1923).

Nitrification is a rapid process, so that ammonia and nitrous acid are used as soon as they are formed. As a result, these substances do not usually occur in high concentrations in the soil, provided that the amount of ammonia formed is not excessive, and that there is no oxygen deficiency nor any other conditions limiting the activity of the bacteria. Within a certain range, the rate of turnover is therefore determined by the rate of production of ammonia.

Where large amounts of ammonia are formed, for example in places where urea accumulates, particularly if aeration is inadequate, nitrification cannot keep pace, and some of the ammonia is lost as gas.

In peat bogs and raw humus soils, ammonification is inhibited. As a consequence, nitrification is limited not only by lack of oxygen but also by lack of ammonia, and the vegetation may experience deficiency of nitrogen in spite of the large amounts present in the soil.

In previous centuries, nitrification was the process responsible for the supply of saltpetre (KNO_3) to industry, for example, to potteries. The raw materials, soils with a high content of saltpetre, were so important in Sweden, and needed in such large quantities by industry, that special laws were introduced. Gustav Vasa, in the sixteenth century, requisitioned the saltpetre soil near and under farm buildings and obliged the farmers to supply such soil and to help with extracting the saltpetre. Hence, at a time when agricultural production was severely limited by deficiency of mineral nutrients in the soil, nitrogen was removed for other purposes and agricultural nitrogen reserves were exploited as a source of raw material for industry.

In the nineteenth century, particularly in the second half, Chile saltpetre came on the market; and at the beginning of the twentieth century a method of fixing nitrogen from the air was developed. As a result, industrial consumption of nitrogen obtained from the soil came to an end, and movement in the opposite direction began: nitrogen produced by industry was added to the soil. The nitrogenous fertiliser industry has made one of the most important contributions to the rise in crop yields which has taken place over the last hundred years, beginning in the middle of the nineteenth century when mineral fertilisers were first applied.

Denitrification

In the absence of oxygen, nitrate may act as an oxidising agent towards other substances, e.g. lactic acid. The nitrate acts as a hydrogen acceptor and is reduced to nitrite.

$$CH_3.CHOH \cdot COOH + KNO_3 \longrightarrow CH_3.CO.COOH + H_2O$$

lactic acid pyruvic acid

$$+ KNO_2.$$

Some of the widely-occurring organisms in the soil, bacteria, actinomycetes and mould fungi, make use of the tendency of nitrate to give up oxygen. Under aerobic conditions the oxidation processes in their metabolism utilise oxygen from the air, but when aeration is poor oxidation proceeds in conjunction with the reduction of nitrate. Some organisms are specialists in various parts of

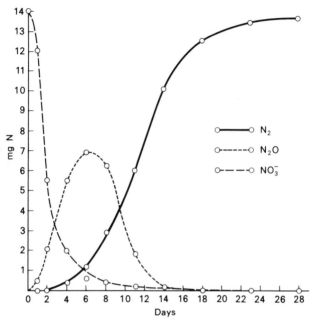

FIG 36. The course of denitrification in a heavy clay to which nitrogen in the form of KNO_3 had been added. (After Nömmik, 1954.)

this process: some reduce nitrate to nitrite, others convert nitrite to ammonia.

If there is a good supply of nitrate and of organic material, but aeration is variable and conditions fluctuate between aerobic and anaerobic, the organic material is subjected to corresponding fluctuations between oxidation and reduction. In these conditions, ammonia may be oxidised in such a way that free nitrogen is formed.

$$2\,NH_3 + O_2 \longrightarrow N_2 + 2\,H_2O.$$

All these processes form part of the denitrification aspect of nitrogen turnover in the soil. The first phase of denitrification is an oxidation, by which various organic compounds, especially alcohols and fatty acids, are oxidised and serve as hydrogen donors. Nitrate, the hydrogen acceptor, is reduced to nitrite or ammonia, depending on the conditions. If the pH is high and aeration is good, the reduction stops at the nitrite stage. If the medium is acid and poorly aerated, the process proceeds to the ammonia stage (Rokitzka, 1949).

Denitrification may be summarised as follows, where H_2A is the hydrogen donor.

$$4\ H_2A + HNO_3 \longrightarrow 4\ A + NH_3 + 3\ H_2O.$$

The course of the process depends on the organisms involved and on the external conditions. If these are suitable for the continued oxidation of ammonia, gaseous nitrogen may be released (Wikén, 1942). Denitrification can also proceed to the release of gaseous nitrogen by other, more direct pathways: nitrate is first reduced to nitrous oxide (N_2O) and then to gaseous nitrogen (Rokitzka, 1949; Nömmik, 1954) (Fig. 36).

$$5\ C + 4\ KNO_3 + 2\ H_2O = 4\ KHCO_3 + CO_2 + 2\ N_2.$$

This type of denitrification is very important in cellulose decomposition. One prerequisite is the absence of any organic nitrogen compounds (Rokitzka, 1949). The reaction brings about an increase in pH, since potassium bicarbonate is formed.

In the presence of amines or amides, denitrification may lead to the release of amino nitrogen (Rokitzka, 1949).

(a) $R - CH\ NH_2 . COOH + HNO_2 = R - CH\ OH . COOH$
     amino acid

$$+ H_2O + N_2.$$

(b) $R - CONH_2 + HNO_2 = R - COOH + H_2O + N_2.$
     amide

The pH optimum of denitrification is about pH 7 to 8, which is also favourable for the activity of nitrifying bacteria. However, denitrification and nitrification are not affected in the same way by changes in aeration. Nitrification is most rapid when there is a good oxygen supply, whereas denitrification requires a limited oxygen supply or an oxygen-free medium. Conditions may therefore be suitable for denitrification in waterlogged soils, when the entry of oxygen is prevented by water, or in places where there is an accumulation of organic matter, such as manure heaps. In such places, the oxygen is used up in the uppermost layers and there is a lack of oxygen in the deeper layers. Hence very little saltpetre is formed, and no large quantities of nitrate, and this limits denitrification.

It is apparent, therefore, that nitrification and denitrification make different demands on the environment. In conditions favourable to denitrification there are generally only limited supplies of nitrate. The deleterious effects of denitrification are therefore not usually as extensive as might be expected from knowledge of the widespread occurrence of the process and of the large numbers of denitrifying organisms.

In agricultural practice, denitrification is prevented or minimised by draining waterlogged land and by cultivating the soil so as to increase pore spaces and aeration. The addition of nitrate to such soil is avoided, to discourage denitrification; instead, ammonia or organic fertiliser is used.

Other nitrogenous gases are probably formed as a result of mineralisation of litter. Gaseous oxides of nitrogen are common in the soil and in the atmosphere. The highest concentration is in the soil, in which up to one-fifth of the total volume of gases may be oxides of nitrogen. In the air nearest the soil surface, the concentration is only 5×10^{-5} per cent by volume (or $0.5 \text{ cm}^3/\text{m}^3$ air), and it decreases with increasing height above the ground. Higher up in the atmosphere, the gases are broken down photochemically, so that gaseous nitrogen is finally released. Their distribution at these various levels indicates that a state of equilibrium exists between the production of oxides of nitrogen in the soil and their decomposition in the atmosphere (Adel, 1951).

The cycle

As proteins and other nitrogenous compounds are broken down, the nitrogen again becomes available to plants. Some is assimilated by microorganisms in the form of amino acid nitrogen, but probably the major part is broken down to ammonia, some of which is assimilated, some oxidised to nitrate. There may be some loss by evaporation or by leaching. Higher plants can take up nitrogen as nitrate or as ammonia. Probably mycorrhiza are responsible for the uptake in some cases. Melin and Nilsson (1952, 1953) found that a pine mycorrhizal fungus, *Boletus variegatus*, transferred ammonium nitrogen and glutamic acid nitrogen from the external medium to the pine root tissues. This nitrogen was rapidly consumed and converted to amino-nitrogen. The main features of the nitrogen cycle are thereby completed (Fig. 37).

Fixed nitrogen is relatively unstable: nitrogen therefore tends to move along various pathways back to its original gaseous state. The pathways are mainly via the decomposition of proteins and other nitrogenous compounds during humification and mineralisation. The gaseous nitrogen is usually lost from the locality as it is formed, and the organic nitrogen reserves are depleted.

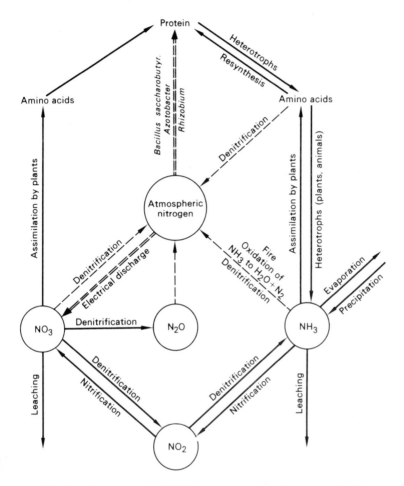

FIG 37. **The nitrogen cycle. Solid arrows represent movement of nitrogen. Dashed arrows represent gain (double dashes) or loss (single dashes) of nitrogen.**

The soil can also suffer a loss of nitrogen in another way. The simpler nitrogen compounds, especially ammonium salts and nitrates, are quite soluble and are therefore liable to leaching, and are carried away in water. In this form, the nitrogen is, of course, not lost to the biological world as a whole, since it is transported to other places where it may be reassimilated, but a particular locality may experience a net loss.

It is apparent that the cycle has various leaks. If the nitrogen is transported away from one locality, but accumulated in another, it is only the position of the reserves that changes. However, when mineralisation proceeds as far as gaseous nitrogen, there is a direct loss from the organic world and the reserves of nitrogen in the soil are depleted. The loss would be considerably greater if it were not for the safeguards in the cycle, which combat the capital losses. One of these is provided by the humic acids and other relatively insoluble nitrogenous intermediates formed as a result of the humification of litter; these intermediates release nitrogen only slowly as mineralisation of the material continues (see page 181). Another is provided by nitrification. If decomposition of amino acids is very rapid, the microphytes cannot consume all the ammonia which is formed, and the excess would rapidly be lost by leaching, or otherwise removed, if it were not used in nitrification. After conversion to nitrate, it is rapidly consumed by the higher plants.

Another safeguard lies in the usually high ratio of carbon to nitrogen in the litter. As a result there is competition between the soil organisms for nitrogen, and the availability of nitrogen is usually a limiting factor, which determines the rate of reproduction and growth of the litter-decomposing organisms. Soil organisms therefore usually suffer from a deficiency of nitrogen, so that any nitrogen is consumed as rapidly as it becomes available through the processes of decomposition. Consequently there is usually no excess nitrogen which can be lost.

Another safeguard depends on the rapid growth and reproduction of the microorganisms. If the nitrogen supply increases, the number of soil microphytes increases, too, so the nitrogen is consumed and no excess accumulates. In unfavourable conditions, for example a poor supply of nitrogen, the microphytes enter a resting stage, often as a particular type of spore, and then reproduce themselves again as soon as conditions improve.

In this way, the microphytes act as a caretaker, taking up the nitrogen which is made available during humification or which enters the inorganic state and therefore becomes liable to be lost.

Because of these safeguards, loss of soil nitrogen is rare, occurring only when the nitrogen supply is optimal or supraoptimal, for example, in a substrate rich in nitrogen, but poor in carbon. Nevertheless, loss of nitrogen would be deleterious over a long period if it were not balanced by fixation of gaseous nitrogen from the air.

Nitrogen-fixing processes

Atmospheric nitrogen can be converted to a bound form both physically and biologically. The first type of process includes electrical discharges in the atmosphere, by which gaseous nitrogen is oxidised. The resultant oxides of nitrogen are then carried down by wind and rain. At great heights, there may also be some photochemical nitrogen fixation (Virtanen, 1952). The amounts involved are discussed below (page 163). Biological fixation plays a much more important part in the nitrogen economy of living organisms.

Clostridium; Azotobacter

The atmosphere, containing about 79 per cent nitrogen, or about 77 000 Mg per hectare of surface, is the primary source of nitrogen for plants. During the course of time some of it has been bound into organic matter by plants. This makes up the nitrogen reserves of the biological world.

Only a few species of plants are able to fix molecular nitrogen. The process was first demonstrated in 1862 by the Frenchman, Jodin, but it was another thirty years before the organisms responsible were identified. In 1893, a microbe which fixed nitrogen was isolated by the Russian bacteriologist Winogradsky. It was found to be an anaerobic butyric acid bacterium, which Winogradsky called *Clostridium pasteurianicum*. It was subsequently confirmed that the ability to fix nitrogen was general among anaerobic butyric acid bacteria. These species were first given the collective name of *Bacillus amylobacter* and then *Clostridium butyricum*.

Fermentation of carbohydrates by these bacteria leads to the formation of butyric acid, acetic acid, carbon dioxide, butyl-, ethyl- and isopropyl-alcohols, acetone, hydrogen and other compounds. At the same time, atmospheric nitrogen is assimilated and converted to an organic form. Such bacteria are thus autotrophic for nitrogen, but not for carbon.

Winogradsky found that nitrogen fixation was more or less proportional to the amount of carbohydrate (glucose) consumed. Fermentation of 1 g carbohydrate was equivalent to fixation of 3 to 7 mg nitrogen by the butyric acid bacteria.

The butyric acid bacteria are present in all types of soils and in water. Their distribution is certainly limited by their anaerobism, but in symbiosis with pronounced aerobes they can grow even in the surface layers of the soil. They are not particularly sensitive to acids: they produce acids themselves. Even acid soils of pH 5·5 contain large numbers of butyric acid bacteria.

Another nitrogen-fixing bacterial species, *Azotobacter chroococcum*, was isolated in 1901 by Beijerinck. This is again autotrophic for nitrogen, but heterotrophic for carbon. It ferments hexoses, disaccharides, starch and other carbohydrates. However, it cannot itself utilise cellulose, but lives side by side with cellulose-decomposing species and attacks the intermediates produced by their activity. *Azotobacter chroococcum* is an aerobe and requires good aeration and a good supply of nutrients. It is sensitive to acids and rarely grows at pH values less than 6. Nitrogen fixation is rapid. Fermentation of 1 g carbohydrate is equivalent to fixation of 10 to 12 mg nitrogen.

Other species of *Azotobacter* are known to fix nitrogen, e.g. *A. Beijerincki*, *A. Vinelandi*, *A. agilis*, etc. The way in which nitrogen fixation occurs has not been worked out in detail. Frey Wyssling (1945) points out the similarity between nitrogen fixation by bacteria and by the Haber process, in which hydrogen and nitrogen react to form ammonia when they are heated to 400 to 500°C at a pressure of 200 bars and in the presence of metal oxides as catalysts. Although the synthesis is weakly exothermic, the tendency of the two elements to react with each other is so slight that the process is only started with great difficulty.

$$N_2 + 3 H_2 \longrightarrow 2 NH_3 + 49.8 \text{ kJ.}$$

The bacteria solve the problem physiologically, without the need for high temperatures and high pressures. The hydrogen released as a result of oxidation of carbohydrates in the cytoplasm of the bacteria is transferred to gaseous nitrogen taken up by the cells. It is possible that this process may be catalysed by metals. It has been found that traces of molybdenum and vanadium promote nitrogen fixation, and the presence of molybdenum is thought to be essential before nitrogen fixation can begin. The presence of humus has a favourable effect, thought to depend on its iron content.

The nitrogen autotrophy of these bacteria is only faculative. Any ammonia or nitrate present is assimilated preferentially, ammonia first and then nitrate. Only when these are used up is gaseous nitrogen assimilated. Nitrogen fixation therefore seems to come into operation only when no other possibilities remain. It ceases immediately if ammonium salts are added to the system. The bacteria then use the ammonium ions, whereupon the trace elements cease to be necessary. The change from the heterotrophic type of assimilation to the autotrophic, and vice versa, occurs easily and there seems to be no intermediate adaptation stage.

Azotobacter and *Clostridium* are widely distributed in soil and in salt and fresh water. In soil, *Azotobacter* are found down to a depth of 0.5 m. Because of their different sensitivity to the acidity of the medium, *Azotobacter* predominate on neutral and alkaline soils and *Clostridium* on acid soils and soils with a low lime content.

Symbiosis between higher plants and nitrogen-fixing microphytes

Theophrastus (died 285 BC), Cato II (died 149 BC) and Pliny (died AD 79) knew that leguminous plants (*Papilionaceae*) could increase soil fertility. The farmer had learned by experience that the yield of grain, for example, was higher if there had been a previous leguminous crop. However, it was not until the nineteenth century that the reason for this became clear. In 1838 Boussingault found that the legumes in some way increase the supply of nitrogen in the soil by using nitrogen from the air. Later, in 1888, Hellriegel and Willfarth showed that the nitrogen-fixing agent was in the nodules which occur on the roots of legumes. Also in 1888, Beijerinck isolated the agent and found that it was a bacterium, which he called *Bacterium radicicola*. However, the nitrogen-fixer is not a single species, but several species or forms now known by the generic name *Rhizobium*. Each of these is specific for a particular species of legume, and cannot grow at all, or only to a limited extent, on any other species.

The bacteria also live free in the soil and they infect the host plant by entering the root hairs. Their presence in the root hairs causes abnormal growth of parenchyma at the site of infection so that nodular outgrowths are formed. A symbiotic relationship develops between the bacteria and the host plants: the bacteria obtain organic substances and mineral salts from the hosts, new compounds are synthesised from these materials and gaseous nitrogen, which diffuses through the cell membranes of the root parenchyma into the cells themselves, is assimilated. Some of the nitrogenous compounds are lost from the bacteria and are utilised by the host. The exchange of materials is facilitated by special vascular tissue which develops in the nodules.

At the end of the growing season, the nodular cells die, and the bacteria and other cell contents are incorporated into the soil so that its nitrogen content is increased. According to Virtanen and Laine (1939) there is also an enrichment of the soil nitrogen before this stage, by outward diffusion of amino acids from the cells in the nodules. The amount involved may be equivalent to 60 to 80 per cent of the fixed nitrogen. Cereals and grasses which are grown together with legumes are able to use this amino nitrogen.

Nitrogen fixation by *Rhizobium* species is relatively extensive and may be as much as 100 mg nitrogen per gramme dry weight of root nodules, so that the nitrogen reserves in the soil may be increased considerably. A field of lucerne or clover may fix 150 to 400 kg of gaseous nitrogen per hectare in a growing season (Virtanen, 1952). Enrichment of the soil is particularly large if the legumes are left in place to humify. The use of legumes for green manuring is based on this.

Green manuring is mostly suitable in regions with a mild, wet winter, where the climate allows two crops in one year. The winter crop can then be used for green manure without competing against the crop which is to be harvested. In the south-eastern states of the USA it is possible to grow maize or cotton as a summer crop and legumes as a winter crop to be ploughed in to replace the nitrogen removed in the summer harvest.

In coffee or tea plantations in the tropics, leguminous trees, e.g. *Erythrina* and *Albizzia*, are grown between the bushes so that their litter, particularly the leaves, acts as a supply of nitrogen to the soil. This can be increased by cutting off branches and twigs and spreading them on the ground as green manure.

Green manuring is also advantageous on saline soils, since the plant material used as manure also hinders evaporation from the soil and so reduces salt accumulation at the surface.

Specialisation of the bacteria to a particular host may cause difficulty to the farmer since soil sown with a particular legume may not contain the appropriate strain of bacterium. This often happens with newly-cultivated land or with introductions of new species of legume. The deficiency can be rectified by transferring some soil from fields in which the legume has previously been grown, which will be infected with the right strain of bacterium. Alternatively the soil may be inoculated with special pure cultures prepared in central laboratories.

Numerous investigations of the conditions in which *Rhizobium* species grow have shown that the symbiotic arrangement is necessary for nitrogen fixation to take place: there is no nitrogen fixation in a pure culture of the bacteria.

Again, autotrophic nitrogen fixation is only faculative in *Rhizobium*. Any nitrate nitrogen in the soil is utilised before the bacteria will start to fix nitrogen; and if the nitrate content increases, nitrogen fixation gradually ceases.

The presence of calcium, phosphorus and molybdenum is essential. If the molybdenum content of the soil is too low, molybdenum accumulates in the root nodules so that its concentration there may be five to fifteen times as high as that in the other parts of the root (Jensen and Betty, 1943).

Nitrogen fixation is fairly tolerant of soil acidity and continues even if the pH is as low as from 4·2 to 4·5. The host plant is still more tolerant and can live at pHs low enough to have caused the cessation of nitrogen fixation (Virtanen, 1928).

The respiration rate in the root nodules is higher than in the other parts of the roots, indicating that the bacteria have a high metabolic rate. The pigment haemoglobin occurs in the nodules, but its role in nitrogen fixation is still somewhat speculative (see Stewart, 1966).

By using an isotope technique (^{15}N), it has recently been possible to show that gaseous nitrogen is also assimilated in several other angiosperms with root nodules, for example, in *Alnus glutinosa* and *Hippophaë rhamnoides*, in which the nitrogen-fixing organisms are actinomycetes, and also in *Myrica gale*. The nitrogen is fixed and accumulated in the nodular roots and is transported from them to all parts of the plants (Bond, 1957; Stewart, 1966).

Nitrogen-fixing algae and fungi

There is probably some nitrogen fixation on a small scale by some moulds and yeasts, but its significance has been only little studied (Stewart, 1966, page 83).

However, soil algae play an important part in the nitrogen economy. They occur on the soil surface, and in the soil itself, sometimes to a depth of several centimetres. Their main growing periods are spring and autumn, when the soil water content is high. The ability to assimilate molecular nitrogen is characteristic of some of the blue-green algae, especially species of *Nostoc*, *Anabaena* and *Cylindrospermum*. As these algae can also assimilate carbon dioxide, and are therefore both carbon-autotrophic and nitrogen-autotrophic, they can live on a purely mineral substrate and are therefore more independent of the organic world than most other living organisms.

Nitrogen fixation by soil algae does not itself need light and so might be expected to occur even in individuals living beneath the soil surface. However, there is probably no nitrogen fixation, or only an insignificant amount, below the soil surface, since the presence of carbohydrates to which the nitrogen can be transferred is required. It has been established that nitrogen fixation is rapid in algae which have a good supply of mineral nutrients and good illumination, so it is probably limited to algae above the soil surface (see Paech, 1940).

Nitrogen fixation by soil algae is again facultative. If nitrate or ammonium compounds are available, there is no nitrogen fixation until they have been used up. Addition of nitrate will stop any nitrogen fixation which is taking place.

The similarity to the processes in other nitrogen-fixing organisms is shown by the requirement for molybdenum. The process is most rapid in neutral or alkaline media, within the pH range *c.* 7·0 to 8·5.

The nature of the nitrogen economy of blue-green algae is only partly known. There are probably ecologically and physiologically important features, for example, concerned with symbiosis. There are several *Nostoc* species which live endophytically in the leaves, stems and roots of aquatic plants; others which occur in the leaves of liverworts; and *Anabaena* is an endophyte in the leaves of *Azolla* (a fern). The relationship between *Anabaena* and *Azolla* indicates some sort of symbiosis between the two: *Azolla* can live on nitrogen-free media, if *Anabaena* is present. *Anabaena* seems also to be dependent on some other living organism, and cannot be grown in pure culture (Pirson, 1946).

Nitrogen fixation by blue-green algae is probably especially important in tropical soils. Watanabe *et al.* (1951) isolated thirteen species of blue-green algae from a Japanese rice field, all of which assimilated gaseous nitrogen. Singh (1942) studied rice field soils in the East Indies and found that the film of algae covering the water and the wet soil consisted of blue-green nitrogen-fixing species. He found *Anabaena* species which could secrete more than 40 per cent of the nitrogen taken up, in the form of soluble organic compounds. This phenomenon doubtless explains why tropical rice fields can bear a crop year after year without any addition of nitrogenous fertilisers.

In the crust which forms on some desert soils as a result of the binding of the soil particles by algae, fungi and rain water, the nitrogen content is up to four times higher than in the layers beneath. Since there are nitrogen-fixing species in the algal population, it is probably because of their activity that the nitrogen accumulates (Fletcher and Martin, 1948).

The fact that blue-green algae are most common in the tropics (Chapter 6) may be explained by their high pH optimum, since tropical soils are generally alkaline. Whether these algae are of any importance in temperate or cold regions is not known. In these regions it is the green algae, not the blue-green, which are dominant among soil algae. The relatively low pH optimum of the green algae is most suitable for the subneutral and acid soils in these regions.

Hitherto, no green algae have been found to fix nitrogen. However, there are known cases of symbiosis between green algae and nitrogen-fixing bacteria. For example, *Chlorella* can grow in tap water if *Azotobacter* is present. The two make up a unit which is autotrophic for both carbon and nitrogen.

The soil nitrogen balance

The nitrogen content of the soil is variable: in forest soil it is often 0·1 to 0·3 per cent (Wilde, 1946); in arable soil up to 0·15 per cent; in pastures 0·3 per cent; in bog, fen and prairie soils even higher; and in raw humus 1 to 4 per cent (Lutz and Chandler, 1946).

Only about 2 to 3 per cent of soil nitrogen is in the form of nitrate or ammonia. The rest is in stable organic compounds, a third to a half of which is probably protein nitrogen, and the remainder amino sugars, chitin and lignin.

The nitrogen cycle is not a closed cycle, but is affected by losses and gains of nitrogen to and from the atmosphere. Whether these are in equilibrium with each other cannot easily be determined, since to do so it would be necessary to distinguish between the nitrogen reserves in the organic world as a whole and those in the particular locality.

The movement of nitrogen from one locality to another

Some of the nitrogen which is lost from one locality is not lost from the organic world as a whole, but merely moved from one place to another. This may occur in several ways.

1 Leaching of nitrate, ammonium salts and other soluble nitrogenous compounds occurs in places where there is water movement (in running water, ground water or surface water). The amount lost depends on the quantity of dissolved substances, which, in turn, depends on the balance between their production and consumption in the locality. Since consumption is usually high, nitrate is absorbed rapidly by the higher plants and ammonium salts by the soil microphytes. Generally, then, the nitrogen concentration in the soil water is so low that very little is lost. However, leaching increases if the natural equilibrium in the cycle is disturbed, for example, by changing the composition of the plant community. In a forty-five year experiment at Rothamsted, soil with no plant cover lost an average of 30 kg nitrogen per hectare and year (Russell, 1952).

2 In some conditions, there may be so much ammonia released as a result of humification that some escapes in the gaseous form. This is transported by the wind, but eventually dissolves in rain water so that it is returned to the soil. Such conditions arise if ammonia production is greater than consumption, for example, in places where large quantities of material rich in nitrogen, such as fertilisers, urine, or animal waste, accumulate. Loss of nitrogen as ammonia is greatest in the tropics, where temperatures are high

and soils are usually alkaline, so that both physical and chemical conditions are suitable for the release of ammonia. In rice fields, where the high soil moisture content favours the formation of ammonia, loss of nitrogen in this way is probably of some significance.

3 Removal of harvested crops from arable land, movement of grazing animals from one place to another, removal of timber from forests, and other processes involving loss of nitrogen-containing organic material from the place where it was produced, all disturb the natural nitrogen balance.

Where animals or men live in communities there is an accumulation of nitrogen and other nutrients. A constant stream of organic material passes into towns and villages, then, after transitory use—as food, as part of living organisms, as clothes, etc.—it is mineralised in sewers, sewage works and water courses, from which it passes into lakes, the sea, or the atmosphere. This enriches the

TABLE 33 Gain in nitrogen in an arable soil at Rothamsted, England, which last bore a wheat crop in 1882 and was then allowed to revert to natural prairie. (After Russell, 1952.)

	CARBON %		NITROGEN %	
	1881	1904	1881	1904
1st 9 in.	1·14	1·23	0·108	0·145
2nd 9 in.	0·62	0·70	0·070	0·095
3rd 9 in.	0·46	0·55	0·058	0·084

nitrogen reserves in lakes and the sea, at the expense of the reserves of the land. Besides increasing soil nitrogen deficiency, this has a deleterious effect in increasing silting up and pollution of lakes and water courses. In the long term, accumulation in lakes and rivers is only temporary and the final result of all this movement is an increase in the nitrogen reserves of the sea.

The nitrogen content of arable land is decreased by removal of the harvested crops—unless the nitrogen is replaced by supplying fertilisers—and by leaching, which is accelerated by removal of the natural vegetation. The loss is greatest during the first year of cultivation. Jenny (1933, cited by Millar, 1955, page 96) found that a prairie soil lost 25 per cent of its nitrogen reserves during the first twenty years of cultivation, 10 per cent in the second twenty years, and 7 per cent in the third twenty years. There are similar figures for other soils (see literature cited by Millar, 1955).

The process can also operate in the opposite direction. If arable soil becomes depleted of nitrogen, the nitrogen can be regained by ceasing cultivation and allowing the land to be recolonised by a natural plant community. The carbon content also increases. Table 33, illustrating a Rothamsted experiment, is an example. The gain in nitrogen is about 17 kg per hectare and year.

The loss of fixed nitrogen

The nitrogen reserves in the organic world, which have been mainly built up by the organisms themselves, are depleted by any conversion to molecular nitrogen. The following processes are the most important causes of depletion of the reserves.

1 Denitrification and conversion of organically fixed nitrogen to molecular nitrogen in some of the reactions involved in humification, with ammonia and nitrous oxide as intermediates.

2 Combustion of organic material in fires, leading to release of molecular nitrogen. The largest losses result from the use of wood or peat for fuel, and from forest and prairie fires. However, the magnitude of this loss is difficult to assess. Surface burning of a thick layer of raw humus causes only moderate losses, according to Uggla (1958).

3 Some fixed nitrogen remains out of circulation for long periods in peat bogs, at the bottom of the sea or of lakes, and in other places where organic substances accumulate.

The gain of fixed nitrogen

The nitrogen reserves in the soil are increased in several ways.

1 Nitrogen-fixing organisms. Many microphytes probably have the ability to fix atmospheric nitrogen, but as far as is known only a few do so to any great extent. The most important have already been named. They are species of the bacterial genera *Clostridium*, *Azotobacter* and *Rhizobium*, and the algal genera *Nostoc*, *Anabaena* and *Cylindrospermum*.

2 Oxides of nitrogen are produced as a result of electric discharges in the atmosphere and are carried down into the soil in rain water, in the form of potassium nitrite and nitrate. This source may provide 2 to 6 kg nitrogen per hectare and year in Sweden (Egnér, 1954), about 7 kg per hectare and year in England (Russell, 1952, page 304), and about 3 kg per hectare of ocean surface (Steeman Nielsen, 1944, page 8). However, about 2 kg of this nitrogen is in the form of ammonia, some of which has come from the soil and is therefore of organic origin (see page 453).

3 Nitrogenous fertilisers. Atmospheric nitrogen is converted industrially to nitrate or ammonium salts, which are used as nitrogenous fertilisers.

In nature, there is probably a balance between these processes of gain and of loss, so that the amount of nitrogen in the vegetation, in animals, and in the soil is more or less constant. These nitrogen reserves represent the net gain made from these opposing processes, which have been going on for thousands of years. Since these reserves have accumulated, nitrogen fixation must at some time have exceeded loss of fixed nitrogen. However, as nitrogen reserves increase, loss of fixed nitrogen also increases: the deeper the layer of accumulated organic material, the more extensive the anaerobic decomposition—so that equilibrium automatically becomes established.

As long as a plant community remains undisturbed, this equilibrium is maintained: if the balance is upset, the losses increase. Cultivation, intensive grazing, forest clearance, and burning, remove or change the vegetation in such a way that the production of litter increases abnormally and temporarily, stable humus is destroyed, the composition of the microphyte flora changes, the nutrient salts in the organic material are prematurely released, and so the conditions for both accumulation and loss of fixed nitrogen are changed. The amount of nitrogen fixed industrially has been doubling about every six years, and together with nitrogen fixed by cultivated legumes, it now exceeds (by about 10 per cent) the amount of nitrogen fixed in nature. Fertilisers and nitrogenous wastes must therefore be carefully managed in order to keep the nitrogen cycle in reasonable balance. The nitrogen concentration already exceeds acceptable levels in some rivers, lakes and groundwater systems (Delwiche, 1970).

The sulphur cycle

Sulphur occurs in several minerals, including iron pyrites (FeS_2). Sulphates are formed as a result of weathering and oxidation. Volcanic gases are another source: they include large quantities of hydrogen sulphide and sulphur dioxide. Near volcanoes and sulphur springs, sulphur occurs in the free state. These forms may also be oxidised to form sulphates.

In comparison with common soil constituents, such as silicon, iron, aluminium, etc., the content of sulphur in most soils is only small, about 0·05 to 0·07 per cent, of which about 90 per cent is in an organic form. In bog peat the content has been estimated at 0·05 to 0·3 per cent; in fen peat it is 0·2 to 1·5 per cent. Gyttja and gyttja-containing soils are rich in sulphur and may contain up to 2 per cent or more (Wiklander, 1957). Whether these amounts are sufficient for plant requirements does not seem to have been investigated. Many arable soils are somewhat deficient in sulphur. About as much sulphur as phosphorus is removed in harvested crops, but rather more sulphur than phosphorus is lost from the soil through leaching, since sulphur compounds are more soluble and therefore more prone to it. Increase in the supply of sulphur to arable soil has increased the yield in some areas, for example, in some agricultural districts or the United States (see Gilbert, 1951).

Plants take up sulphur in the form of sulphate and use it both in metabolism and growth. It is a constituent of proteins and other organic compounds. The amino acid cysteine contains about 1·5 per cent sulphur. Since, in this form, the sulphur has a valency of 2, whereas in sulphate its valency is 6, it has been reduced in the plant cells. The sulphur becomes inorganic again when the cells die and become litter, and go through the oxidation and reduction processes of humification. It is released as hydrogen sulphide. However, this is not the end of its biological activity: it is oxidised by soil microorganisms, and can then again take part in their metabolic and synthetic processes in various ways.

Sulphurification

When considering the oxidation of ammonia, nitrification and the nitrate bacteria were discussed. The corresponding processes for hydrogen sulphide may be given the name *sulphurification*, and involve the sulphur and sulphate bacteria, but they differ from the nitrifying bacteria in some respects. They are not such a uniform group, and the process of sulphurification is more varied than nitrification. It includes the oxidation of free sulphur, sulphides, thiosulphate and other sulphur-containing compounds.

The oxidation of hydrogen sulphide is important biologically: partly because it is a poisonous substance occurring in relatively large amounts, so that its accumulation would be inhibitory and harmful to living organisms; partly because the oxidation of hydrogen sulphide, like that of ammonia, completes the circle for the organic cycle of the element.

The sulphurification of hydrogen sulphide is so rapid that the gas does not usually accumulate in the soil to any extent, except in anaerobic conditions in which the hydrogen sulphide formed is mostly the product of microbial reduction processes. These may be termed *desulphurification*, and are analogous to denitrification.

Some sulphur bacteria oxidise hydrogen sulphide in stages, so that free sulphur is released, and then oxidised.

Free sulphur is released within the cells of some aerobic sulphur bacteria. It is oxidised subsequently, for example, when the supply of hydrogen sulphide diminishes, and so it acts like a reserve store of fuel.

$$2\,H_2S + O_2 = 2\,H_2O + 2\,S + 511\text{ kJ}$$
$$2\,S + 3\,O_2 + 2\,H_2O = 2\,H_2SO_4 + 1181\text{ kJ}.$$

The sulphur bacteria include *Spirillium volutans*, and species of *Beggiatoa*, *Thiotrix* and others which occur in stagnant water where decomposition is going on and hydrogen sulphide is produced. This may be in some bays at the edge of the sea or lakes and in other bodies of water where large amounts of organic matter collect. Stones or other material in the water may be covered by a yellow-white mass of bacteria, so coloured because of the large amounts of sulphur in the cells. The same species of bacteria occur in sulphur springs and in the soil.

Some sulphur bacteria, e.g. *Thiobacterium* (*Thiobacillus*) *thioparus*, secrete sulphur, and give rise to deposits of free sulphur. Some denitrifying forms are of this type. They live anaerobically, with nitrate as a source of oxygen, e.g. *Thiobacillus denitrificans*.

$$6\,KNO_3 + 5\,S + 2\,CaCO_3 = 3\,K_2SO_4 + 2\,CaSO_4 + 2\,CO_2$$
$$+ 3\,N_2 + c.\ 2760\text{ kJ}.$$

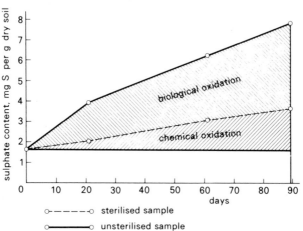

FIG 38. The formation of sulphate through chemical and biological oxidation (probably by *Thiobacillus* species) of sulphur in gyttja clay. (After Wiklander, Hallgren and Jonsson, 1950; Wiklander, 1957.)

Other forms oxidise sulphur and sulphur compounds, e.g. the iron sulphide in gyttja clay to sulphuric acid, directly or in steps (Wiklander, 1957).

$$FeS + H_2O + \tfrac{1}{2}\,O_2 = Fe(OH)_2 + S + 151\text{ kJ}$$
$$2\,S + 2\,H_2O + 3\,O_2 = 2\,H_2SO_4 + 1001\text{ kJ}.$$

Thiobacillus thioxydans is a strict aerobe of this type. It is very tolerant towards acids: the pH optimum is 3 to 4 and the minimum as low as pH 0·6.

The production of sulphuric acid is important to soil processes, since it accelerates the weathering of minerals and the solution of phosphates with low solubility.

Carbon assimilation of sulphur bacteria may take place in one of the following ways.

Chemosynthetic, carbon-autotrophic sulphur bacteria. Some sulphur bacteria, e.g. *Beggiatoa* and *Thiotrix*, use the energy released by oxidation for the chemosynthetic assimilation of carbon.

This differs from the photosynthesis of green plants not only in the source of energy but also in the hydrogen donor, which is hydrogen sulphide. In cases where the carbon dioxide assimilation of the bacteria involves a reduction process as well as carboxylation, the hydrogen comes from the hydrogen sulphide, the sulphur being released as free sulphur. Thus, the hydrogen donor in carbon dioxide assimilation can be either water or hydrogen sulphide.

$$2\,H_2O + CO_2 + 469\,kJ \begin{array}{c} \text{solar} \\ \text{energy} \end{array} \xrightarrow[\text{photosynthesis}]{\text{green plants}} (CH_2O) + H_2O + O_2$$

$$2\,H_2S + CO_2 + 469\,kJ \begin{array}{c} \text{chemical} \\ \text{energy} \end{array} \xrightarrow[\text{chemosynthesis}]{\text{sulphur bacteria}} (CH_2O) + H_2O + 2\,S.$$

It is the sulphur released as a result of chemosynthesis which is deposited within the cells, secreted, or oxidised to sulphuric acid.

Carbon-heterotrophic or faculatively carbon-autotrophic sulphur bacteria. Some sulphur bacteria oxidise sulphur compounds such as thiosulphate, but cannot assimilate carbon in an inorganic medium. They are therefore considered to be carbon-heterotrophic. The oxidation of the sulphur compounds is only partial and would hardly provide enough energy for chemosynthetic carbon assimilation. The reaction is probably:

$$2\,Na_2S_2O_3 + H_2O + \tfrac{1}{2}\,O_2 = Na_2S_4O_6 + 2\,NaOH + 88\,kJ.$$
$$\text{sodium}$$
$$\text{tetrathionate}$$

Photosynthetic carbon-autotrophic sulphur bacteria. Some of the purple bacteria (*Thiorhodacae*) oxidise sulphur and sulphur compounds, including hydrogen sulphide. They differ from the other sulphur bacteria in their pigments and photosynthesis. The pigment, which probably functions like chlorophyll in higher plants, may be red (bacteriopurpurin) or green (bacterioviridin).

The purple bacteria occur in soil or water, but spread most

rapidly in water in which hydrogen sulphide is produced. In the laboratory purple bacteria can be grown on inorganic substrates containing hydrogen sulphide, provided that oxygen is excluded and that the culture is illuminated. The bacteria can then assimilate carbon dioxide with light as the energy source.

Hydrogen sulphide is oxidised as carbon dioxide assimilation proceeds. In the green species sulphur is released in the free form and in the red species as sulphuric acid. The source of energy is the same as in green plants, but the hydrogen donor, hydrogen sulphide, is different.

Desulphurification

Hydrogen sulphide is produced from decomposing organic material at the bottom of lakes and other bodies of water, often in such large amounts that the water is evil-smelling. The major part is derived from sulphate and other oxidised forms of sulphur, which are reduced by anaerobic organisms, of which *Vibrio desulfuricans* is probably the best known. The reduction takes place at the expense of various organic compounds, which are oxidised in such a way that they serve as hydrogen donors. The principle of the reaction is:

$$CH_3 COOH + 2 H_2O \longrightarrow 2 CO_2 + 8 H$$
$$H_2SO_4 + 8 H \longrightarrow H_2S + 4 H_2O.$$

Sulphate reduction of this type can probably also take place in waterlogged soil, e.g. arable soil to which sulphate fertilisers have been applied. This desulphurification process is analogous to denitrification and to the reduction of carbon dioxide to methane. Wikén (1942) summarises the three types of reduction process in the following way, where H_2A represents the organic compounds, e.g. alcohols, fatty acids, etc., which act as the hydrogen donor. Methane fermentation:

$$4 H_2A + H_2CO_3 = 4 A + CH_4 + 3 H_2O.$$

Denitrification:

$$4 H_2A + HNO_3 = 4 A + NH_3 + 3 H_2O.$$

Desulphurification:

$$4 H_2A + H_2SO_4 + 4 A + H_2S + H_2O.$$

The cycle

Except for the sulphur bacteria, plants take up sulphur in the form of sulphate (valency 6) and return it to the soil as hydrogen sulphide (valency 2). The sulphur obtained by animals from plants is also returned as hydrogen sulphide, and it is hydrogen sulphide which is the product of bacterial desulphurification. Nevertheless, hydrogen sulphide does not usually accumulate, because the sulphur bacteria oxidise it to sulphuric acid so rapidly. This oxidation completes the cycle, in the same way as oxidation by the nitrate bacteria completes the nitrogen cycle (Fig. 39).

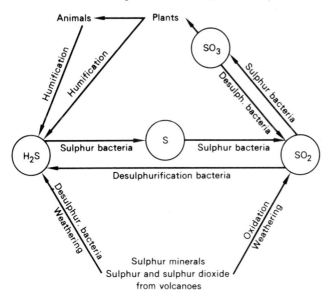

FIG 39. The sulphur cycle.

TABLE 34 A comparison between the nitrogen, sulphur and carbon cycles.

NH₃	H₂S	CH₄
Formed as a result of humification	ditto	?
Formed as a result of denitrification (by denitrifying bacteria)	desulphurification (desulphurification bacteria)	methane fermentation (methane bacteria)
Oxidised by specialised bacteria (saltpetre bacteria)	ditto (sulphur bacteria)	ditto (*Bacillus methanicus*)
Oxidation rapid	ditto	ditto
Oxidation yields an acid, the anion of which is assimilated by the plants	ditto	ditto
Oxidation yields energy, used for chemosynthetic assimilation of carbon	ditto	ditto
Occurs in relatively low concentrations	ditto	ditto

The biological advantages of the cycle are twofold: first, the sulphur economy of the organic world can be maintained using only a relatively small amount of sulphur, and, second, the accumulation of poisonous quantities of sulphur or hydrogen sulphide is avoided.

As is apparent from Table 34, there are many similarities between the sulphur, carbon and nitrogen cycles: there is a balance between oxidation and reduction processes, and there is resynthesis of the products of humification, so that only relatively small amounts of the elements are required for the maintenance of the biological world.

The phosphorus cycle

The decomposition of organic phosphorus compounds

The phosphorus requirements of plants and animals are met by the sparsely occurring minerals apatite and phosphorite, which contain phosphorus in the form of calcium phosphate. By a reaction with carbon dioxide, this salt is converted to soluble hydrogen phosphate:

$$Ca_3(PO_4)_2 + 4\,H_2O + 4\,CO_2 \longrightarrow Ca(H_2PO_4)_2 + 2\,Ca(HCO_3)_2.$$

The acids produced as intermediates in the humification of carbohydrates, e.g. butyric acid, lactic acid, acetic acid and citric acid, have a similar action, as do nitric acid and sulphuric acid produced by bacteria or in some other way. The soil organisms contribute to the solution of the mineral through their production of acids, and also to the solution of the relatively insoluble triphosphate, which tends to precipitate during phosphate turnover in the soil.

Phosphate ions are taken up by plants. Phosphate is a constituent of several organic substances, of which adenosine, di- and triphosphate, nucleic acids, lecithin and phytin are probably the most important. These substances are vitally important in energy transfer, as constituents of cell nuclei and the plasma membranes of the cell, and of animal nervous systems. Phosphorus deficiency has serious results, including inhibition of growth and development.

When cells and organs die, phosphoric acid is released, and becomes part of the soil phosphates again. This is partly the result of autolysis, partly of microbial activity (Steiner, 1948). Autolysis in plants begins while the organs are still living. Thus, the yellowing of leaves is accompanied by a decrease in total phosphorus and a rise in autolytically separable phosphorus. When the leaves die, some of the phosphorus compounds split, and phosphorus is released as phosphate or other inorganic forms, and is leached out of the litter. It is only later that microorganisms play a part.

The microorganisms attack the rest of the organically bound phosphorus in the litter, mainly that contained in phytin (inositol hexophosphate), nucleic acids and nucleotides. The detailed course of decomposition is not known. It is probable that many soil organisms split the organic phosphorus compounds. Some produce an enzyme (phytase) which hydrolyses phytin to inositol and phosphoric acid.

$$\text{phytin} + 3\ H_2O \xrightarrow{\ \text{phytase}\ } \begin{array}{c} CHOH \\ \diagup \quad \diagdown \\ HOHC \quad\quad CHOH \\ | \quad\quad\quad | \\ HOHC \quad\quad CHOH \\ \diagdown \quad \diagup \\ CHOH \end{array} + 6\ H_3PO_4$$

inositol

There are early reports of microorganisms which reduce phosphate to phosphite, hypophosphite and phosphine (PH_3) (Stephenson, 1949).

Higher plants may also play a part in the process of decomposition. Nilsson (1950) considers that the slimy mass surrounding the root tips of maize, for example, contains enzymes (exo-enzymes) which are produced by the roots and which attack nucleic acids.

The cycle

The amount of phosphoric acid in the soil varies between 0·03 and 0·3 per cent, and is usually about 0·1 per cent. The amount in solution in the soil water is only 1 to 3 ppm, which would be expected to be less than the optimum for plant requirements, and industrially produced phosphorus fertilisers are therefore necessary for crop production. The ability of plants to increase their phosphorus reserves is small in comparison with their ability to increase nitrogen, for example. The supply of phosphorus is considerably less than that of nitrogen, but the difficulty of obtaining phosphorus is balanced by an effective system for storing and retaining what there is. When phosphoric acid is released, phosphates are formed, and since these are relatively insoluble they are not subject to loss by leaching or evaporation, unlike nitrates. If there is a good supply of calcium in the soil, and thus high pH, insoluble calcium phosphate is formed. In acid soils, with little calcium, phosphoric acid combines with iron aluminium or magnesium, becoming a part of the colloidal hydrates formed by these metals, which are precipitated. In either case, the phosphorus is

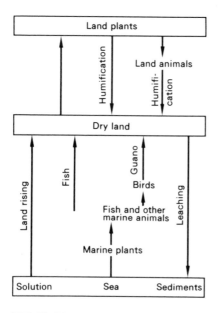

FIG 40. The phosphorus cycle.

retained in the soil and leaching is insignificant compared with that of nitrate or sulphate. Phosphorus therefore accumulates in areas inhabited by man: in towns, villages and farms (Arrhenius, 1931).

Phosphate leaching is not prevented completely. In the long period of time during which rain water has filtered through the soil layers of the earth's crust, large amounts of phosphate have been leached out and carried in streams and rivers to the sea. Steeman Nielsen (1944) calculated that the total content of phosphate in the sea was about 100×10^9 Mg (or about 80 mg/m^{-3}) to which about 2×10^6 Mg is added every year.

The construction of towns and cities and the development of modern methods of sanitation have intensified this movement of phosphate to the sea, although some of it is regained in the form of guano from sea birds and seals, which feed on fish and other aquatic animals. Guano is rich in both phosphorus and nitrogen. During the nineteenth century it was used as fertiliser, especially in Europe and America; and in South America it was used earlier than that. Guano is produced along coasts and on islands, but it accumulates only where the climate is so dry and warm that decomposition by microorganisms is prevented. In humid and semi-arid regions the guano is rapidly broken down, so that the vegetation along the coasts of such regions receives significant amounts of extra nitrogen and phosphorus.

Some phosphate is regained in the fish transported by man or animals from the sea to the land. In the course of time, more becomes available again in the considerable areas of the sea bottom which have become dry land, in places where the land is rising. The gyttja and other sediments on the sea bottom probably contain some phosphorus retained there in the form of insoluble phosphates or other compounds.

Finally, some phosphorus is transported in rain and wind, from the sea and from wind-eroded land. Tamm (1953) found 0·12 to 3·5 g/m^3 phosphorus in rain water.

The iron cycle

Although iron is essential to the metabolism of all organisms, it is used only in small amounts. Plants take up iron from the salts dissolved in the soil water or adsorbed onto the soil particles. In the cells, iron is a constituent of organic compounds such as haemoglobin and tannic acid, or remains in the inorganic state. When the cells die, the organic iron compounds are attacked by heterotrophic bacteria, whose oxidation processes lead to the release of iron in the form of ferric hydroxide.

However, the major part of the iron hydrate occurring naturally has a different origin. In minerals, iron occurs in the bivalent and trivalent forms, which are released into the soil by weathering. If the substrate is deficient in oxygen, the trivalent form is reduced to the divalent. According to Bromfield (1954), this is because some bacteria (*Bacillus polymixa*, *B. cirkulans*, *B. megatherium*, *Aerobacter aerogenes*) use the system $Fe^{+++} \leftrightarrows Fe^{++}$ as a hydrogen acceptor in the oxidation of organic compounds. The electron removed is taken up by Fe^{+++}. In the presence of carbon dioxide, ferrous iron forms neutral or acid carbonates.

Ferrous compounds are unstable. As soon as they come into contact with oxygen, they are oxidised to ferric compounds. Ferrous carbonate is oxidised to ferric carbonate, which is immediately hydrolysed so that ferric hydroxide, which is stable and insoluble, is precipitated.

$$4 \, FeCO_3 + 6 \, H_2O + O_2 \longrightarrow 4 \, Fe(OH)_3 + 4 \, CO_2 + 167 \, kJ.$$

Ferrous iron is therefore present in the soil mainly in places to which gaseous oxygen does not penetrate, or where the oxygen is used up by bacteria. However, it may also be formed as a result of reduction processes brought about by aqueous extracts of litter of various kinds, in which the iron becomes part of organic ferrous complexes (Bromfield, 1954).

The oxidation of ferrous iron is a weakly exothermic process. The so-called iron bacteria use the energy released for the chemosynthetic assimilation of carbon dioxide. Some of these bacteria are carbon-autotrophic, others only partly so. Whether the autotrophy is maintained under natural conditions has not been established. Possibly the bacteria use both methods; Pringsheim (1949) considers that the iron bacteria *Gallionella*, *Leptotrix* and *Sphaerotilus* are probably amphitrophic.

The morphology, physiology and ecology of the iron bacteria is insufficiently known to allow a summary of their way of life to be given. They occur in water which contains iron. Some are free-floating in the water (e.g. some forms of *Leptotrix*); others grow on plants, stones or other objects, and these are generally filamentous bacteria, forming rows of cells, which may be branched (*Clenotrix*) or unbranched (*Leptotrix*). The morphology is partly related to the way in which the iron is secreted from the cells. The iron bacteria take up ferrous carbonate, which is very soluble, and catalyse its oxidation, with the release of ferric ions, which are precipitated as ferric hydroxide. In the filamentous forms, the ferric hydroxide is deposited in the slimy sheath around the threads. The sheath swells so that the width of the threads is increased. If dilute hydrochloric acid is added, the iron dissolves and the sheath is freed.

The deposition of the iron may cause other forms to be developed. *Gallionella ferruginea* is a kidney-shaped bacterium, which is not filamentous. Ferric hydroxide is deposited on one side, so as to form a stalk with the bacterium on top.

The iron bacteria are aerobic, but since the ferrous carbonate on which they live is formed mainly under anaerobic conditions, they grow best along the narrow zone between the aerobic and anaerobic substrates. Not even here are conditions always favourable, since the bacteria have other requirements: they do not tolerate low pH, nor high concentrations of organic substances or iron (Pringsheim, 1949), they also require manganese. Hence, there are few places suitable for their extensive development. Pringsheim (1949) distinguishes three groups of such places.

1 Mineral springs, with water rich in carbon dioxide and in iron, usually contain ferrous bicarbonate. On contact with air, the iron is oxidised to ferric hydroxide, which is deposited on the surface of the water as an irridescent film, particularly in spring. The

iron bacteria, especially *Gallionella*, catalyse this oxidation. When the bacteria die, their sheaths or stalks remain as rust-coloured gelatinous masses and slimy casings on the bottom and around any objects in the water. In time, solid layers of ferric hydroxide may build up.

2 Humus-rich water from bogs and fens is relatively rich in ferrous iron as a consequence of the reducing effect of the stable humus and the tendency for iron to form complexes with organic compounds such as proteins, carbohydrates, humates, etc. However, the hydrogen ion concentration is usually too high for iron bacteria to flourish. If the water runs through base-rich media, such as limestone soil, the acidity is reduced. Then iron bacteria appear, especially *Leptothrix* species, and oxidise iron to ferric hydroxide, which is deposited in thick layers. These iron deposits have long been exploited, and are known as bog iron ore.

The capacity of the bacteria to convert such large amounts of iron is related to some extent to the nature of the oxidation process. The yield of energy is relatively small. The bacteria must oxidise about 500 g of ferrous hydroxide in order to produce one gram of new cell material.

3 In polluted water, such as ponds and lakes polluted with sewage or animal excrement, there are large numbers of micro-organisms in the water itself and in the gyttja, including iron bacteria and the bacteria responsible for sulphurification and de-sulphurification. In a reducing medium such as gyttja, ferrous compounds, hydrogen sulphide, and then ferrous sulphide are formed. The gyttja is coloured black by the ferrous sulphide (see Chapter 10). In the presence of carbon dioxide, the sulphide is converted to ferrous bicarbonate and hydrogen sulphide, which are oxidised in the better aerated parts of the body of water, by iron and sulphur bacteria. The activities of these two groups of bacteria are therefore interrelated to some extent, and they are often found together (Pringsheim, 1949).

4 Moist soil on a substratum containing iron and with a high water table. Ferrous carbonate is carried upwards to the highest level of the water table, where it is oxidised and precipitated as ferric hydroxide, forming a horizon rich in iron (the gley horizon, see Chapter 14). Similar processes take place in water pipes in which water is allowed to stand for some time: *Crenotrix polyspora* is especially active in these conditions, and produces large quantities of rust.

Oxidation by iron bacteria is analogous with that carried out by nitrate and sulphate bacteria. The corresponding reduction processes (see page 173) are insignificant in extent, so that iron accumulates as ferric hydroxide. This has no harmful consequences for the organisms because the iron compounds are not poisonous, unlike ammonia or hydrogen sulphide. Nor does the loss of iron from some sites readily result in iron deficiency, because minerals are rich in iron and because plant requirements are only small. After oxygen, iron is the plant nutrient most common in nature. Even so, iron deficiency does occur. Hewitt (1963) found that soil which contains large quantities of certain metals, such as manganese, copper, zinc, nickel, cobalt, etc., may bring about deficiency symptoms attributable to iron deficiency. The mechanism of these effects is not known.

The manganese cycle

Manganese occurs mainly as oxides, of which pyrolusite, a black crystalline solid which is the dioxide, is the most important. Its solubility is mainly dependent on soil pH. In acid soils the manganese is reduced, and forms salts of the lower valencies, particularly the bicarbonate and sulphate of bivalent manganese. This reduction is brought about by bacteria which utilise the oxygen from the manganese compounds. It is analagous to denitrification and desulphurification. Acid soils therefore contain manganese in a form readily available to plants (Mattson *et al.*, 1948).

In neutral and basic soils, usually those rich in calcium and other bases, the manganese is oxidised to less soluble high valency forms:

$$2\,MnCO_3 + 2\,H_2O + O_2 = 2\,MnO(OH)_2 + 2\,CO_2 + 452\,kJ.$$

The hydrate decomposes to water and the stable oxide MnO_2. The process is catalysed by bacteria, including iron bacteria, in the same way as iron oxidation. In calcium-rich soils, therefore, the manganese is in a form not readily available to plants. The lack of available manganese is enhanced by the fact that the iron bacteria are aggregated around plant roots, so that any dissolved manganese salts are prevented from entering the roots.

Cycles of other mineral nutrients

Potassium, calcium and other cations which are important in plant nutrition, cycle between the plants and the soil and at times take part in the metabolism of microorganisms. After uptake by the roots and transport to the above-ground organs, where they remain for a period, they return to the soil again in various ways:

1 The nutrients are returned to the soil through the humification of the litter. Even while humification is proceeding some salts

are taken up by higher plants and enter the cells again. The cations, especially calcium, are bound by other substances, including the acidoids which are produced in metabolism (Mattson and Koutler-Andersson, 1945). More of the salts are used by soil microorganisms and for a time become part of their cell substance. The ash of microorganisms may contain up to 37 per cent potassium. In this way, not inconsiderable amounts of plant nutrients are stored. These might otherwise be carried away in water, but are taken up by microorganisms because of the rapid increase in their numbers resulting from increase in the supply of litter.

2 Living leaves and other organs, especially of trees, may lose salts by secretion and leaching in large enough amounts to be significant ecologically. The earlier information was conflicting (see e.g. Bojarski, 1948; Tamm, 1953, page 85), but more recently large losses have been recorded (Tukey, 1970). K, Ca, Mg and Mn are usually leached in the greatest quantities.

Regardless of the pathway taken by the salts during their cycle between plants and soil, in humid climates they are exposed to the risk of leaching every time they are at the stage of being in the soil. This risk is particularly large for the soluble salts of the alkali metals. However, this loss is balanced by the constant supply of new salts from the soil minerals. The greater this supply, the more of the salt is available for leaching. Hence an equilibrium develops between supply and loss by leaching, provided that the plant community is not disturbed. If a plant community is destroyed by forest or prairie fires, for example, salts are released in large quantities and leaching is abnormally high. This occurs as a result of previous management practices involving shifting cultivation followed by burning. (See Chapter 17 for a discussion of other aspects of salt transport.)

9

Humus as a factor in plant ecolo

The formation of humus

According to the definitions of litter and humus given previously (Chapter 7), the boundary between these two types of material corresponds to the degree of decomposition at which it is no longer possible to identify the constituents morphologically with the naked eye. For example, when parts of a leaf cannot be recognised as such, the organ has become humus. Since the definition does not require a distinction between the macroscopic and microscopic parts of the material, the humus fraction includes microorganisms and various microscopic organs, such as pollen and spores, even though they may seem unchanged under the microscope. They are included in the term *humus* because it is difficult or impossible in practice to distinguish microorganisms or their remains from the products of decomposition of larger organisms. The so-called humus may therefore contain significant amounts of various kinds of living or dead microorganisms. There may be up to 2 kg of living microorganisms per square metre of soil 20 cm deep (Thorp, 1949).

When decomposition of the litter from higher organisms has reached the humus stage, there are generally only remains of the organs left. First, those parts which break down autolytically are utilised by the microorganisms, together with simple compounds which can be absorbed directly or after a single splitting of the molecules, for example, sugars and carbohydrates in general. Cellulose is also used fairly quickly, so that cellulose and sugars are almost completely absent from the humus complex. The respiration of the microorganisms rapidly converts some of these substances into carbon dioxide, water and salts, especially in aerobic conditions.

However, a large proportion of the litter consists of more resistant substances, which are decomposed to give various intermediates, which accumulate in the soil to an extent which depends on the rapidity of decomposition. A small proportion of such products consist of organic acids, alcohols, esters and amino acids which can rapidly be consumed by microorganisms or combine with other compounds present in the soil, so that they do not accumulate. It is even open to doubt whether these substances, shown by analyses, really occur free in the soil, or if they are released by the action of the chemicals used in the analyses. Anyway, the amounts in the humus are probably insignificant.

Other intermediates, such as compounds between proteins and lignin, are more resistant to the soil organisms and the chemicals in the atmosphere and the soil. The nature of these intermediates is generally incompletely known.

Animal excrement is another type of intermediate. Since the most easily broken-down parts of the animal food are taken up in the alimentary canal, what remains is the most chemically resistant part of the material, which leaves the body more or less unchanged. During the passage through the alimentary canal these materials are exposed to mechanical and chemical treatment which make subsequent decomposition by microorganisms easier. The formation of this type of intermediate certainly does not cause a delay in decomposition, rather an acceleration, but it leads to the separation and accumulation of the constituents which require the longest time for humification, because of their chemical nature.

Also in the humus are various resistant parts of the litter which are only attacked by specialised microorganisms, so that they take a long time to decompose and therefore accumulate even in their unaltered form. Suberin, wax, tannins, etc., belong to this group.

Humus is thus a mixture of living and dead microorganisms and plant and animal remains which are left after partial mechanical and chemical decomposition (humification, fermentation, mechanical and chemical digestion in animal alimentary canals). Among these remains are intermediates resulting from chemical decomposition, compounds between such intermediates, and chemically unchanged fragments of the material.

In general, the humus intermediates are derived from two types of reaction sequences, those leading directly to the mineralisation of the organic material, and others which lead only indirectly to this goal, and which may include some processes of synthesis. Some of the litter consumed by soil organisms is assimilated by them and enters a cycle within the humus complex itself. Figure 41 shows that the cycle may proceed in various ways. A particular part, e.g. a carbon atom in a leaf, may be transferred to the atmos-

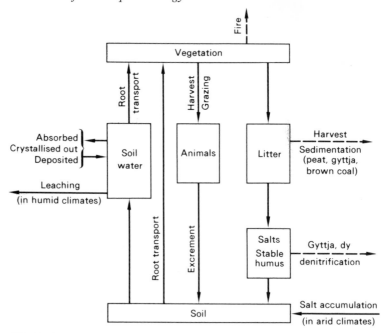

FIG 41. Survey of the decomposition of litter and the conversion, movement and loss of the products.

phere immediately, in the form of carbon dioxide, as a result of the humification of the leaf, or it may enter a cycle within the humus complex and be retained there for some time.

The chemical fractions

There have been numerous attempts to determine the chemical character of the substances which comprise the humus. One of the first workers in this field was Sprengel (1826), who divided the organic material in the soil into 'acid humus' and 'mild humus'. Acid humus is relatively stable and formed in places with a low content of basic material, and therefore occurs in peat. Mild humus is more easily broken down to carbon dioxide and water and mainly occurs in more fertile areas. Sprengel isolated an acid fraction which forms salts with bases, releases carbon dioxide from carbonates and silicic acid from silicates, and therefore was assumed to consist of acids, the so-called *humic acids*.

Later work has led to the division of the humus into other fractions, the composition of which is, however, partly dependent on the solvent used. Humus is almost insoluble in water, part dissolves in acids, other parts in alkalis, alcohols or other solvents.

Laatsch (1954, page 132) distinguishes the following main constituents:

1 Soluble in 1% sodium fluoride or sodium pyrophosphate solution.
  (a) Fulvic acids. Fraction:
      (i) Oxyacids and other substances not adsorbed by carbon.
      (ii) Precursors of humic acids.
      (iii) Soluble polysaccharides made up of galactose, glucose, mannose, arabinose, xylose, gluconic acid.
  (b) Humic acids
      (i) Non-hydrolysable acids with properties of tannins.
      (ii) Hydrolysable acids which yield amino acids.
2 Insoluble in 1% sodium fluoride or pyrophosphate solution.
  (a) Hydrolysable, including polysaccharides, chitin, cell proteins, etc.
  (b) Non-hydrolysable, including waxes, resins, lignin and lignin derivatives, humic acids combined with mineral colloids, etc.

Litter contains a large quantity of lignin, which is therefore regarded as one of the main sources of humus. During the humification of litter, the lignin is oxidised, and there is an increase in the number of carboxyl groups in the molecule, or in the parts into which it splits. This oxidation is facilitated by soil microorganisms when they attack and remove the cellulose in which the lignin is embedded (Laatsch, 1954, page 121). The further the oxidation proceeds, the more acid the product, as long as no neutralisation takes place. If the substrate has a low base content, the oxidation leads to increased charges on the colloidal anions, which therefore disperse in the soil water and filter down to the deeper layers, where there is usually a larger supply of bases and where precipitation can take place. As lignin is oxidised, ammonia is taken up and incorporated into the humus complex. This explains why the nitrogen content of many soils, including arable soils, increases with depth.

During oxidation, the lignin molecules are broken up into smaller molecules, which combine, in conjunction with the uptake of ammonia, and form acids of high molecular weight—humic acids—which are neutralised by the bases in the soil and form humates. The humic acids are brown colloids which are only slightly dispersed in water. The humates are of variable solubility; sodium humate is readily soluble in water, calcium humate is insoluble.

The oxidation of lignin seems to be dependent on the soil

acidity, or, rather, on its base content, particularly lime. In material poor in bases and generally acid, oxidation is slow, whereas it is rapid in alkaline media, provided that oxygen and ammonia are available in sufficient amounts.

The soil microorganisms and their metabolic products are another of the major constituents of the humus complex. The nitrogen contained in the humus fraction is of especial interest, since it is bound considerably more firmly than that in other nitrogenous compounds. Nitrogen in the litter from higher plants can be released relatively easily through humification to form ammonia or other compounds, but the nitrogen which has been assimilated by microorganisms is strikingly resistant. This microbial nitrogen, however, makes up almost all of the humus nitrogen, only 2 to 3 per cent of which consists of nitrate and ammonium compounds. When the microbial litter is humified, the microbial nitrogen becomes part of the humic acids.

Because the humic acids and humates are fairly insoluble in water, a quantity of nitrogen and soil bases is bound, so that it cannot be rapidly consumed by plants, sometimes in excess of requirements, or be leached away.

The content of nitrogen in plants is insignificant compared with the lignin content. The nitrogen in plant litter would not alone be sufficient for the humification of the lignin. It has therefore been suggested (Laatsch, 1948) that the humus nitrogen is enriched by animal litter, e.g. excrement. In the series of processes which the organic substances go through in the animal body, as a result of respiration the carbon content decreases more than the nitrogen content, and nitrogen is accumulated in the body and in the faeces.

A large amount of litter and humus is eaten by worms, particularly earthworms, so it is likely that some lignin oxidation takes place in the alimentary canals of these animals. Nitrogen is plentiful there, including ammonia, and there is a good supply of calcium, so that the humic acids can be neutralised.

When plants themselves produce humic acids, it is with the aid of microorganisms which form polyphenols and quinonelike carbon rings, which are components of the stable humus, metabolically.

Hence, the major part of the stable humus consists of the following two components:

1 A lignin component, formed as a result of oxidation and simultaneous combination with ammonia. Probably this process occurs mostly in the alimentary canal of saprophagous animals, especially earthworms.

2 A protein component, probably mostly microbial protein, together with carbohydrates, especially pentose and hexose residues, partly in the form of uronic acids.

Waksman and Stevens (1930, page 112) analysed a chernozem profile and found the following proportions (expressed as a percentage of the total organic material). See page 303 for the explanation of the terms A horizon, etc.

	In A horizon	In B horizon
Ether-soluble substances	2·72	2·18
Soluble in cold water	1·16	1·01
Soluble in hot water	1·31	1·24
Alcohol-soluble substances	0·98	0·91
Hemicelluloses	1·81	2·75
Celluloses	6·12	4·86
Lignin humus	47·70	45·82
'Soil protein' or organic nitrogen compounds	30·27	27·08

The physical and chemical properties of the humus

Humus is brown, black or grey in colour. When moist, it is greasy and coherent or of porridgelike consistency, depending on the variable colloidal state of the constituents. When dried, it crumbles to a brownish-black powder.

Its composition is complex, not only because of the different organs, parts of organs, and living and dead organisms which form part of it, and the large number of chemical compounds which form the basis of the complex, but also because of the chemical processes taking place, the oxidations, reductions, hydrolyses and syntheses, and the biological processes including everything involved in the growth and reproduction of the organisms, metabolism, competition, etc. Humus is a living system: it is the habitat for the soil microphytes and it is the most dynamic component of the soil, since it has the potential for greater activity and adaptation than any other part of the soil, both in its reaction in response to climatic or edaphic factors or to living organisms. In the interplay between organisms and environment, the humus is a link between them. Some of its properties are of particular importance for this, particularly those discussed below.

Colloidal properties

Stable humus consists of typically hydrophilic colloids; because they swell in water without dissolving, they can store a large amount of water. So, together with clay, humus is the most important store of water for plants. Swelling and shrinkage are reversible, and like the dispersion and flocculation of the humus, they play a part in determining the structure of the soil. This is the more significant because the stable humus is able to combine with lime and with clay to form a clay-humus complex, and the properties of the clay and other mineral fractions are dependent on the humus constituents. The structure of clay soil is determined primarily by its physical properties; these affect the tendency to flocculate and form the particles on which the crumb structure of the soil depends. However, the presence of humus facilitates the development of this structure: flocculation occurs more readily because of the precipitation of the dispersed stable humus on the clay particles, which join together into larger aggregates as a result (Laatsch, 1948).

Base exchange capacity

Partly because of its many carboxyl groups, the stable humus combines with bases, which are thereby retained in the soil instead of being leached away. Roots can take up cations from these compounds, in exchange for hydrogen ions. In this way, the stable humus acts as a reserve of mineral nutrients. The size of this reserve, the base exchange capacity, depends, among other things, on the number of carboxyl groups, which is, in turn, connected with the degree of oxidation of the lignin component. According to Mattson and Koutler-Anderson (1943), chernozem humus has a higher base exchange capacity than podsol humus, since it is formed under conditions of better supply of bases, has attained a greater degree of oxidation, and has a higher nitrogen content than podsol humus, which is formed under more acid conditions. This difference is, in turn, connected with, and explained by, the difference in the kinds of animals in the two kinds of soil. Because of its higher base content chernozem has more soil animals, especially the larger ones, and there is therefore more rapid turnover, including more rapid oxidation. The organic material in podsols is deficient in bases, and if the podsol has a high lignin content, humic acids are formed in large quantities, but neutralisation is relatively insignificant so that the soil becomes acid and the animal life and oxidation is inhibited.

The oxidation and reduction system

The humus complex may be regarded as an oxidation-reduction system, since both processes are going on within it. It reacts to changes in the partial pressure of oxygen in about the same way as living organisms do: a deficiency of oxygen causes the inhibition of certain processes and the stimulation of others. Decomposition changes in character if the oxygen supply changes, but it continues in any case, even at great depths in the soil. In other words, the humification process is adaptable to changing conditions brought about by variations in oxygen content of soil and water.

The balance between oxidations and reductions affects the base exchange capacity, among other things. Bases, bound as a result of oxidation and the formation of carboxyl groups, may be released if the carboxyl groups disappear as a consequence of reduction.

The carbon/nitrogen ratio

The proportions of the various substances in the litter change during humification to the values characteristic of humus. This is particularly true of carbon and nitrogen. At first, these components are mineralised to a different extent, so that the quantitative relationship between them (the carbon content/nitrogen content, or the carbon/nitrogen ratio) changes. The progress of mineralisation depends primarily on the type of litter. In some types the C/N ratio is large: in others small. Dvôrák (1912) and Wittich (1952) have given several examples: in leaf litter of poplar the C/N ratio is 63, of pine 66, of larch 113.

Litter with a high nitrogen content, with a C/N ratio less than 25 to 30, is broken down rapidly by the animals and microorganisms in the soil; this releases ammonia and the ratio rises. If the litter has a low nitrogen content, and a high C/N ratio, the available nitrogen is used up by the microorganisms, so it becomes part of the living constituents of the soil for a time. Consequently, the C/N ratio decreases.

This relation between the nitrogen economy and the supply of litter has important consequences for the vegetation. Nitrogen is released from litter rich in nitrogen, and is immediately available to the higher plants, but some is subject to loss by leaching or volatilisation, if not taken up sufficiently rapidly.

If the litter is poor in nitrogen then the higher plants may suffer a deficit of nitrogen, at least temporarily, because the microorganisms use up the nitrogen released by humification and maybe also that which was present in the soil in the form of ammonia or nitrate.

As a consequence, the addition of large quantities of litter with a low nitrogen content (straw, C/N ratio 80 to 100; sawdust, etc.) to arable soil may cause a temporary nitrogen deficiency. In these conditions, nitrogen fertilisers may be ineffective, because they are also used up by the microorganisms. However, the fertiliser treatment promotes the development of the microorganisms and the formation of microbial protein and thereby increases the total nitrogen reserves in the soil. When the microorganisms subsequently become humified and the microbial protein is mineralised, the nitrogen becomes available to higher plants.

Changes in the carbon/nitrogen ratio also depend on the mineralisation of the carbon compounds. Compounds with a relatively low carbon content, such as sugars, starch, hemicellulose and cellulose, are used up first, so that the humus becomes richer in compounds which contain more carbon, such as lignin and protein, mostly in the form of microbial protein. A summary of these changes is provided by Tables 29 and 30, pages 133 and 134.

In the first phase of humification the C/N ratio may increase or decrease, depending on whether the mineralisation of carbon or of

TABLE 35 The carbon and nitrogen contents and the C/N ratio of some soils at Rothamsted, England. (After Russell, 1952.)

	C %	N %	C/N
Old pasture (7–17 cm)	1·52	0·160	9·5
Old woodland (12–17 cm)	2·38	0·250	9·5
Arable land (Broadbalk) in 1893 after 50 years' continuous wheat, 0–22 cm			
No manure since 1839	0·89	0·099	9·0
Complete minerals and sulphate of ammonia, most years since 1843	1·10	0·12	9·0
Farmyard manure annually since 1843	2·23	0·22	10·1

nitrogen has the upper hand. The substances pass from one heterotrophic organism to another, losing carbon on the way, primarily because of respiration; whereas nitrogen, in contrast, is released mostly in the form of ammonia and is reassimilated, so as to reenter the cycle. As a consequence of these processes, the C/N ratio gradually decreases until it reaches a minimum, which is generally 9 to 12. The minimum is reached at about the point when an equilibrium between production and decomposition of the litter is attained (Table 35), and after that the value remains more or less constant, probably depending on the C/N ratio in the lignoprotein complex, and on the dominant position of this complex among the organic constituents of the soil.

The final value of the ratio is independent of that of the litter, but is affected by several other factors, some of which will be mentioned.

L. Meyer (1942) found that earthworms fed with rye straw mixed with powdered basalt reduced the C/N ratio from 23 to 11 during the course of two years, whereas during the same period other soil organisms only reduced it to 18. The value of the ratio is therefore affected by the type of organisms present.

The ratio is also affected by the acidity of the soil, or perhaps more precisely by its content of lime and other bases. It is higher in acid soils poor in bases, such as raw humus and peat, than in less acid or neutral brown earths or chernozem. There are more animals in these last two; oxidation of organic material proceeds further and more carbon dioxide is released. Hesselman (1926) showed that the nitrogen content of raw humus (coniferous forest) increased with increasing pH value, and Wilson and Staker (1932) demonstrated a relation between the C/N ratio in peat and the concentrations of calcium ions and hydrogen ions. The ratio showed a higher correlation with calcium than with pH.

Depth below soil surface is also significant. The C/N ratio is highest at the surface and decreases with increasing depth. This is not because there is more intensive oxidation at greater depths, but because there is a supply of humic acids and other oxidised material from the layers above (see page 303).

Then the ratio is affected by procedures in agriculture or forestry which bring about an increase in oxidation of the humus material, e.g. tilling the soil generally, and drainage of wet land.

Humus as a plant nutrient

Through mineralisation, the plant nutrients contained in the litter, carbon dioxide, water, and salts, are returned to the soil and the atmosphere. The salts are of particular importance to the vegetation because many of them occur in the soil only in suboptimal amounts. The nutritional effect of the humus decreases as humification goes on. Litter has an even larger nutritional effect before it is humified (Table 36). Nevertheless, humus is a better substrate for plants than litter, because it can take up more water and salts, released or supplied from the environment in some other way. Together with clay, humus is the most important store for water and salts in the soil.

In agriculture and horticulture, the formation of humus is accelerated by making compost from litter. Leaves, straw and other plant debris is kept for some time piled together in a heap, so that loss of heat from the innermost parts is minimised. Heat is produced during humification, the temperature inside the heap rises, and the rate of decomposition rises. When the heap is really hot, the litter has been changed to humus. For the same reason, the temperature in a heap of manure rises. However, composting material results in some losses, particularly of carbon dioxide and nitrogen, for example, as ammonia. The fertiliser effect is therefore lower after composting (Table 36). In spite of this, the procedure

TABLE 36 The fertiliser effect of straw, as fresh litter and after composting. (Russell, 1952.)

YIELD	FIRST YEAR		SECOND YEAR	
	COMPOST	STRAW PLOUGHED IN	COMPOST	STRAW PLOUGHED IN
Potatoes	3·17	3·80	3·00	3·22
Sugarbeet	4·64	5·07	4·55	4·77
Barley (grain)	3·46	3·91	3·32	3·50

has its advantages. Transitory supplies of plant litter can be accumulated and kept for a while until needed for some particular purpose. Also, the high temperature may destroy a number of plant and animal parasites, and weed seeds.

Types of organic soil

Terminology

The land consists of solid bedrock and the loose particles which cover it and which may be in definite layers or mixed together. There are several terms applied to this material, such as *soil, soil type, soil layers, solum*, etc. There has been an attempt to develop a definite terminology in soil science, so that the same phenomena have the same name. However, there is as yet no generally accepted nomenclature, and it is doubtful whether this is necessary or even desirable as long as the phenomena to which the names would apply are incompletely understood, as is the case with most soil processes and soil constituents. Hesselman (1926) has pointed out the dangers of labelling incompletely understood phenomena with a fixed terminology; there is a tendency to accept the names as an explanation of the phenomena, and this hinders rather than helps in stimulating further research.

In this book the following terms will be used (derived from Swedish terms defined by Hesselman, 1926, and Ekström, 1951).

The peripheral part of the earth's crust is bounded by the atmosphere or the hydrosphere and is called either *dry ground* or *water-logged* or *water-covered ground.*

Soil is a collective name for the loose parts of the earth's crust. It has two main types of component, that of mineral origin, so-called mineral soil (such as sand and clay) and that of organic origin, called organic soil (such as raw humus and dy). These terms refer to the materials themselves, which are usually mixed together to form the different soils.

Soils are the particular mixtures of soil materials found in certain environments. The name of the soil generally depends on the material dominant in the mixture, sandy soil if sand is the most obvious component, raw humus soil if raw humus dominates, etc.

Soil layer, for example, sand layer, clay layer, etc., are terms used when stratification is distinguishable, and there are layers in which sand, clay, etc., are dominant.

The *solum* is that part of the earth's crust near the surface which is influenced by climate and vegetation, i.e. by chemical, physical and biological processes, so that its properties become different from those of the underlying layers.

The substratum of the solum is not affected by climate or vegetation to a significant extent; it is usually the bedrock. Understanding the properties of the solum requires geological knowledge of the substratum.

A *natural solum* is one which has developed undisturbed by the influence of man, in contrast to cultivated soil.

Cultivated soil is a solum which has been formed with considerable influence of man, e.g. arable soil and garden soil.

The solum consists of various types of soil material, together with water, gases and living organisms. Many of its materials, forces and processes are significant to plants and comprise the edaphic environmental factors. The number of edaphic factors is large, because every process and property of the substratum, the climate and the soil organisms usually affects the solum in some way or other.

The *soil profile* is the name given to the vertical section of soil from the surface through the solum to the bedrock. It therefore includes the solum and the substratum. Arable soil profiles are usually divided into topsoil, subsoil and substratum. The *topsoil* is the uppermost layer, 15 to 25 cm deep mixed by the cultivating implements; it usually contains humus, derived from rotting roots, stubble, manure, etc. The *subsoil* is the lower layer of the original solum, which was there before the land was cultivated. The *substratum* is under the subsoil.

The *bedrock* plays a part in the formation of the solum because the products of its weathering are the mineral components of the soil. Changes in bedrock cause changes in the solum.

The *climate*, both macroclimate and microclimate, affects the solum in several ways. The temperature of the solum rises or falls, varying in magnitude and duration diurnally and annually, and with latitude. These temperature variations change the structure of the solum, particularly if there is freezing and thawing, since the soil particles aggregate and the space in between increases. Precipitation affects weathering and humification processes and hence nutrient production and accumulation. It may also leach the solum so that nutrients are carried away: the balance between these two processes depends on the duration of precipitation, the amount, run-off and evaporation.

Water and *wind* cause soil erosion, carrying away some of the soil constituents and depositing them elsewhere. The soil may be sorted according to size and weight of the particles, or these may be mixed up.

The *organisms*, both those living wholly or partly above the soil surface and those living under the surface, form humus after their death. The products of their decomposition become part of the solum and constitute its organic material. The solum may be wholly organic, e.g. peat. The roots of plants permeate the soil; after their death, the root channels remain and facilitate the infiltration of water. In this way, plants affect mineral weathering, water economy in the solum, etc.

Soil science

Geologically, the solum and the loose soil layers are relatively young. The constituents of the solum are constantly re-formed and used up. Chemical, physical and biotic forces are responsible for these processes of formation and destruction and bring about mixing, sorting and movement of materials so that the structure is changed. The forces involved, and the way in which they combine, together with the geological substratum and the supply of organic material, affect the way in which the solum develops. The study of this dynamic system of materials, forces and processes, together with the nature and properties of the soil layers, is soil science, or pedology.

Although soil science in its modern form is a young science, it has a long history. The secrets of the soil aroused the interest of prehistoric hunters and medicine men; and the first farmers probably encountered problems still important today. It was primarily the need of agriculture for knowledge of the soil which stimulated the first real studies of soil, some of which have been mentioned in the introduction. Soil science therefore originated as a branch of agricultural research. A historical survey of the development of soil science has been made by Lutz and Chandler (1946) and Wilde (1946).

Since soil science is concerned with the analysis of soil factors it opens up possibilities for the study of the relationship between these factors and the vegetation, and it is therefore one of the sciences closely connected with ecology.

Organic soil material

Litter and humus are sometimes separate in the soil, sometimes mixed together. In both fractions there are many animals large enough to be visible to the naked eye (living microorganisms are considered to be part of the humus itself, as mentioned previously). The humus, the litter, and the living soil organisms comprise the organic material in the soil, the main constituent of the organic soils, whether they are purely organic or mixtures with mineral material as a minor constituent. Some types of organic soil materials have a particular morphological character and have therefore been given separate names, such as *raw humus, peat* and *dy*. A more detailed description of these begins on page 203.

There is as yet, unfortunately, no definite terminology for organic soil materials. The term *humus* is sometimes used for such material and sometimes the term *mull*. However, humus is only one of the components. (For a discussion of mull, see page 199.)

The proportions of the various organic components vary with depth below the soil surface. The uppermost layer is mostly litter such as leaves, twigs, etc., but contains small quantities of humus, for example, products of autolysis of dead organs, dead fungal hyphae, insect excrement, etc. The humus component increases with depth and the litter component decreases. However, no precise separation of the two is possible: even in the deepest layers there is litter, in the form of remains of roots, pieces of bark and other resistant materials either produced *in situ* or transported by soil animals.

The content of organic material in the solum is variable. It varies according to the relation between the productivity of the habitat and the rate of humification and mineralisation. If mineralisation is slow in relation to production, then organic material accumulates until there is so much that the rate of total decomposition is equal to the rate of litter production. The relationship between production and decomposition is, in turn, dependent on the environmental conditions. In some conditions, decomposition is so rapid that the solum contains only an insignificant amount of organic material and is almost entirely mineral soil. In other conditions, the rate of accumulation is so high that the solum is entirely made up of organic material. The organic content usually varies between these two extremes and the solum is usually a mixture of mineral and organic materials.

Factors causing accumulation of organic material

By measuring carbon dioxide production from the soil, a relative estimate of the rate of humification and mineralisation can be obtained. Decomposition is mainly an oxidation process and the quantitatively most important end products are carbon dioxide and water. Thus, carbon dioxide production and oxygen uptake of the soil (soil respiration) are an expression of the rate of decomposition or of the biotic activity of the soil (Tamm, 1940).

Little is known about this activity in different plant communities and different habitats. Romell (1928) found, for example, that in a moss-rich pinewood in Central Sweden in summer an average of 0·2 litre of CO_2 was produced per hour and square metre of soil surface. Similar figures were published by Meinecke (1927). Romell calculated that a 1 cm thick layer of raw humus contains about 1 kg/m^2 carbon, which is equivalent to 1·97 m^3 CO_2 (15°C, 760 mm Hg) and that such a layer would be mineralised in three summers at that site.

In general, the biotic activity depends on the composition and numbers of the soil microflora and fauna, and on the conditions affecting the organisms.

Climatic and edaphic factors

1 Temperatures between *c.* 30°C and 45°C are most favourable for humification. The temperature optima of most soil organisms lie within or near this range. As previously mentioned (Chapter 6), there are also microorganisms with higher optima, which mostly occur in warm localities and warmer regions, as a result of selection; and organisms with lower optima, mostly found in colder regions. In warmer places, humification and mineralisation is rapid so that the organic content of the soil is low, while in colder places, organic material accumulates. However, it is not only the effect of temperature on the soil organisms which determines how much organic material accumulates, but the balance between decomposition and litter production. In warmer regions, a rise in temperature usually brings about a more rapid increase in rate of decomposition than in rate of production, whereas the opposite is true of colder regions. The reason for this is the difference in the way in which microorganisms and higher plants react. The lower the temperature, the smaller the number of species of both soil organisms and higher plants. The result of this is to decrease production relatively less than decomposition. A single higher plant, a coniferous tree, for example, produces large quantities of litter, and a single tree species may cover a wide area. However, a single species of soil organism cannot bring about mineralisation: several species in conjunction are required, and if one is excluded because of low temperature, the whole process of decomposition is slowed down. It is probably also significant in this respect that the numbers of larger soil animals decrease rapidly in colder latitudes. Decomposition is also delayed as a result of leaching of the litter in

cold, damp climates, and the consequent loss of some of its nutrients.

2 Good aeration and a suitable moisture content promote decomposition. If the litter lies in very thick layers, the supply of oxygen is insufficient and aerobic processes are inhibited. On the other hand, the anaerobic phase is favoured, because the moisture content of such thick layers is more uniform. A thin layer of litter tends to dry out rapidly. A porous structure which allows a high rate of gas exchange is most favourable for decomposition as long as it does not allow too rapid drying-out.

A water deficit causes the microbes to dry up and inhibits their activity. On the other hand, an excess of water restricts the oxygen supply, unless it is running water, with a large surface area in contact with air (very small brooks, or waterfalls). If still water lies over organic material, an oxygen deficit develops rapidly. In gyttja, all the oxygen is absorbed by a layer only a few millimetres thick immediately under the water/gyttja boundary (Alsterberg, 1922). Decomposition of the material in the gyttja is therefore mostly under anaerobic conditions and a number of intermediates of various types are formed. In the absence of oxygen, it is really only the methane bacteria which attack these intermediates. Hence, conditions for decomposition of gyttja are generally unfavourable, with the result that these intermediates and other forms of humus accumulate and the gyttja becomes a thick layer before the onset of equilibrium between production and decomposition (Wikén, 1942). In the same way, oxygen deficit is a significant factor in peat and dy formation.

In time, the anaerobic processes slow down until they proceed only very slowly, or not at all. The anaerobic products in gyttja, dy and peat are therefore in a relatively stable state (Tamm, 1940).

3 In addition, a sufficient supply of assimilable nitrogen is an important requirement for the growth of microorganisms. If the material is poor in nitrogen, the microorganisms use up the nitrogen reserves available from the environment, and decomposition is then dependent upon this supply. The soil animals are also affected by nitrogen content. For example, the consumption of wood litter by earthworms decreases as the C/N ratio of the material increases.

4 Acidity may affect the rate of decomposition through its selective effect on the microphytes and soil animals, although little is known about the extent of such an effect. The study of this question is complicated because in many cases the effect of acidity on decomposition is probably only apparent, and the reaction of the microorganisms is really caused by the supply or lack of bases.

5 Finally, mineral salts have a multiple effect on both de-

composition and production processes. Salts are an important factor affecting higher plant vegetation and therefore the qualitative composition of the litter, and play a part in the decomposition processes themselves, for example in the neutralisation of humic acids by lime, and consequent change in pH. The rate of decomposition of litter is more rapid on base-rich soils, particularly those with a high lime content, than on soils poor in bases (see Table 17, page 85) (Mattson and Koutler-Andersson, 1941).

The biotic origin and characteristics of the litter

The characteristics of the litter are another group of factors which affect humification. Some parts break down rapidly, others are chemically resistant. Lignin is the most important of the chemically resistant components, partly because it is present in large amounts, partly because of the resistant compounds formed between lignin and other humus constituents, especially nitrogenous compounds. Nitrogen in combination with lignin is not readily available to the microorganisms (see page 182). Because of such conservation effects, lignin is a direct hindrance to decomposition of the litter. The more lignin in the litter, the slower it breaks down. Hence older tissues are more resistant than younger, since lignin content increases with age. Young seedlings, leaves and organs decompose most rapidly.

In contrast with litter from deciduous broad-leaved trees and herbs, that of coniferous trees is more resistant, probably because of the lower content of bases and the higher content of acid-forming substances, particularly lignin, and also because of the resins, which are particularly resistant to microorganisms. Lindeberg (1944, page 170) found, for example, that *Marasmius* species broke down 80 to 90 per cent of the cellulose in the leaves of broad-leaved deciduous trees, but only 30 per cent of that in conifer needles, in the same time.

Leaves of different broad-leaved deciduous trees are also broken down at different rates. Julin (1948) analysed a leaf fall and found that after nine months (November–August) there were no rowan leaves left; but 4 per cent of the hazel leaves, 7 per cent of alder, 14 per cent of birch, 36 per cent of oak, 45 per cent of aspen, and 50 per cent of willow leaves were left.

Herbivorous animals prefer broad leaves and herbs to coniferous needles, dwarf shrubs and mosses, and this is also true of animals in the soil, the earthworms, snails and slugs, and insect larvae. The larger the quantity of broad leaves and litter from herbs, the larger the numbers of microorganisms in the soil. In coniferous forests, with different types of vegetation under the trees, it is mainly the understorey vegetation which determines the species and numbers of soil organisms (Table 37).

There is therefore more rapid turnover in the soil of a mixed wood than in a coniferous wood, and where there are herbs rather than mosses. *Sphagnum* litter is extremely poor in salts and is very unsusceptible to attack by microorganisms.

TABLE 37 The relation between the content of bacteria and the composition of the understorey vegetation in coniferous forest. (Swinhufvud, 1937.)

FOREST	NOS. OF BACTERIA (per g soil)
Oxalis-Maianthemum-type	13 501 000
Oxalis-Myrtillus-type	7 926 000
Myrtillus-type	5 337 000
Vaccinium-type	4 871 000
Cladonia-type	3 095 000

The biotic environment of the soil

The type and number of soil organisms, or the biotic environment of the soil, is most important in determining the breakdown of litter. However, the soil organisms are themselves dependent on the climatic and edaphic conditions, and on the nature of the litter. If these conditions are favourable, there are large numbers of species and individuals, and the litter is broken down comprehensively and rapidly. If, on the other hand, conditions are extreme in some respect, only specially adapted species can exist, and these can only bring about a partial breakdown. If the medium is poor in oxygen, animals and all aerobic organisms are excluded; if there is a lack of bases, and humification produces acid substances, basophilic organisms are excluded. In such cases, turnover is less rapid and organic material accumulates.

Decomposition is also dependent on the relationships between the various organisms; there may be cooperation or antagonism. The end product of one species' activity is often food for another. *Bacterium coli*, *B. aerogenes*, and others, release nitrogen compounds which are used by cellulose-decomposing *Clostridium* species. Cellulose-decomposing bacteria therefore operate more effectively when together with other bacteria than they do in pure culture (Janke, 1949). The larger organisms mix the material so that it becomes infected throughout by the smaller forms, which move insignificant distances in the soil. *Bacillus radicicola*, for example, moves about 1 mm in an hour.

Microorganisms may also be antagonistic. One species may produce substances (antibiotics) poisonous to another, and therefore inhibitory to breakdown. The activity of the cellulose-decomposing bacteria, for example, may be inhibited by these toxins (Janke, 1949). However, decomposition is not hindered by the type of antagonism between predators and the plants and animals on which they feed, for example, the series bacteria—amoebae—nematodes—beetles. The material is exposed to repeated exploitation in this way, and mineralisation is favoured.

For normal decomposition, some balance between the soil organisms is necessary. If this is upset, perhaps by drastic interference with the plant community, some organisms are favoured at the expense of others and the rates and direction of the various processes are changed.

The phenomenon known in agriculture as soil sickness, when the yield of an arable soil which has borne the same crop for too long falls off, may be attributable to a disturbance of the balance between the soil organisms, brought about by one-sided soil treatment and a one-sided type of litter.

Chemical and biotic changes in the material during humification

Litter decomposition is relatively rapid at first, when sugars and other substances preferentially consumed by the microorganisms are used up. Subsequently, the process goes more slowly: the substances are more complex, decomposition proceeds in stages, and several species of microorganisms are involved, so that there may be delays before the habitat becomes suitable for them. Also, during decomposition the material loses energy and the proportion of more resistant substances increases, so that the rate of turnover gets slower and slower, since it is the potential energy of the material which keeps the reactions going.

Since decomposition is mainly brought about by heterotrophic organisms, it is probable that some of the important constituents of the material, the vitamins, for example, may be rapidly used up, and this again would cause slower subsequent decomposition. If fresh litter is supplied, the biotic activity is stimulated, suggesting that this does in fact occur. Broadbent and Norman (1947, 1948) found that an addition of fresh litter to humus caused increased turnover in the humus. It has become apparent that green manuring, in which a mass of green material is added to the soil, does not increase or even maintain the already existing humus content, but brings about more intensive utilisation of the material already in the soil. The effect of green manuring lies not only in the nitrogen and other substances which are supplied to the soil, but also in the mobilisation of older organic material which it makes available.

Decomposition is also delayed by the assimilation by microorganisms of the products of decomposition, which consequently enter the cycle again. When this material again becomes litter, it may be a more resistant form than it was previously. In any case, mineralisation is delayed by the entry of the substances into new syntheses, new cell formation and new decomposition processes. Nitrogen and the substances found in ash are particularly likely to reenter the cycle repeatedly in this way before they finally reach the mineral stage. For instance, a section of a protein molecule may be assimilated by a bacterium, then become part of an amoeba, then a part of a whole series of animals which eat each other, nematodes—beetles—moles, for example. Subsequently, it may become part of the body of the microorganism which releases it during the humification of one of the animals, assimilates it and makes it part of a new organic complex.

Through these processes, some of the components of the primary litter are retained within the soil organisms for a short or long period of time; they are held within the living system, and may be said to have undergone biotic absorption.

In this way cell material may cycle several times, but not an unlimited number of times. Each new turn of the cycle involves the loss of some energy in the form of work or heat, and the loss of some material, accompanied by qualitative changes as well. Most carbon is lost in the form of carbon dioxide (soil respiration), and the microphytes assimilate some of the nitrogen. The net result is a decrease in the carbon/nitrogen ratio of the material.

As a result of these chemical and biological changes in the material itself, the rate of decomposition slows down. By the time the stage of lignoprotein formation is reached the material is in a relatively stable state. The subsequent pathways of decomposition are not known in detail, but probably saprophagous and coprophagous animals play an important role, because of the chemical–biological treatment undergone in the animals' alimentary canal.

Organic soil materials and soils

Organic soil material has already been defined as a mixture of humus, litter and living soil organisms. The composition of the plant communities and the characteristics of the habitat determine the composition of this mixture, which varies in biological, chemical and physical properties and therefore in its ecological effects.

The various organic soil materials are practically never found in isolation, but are mixed with each other and with mineral materials. The organic soils are usually made up of such mixtures. The term organic soils will be applied to mixtures of materials in which the organic constituents are present in such large quantities that they dominate the mixtures in ecological and physico-chemical respects. Thus, a peat soil has peat as its characteristic constituent, but also contains small amounts of gyttja, clay, etc. If the amounts of these are larger the peat soil loses its characteristic peat properties and becomes some other type of soil, such as dy soil, gyttja soil, clay soil, etc. In the same way that mineral materials are minor components of organic soils, organic materials are minor components of inorganic soils. Only exceptionally are soils made up of a single component; an example is the soil found deep down in a glacial moraine (see Chapter 2).

There are, of course, transitional types between the organic and mineral soils, and between the various soils within these two large groups. Soils usually differ from one another quantitatively and only rarely qualitatively.

The most important groups of organic soils are the mull soils, raw humus soils, peat soils, gyttja soils and dy soils.

Mull and mull soils

The term 'mull' was used by P. E. Müller (1879) to denote a mixture of humus and mineral soil. Mull soils can be defined as terrestrial mixtures of organic and mineral material, mainly formed under aerobic conditions, with large numbers of bacteria and animals, including earthworms, with a relatively high pH and a large enough content of organic material to affect the properties of the mixture.

Mull is found where the minerals are relatively rich in nutrients, especially calcium, and where soil moisture and other environmental conditions are favourable for higher plants. A more demanding flora establishes itself in these places, consisting of herbs and broad-leaved deciduous trees, especially the species with highest nutritional requirements (see Tables 17 and 18, pages 85 and 86). Deciduous woodland usually grows on typical mull; and it is also true to say that the vegetation itself gives rise to the mull. The plants produce a base-rich litter, because the habitat is rich in nutrients, especially in bases, and because the flora makes relatively large demands upon these nutrients. Such litter fulfils the conditions for the formation of neutral or nearly neutral humus, which provides the best environment for bacteria and the higher soil animals which are present in such numbers that they dominate the total activity of the soil organisms. Well-developed mull always contains many soil animals, particularly earthworms. The conditions for fungi are not so favourable, partly because the pH is too high; and also because the hyphae are so much damaged or eaten by the animals in the soil that fungal growth is limited (Romell, 1935).

FIG 42. Deciduous woodland on Öland. The vegetation is in several layers. (See page 67.)

Thus bacteria and the larger soil animals strongly influence mull formation. The production of humic acids is probably less when bacteria are the litter-decomposers than when fungi are. An indication of this is that the aerobic cellulose-decomposing bacteria do not produce acids (Waksman and Reusser, 1930). The effective aeration which results from the activity of the larger soil animals also favours the aerobic bacteria and the more complete oxidation of the material. If acid substances are formed, they are neutralised by the basic substances present. Because of the high pH, the humus colloids flocculate and cement together the soil particles both in the soil and in the alimentary canals of the animals. Mull therefore has a granular and porous structure.

Although mull formation is influenced by soil organisms and by the properties of the litter which they exploit, the primary requirement relates to the mineral composition of the habitat and in particular its base-richness.

Mull formation takes place only if there are no other environmental conditions limiting or preventing it, such as temperature. In many cases, there are such limitations. Mull soils are only found in regions where the temperature is sufficiently high to allow the necessary biological processes in the soil, but not so high that the organic matter is reduced to an insignificant amount. In Sweden, mull formation is common in the south and central parts of the country, but rarer in the north where it occurs only in especially

TABLE 38 The content of organic matter in some American soils. (After Romell and Heiberg, 1931, page 589.)

	THE DISTRIBUTION OF THE SOIL SAMPLES IN PERCENTAGE CLASSES OF ORGANIC MATTER CONTENT				
Percentage of organic material in the sample	0–10	11–20	21–30	31–40	41–50
Mull soils	11	25	12	5	7
Raw humus soils			1	3	5
Percentage of organic material in the sample	51–60	61–70	71–80	81–90	91–100
Mull soils	3	1	2	3	—
Raw humus soils	8	12	15	32	58

favourable localities, such as south-facing slopes, and usually in places where there are springs of water which is base-rich and well-oxygenated.

The proportion of organic and mineral components in mull soils is variable. In American mulls, Rommell and Heiberg (1931, page 259) found that the organic content was generally between 10 and 30 per cent, but occasionally as high as 80 per cent (Table 38). Köie (1951) found similar variations in a study of 148 Danish oak woods: the loss on ignition was between 4 and 81 per cent.

As the organic component decreases, the characteristic properties of mull become less marked, the mineral component becomes dominant, and the mull soil changes to a mineral soil.

Because of the brown-black colour of mull, mull soils are called brown earths or black earths (or chernozem) (see Chapter 14).

The intermingling of organic and mineral constituents is a result of transportation by soil animals, particularly by the larger worms; and of downward filtration of dissolved or dispersed humus from the uppermost layers of soil; and of the movement or growth of living organisms or organs in the soil, to the places where they eventually die, form litter and undergo humification.

There is no distinct boundary between the mull soil and the underlying mineral soil, but a gradual change as the organic fraction decreases with depth. In the transition zone between the two, various processes cause blending: material filters down only in some places, worms burrow to different depths, root depth is different, etc. (Tamm, 1940).

The structure is usually granular or lumpy, so that it is porous whether it is wet or dry. It can also be more finely granular and it is then relatively solid when dry and porridgelike when wet. The differences depend on the way in which the two constituents are mixed and affected by organisms and by atmospheric conditions. Some of the humus consists of colloids which are dispersed or precipitated as a result of changes in acidity, base content, temperature and precipitation. The precipitated humus colloids act as a cement between the mineral particles, which form clumps, the size of which depends on the stability of the precipitated colloids. The more stable the colloids, the better they withstand the effects of wetting or drying-out of the soil. If they are unstable, the clumps break up when the soil takes up water after rain. When the colloids are concentrated as a result of drying-out, they again exert a cementing effect. The same changes are brought about by freezing and thawing, so that alternation of wet and dry periods, and frost and warmth, is important for the development of the soil structure. Aggregation of soil particles also takes place in the alimentary canal of soil animals, especially earthworms (Lindquist, 1938; Wilde, 1946).

Because of their high fertility, mull soils have usually been exploited for crop production and used as arable land or meadows. Those still unused are mostly on land which it would not be economic to cultivate because of stoniness, topography, position, or other such reasons, and on such soils there may still be a more or less undisturbed mull flora. However, even these remnants are probably subject to some exploitation, such as removal of timber, or grazing. Deciduous woodlands, which usually have mull soils, are usually affected in these ways.

Exploitation of mull soil, like any other utilisation of soil for cultivation, takes a toll of the humus and mineral nutrient reserves, so that the conditions for the development of this type of soil—good supplies of mineral nutrients and litter—are less likely to be maintained. If there is no replacement by manuring, or if replacement is insufficient, quantitatively or qualitatively, the mull soil loses its character. For example, if a meadow soil is ploughed, the content of organic material may sink from 34 per cent to 2 per cent within fifty years, without manuring (Franck, 1951). After sixty years, arable soil in South Dakota had lost about 40 per cent of its organic material (Millar, 1955). Decreased base content causes changes in acidity, which affect the composition of the flora, the quality of the litter, the activity of the soil organisms, the soil structure, etc. Romell (1935) cites Winogradsky's studies of cellulose-decomposing bacteria as an illustration of this type of change. In neutral or weakly acid conditions, and with a good supply of ammonia, these bacteria can work so quickly that they have no competitors. However, in acid conditions they are suppressed by fungi, which break down the cellulose more slowly, but which are more tolerant of acidity.

Such changes cause the humus soil to lose its character. Since this involves the loss of ecologically valuable properties, and decrease in productivity, the soil is said to degenerate. Mull may also undergo degeneration in the absence of interference by man, for example, if severe drought decimates the soil fauna. The sensitivity of earthworms to drought was pointed out on page 102. During degeneration, as well as during development, there are transitional stages between the highly productive mull soil and other soil types. Some of these intermediate stages are so characteristic that they are distinguishable from one another and can be classified as subdivisions of mull soils (see, for example, Romell and Heiberg, 1931; Lindquist, 1938).

Raw humus soils

1 Raw humus, or mor, is a terrestrial, mainly aerobic, organic soil material, rich in fungi and small soil animals, with a relatively low pH.

Raw humus consists of litter, remains of litter and living material. Besides hyphae, roots and underground stems, there are usually bits of mosses, shoots, bark, coniferous needles and cones from the tree layer, in significant quantities, so that there is a large litter component in the organic soil. However, there is only a small amount of animal excrement and of other humus. Raw humus is also relatively poor in colloids and consequently it is a dry and poor type of soil, supporting a vegetation usually made up of xeromorphic species. The structure and appearance of the mor is affected by the composition of the plant community growing on it. In a spruce forest with mosses or dwarf shrubs, the mor is usually densely permeated by living and dead hyphae, roots and stems. Consequently, it is tough and bound together so that it can sometimes be lifted off the layer below like a mat. In other cases, as on some pine heaths, it can be loose and crumbly, and all the intermediates between these types can be found (Tamm, 1940; Romell and Heiberg, 1931). The permeability of some types of raw humus is so low that they inhibit the downward penetration of water into the soil more than any other layers of the soil profile.

The conditions for mor development are mainly a poor mineral supply, but the composition of the plant community is also important. On poor land, especially that with a poor supply of bases and a poor moisture supply, a flora of coniferous trees, dwarf shrubs, mosses, lichens, etc. develops; and these are plants with a low optimum mineral nutrient supply and pH (see Tables 17 and 18, pages 85 and 86) and some degree of drought resistance. Because of the soil type, these plants produce litter which is poor in bases but generally rich in lignin. Conditions for humification of this litter are relatively more favourable to fungi and the smaller soil animals than they are to bacteria and larger animals. Decomposition is therefore by different pathways from that of mull litter. The humification of this base-poor and acid litter gives rise to acid substances, like humification of any litter, but probably to a significantly greater extent, because of the high lignin content. Because of the low base content, the acids are incompletely neutralised and the pH value of the substrate is therefore low. Hesselman found that the pH value of mor was generally between 3·6 and 4·3 in coniferous forests with an understorey of dwarf shrubs and mosses.

Because of this high acidity, fungi and small soil animals are dominant among the soil organisms. Romell (1935) considers that the large population of fungi itself causes an increase in production of acids. He cites Falck's (1926, 1930) investigations showing that some fungi break down cellulose to oxalic acid, and also that the brown rot fungi generally produce acids.

The dominant position of the fungi in the microflora is further ensured by the relationship between them and the other vegetation: most trees and dwarf shrubs growing on raw humus form mycorrhiza.

As a result of these conditions, humification of base-poor litter differs from that of base-rich litter and the products have the special characteristics which are diagnostic of raw humus. The chemical nature of these characteristics is largely unknown.

Hesselman (1917a, page 421) found that there was no nitrification of ammonia in raw humus, in contrast to mull, where nitrification is rapid. It is possible that the ammonia formed in raw humus is immediately utilised by the numerous fungi.

2 Raw humus soils are mixtures of raw humus and mineral soil in which the raw humus component is so large that it determines the character of the mixture.

The amount of mineral soil mixed with the mor is very variable (Table 38, page 201). There are raw humus soils made up almost entirely of mor and others which contain as large a mineral component as some mull soils. However, the differences between raw humus soils depend more on the quality of mineral soil than on the quantity. It has already been pointed out that the mineral character determines the development of the organic fraction. The composition of the vegetation depends largely on the composition of the mineral soil; and the development of the organic soil depends on the composition of the vegetation. Therefore the mineral soil is of primary importance for development of raw humus soils, just as it is for mull soils.

Raw humus soils lack the intimate mixture between the organic and mineral fractions which is a characteristic feature of mull soils. The cause of this is related to the fauna. Mor contains many small animals, especially mites and springtails, which probably mainly live on hyphae, but few or none of the larger worms, so that the transport and mixing of material which is carried out by the earthworms in mull soils, and which is important to the character of the mull, does not occur in raw humus soils. However, downward movement of dispersed or dissolved humus from the uppermost layers to the layers underneath, and precipitation of colloids in the upper part of the mineral zone, probably occurs to a considerable extent. Conditions are favourable for this, because of the high content of acids, which cause dispersion of the humus colloids. Another way in which organic material is transported is by the growth of roots, often to considerable depths in the mineral soil, and their subsequent death and decomposition. However, such transport is insufficient to destroy the boundary between raw humus and mineral soil, which exists because the former is built up as a layer on top of the latter. This boundary is a characteristic feature of raw humus soils.

Raw humus soils are more stable than mull soils (Romell, 1935), since the main requirement for stability is a lack of mineral nutrients. A mull soil can be impoverished and change fairly easily to a raw humus soil, but change in the opposite direction proceeds much less readily. As the nutrient reserves are depleted, the demanding mull species disappear and are replaced by less-demanding raw humus species, whose litter is the starting point of the soil processes which gradually change mull to mor. Grazing for long periods without fertilisation can change brown earths to raw humus soils in this way. Leaching may have the same effect. On the other hand, a raw humus soil can be changed to a mull soil if suitable mineral nutrients are added and mull species are enabled to establish themselves.

There are many intermediates between raw humus and mull soils, and between the corresponding plant communities, because these changes can take place, and because the two types differ only in degree. For example, one beechwood may grow on a soil which is predominantly of a mull type, whereas another may be on a soil predominantly of a mor type (Hesselman, 1926); in land with a low clay content in Denmark raw humus is formed in beechwoods (Köie, 1951); in coniferous forest on a lime-rich substrate the soil may be mull, etc.

Raw humus soils cover large areas of the temperate and cold regions of the world. They are the soils most characteristic of coniferous forests. As temperatures get lower, precipitation, leaching and hence the tendency for mor formation increase. It is probable that large areas of the southern parts of the present coniferous forest belt were covered by mull soils and supported a luxuriant vegetation, but that leaching and other impoverishing processes brought about the conversion to raw humus soils with coniferous forests or heaths. In this region the frequency of fungi increases towards the north and that of bacteria and large soil animals decreases (Fehér, 1933).

Peat and peat soils

Peat is a mainly organic soil material, poor in organisms, formed on dry or wet ground, or under water.

The balance between production and decomposition determines the amount of organic matter which accumulates. Peat is formed when the rate of production is high and the rate of decomposition low. If the environmental conditions are such that production and decomposition occur at about the same rate, the gain of organic matter is only as large as the amount which can be mixed into the mineral soil by the various means of transport available. No peat is deposited in these circumstances, but mull and mineral-rich forms of raw humus develop. However, if conditions are more favourable for production than for decomposition, litter accumulates, and there is little or no mixing with mineral material. The organic material which accumulates is called peat.

Since the mechanisms by which the soil is mixed do not operate, the boundary between peat and mineral soil is distinct (Tamm, 1940). In this respect, peat is like raw humus, especially the forms poor in minerals. In other features, too, peat is very like raw humus, and morphologically they are more or less identical; the name given depends more on the thickness of the layer than on anything else. A thin layer is called raw humus: a thick layer— from 10 cm to several metres—is called peat (bog peat, fen peat, forest peat).

'Peat' is an old term. The thick layers of peat have always been of interest to man, partly because of their effect on soil and vegetation, and also because of their many uses. 'Raw humus' and 'mor' are more recent terms, originating from the end of the last century, in the development of modern soil science (P. E. Müller, 1879; see Romell and Heiberg, 1931).

Although peat and mor are similar morphologically, they are not usually identical. Their origins are different in some respects, and there are probably differences in their decomposition processes and chemical composition.

Peat is widespread, especially in cold and temperate regions with moist climates. About 8 per cent of the land in south Sweden is covered by peat bogs (Post and Granlund, 1926), and peat is also formed in marsh and fresh water communities in warmer regions. Lack of oxygen may have such a marked inhibiting effect on humification of litter in marshes, despite the high temperature, that decomposition is slow enough to allow accumulation of material and formation of peat. All peat is rich in lignin. The carbohydrates in the litter are used up during peat formation, whereas lignin and nitrogenous compounds are accumulated (Table 30, page 134).

Peat formation takes place in two main ways:

1 PEAT ON DRY GROUND (FOREST PEAT). In dry places peat forms when humification is very slow, because of low temperatures, for example. There is no difference between this type of peat and raw humus, either in formation or other respects, except that the layer is thicker. It is a terrestrial, autochthonous (formed *in situ*) mode of formation which occurs generally in cold, wet regions, such as the northernmost parts of the north European–Siberian coniferous forest belt, on mountain heaths and in the tundra.

Dry peat is the main constituent of dry peat soils, which are composed almost entirely of peat and differ from raw humus soils only in their greater thickness. Where dry peat occurs, it hinders the growth of higher plant vegetation because of its poverty in nutrients, and because it prevents the access of young seedlings to the more nutrient-rich mineral soil. Hence, difficulties in forest regeneration occur in the colder parts of the northern coniferous forest belt; forest management attempts to overcome these through burning-off the surface layers, partly to reduce the thickness of the peat and partly to increase the supply of nutrients, from the ash to the seedlings.

FIG 43. Raw humus 20 to 30 cm deep, over moraine. Mountain heath, Norway.

2 WET PEAT. In wet places humification may be hindered by lack of oxygen. If production of organic material is relatively high, peat is formed, primarily because of the excess of water which restricts the oxygen supply. The peat is therefore different from that formed on dry land as a result of low temperatures. In dry peat, humification occurs in aerobic conditions, with numerous fungi: in wet peat, conditions are anaerobic and the number of fungi insignificant. Since decomposition is of two different types, the chemical composition of the products can be assumed to be different. However, microorganisms are also active in the decomposition of wet peat, which contains a number of bacteria (Wikén, 1942).

Autolytic decomposition is probably more important in the mineralisation of peat than of any other soil material. However, the relative importance of autolysis and microbial decomposition is not precisely known. Autolysis proceeds slowly, but the time available is unlimited—the deepest layers of a peat bog may be several thousand years old.

An excess of water is the primary requirement for the formation of this sort of peat and if the water is removed by drainage the humification processes are radically altered: decomposition occurs by aerobic pathways, the rate of decomposition is many times faster, and the nutrients stored in the peat are rapidly mobilised. These changes are utilised by man in the cultivation of peat land. Ploughing increases the rates of decomposition and of nutrient mobilisation still further, by increasing the oxygen supply. Nutrient mobilisation in the undisturbed peat is so slow that only plants with very low requirements can survive, but after draining and ploughing it is sufficient to allow harvests for several years even if no additional mineral fertiliser is supplied. However, as a

FIG 44. Water meadow by a river in northern Skåne. Flooded in spring and autumn.

result of this faster decomposition, the peat is rapidly used up, the surface sinks, the ditches must be made deeper, and, finally, the mineral soil is exposed. If the mineral soil is stony till, cultivation is not economic and the land usually becomes afforested again. If it is not so stony, and especially if it is mostly clay or some other fine material, cultivation can be continued.

There are three types of wet peat: lake peat, fen peat, and bog peat.

Lake peat. In shallow lakes and bays, plant communities produce peat which eventually displaces the water and the vegetation encroaches. Plant communities which give rise to peat formation under the surface of the water consist of large grasses, sedges or herbs, so that the water must be nutrient-rich, or eutrophic. The eutrophy of the habitat determines the rate of peat formation and encroachment by vegetation.

Lake peat is a limnological product. It is derived from plants with roots and basal parts growing under water (Malmström, 1928). Different sorts of peat may be distinguished, depending on the dominant plant species.

Phragmites peat consists mainly of roots, parts of stems, and other litter from *Phragmites communis.* Humus, litter and seeds from *Nymphaea, Nuphar, Potamogeton, Menyanthes, Iris,* etc., may also be present. This type of peat is common in south and central Sweden. Where there are large amounts, it is the dominant component of the soil. The other components are mostly gyttja and dy. Such soil is hardly suitable for cultivation because of its loose structure.

Equisetum or *Trichophorum* peat is formed in communities in which these genera are dominant, and such peat is common in northern Sweden in and around lakes. *Equisetum fluviatile* is the most common species.

Some of these communities grow in shallow water or on land which sometimes dries out, so they are not truly aquatic. The peat which they form is intermediate between aquatic peat and semi-aquatic fen peat.

Soils corresponding with *Equisetum* or *Trichophorum* peat usually contain a large amount of dy (Malmström, 1928).

Fen peat. The plant communities known as 'fen' develop on land which is sometimes covered by water. Stream fen may be on level or slightly sloping ground with streams or springs: in rainy periods, the streams flood and the surrounding ground becomes covered by water. The vegetation is semiaquatic: the roots or rhizomes of the plants are in soil which is under water for long periods of the year. Such vegetation develops on low-lying land at the edges of lakes and rivers, where the fen may be flooded when the water level is high. Other low-lying ground may be within the range of fluctuation of the water table and so is sometimes under water. Or the surface run-off from surrounding slopes may cover the low-lying land. This adds to the salts and humus so that the eutrophy of the land increases. Such fens may be called 'ground water fens'.

As it passes through or over the soil, the water comes into contact with many substances which dissolve or are dispersed and are transported to the fen. The water movement ceases, or slows down, in the fen, and material is sedimented, dispersed humus colloids are precipitated and the salts are adsorbed by the soil colloids. When the water level sinks, a thin humus layer can often be seen, deposited by the flood water and covering the plants, litter, etc. The deposit is thickest where the water was deepest so that depressions tend to be filled up and the surface of the fen gradually assumes the evenness which is characteristic of it.

FIG. 45. Hummocks and hollows on a bog in Småland.

In some of the depressions, water stands so long and is so deep that no higher plants grow there. Such areas are usually bare. In other places the surface may be high because of tussock formation, and the tussocks may be so high that even species less tolerant of flooding than the true fen species can grow there. Eventually the tussocks may bear plants which have no direct contact with the water (Vallin, 1925).

Flooding acts like an application of fertiliser, bringing both salts and organic material. In time, relatively large nutrient reserves may be built up, often derived from poor localities. In this case, the change does not result in evening-out of differences, as it does in many of the other dynamic systems in nature, but in building up or accentuating differences.

The large supply of nutrients is one of the reasons for the productivity of fens, which is often very high. Another is the copious water supply, although the high water content means that the fen soil has a low oxygen content, which prevents most plant species from becoming established. Only species which have a special anatomical and histological structure facilitating transport of gases between the submerged or underground parts and the aerial parts

FIG 46. Cotton grass bog at the edge of *Calluna* heath. Halland.

are equipped to survive in such habitats. The lack of oxygen also excludes most microorganisms, so that decomposition of litter is slowed, and organic material accumulates so that peat is formed.

Fen peat is formed in wet conditions, and is both autochthonous and allochthonous, when it comprises sedimentary material. It has a high ash and nitrogen content (Table 41, page 219).

Because of the good nutrient and water supply, fen develops a rich vegetation relative to the immediate surroundings. However, the productivity is variable. In eutrophic areas, especially those containing large amounts of calcium and other bases, the fen vegetation is luxuriant (rich fen), but in oligotrophic areas it is more tussocky (poor fen) (du Rietz, 1949 and literature cited;

Malmer and Sjörs, 1955). There are many intermediates between the two.

Fen litter is humified more rapidly than litter in lakes and bogs because of the alternating wet and dry conditions in fens, and also because of the good supply of mineral nutrients. Hence, a large part of the nutrient reserves in a fen reenters circulation and contributes to increased productivity. Even so, humification is insufficient to prevent peat formation.

A eutrophic fen often bears a species-rich and multi-layered vegetation. There may be birch, alder, pine and spruce in the tree layer; *Salix* species in the shrub layer; a field layer with sedges, grasses and herbs, and sometimes *Equisetum*, or with Ericaceous species in the poorer fens; and, finally, mosses such as *Scorpidium*, *Drepanocladus*, *Campylium*, etc., in the bottom layer, but with no significant amount of *Sphagnum* species. This variable and species-rich flora includes many peat-building species, which are quantitatively the most important constituents of several peat soils. Fens may have special names, depending on the dominant species, e.g. dwarf shrub fen, *Carex* fen. In large fens there may be several types of vegetation, and hence several types of peat, with intermediate types, too (Malmström, 1928).

Among the most important types of fen peat and peat soils are the following:

Broad-leaved tree fen peat. This is common in Sweden. It contains large quantities of twigs, bark, parts of fruits, stumps, roots and pieces of wood, particularly from birch and alder. The soil usually contains some dy, and sometimes some clay. If the fen has developed in the course of encroachment of a lake by vegetation, there may be gyttja in the soil. The soil is rich in nutrients and can often support productive forest, even without drainage.

Carex peat. This mainly consists of remains of litter from *Carex* roots and stems. The soil usually contains a large amount of dy, especially in the wetter parts of the fen. In extreme cases there may be so much dy that it has as much effect in the resultant soil as the peat does. *Carex* fens are common in south and central Sweden. The sedges were formerly mown for hay, and the fen became meadow, as a result, and the composition of the plant communities changed.

Cladium mariscus peat is formed on the island of Gotland and in calcium-rich areas in south Sweden, where there are large amounts of this species. The peat consists mainly of the roots and rhizomes of the plants, often with remains of *Carex* and *Phragmites* intermixed. Both gyttja and clay may also be present in the soil.

Thus, fen peat soils generally consist of a mixture of autochthonous and allochthonous peat, mostly deposited under the water surface, together with dy, gyttja and mineral soil components.

Because of their high productivity, fen soils have often been exploited for cultivation and are of economic importance. They have primarily been utilised for mowing and grazing and in many regions they have been more important for these purposes than have the drier soils. Fen was formerly valued more highly than forest, when the value of estates was estimated. Only the wettest parts, with no trees or bushes, could be mowed without special preparation. Hence, it was mostly the most level parts nearest open water which became mowing meadows.

More recently, most of these meadows have been drained and ploughed; in Sweden, this is particularly true in the central and southern parts of the country. This has destroyed the natural flooding and sedimentation processes; the humus and salts in the flowing water have been transported further; the balance between production and decomposition of the peat has been upset; decomposition has become predominant and destruction of the organic soil has begun. Drainage has also resulted in partial or complete disappearance of many lakes.

Forest fen has also been exploited to a large extent, and drainage has often resulted in productive forests. But again, flooding and peat formation have been prevented and humification accelerated so that the amount of organic soil decreases.

Conversion of fen to arable land or productive forest therefore results in depletion of the nutrient reserves which have been accumulated for centuries. Not only is the interest on the accumulated capital being used, but also the capital itself.

Bog peat, bog soil. Bog peat is mostly derived from mosses, *Sphagnum* species in particular, and is the material of which peat bogs are built. *Sphagnum* species occur in large quantities and in various forms throughout the profile down through a bog. From the uppermost living cover of mosses, the lower dead parts of the stems extend downwards and contribute to a layer of litter, called *low-humified peat*. The degree of humification increases with depth, so that at a depth of a few decimetres under the surface there is highly humified peat. Even at this depth, it is usually possible to distinguish an increase in degree of humification downwards (Sernander, 1918).

Bog peat is a partly terrestrial, partly semiaquatic, mainly autochthonous soil. It is deposited in the same place as at least the main part of the material has been formed, sometimes somewhat above the water table, sometimes at or below it. Even if the water table is relatively low, as it is in a raised bog, humification is slow because of the high water content ensured by the peculiar histological structure of the *Sphagnum* mosses and their consequent capacity for storage of precipitated water. Besides, the peat has considerable capillarity, so it retains water drawn in from the surface or from greater depths. Water may also move in sideways from the laggs where it is stored. Sjörs (1948) considers that the laggs may act as reservoirs of water with a regulatory effect: excess water from the bog during rainy periods is stored in them.

Apart from *Sphagnum*, there are many other species in the flora of peat bogs, some so abundant that they exert a characteristic

FIG 47. The lagg of a bog. Forest at the edge of the bog is on the right, forest on well-drained ground is on the left. Komosse, Småland.

effect on the physiognomy of the plant community. These provide a means of classification of bogs and types of peat.

In the bottom layer, there are often such large quantities of *Sphagnum* that this species alone dominates the composition of the peat, which is then called '*Sphagnum* peat'. The most important type of peat in raised bogs is a *Sphagnum* peat, mostly derived from *S. fuscum* and *S. cuspidatum*.

In the field layer, the dwarf shrubs are particularly characteristic (heather, *Erica* species, *Betula nana*, etc.) and where these are very well developed the bog may be termed a 'dwarf-shrub bog', with dwarf-shrub bog peat.

Eriophorum vaginatum (cotton grass) is the characteristic species of cotton grass bogs and cotton grass peat, which is also called 'fibrous peat' because of its high fibre content. *Trichophorum cespitosum* (deer sedge) is the dominant constituent of the bogs and peat named after it; and, similarly, *Carex* species form *Carex* bogs and *Carex* peat.

Where there is a tree layer, the trees are mostly birch, spruce or pine. These bogs may be termed 'forest bogs', 'pine bogs', etc.,

FIG 48. An experimental area of a peat bog in N. Sweden (Västerbotten) fertilised with wood ash in 1926. The forest grew subsequently to a height of c. 14 m by 1957. (Photo C. O. Tamm.)

according to the species present. The peat contains large amounts of stumps and roots, fallen trunks, branches, bark and cones.

There are transitional forms between all these various types of peat, because there are transitional forms between the various typical plant communities, for example, between a typical *Carex* bog and a typical pine bog. In the same way, there are intermediates between fen peat and bog peat, and between these two and lake peat. Despite this lack of distinct boundaries, the divisions have some scientific and practical value; they allow the information to be classified, and appreciated as a whole.

FIG 49. Tree trunks exposed in peat at a depth of *c.* 1 m. Söderman-land.

Peat bogs are markedly oligotrophic plant communities, poor in nutrients. The vegetation is not in contact with the mineral soil and must depend on the small quantities of salts which are present in rain, an amount which varies according to the distance from the

TABLE 39 The chloride content of rain water in relation to distance from the coast. (From Firbas, 1952, page 189.)

DISTANCE FROM COAST (km)	CHLORIDE IN RAIN WATER (mg/litre)	DISTANCE FROM COAST (km)	CHLORIDE IN RAIN WATER (mg/litre)
0·03–0·1	70–700	100	3–5
1–2	15–30	500	1–2
5–10	6–13	1000	0·5–1·5
50	4–9	2000	< 0·3

coast. The supply of chloride 100 km from the coast may be 35 kg per hectare and year, in an annual precipitation of 700 mm (Table 39). The other salts in sea water are probably distributed in the same way (see Chapter 17). Nutrients are also supplied in dust particles blown from surrounding land and in the animal litter.

Apart from this, the bog plants are dependent on the nutrient reserves already present in the community and which become available again. However, nutrients in this form are not readily available since litter from bog plants is acid and poor in bases and is humified only very slowly. Mattson and Koutler-Anderson (1944) consider that bog plants may be able to take up nutrients from the acid peat because of their own high content of acid constituents.

TABLE 40 Salt turnover in a peat bog (Augstumalmoor) in north-west Germany. (Firbas, 1952.) The calculations are based on analyses of the peat and the bog water, and on the assumptions that the bog grows in height at a rate of 1 mm per year and that annual precipitation is 680 mm, of which 580 mm evaporates and 100 mm runs off.

	1 A 1 MM LAYER OF PEAT CONTAINS (MEAN OF 4 BOGS) (kg/ha)	2 AMOUNT IN 100 MM RUN OFF WATER (kg/ha)	3 1 + 2	4 THE VALUE OF 3 AS A % OF THE ASH
K_2O	0·55	4·1	4·65	10·0
Na_2O	0·63	4·6	5·23	11·3
CaO	1·77	1·1	2·87	6·2
MgO	1·41	0·7	2·11	4·5
$Fe_2O_3 + Al_2O_3$	1·57	0·5	2·07	4·4
P_2O_5	0·38	0·5	0·88	1·9
SO_3	2·21	1·9	4·11	8·9
SiO_2	4·73	9·1	13·83	30·0
Cl	0·21	10·4	10·61	22·8
Ash	12·89	32·9	45·79	—
Nitrogen	6·35	2·7	9·05	19·5

The small income of nutrients to the bog plants is balanced by considerable losses. Some of the nutrients are locked up in the peat, as long as the bog is increasing in depth; some are carried away in the run-off water. There is no continuous flow of water through a bog as there is through a fen: instead there is a slow-moving stream around the edge, the lagg, which carries away salts and humus materials. The water of such streams is often more or less brown, as a result of the large amounts of humus materials in it. Because of the high acidity of the peat, the humus has a high degree of dispersion.

Bog soil consists only of peat, and it is thus a wholly organic type of soil. It is a poor oligotrophic soil, as a result of the low supply of nutrients and of the loss of material (Malmström, 1928). Also, the deficiency of nutrients is even greater physiologically than it is physically, because a large proportion of the nutrients are bound in such a way that they are unavailable to plants. The contents of silicates, chloride and nitrogen are highest, next come sodium, potassium and sulphur (Table 40).

The capacity of bog species to grow despite the low nutrient supply depends on their metabolism and nutrient economy, which are such that the demand for mineral nutrients is relatively low. In *Sphagnum* species, the optimum concentration of nutrient salts is so low that even the normal concentrations in the soil solution are inhibiting. Hence, these species cannot grow at the edges of peat bogs if the water contains electrolytes from the surrounding mineral soil. Partly because of this, and partly because of the reaction of the other flora to the higher nutrient supply, this zone, the so-called 'lagg' of the bog, resembles fen in vegetation,

TABLE 41 The composition of low moor peat (fen peat) and high moor peat (bog peat). (After Waksman and Stevens, 1928.)

	PER CENT DRY WEIGHT	
	LOW MOOR PEAT	HIGH MOOR PEAT
Ether-soluble fraction	0·43	4·34
Water-soluble fraction	1·71	1·61
Alcohol-soluble fraction	1·94	3·40
Hemicellulose	11·03	15·76
Cellulose	0	12·35
Lignin	44·81	44·19
Protein	20·97	4·94
Ash	13·02	1·82
Total nitrogen	3·70	0·81

hydrology and soil (Malmström, 1928, 1940, 1943). There is a sharp vegetation boundary with the more oligotrophic bog proper.

Thus in many respects, bog soil is the opposite of fen soil. It is poor in nutrients; humification is slow; there is no nutrient supply by flooding, but a loss of nutrients in the run-off water; the flora is poor in species and low in productivity. As a result of these differences, there are fewer animals and microorganisms in bog peat than in fen peat. In fen peat, decomposition is rapid, and cellulose and a major part of the hemicellulose is used relatively rapidly by animals and microorganisms; furthermore, there is a high nitrogen content (Table 41). In bog peat, decomposition is slow, cellulose and hemicellulose are left, and protein production is low. The two types of peat resemble each other in one respect: they are rich in lignin, but poor in water-soluble substances (Waksman and Stevens, 1928).

Peat bogs have been, and still are, exploited in various ways. The upper, least humified, layers are used for animal bedding and as packing material. Low-humified *Sphagnum* peat is porous, light and airy, with a high water-absorbing capacity (because of the anatomy of the moss shoots). Peat for fuel is cut from the lower layers which are highly-humified and therefore much denser and heavier. In some areas, in south Sweden, for example, considerable areas have been cultivated. Bog soil consists mainly of bog peat, but also contains larger or smaller amounts of dy and little or no mineral soil. After drainage and ploughing the rate of peat decomposition is increased many times and the plant nutrients which were locked up are released. As a result, cultivated bog soil may yield good harvests for several years after drainage, even without fertiliser treatment. Subsequently a supply of mineral fertilisers and lime is required, and also gravel till, preferably nutrient-rich, to improve the structure. These additions, together with increased humification and neutralisation of the humic acids, bring about the change to mull soil.

Peat bogs are being used more and more in forestry. In southern and central Sweden it is often possible to get productive forest, instead of stunted trees or none at all, merely by draining a peat bog. However, in the north, draining alone is not enough, and a supply of mineral salts is needed as well. Good growth of trees can be obtained after addition of wood ash (Malmström, 1935, 1943). The ash has a direct nutrient effect and an indirect effect through stimulation of mycorrhiza formation (Björkman, 1941).

Whether bogs are exploited for cultivation or for fuel, the peat is used up. During the two world wars in the twentieth century, the supply of imported fuel into Sweden was partly cut off, and peat-burning took place on a large scale. Peat is a natural resource which has taken thousands of years to accumulate; it should therefore be kept in reserve and used only in time of necessity, when requirements for fuel and arable land cannot otherwise be met.

Dy and dy soils

Lagg streams and other water flowing from bogs contain colloidal soluble humus compounds which colour the water brown. There is generally also a large quantity of humus in the run-off water from areas where the soil and the underlying rock is poor in basic constituents. Because of the lack of bases, the neutralisation of humic acids is incomplete and the acids remain in solution and tend to bring about the solution of other types of humus. When these humus materials are carried in the water on to a meadow by a river, for example, they are exposed to the oxygen in the air and water and they are precipitated, forming the type of organic soil known as *dy*. Dy can therefore be described as precipitated, allochthonous humus. Oxidation probably mostly takes place during the process of transport itself, since oxygen is taken up more readily by flowing water than by still water. Dy is precipitated all along water courses, wherever a river flows more slowly or where there is a volume of stagnant water. The salts dissolved out of the

FIG 50. A dry hollow with its bottom covered by dy. (There are crane footprints in the dy.) Komosse, Småland.

mineral soil by the water, or present along the path of flow, for example, in lakes, also contribute to the precipitated material. Oxidation and precipitation also takes place in the laggs of bogs, so that the bottom becomes covered with dy.

Dy is most common in lakes and fens, since it is in such places that conditions bringing about precipitation (oxygen, salts) are encountered. There is also some precipitation in bogs, where the dy originates, at the boundaries between bog and open water. This dy is then called *lake dy*, *fen dy* or *bog dy*, whichever is appropriate. The dy is mixed with peat to form peat-dy, or, if peat is predominant, dy-peat. Dy is usually present in most peat soils.

Dy is brown or black in colour, and forms a loose, porous and light powder when dry.

Dy soils usually contain varying proportions of peat and gyttja, and are usually very oligotrophic. After drainage, they form light, loose, dry soils which are not very good for cultivation.

Gyttja and gyttja soils

Gyttja is an organic material deposited on the bottom of bodies of water, consisting of a mixture of substances from the water and its surroundings, mainly from the communities of higher plants

FIG 51. Vegetation encroaching into a lake. Söderman-land.

growing in or at the edge of the water and from the algal communities in the water. The material may be transported and deposited elsewhere by flowing water (allochthonous gyttja), for example, at river mouths, where it contributes to delta formation. Gyttja is more coherent and more elastic than dy. It contains a lot of material which swells when wet, so it shrinks and cracks on drying-out.

Humification of gyttja takes place in oxygen-deficient conditions, partly because of the large amount of organic material competing for the oxygen available, partly because of the water slowing down the inflow of oxygen. Many of the processes are reductions leading to the formation of hydrogen sulphide, ammonia and other un-pleasant-smelling gases. For this reason, this type of soil is some-times called *sapropel* (see Kubiëna, 1953, for example). Iron and sulphur compounds are particularly affected by reduction pro-cesses: ferrous iron and hydrogen sulphide are formed, and ferrous sulphide is deposited, colouring the gyttja black (see Chapter 8, page 175).

Gyttja soils contain a large amount of sulphur, sometimes as much as 2 to 3 per cent. If the soil is drained and cultivated, this is oxidised to sulphuric acid, which is gradually leached out or neutralised (Wiklander *et al.*, 1950).

Gyttja can be divided into two types, detritus gyttja and algal gyttja, depending on how it is formed.

Detritus gyttja consists mostly of the litter from the communities of higher plants at the water's edge, made up of leaves, shoots, parts of fruits, fruits, seeds and whole plants, from both aerophytes and hydrophytes, together with the humus formed as a result of decomposition of the litter. If the litter is predominant, the gyttja is called coarse detritus gyttja: if the humus fraction predominates it is fine detritus gyttja. In both, there are many organic materials such as pollen, animal excrement, small amounts of algae, etc. Detritus gyttja contains chlorophyll derivatives and carotinoids which are stable for many thousands of years under the prevailing conditions (Gundersen, 1956) and it is therefore grey green to greenish brown in colour.

Algal gyttja is formed from litter derived from aquatic algae; e.g. *Vaucheria* gyttja, with *Vaucheria* the dominant alga (the so-called paper gyttja, tough and bound together by the coarse threads of the alga); Kieselguhr (diatomaceous earth), a deposit of diatom frustules; *Scenodesmus* gyttja, etc. Algal gyttja is usually red brown, and elastic.

The distribution of gyttja is dependent on the distribution of the communities from which it is derived. Detritus gyttja is mostly formed along the edges of water, and in shallow water, whereas algal gyttja is deposited everywhere under open water, even out at sea. At the boundaries between these main sites transitional forms occur between the two types. These contain some dy and peat.

There are several sorts of gyttja soils, made up of gyttja together with other organic and mineral materials. These may be exposed if lakes are drained, or when shallow beaches are exposed at ebb tide. Their composition depends on the characteristics of the surrounding area. In places with a large amount of finely-divided mineral soil, such as fine sand, silt or clay, or with water rich in lime or iron, and depositing calcium carbonate or iron ochre, dylike material is carried down in the water and deposited with the gyttja. If there is so much of this material that it makes an important contribution to the soil properties, the resulting soil is called clay gyttja or chalk gyttja, for example. If the clay is predominant, the soil is termed gyttja clay. In this way, the gyttja soils include transitional forms to mineral soils, like the mull soils.

Agriculture plays a part in the formation of clay gyttja and gyttja clay. Considerable amounts of clay are lost annually from open arable land and are carried in drainage water to some area of still water where gyttja formation is going on. There is some movement in the opposite direction, although on a smaller scale, if a material such as chalk gyttja is added to agricultural soil as a means of improving it. Such gyttja often contains large numbers of snail shells.

As a result of the high content of finely divided, partly colloidal material, both organic and mineral, gyttja soils and mineral soils containing gyttja have a high water- and salt-holding capacity. They are therefore some of the most productive soils. The high fertility of delta soils is due to the accumulation of such material and the continuous supply carried by the river. Considerable areas of shallow sea bays have been reclaimed as fertile agricultural land through damming and drainage.

Summary of soil types and their composition

The solum consists of various mineral or organic soils. A soil is termed a *mineral soil* when its properties are primarily determined by the mineral constituents (e.g. sand or clay) and an *organic soil* when the organic constituents (e.g. raw humus) are dominant.

Organic constituents and organic soils

Organic soil material comprises humus, litter and living soil organisms.

Organic soil material occurs in nature in several forms, which differ in biological, chemical and physical characteristics.

1 Mor or raw humus—terrestrial, mainly aerobic organic soil material, rich in fungi and small soil animals, and with a relatively low pH.

2 Peat—terrestrial, semiaquatic or aquatic, mainly anaerobic soil material, poor in bacteria, fungi and animals.

3 Gyttja—aquatic, anaerobic soil material, poor in fungi and bacteria, but relatively rich in animals in the upper layers.

4 Dy—precipitated, allochthonous humus.

5 Mixtures of these soil materials, e.g. peat and gyttja, peat and dy, etc.

Organic soils consist of organic substances with or without a mineral component.

1 Mull soils—mixtures of organic and mineral material, e.g. brown earths, black earths.

2 Raw humus soils—mixtures of raw humus and mineral material.

3 Peat soils.

(a) Terrestrial peat soils—terrestrial, autochthonous humus or peat together with mineral material.

(b) Bog peat soils—mainly terrestrial, autochthonous peat with an admixture of dy.

(c) Fen peat soils—mixtures of allochthonous and autochthonous mainly semiaquatic peat with an admixture of dye and mineral soil.

(d) Lake peat soils—mainly aquatic, autochthonous peat with an admixture of dy and mineral soil.

4 Gyttja soils—mainly gyttja with an admixture of dy and mineral soil.

5 Dry soils—mainly dy with an admixture of peat and gyttja.

Intermediate types of organic soils, e.g. intermediates between bog peat soil and fen peat soil, between peat soil and dy soil, etc.

Intermediates between organic and mineral soil materials, e.g. mull; and between organic and mineral soils, e.g. gyttja soils and clay soils, etc. (For further discussion of soil classification see Kubiëna, 1953.)

Mineral materials and mineral soils (see Chapter 11).

Mineral materials consist of:

1 Crystals, fragments of crystals and crystal conglomerates of the crushed rock minerals.

2 Salts precipitated from the soil solution, e.g. calcium carbonate, limonite, vivianite, etc.

Mineral soils consist of mineral material with or without an organic component. The most important groups are boulder soils, stony soils, gravel, sand, fine sand, silt and clay soils.

Mixtures of these soils are till soils (see Chapter 11).

The mineral constituents of the soil

The composition of minerals and rocks

The lithosphere, or earth's crust, is the major source of mineral nutrients for the plant world and the majority of plants obtain their nutrients from the mineral fraction of the soil. Mineral soil is formed as a result of weathering of rocks and is therefore made up of the same materials as the rocks. However, this relationship applies only on a very large scale, and not to small areas, which may have a mineral soil which was formed elsewhere and has been transported to the site in some way.

The mineral soil satisfies the primary requirements of plants from the soil. Its ecological value depends mainly on the chemical composition of the mineral particles, their size and ease of weathering. In turn, these characteristics depend on the composition of the parent rock and the ease with which it weathers. The parent rock may be very variable in this respect, because it comprises different types of rock, each made up of one or several minerals. Some minerals contain large amounts of valuable plant nutrients, others very small amounts; some are easily weathered, others are more resistant, and so on.

A brief survey of the minerals which are most important ecologically is given below.

Minerals

Potassium felspar (orthoclase) Si_3O_8AlK, is a common red mineral, frequent in rocks and mineral soil. For example, Tamm (1940) gives the amount in north Swedish forest soils as *c*. 20 per cent by weight. It is resistant to weathering and the large amount of potassium which is present in the mineral (about 17 per cent K_2O) is therefore only available to plants to a very small extent.

Sodium felspar (albite) $Si_3O_8Al\ Na$, also occurs frequently, and is resistant to weathering.

Calcium felspar (anorthite) $Si_3O_8Al_2Ca$, weathers fairly easily and is therefore ecologically important, because of its calcium content. This mineral is usually found together with sodium felspar, in a mixture called 'plagioclase'.

The felspars, which usually occur in mixtures in the rock, make up about 60 per cent of all minerals and are therefore of fundamental importance to the formation and properties of soils.

Mica, like the felspars, has many forms, because of variations in its composition. The formula may be shown diagrammatically as follows:

$$(SiO_4)_3\ Al_2\ R_1\ R_{11}$$
$$Fe\ \ Fe\ \ K$$
$$Mg\ \ Na$$
$$Ca\ \ Li$$
$$F$$

Mica contains several elements important to plants, and since it weathers easily it is one of the most important minerals ecologically. Because it is soft, it has been extensively ground up by glaciers into fine particles which are one of the main constituents of clays. Although mica has a lower percentage content of potassium than potassium felspar, it is more important as a source of potassium because the clay particles weather easily, and Tamm (1940) considers that mica is the most important source of potassium for wild plants. The mica particles also give the clay plasticity and are therefore important in influencing the physical properties of clay soils.

Augite and *hornblende* are two greenish-white and brown-black groups of minerals, consisting of Mg—Fe—Ca silicate. The content of Ca and Mg may be up to 30 per cent. They weather easily.

Olivine is Mg—Fe—Mn silicate, and Ca—Mg silicate.

Hypersthene is a mixture of $MgOSiO_2$ and $FeOSiO_2$.

Calcite, $CaCO_3$, is the most important mineral in limestone.

Dolomite, $(CaMg)CO_3$, is closely related to calcite.

Apatite, $Ca_5(PO_4)_3(F, Cl)$, occurs in small amounts (0·1 to 0·2 per cent) in the common types of rock and in the soil and is very important ecologically since it is the most important source of phosphate for plants. Because apatite is infrequent, and weathers relatively easily, there is a possible risk that the biological world is in danger of suffering from phosphate deficiency. In many cases, phosphate deficiency does exist, but to a lesser extent than would be expected, because phosphate is not readily soluble and so tends to be retained in the soil.

Phosphorite is closely related to apatite. It is the raw material of the phosphate industry.

Pyrites, FeS_2, is a brassy yellow mineral found in small quantities in many types of rock. It is easily oxidised to sulphate.

Iron sulphide, FeS, is formed where sulphur and iron compounds occur together in reducing conditions.

Siderite, $FeCO_3$, is frequent where there are minerals containing iron. It is precipitated in peat, for example. It is oxidised to limonite (see pages 173 and 223).

Limonite (bog iron ore, iron ochre) $Fe_2O_3 \cdot 3\ H_2O$, is a red brown, secondary product deposited from iron compounds dissolved in water, such as iron carbonate. It occurs as solid layers on the bottom of lakes and in bogs, and also as a constituent of soil (see page 175).

Vivianite is an iron phosphate, sometimes precipitated in peat and clays.

Quartz is SiO_2.

Mohr and van Baren (1954, page 114) have given a survey of the chemical composition of the most important minerals, how easily they weather, and the products of weathering.

The constituents of the minerals listed above include all the quantitatively most important plant nutrients, with the exception of nitrogen. The earth's crust contains about ninety elements, which all occur in the surface layer of soil.

The majority are found only in small quantities and in rare minerals which are mixed with the other minerals in the rocks and the soil. They include elements important to plants, e.g. copper, zinc and boron. However, these elements are physiologically active at such low concentrations that even a trace in the soil solution is sufficient for optimum growth, and they are therefore known as *trace elements*, or *microelements*.

Types of rock

Rocks are usually mixtures of several minerals and are therefore of many different forms. Some of the ecologically important types are listed below.

Quartzite, *sparagmite* and *sandstone* are made up of quartz or a mixture of quartz and felspar. They form poor soils for plants because the minerals are resistant to weathering and because there is a high content of SiO_2 (Table 43 provides an example of the composition of sandstone).

Granite or, more correctly, the granite group, contains quartz, potassium felspar, sodium-rich plagioclase, magnesia mica (biotite), sometimes potash mica (muscovite) or hornblende, and small quantities of apatite (see Table 42).

Gneiss is made up of the same types of mineral as granite, but often contains other minerals as well (Table 42). The gneisses and granite are valuable in forming soil mainly because of their content of such minerals as hornblende, apatite and felspar. The higher the proportion of these, and the smaller the proportion of quartz, the richer is the derived soil (Tamm, 1940).

Mica schists contain quartz and mica in varying proportions, which determine the properties of the resulting soil.

Hyperite (black granite) consists of calcium-rich plagioclase, augite, olivine and hypersthene.

Diorite contains mainly plagioclase and hornblende (Table 42).

Gabbro and *diabase* contain calcium-rich plagioclase, augite, olivine and apatite.

TABLE 42 Example of the percentage composition of some of the common types of rocks. (Sörlin, 1943.)

	GRANITE	GNEISS	DIORITE	DIABASE	GABBRO
SiO_2	70·70	63·41	62·05	52·29	48·57
TiO_2	0·03	0·65	0·67	0·24	0·21
Al_2O_3	13·13	19·22	17·16	14·99	18·48
Fe_2O_3	2·73	2·74	1·74	6·77	0·67
FeO	0·69	4·00	4·23	3·70	6·21
MnO	0·13	0·05	0·05	0·50	0·07
MgO	0·49	2·84	1·80	5·95	9·56
CaO	1·15	0·65	6·92	7·62	12·23
Na_2O	4·94	1·21	2·96	4·16	3·22
K_2O	4·41	4·36	1·90	2·26	0·30
H_2O	1·49	0·60	0·82	1·65	0·81
P_2O_5	—	—	0·17	—	—

Amphibolite is made up of plagioclase, hornblende, and small quantities of apatite.

Basalt is a dark volcanic rock containing calcium-rich plagioclase, augite and magnetite.

Greenstone is a term covering hyperite, diorite, diabase, gabbro and basalt, the components of which are mostly dark-coloured minerals, especially augites and hornblendes. The presence of these types of rock is of particular importance to the soil, because they contain several elements important in plant nutrition, including calcium, and because they are easily weathered; and it is often reflected in the flora.

Limestone has many forms. It is mostly made up of calcium carbonate, either in a crystalline form (calcite) or sedimented (Table 43 gives an example of its composition). The sediment consists of whole or crushed shells of animals, together with varying amounts of clay. Depending on how they were formed and on their position in the geological strata, Silurian, Devonian, Carboniferous, Jurassic and other limestones can be distinguished. Most of the layer of Silurian limestone which formerly covered the Scandinavian peninsula has been weathered and eroded away, particularly by glaciers, so that only small remnants are left. These remnants, especially the soil derived from the limestone, are important reserves of nutrients for the flora. There are also small amounts of phosphorite in limestone.

Shales are clays which have been hardened and changed in other ways during the process of time. The composition varies according to the type of clay. Some shales, alum shale for example, contain organic remains (carbon and bituminous substances) and iron

TABLE 43 An example of the composition of some Cambro-Silurian rocks: sandstone, covered by alum shales, covered, in turn, by limestone. (After Sörlin, 1943, page 25.)

	SANDSTONE	ALUM SHALES	LIMESTONE
SiO_2	97·7	45·22	8·30
Fe_2O_3	0·1	7·26	2·30
Al_2O_3	1·3	13·60	1·60
CaO	0·2	0·75	49·12
MgO	0·1	0·71	0·81
Na_2O	—	0·40	—
K_2O	—	4·74	—
Loss on ignition	0·6	25·83	38·60
Total	100·0	98·51	100·75

pyrites (Table 43). The content of potassium and other plant nutrients is large because of the capacity of clay to absorb salts. Shales give rise to nutrient-rich and productive clay soils, after crushing and weathering.

The mechanical weathering of minerals

The bedrock, and to some extent the mineral constituents of the soil, are subject to mechanical weathering brought about by the crushing and grinding action of the glaciers, by temperature changes, especially frost, and by the pressures exerted by roots and other growing parts of plants. The products of weathering in previous epochs have been sorted by water and by the wind into layers, or have become compacted and cemented together by precipitated or crystallised salts. In this way the material may again

FIG 52. Diabase outcrop with luxuriant lichens. (Photo H. Vallin.)

FIG 53. Silurian limestone cut through by a stream forming a gorge.
Billingen, Västergötland. (Photo. A. Holst.)

form solid bedrock in the form of layers of sandstone, schists or limestone. More recently, soil has been formed as a result of weathering of these sedimentary rocks as well as of the primary rocks.

Before the last glaciation, large parts of the northern hemisphere, including Scandinavia, were covered by weathered material, which was then largely carried away by the ice. At the same time, more rock was crushed, forming the mineral soil which now covers these areas.

Hard rocks, such as granite or gneiss, with a large proportion of quartz, were relatively resistant, and they were broken up by the ice into relatively coarse material including large numbers of boulders and stones. Gneisses formed somewhat finer deposits, with fewer stones, than granites.

Greenstones were less resistant and were crushed into finer material.

Shales and limestone were crushed into such fine particles that they formed fine earth (< 2 mm) and clay (Tamm, 1940).

The greater part of the mineral soil which now covers these formerly glaciated areas is this glacial material. In comparison the amount of mineral soil formed since the last glaciation as a result of weathering is insignificant.

Transport, mixing, sorting and deposition of the weathered material

It is unusual for soils which are the products of weathering to remain in the places where they were formed, i.e. to be autochtonous. They have usually been transported by ice, water, wind or gravity, and are therefore allochtonous.

Material transported by the ice (glacial material) is deposited when the ice melts, forming moraines. The particles are mixed, of various sizes, with uneven and angular surfaces, and of varying density.

Other material is transported by water (alluvial or diluvial material). Glacial rivers deposit boulders, shingle, gravel, sand, silt and clay. During transportation, the material becomes ground and polished, so it differs morphologically from that in moraines. The particles are deposited according to size, and they therefore become sorted to some extent. Clay is deposited where the water flows most slowly, in lakes and in the sea. Elevation of parts of the earth's surface has lifted up some of these clay deposits above the surface of the water, and now they form clay plains on dry land. In some cases there was so much limestone among the clay that the material (marl) has been used as a fertiliser. The lime content of this type of clay is between 8 and 45 per cent: the clay content between 8 and 60 per cent.

Presentday water courses still transport mineral soil, mainly clay, which is deposited at the bottom of lakes and in river estuaries,

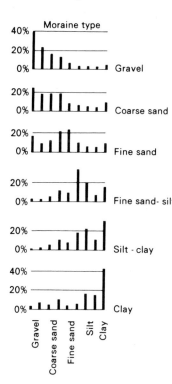

FIG 54. **Distribution of particle size in different types of moraine. Each fraction is divided into coarse and fine parts. The shift in height of the columns from left to right results from crushing of the material. (After Magnusson, Lundqvist and Granlund, 1957.)**

where it may cause delta formation. Some of the material is from the river banks, eroded by the flowing water. Some is carried in run-off water from the surface of the land.

High precipitation, especially torrential rain, washes away bare soil: exposed arable soils, and other soils denuded of vegetation, for example, severely overgrazed and trampled pasture, are eroded by heavy and persistent rain in many regions. The finest material is carried away first, and is washed down from higher levels into the hollows. On sloping ground, the surface run-off water forms runnels which excavate deeper and deeper into the soil so that the water flow in them increases and erosion is accentuated. Coarser material begins to move and is carried down to the valley bottoms and the larger rivers. If this process continues, the land is dissected

FIG 55. Desert, with bushes, on a beach of wave-sorted material. Grogarnsberget, Gotland.

FIG 56. Fen with a bottom of deposited calcium carbonate. Gotland.

into hillocks with deep channels between, and the highest ground may be denuded of soil (see Chapter 18).

Waves and currents in the sea or in lakes also cause erosion: particles are transported and deposited, sorted into shingle, gravel, sand and clay fractions.

FIG 57. Quarry in dunes of blown sand, now afforested. Västergötland.

Water also transports small quantities of material in solution or as dispersed colloids. When these are precipitated and deposited, they form special types of soil material, such as calcareous tufa, which is precipitated calcium carbonate, deposited in springs; calcium carbonate deposited at the bottom of lakes; limonite (bog iron ore), which is precipitated iron hydroxide, etc. Some clay fractions also belong to this group.

Windborne material consists of sand and other soil particles which are usually blown away either from layers of particles already sorted as a result of water transport and deposited on beaches by the sea, lakes or rivers, or from the weathered material in areas of sandstone. Windborne material is also blown from arable fields, roads, and other bare soil surfaces where the wind is in direct contact with the particles (wind erosion, see Fig. 57 and 58).

FIG 58. Dunes stabilised by fencing. Halland.

Material moved by gravity is found in some places, but is comparatively limited in extent. Steep cliffs, particularly those facing south, are often covered with boulders, stones and grit, which have been broken off by frost action or other mechanical weathering and have accumulated on the lower ledges. Such scree soils are usually made up of coarse material, but they often support a rich flora different from that in the surrounding area. The reason for this has often been the subject of discussion. An interesting hypothesis, which needs to be tested, has been put forward by G. Weimarck (1952, page 118), who suggests that the high productivity of such soils is because they were unsuitable for shifting cultivation following burning, and therefore escaped the impoverishment and decomposition suffered by more suitable land during the period when this practice was popular (see Chapters 13 and 18). However, Selander (1955, page 341) considers that the flora of screes under south-facing cliffs is peculiar because of the unstable nature of the ground, which prevents dwarf shrubs from establishing themselves and hinders the formation of raw humus.

FIG 59. Soil profile of strongly weathered limestone. The fine earth (black) has run down through cracks in the rock from the thin layer of soil on the surface. Roots have penetrated the cracks. The part of the profile shown is 1·5 m deep. Fiuggi, central Italy.

The size and physical properties of the mineral particles

The mineral particles in the soil consist of crystals, fragments of crystals and conglomerates of crystals, generally of several minerals. This material has been sorted into fractions to some extent, depending on how much it has been subjected to the action of wind or water, and there are long-standing names for these fractions, such as gravel, sand, clay, etc. These names apply nowadays not only to the material itself, but in many cases to soils in which the particular materials are important components. Hence 'sand' means mineral material of a certain size, and also a soil containing a large amount of sand.

In order to define types of soils and materials more exactly, a classification of the fractions in terms of the diameter of the particles has been worked out. The names and size ranges of the fractions are listed in Table 44.

The physical, chemical and ecological character of the material depends on the particle size: the smaller the particles, the larger is the total surface area (Table 44). Chemical weathering increases with surface area, and the rate of production of plant nutrients is

TABLE 44 The classification of mineral particles according to size and surface area. (After Ekström, 1951.)

SOIL	PARTICLE DIAMETER D	NO. OF PARTICLES IN 8 LITRES MATERIAL	TOTAL SURFACE IN 8 LITRES MATERIAL, CALCULATED AS IF EACH PARTICLE WAS:	
			A CUBE WITH SIDES D m^2	A SPHERE WITH RADIUS $\frac{1}{2}D$ m^2
Boulders	> 200 mm	1	0·24	0·1256
Stones	200–20 mm	$1–10^3$	0·24–2·4	0·1256–1·256
Gravel	20–2 mm	$10^3–10^6$	2·4–24	1·256–12·56
Coarse sand	2–0·2 mm	$10^6–10^9$	24–240	12·56–125·6
Fine sand	200–20 μ	$10^9–10^{12}$	240–2 400	125·6–1 256
Silt	20–2 μ	$10^{12}–10^{15}$	2 400–24 000	1 256–12 560
Coarse clay	2–0·2 μ	$10^{15}–10^{18}$	24 000–240 000	12 560–125 600
Fine or colloidal clay	< 0·2 μ	$> 10^{18}$	> 240 000	> 125 600

TABLE 45 The relation between particle size and the rate and height of the capillary rise of water. (Tamm, 1940.)

SOIL PARTICLE DIAMETER (mm)	CAPILLARY RISE IN 24 H (cm)	MAXIMUM CAPILLARY RISE (cm)
5–2	2·2	2·5
2–1	5·4	6·5
1·0–0·5	11·5	13·1
0·5–0·2	21·4	24·6
0·2–0·1	37·6	42·8
0·1–0·05	53·0	105·5
0·05–0·02	115·3	c. 200
0·02–0·01	48·5	400
0·01–0·005	28·5	820
0·005–0·002	14·3	c. 1700
0·002–0·001	5·5	c. 7000

therefore greater; the water-holding capacity also increases, because this is partly dependent on the area of wettable surface, and partly on the capillarity. Finer material has smaller spaces between the particles and therefore greater capillarity (Table 45), but at the same time is less permeable to water and to air. This affects the transfer of both heat and water in the soil and hence influences the soil temperature and availability of water to plants.

Gravel is a poor, dry soil material. Weathering is slow; water-holding capacity and capillary rise of water are insignificant; aeration is good; evaporation is slow and conduction of heat is poor, so gravel soils tend to be warm. An admixture of gravel in soils which lack these properties, clay soils, for example, has an ecologically favourable effect since the soil is rendered lighter, warmer and better aerated.

Coarse sand has properties like gravel, but more accentuated, and productivity is therefore higher. Coarse sand is more effective than gravel in improving clay.

Fine sand. The coarser particles of the fine sand fraction have a higher water-holding capacity than coarse sand, but nevertheless are a dry soil material. The rate of weathering is insufficient to support the more demanding species of plants: fine sand usually supports a vegetation which has relatively low nutrient requirements and which has xeromorphic characteristics enabling it to tolerate a water deficit during dry periods.

Gravel, coarse sand, and the coarser part of the fine sand fraction are poor, dry materials. Productivity is increased considerably if they are mixed with fine materials such as clay and humus. The lack of nutrients and low water-holding capacity of the coarser fraction is counteracted; the aeration, permeability and soil temperature of the finer fraction is increased. The productivity of the coarse mineral soils may also be increased by a good supply of water. There is often a rich flora in places where gravel and coarse sand soils are kept moist by running water.

Table 44 shows that the dividing line between macroscopic and microscopic particles lies within the fine sand fraction. Within the size range represented by this fraction, there is a rapid rise in capillarity with decrease in size, and the rate of capillary rise reaches its maximum (Table 45). These factors, and their important ecological effects, are the reason for distinguishing between the coarser and finer parts of the fine sand fraction; the usual dividing line corresponds to a particle diameter of 0·06 mm.

Ecologically important phenomena such as solifluction and frost heaving occur to some extent in the coarser fraction of fine sand, but are more apparent in fractions finer than this. Solifluction is soil movement brought about by large amounts of water in the soil: it occurs only when the soil particle size is such that high water-holding capacity is associated with high capillarity and a rapid rate of capillary rise. These properties are especially characteristic of the finest sand and silt. In conditions of abundant water supply, in periods of high rainfall, or when snow and ice are melting, so much water is taken up by these soils that on slopes they become unstable and slide downwards, so that the vegetation is disturbed or destroyed. Trees are displaced, and their trunks are crooked (Tamm, 1940). Dense vegetation lessens the risk of solifluction, because the plants reduce the water supply to the mineral soil. The risk increases with increasing precipitation. Solifluction is most common in cold and arctic regions, where conditions so favour it that other types of soils may also be affected. In more temperate regions it occurs only in places where the vegetation has been removed and where the mineral soil is mostly fine sand, silt or clay. Such soils pose difficult problems agriculturally since drains, either open or closed, are prone to silt up as soil tends to flow in from the sides. (See also page 291.)

In soils subject to solifluction, the plants are also more liable to be affected by frost heaving than on other soils. Frost heaving is caused by ice formation deep in the soil so that the more superficial parts are lifted upwards. It occurs only if ice formation starts deep enough, and in places where the water supply is sufficient for formation of enough ice to lift up the soil lying over it. There may be a continuous layer of ice, which lifts up the soil above it because of the increase in volume of water on freezing. Alternatively, ice formation may be concentrated at various points, lying close to each other, from which prisms of ice grow up, often to a height of several centimetres. These crystals together form a layer of ice with a spiny and uneven surface, on which soil is held. The surface moves upwards as the ice prisms grow, and the soil is lifted as a result. Both these types of ice formation have a water supply from below, through capillary rise of water, which adds to the layer of ice. In the finest sand, in silt and in the coarsest clays, the capillary forces and the rate of capillary rise are large enough to allow rapid formation of thick layers of ice.

Frost heaving is most frequent in places with a high water table; it can therefore be prevented by drainage. It is more common in soils with sparse vegetation and a thin humus layer than in soils supporting a dense vegetation, since vegetation and humus decrease heat loss from the mineral soil.

Finest sand. The finest sand fraction (Tables 44, 45) is characterised ecologically by its liability to frost heaving and solifluction. However, its water-holding capacity and rate of weathering are sufficiently high to support coniferous forest, for example, and it forms a good forest soil. Pine heaths in northern Sweden are often on this type of soil. The soil surface is rough as a result of frost heaving (Tamm, 1940), which is the more common because the bottom layer of the vegetation, and the humus layer, are sparse.

Silt (Tables 44, 45) is more prone to solifluction and frost heaving than fine sand. Its water-holding capacity is large, often too large for plants such as forest trees. Silt soils tend to be cold, but often a high proportion of arable land is silt, for example along the coast of northern Sweden.

Coarse, or light clay is like silt in its physical characteristics (Tables 44, 45). As arable land, it is difficult to drain because the soil blocks the drains so easily, and natural drainage by crack formation does not occur to any large extent. Cracks result from the swelling and subsequent drying out of the material.

Fine, or heavy clay has a high proportion of colloidal particles, to which large amounts of water are bound. When the soil dries out, a system of cracks develops, and these cracks often remain even after the water supply has been renewed. These stable cracks act as a natural drainage system. Water uptake is slow, but because of the large capacity to store water, both as colloidally-bound water and as capillary water, clays provide a relatively constant supply of water. Since chemical weathering is rapid, there is a high nutrient content, and high productivity.

Quaternary mineral soils

As the ice retreated, the material it had ground up and transported was deposited on land and in water. In previously glaciated areas, the major part of the mineral soil material which now makes up the soil solum, together with organic soil materials, was derived in this way.

Moraine soils, or till

The unsorted material, till, which was deposited on land and on the bare bedrock, consisted of uneven, angular particles, and formed layers of soil between *c.* 0·5 m and 50 m thick. Depending on the fraction most in evidence, stony, gravelly, sandy, silty or clayey till can be distinguished, and the soils made up of a mixture of till and organic material can be categorised in the same way. Examples of the composition of some of these soils are given in Table 46 and Fig. 54.

Till soils are common in formerly glaciated areas: they cover about three-quarters of the surface of Sweden (Magnusson *et al.*, 1957). The value of till soils depends mainly on their chemical composition, particle size, and hardness.

Hardness depends on the conditions of deposition. The material on the surface or in the upper part of the ice was deposited without being compressed to any great extent and now has a loose structure favourable for plant growth, whereas material towards the bottom

TABLE 46 The composition of some typical mineral soils as determined by mechanical analysis (sieving, suspension in water, sedimentation). Values are percentages. (After Ekström, 1951.)

MINERALOGICAL MATERIAL	STONES	COARSE GRAVEL	FINE GRAVEL	COARSE SAND	MEDIUM SAND	FINE SAND	VERY FINE SAND	COARSE SILT	FINE SILT	CLAY
Particle diameter	> 20 mm	20–2 mm	20–2 mm	2–0·2 mm	2–0·2 mm	220–60 μ	60–20 μ	20–2 μ	20–2 μ	< 2 μ
Sorted mineral soils										
Gravel	12	26	31	29	2					
Coarse sand			4	73	21	2				
Medium sand				28	55	16	1			
Fine sand					1	85	12	1	1	
Very fine sand						20	58	17	2	3
Silt						4	30	42	21	3
Fine sandy-clay						33	25	14	9	19
Silty-clay						2	31	35	14	18
Medium clay						5	26	21	10	38
Heavy clay						2	7	19	15	57
Very heavy clay						1	2	3	6	88
Unsorted mineral soils										
Sandy till	27	6	6	11	25	12	4	3	2	5
Fine sandy till	10	8	12	8	13	17	15	11	2	4
Slightly clayey sandy till	14	4	7	10	20	16	8	7	4	10
Light clayey till	7	3	7	11	17	12	9	10	7	17

of the layer of ice was greatly compressed and remains very resistant to penetration by plant roots and soil organisms.

Chemical composition and particle size are largely determined by the nature of the parent material. Bouldery, stony or gravelly tills are derived from the hardest and toughest types of rock and they are therefore found in or near areas of fine-grained granite or of certain sandstones. Less stony, more sandy tills are derived from coarse-grained granite and gneiss. These sandy tills cover large parts of north Sweden and the high ground in south Sweden. All these soils are suitable for forest and are mainly covered by coniferous forest, in the form of pine heath on the dryest and poorest moraines (Magnusson *et al.*, 1957).

Sandy tills with few stones are formed from greenstones, which are rich in bases. To some extent, this compensates ecologically for the lack of fine material. Similar tills are derived from some shales. However, shales (for example, clay shales) may be crushed to a greater degree, to form tills with a high content of silt or clay. These soils are also mainly forested; in north Sweden they usually bear moss-rich spruce forest. Productivity is higher than that of tills with larger particles. If the ground is not too rocky, or topography unsuitable, silt or clay tills are usually cultivated, as in parts of south Sweden. Some of the resulting arable soils contain many small stones.

Clayey and lime-containing tills with few stones are formed from shales and limestones. Clay till is ecologically favourable in several respects, because of the suitable proportion of clay to coarser material and because of the high base content. It is nutrient-rich, has a high base-exchange capacity, is not subject to soil slip like silts are, and is warmer and better drained than the sedimented clays. Capillary forces are large (Table 45) and the water-holding capacity is high. The hygroscopicity (the amount of adsorbed bound water, as a percentage of the volume) is *c.* $1 \cdot 0$ for gravelly till; $1 \cdot 2$ for sandy till; $1 \cdot 7$ for fine sandy till; $2 \cdot 1$ for silty till; $3 \cdot 4$ for clayey till and $3 \cdot 5$ to $10 \cdot 0$ for pure clay till (Magnusson *et al.*, 1957).

Clayey tills are usually cultivated, and they form some of Sweden's most fertile agricultural land, such as the extreme south (south Skåne), the islands of Öland and Gotland, etc. In southwestern Skåne, the clay content is so high that the clay tills are medium or heavy clays, whereas others are usually light clays. Figure 54 shows the particle sizes characteristic of the various tills.

In the post-glacial period, some tills have been subject to wave action by water, causing some sorting of the particles. These include coastal tills which were formerly water-covered but are now dry, because the land has risen. They form a coastal belt of variable width, situated below the highest former coast line (which is

200–295 m above presentday sea level in north Sweden and *c*. 30 m in south Sweden). Tills at the edges of lakes dammed by the ice have also been subject to wave action. The finer particles were carried away from beaches exposed to waves and deposited in deeper water, to form mineral soils of various types, gravel, sand and clay. This so-called post-glacial clay is not stratified like glacial clay, and it contains fossils and organic material. The coarser material has been left near the beach as shingle. The remaining uppermost layer contains many boulders and stones, and is ecologically very different from the original material. It is also different from till with stones and boulders pushed to the surface by frost heaving, which has different properties.

Fluvioglacial sediments

Stones, sand and other materials carried by the glacial rivers were sedimented in relation to the size of the particles and the rate of flow of the water, and formed 'sorted' soils (Table 46).

The coarsest material was deposited at the mouths of the rivers, as ridges of rounded and polished particles and stones. These soils have a low water-holding capacity and a low content of dissolved nutrients. Except where they have a high lime content, they are dry, poor soils, usually covered by coniferous forest, mostly pine.

Clay and other fine material was retained in the water and deposited when the river water flowed more slowly or stopped flowing altogether, in places such as bays of the sea or in lakes. In any one place, the particle size and organic content of the clay varied with the season of the year, so that the sediments are stratified: glacial clay or silt deposited in the sea or in lakes is therefore known as stratified clay or stratified silt (varves). Because of the release of water from lakes dammed by ice, and because of the rise in land level, many of these lake bottoms or bays of the sea are now dry ground, forming many of the presentday clay plains. Glacial clay is quantitatively the most important of the Swedish clays. There are coarser (light) and finer (heavy) types. When material was derived from hard rocks, thin layers of silts and light clays were deposited. The clay shales, on the other hand, gave rise to heavy clays in thick (2–5 cm) annual layers. Clay marls are derived from material containing limestone. Because they are nutrient-rich, have a high water-holding capacity and other favourable properties, they are now used as arable land.

As the land rose, both the ridges of rounded stones and the stratified clays were affected by wave action, in those areas which lay beneath the coast line. Sand and gravel was carried away from the exposed parts of the ridges and deposited further out to sea, forming sandy areas which now bear poor pine heaths or other types of heath. However, if there is a good supply of spring water, a more mesophytic type of plant community may develop. The stratified glacial clay which was carried away was re-deposited as post-glacial clay, like the loosened and redeposited clay till. This often forms a layer on top of the original glacial clay, tills, or sand deposits, and may, in turn, be covered by even younger clays and sands.

Alluvial soils

Rivers and other flowing water erode the land, and the finer material is carried away and deposited where the rate of flow slows down, e.g. in broad valleys, in lakes, and at the mouths of rivers, as delta soils. Parts of the river bottoms are covered by the material, which builds up an even surface. Along the lower reaches of the river there are now flat areas in many places, forming terraces above the presentday water line. Erosion continued as the land rose, and the rivers cut their way down through their own former sediments, often forming steep cliffs along the water's edge. The terraces may project into the water as tongues of land. The soil on the river terraces is mainly sand and fine sand, sometimes with some humus derived from underwater litter, and now bears forest or meadow. The fertility mostly depends on the particle size, and the finest soils are cultivated. The buildings along river valleys are often concentrated on the river terraces.

Windborne soils

Windborne particles of silt or sand have accumulated to form plains, broken here and there by rounded dunes. Loess soils are thought to be of Aeolian (windborne) origin: they were probably derived from surrounding areas of desert or from moraines exposed to winds blowing off the ice-covered land.

Blown sand is generally found in places where the mineral soil is already sorted, such as on beaches of the sea, rivers or lakes. The alluvial material on the beaches is still mixed to some extent (Table 46). The wind brings about more precise sorting, and the Aeolian fractions are therefore more uniform and consequently looser and more mobile than other soils. Mobile blown sand can be stabilised by covering it with small branches or other material which decreases the effect of the wind and allows vegetation to develop; or suitable species can be planted. Such sand forms an extremely dry and poor soil, supporting desert or heath vegetation.

Chemical weathering of the minerals

The chemistry of weathering and the products of weathering

The mineral salts which are plant nutrients are derived from minerals brought into solution by atmospheric influences and by the influence of plants themselves.

In general, the minerals are the salts of strong bases (K, Na, Ca, Mg, etc.) and weak acids (silicic acid, aluminium silicic acid, carbonic acid). Solubility is therefore low and variable. The crystals are eroded by water and other substances in the soil so that some of the components dissolve and some remain undissolved, or form colloidal systems. This process of solution of the mineral particles is called *weathering*.

The effect of water on felspar may be cited as an example of chemical weathering. The hydrogen ions in the water replace the potassium ions in the felspar crystal, the potassium ions go into solution, the molecule breaks up and some of the silicic acid is released, and the rest of the molecule forms kaolinite (aluminium silicic acid) which remains undissolved.

$$\underset{\text{felspar}}{KAlSi_3O_8} + \underset{\substack{\text{from} \\ \text{water}}}{\underbrace{OH^- + H^+}} \longrightarrow \underset{\text{kaolinite}}{H_4Al_2Si_2O_9} + K^+ + OH^- + SiO_2.$$

Finely divided silicate material, dispersed in water, therefore has a weakly alkaline reaction.

This type of weathering is a hydrolysis. If an acid is present the following reaction may proceed:

$$\underset{\text{felspar}}{KAlSi_3O_8} + 2H^+ + CO_3^{--} \longrightarrow \underset{\text{kaolinite}}{H_4Al_2Si_2O_9} + 2K^+ + CO_3^{--} + SiO_2.$$

If the mineral is a carbonate:

$$\underset{\text{calcite}}{CaCO_3} + 2H^+ + CO_3^{--} \longrightarrow Ca(HCO_3)_2.$$

It is hydrogen ions which are active and which replace the cations in the mineral. In the first type of weathering the hydrogen ions come from water: in the second from carbonic acid. All acids and other substances in the soil which can act as donors of hydrogen ions contribute to chemical weathering. In addition, weathering is affected by numerous other factors, particularly the type of mineral, the particle size and the temperature.

Factors affecting chemical weathering

1 THE TYPE OF MINERAL. Minerals vary in their resistance to weathering. Tamm (1940) groups them as follows: the most readily soluble are limestones consisting of calcite, apatite and phosphorite; next come the greenstones, greenstone shales and clay shales, which contain calcium-rich plagioclases; then come granites, gneisses and porphyrites, and mica-rich mica schists, all of which contain potassium felspar and sodium-rich plagioclases; then porphytes, leptites and mica-poor mica schists; the most resistant are quartz-rich sandstones and quartzites. However, this sequence is only valid for particles larger than the clay fraction. The differences become less as the particles get smaller, because the tendency to weather is related to the total surface area. Table 44 shows the relationship between the diameter of the particles and their surface area. For particles as small as clay, the surface effect is dominant in affecting the rate of solution, so that weathering is more or less independent of the nature of the material. Gravels and sands generally weather slowly, forming poor types of soil, whereas clays weather more rapidly and are therefore nutrient-rich.

The significance of particle size to plants is sometimes reflected by the relation between distribution of a species on a particular soil type and the size of the soil particles. Köie (1951) investigated a number of Danish oak woods and showed that oak was confined to soils with a certain minimum content of silt or clay particles below 0·02 mm diameter. Eighty per cent of the oaks grew on soils with 4 to 12 per cent of silt and clay particles. Similarly, there were no stands of beech on soils containing less than 8 per cent of silt or clay particles; and there were no *Corylus* or *Crataegus*, and usually no *Malus sylvestris*, *Tilia cordata* or *Viburnum opulus*, on soils with less than 4 per cent silt or clay. The numbers of these trees increased with the content of small particles in the soil.

2 TEMPERATURE. Like other chemical processes, weathering is affected by temperature. It is more rapid in hot climates, and more rapid in summer than in winter. The soils of polar regions are only insignificantly affected by weathering, whereas tropical soils are changed considerably, down to depths of several tens of metres. This difference is a consequence not only of differences in temperature but also of differences in age of the soils. In comparison with tropical soils, which were unaffected by the last glaciation, soils of the polar regions and all other glacial soils are young and relatively unaffected by weathering. Viro (1953) calculated that the rate of weathering in Finland is equivalent to the loss of a layer of mineral soil 3 mm thick since the end of the last glaciation.

3 THE PRODUCTION OF ACIDS IN THE SOIL. Several types of acids play a part in weathering.

(a) *The contents of acids and bases in the soil minerals.* Acid residues are formed as a result of weathering, in amounts depending on the composition of the parent minerals, some of which have more acid constituents than others (Table 47). Silicic acid is frequently released, and makes up as much as 25 per cent of the products of weathering in forest soils (Tamm, 1920). In addition, other less acid compounds are formed by other processes, e.g. SO_2 from combustion of coal or smelting of minerals.

(b) *Roots and rhizoids.* Roots and rhizoids produce carbon dioxide and other acids, such as malic acid; these are particularly effective in accelerating weathering, because the roots and root hairs are in intimate contact with the soil particles. The thin, elastic and flexible walls of the root hairs lie alongside the mineral surfaces so that ion exchange can take place directly, and not via the soil

TABLE 47 The content of acid constituents in
various rocks (from Sörlin, 1943.)

	PER CENT SiO_2
Quartzite, sparagnite	85–80 acidic
Granite, gneiss	80–65
Diorite, amphibolite	65–52
Gabbro, diabase	52–42
Olivine stones, serpentine rocks	45–35 basic

solution. The acids are undiluted at the points of contact so that the mineral surface is attacked by the highest possible concentrations; the acids are not immediately neutralised by the base cations in the soil solution, because of the proximity of the surfaces. This is particularly important to plants growing in neutral or alkaline soils, since the acid root exudates can act on the soil minerals and increase the supply of mineral salts, in spite of the large amounts of bases in the surrounding soil solution.

The effect is further enhanced by the acids produced by the microorganisms in the rhizosphere. The concentration of fungal hyphae is greater at the root surfaces than elsewhere in the soil, and the numbers of bacteria and fungi may be many (seven to eighty-one) times higher there (literature cited by Lundegårdh, 1949). The microorganisms can make use of the various substances secreted from the roots, such as phosphatides (Melin, 1924), thiamine, biotin and amino acids (West and Lochhead, 1940).

There is thus a kind of symbiosis between the soil organisms and the higher plants in accelerating the rate of chemical weathering.

Lichens on rocks, and the mosses which may succeed them, are also in direct contact with the mineral substratum by means of their rhizoids. However, little is known of the extent to which the lichen acids and the other acid substances produced by these plants take part in the process of solution of the minerals.

(c) *Microorganisms.* Acid production by microorganisms was discussed in Chapter 7. Acid products include carbon dioxide, nitric acid, sulphuric acid and a number of organic acids, such as fumaric, acetic, malic, and benzoic acids, together with humic acids. Lundegårdh (1949) calculated that between two-thirds and three-quarters of the soil respiration in a crop-bearing arable soil is attributable to microorganisms.

(d) *Basic and acidic buffers in the litter.* Litter contains considerable amounts of basic and acidic buffers. Acids produced in moist soils are dissolved or dispersed in the soil solution. These acids accelerate weathering only if they remain unneutralised. However, there is some neutralisation by bases released from the minerals and from humification processes, so that the nature of the litter has an effect on weathering. Some litters have more acid constituents than others (see Tables 17 and 18, pages 85 and 86). An acid soil solution, effective in accelerating weathering, tends to be derived from acid litters, and vice versa. Hesselman (1925) pulverised and made extracts from fresh leaf litter of various species, and measured the pH and the buffer content. The basic buffer content was determined by measuring the change in pH after the addition of increasing amounts of an acid to a suspension of the material in water: the content of acidic buffers was determined similarly, by adding an alkali (Fig. 24, page 84. Also see page 83 for discussion of the term 'buffer').

From titration curves similar to those in Fig. 24, it can be seen that the litter of different species has different degrees of acidity, both actual and potential. These different types of litter are characteristic of certain plant communities. The litter from the trees, dwarf shrubs and mosses typical of coniferous forests has a high content of acidic buffer substances, and so the soil solution has a low pH value (Table 48). These acids accelerate weathering, but humification tends to produce mor (see Chapter 10). However, the low content of basic buffers in the mor does not affect the mor species to any great extent, since it is compensated by the effective weathering of the minerals.

In contrast, litter from broad-leaved deciduous trees and herbs has a high content of bases (Tables 49 and 50) and mull and brown earths develop. Other plants, e.g. sycamore, oak and larch, produce

TABLE 48 Types of litter with a high content of acid buffers and a low content of basic buffers. The acidity of extracts from the litter is listed. These are all species of coniferous woodland and form raw humus. (After Hesselman, 1926.)

	pH		pH
Spruce	3·8–4·2	*Vacc. vitis-idaea*	3·7–3·8
Pine	4·0–4·2	*Empetrum*	4·2
Juniper	3·8	*Calluna*	4·4
		Vacc. myrtillus	4·0–4·5
		Vacc. uliginosum	4·7

TABLE 49 Types of litter with a moderate content of acid buffers and a fairly high content of basic buffers. The acidity of extracts from the litter is listed. These are generally species of deciduous woodland and meadows. They generally form mull; there is little or no raw humus formation. (After Hesselman, 1926.)

	pH		pH
Birch	5·0–6·1	*Hylocomium* spp.	4·6
Alder	4·6–6·3	*Maianthemum bifolium*	5·6
Aspen	5·3–6·1	*Trientalis europaea*	6·1
Willow	5·6–6·1	*Athyrium filis-femina*	5·9
Beech	5·3–6·6	*Dryopteris* spp.	5·5–6·3
		Melica uniflora	6·0
		Mercurialis perennis	7·4

TABLE 50 Types of litter with a low content of acid buffers and high content of basic buffers. Generally plants of deciduous woodland, forming mull. (After Hesselman, 1926.)

	pH		pH
Elm	7·3	*Mulgedium alpinum*	6·9
Hazel	6·6	*Stachys sylvatica*	6·?

litter with a high content of both acidic and basic buffers or with a low content of both, e.g. *Deschampsia flexuosa* (Hesselman, 1926).

According to Blomfield (1954), aqueous extracts from litter tend to dissolve both iron and aluminium compounds by formation of organic complexes and reduction of iron to the ferrous state.

4 SOIL ANIMALS. Soil animals may also accelerate weathering. The mineral particles are worn down in the alimentary canal, particularly in earthworms, and cleaned of the products of weathering, such as silicic acid, which otherwise remain on the surface and prevent contact between mineral and weathering agent.

The ecological significance of chemical mineral weathering

This is twofold. The plant nutrients, K^+, Ca^{++}, Mg^{++}, SO_4^{--} etc., are dissolved out of the minerals; and weathering contributes to soil formation, both by the particles formed and by the hydrated oxides ($Al_2O_3 . n\ H_2O$, $SiO_2 . n\ H_2O$, $Fe_2O_3 . n\ H_2O$) and other substances which are precipitated in the soil directly or are released and dispersed in the soil solution, carried to lakes or the sea, and precipitated and sedimented there.

12

Soil colloids

Types of colloids

As a result of humification processes and physical and chemical weathering, particles of very different sizes, ranging from boulders and stones down to molecules and ions, are formed (Table 44, page 238). The finest particles are in true solution and the coarser ones may form suspensions in water. The colloids are intermediate in size, in some respects behaving as solutions, in others as suspensions. There are no distinct boundaries between these categories. Colloidal particles generally have diameters of 0·1 to 0·001 μm. Inorganic colloids are products of weathering; and organic colloids are products of humification.

Some inorganic colloids are formed by mechanical weathering, since this is sometimes intensive enough to grind the minerals into particles of colloidal dimensions. These are called primary mineral colloids, and the particles are fragments of minerals, crystals or parts of crystals, mostly silicates, like the minerals themselves, and usually complex iron aluminium silicates in combination with various bases.

Other mineral colloids are made up of particles which are products of chemical weathering, combined in groups, perhaps because of the influence of some of the compounds in true solution. These are called secondary colloids. Some of them are deposited in sediments, forming secondary minerals (clay minerals). Their main constituents are silicic acid $(SiO_2 . n\ H_2O)$ and hydroxides of iron and aluminium $(Fe_2O_3 . n\ H_2O;\ Al_2O_3 . n\ H_2O)$ or of manganese and other heavy metals. At least at first, the structure is amorphous.

Clays, particularly the finest fraction, are mostly mineral colloids. Because of their important effects on many chemical, physical and biological processes, even small amounts of clay colloids affect soil properties.

The organic soil colloids are mainly stable humus, with colloidal properties because of the large size and complexity of the molecules. In some organic soils, e.g. highly humified peat, the humus colloids may be the most important constituents. The properties of peat and other humus-rich soils are determined primarily by the humus colloids.

Other soil colloids are formed by association of organic and inorganic particles into large complex colloids.

The pedological and ecological importance of the soil colloids is related to their enormous surface area (see Table 44). All processes which involve surface reactions are more rapid in soils with a high colloid content. The electrical charges, dispersion, flocculation and adsorption properties associated with colloids (described below), have effects of fundamental ecological significance. The retention of water and nutrient salts by the soil, the soil structure and the processes affected by soil structure, such as gas exchange, water turnover and biological activity, are properties largely dependent on colloids.

The charge, dispersion and coagulation of colloids

Acidoids, basoids, ampholytoids

Colloidal soil particles are usually made up of several molecules, each comprising several elements, together with molecules and ions adsorbed by the particles from the surrounding solution. When in contact with water, the particles release cations or anions, or both, and thereby acquire an equivalent charge. Ions attracted by the charged particles form a shell around them, so that the particles are said to be surrounded by an electrical double layer. The diagram in Fig. 60 shows the situation in which the shell is made up of cations.

Colloidal particles which release only cations are negatively charged and may be regarded as giant anions. The amount of the charge varies in relation to the numbers and valencies of the cations. Because of the electrostatic attraction, the cations are bound to the colloidal anions and cannot diffuse away, although other more distant ions are free (Fig. 60). If the positive and negative charges were exactly balanced, the particles and their ionic shells would be units with no net charge. However, the cations cannot balance the negative charge of the colloidal anions, because of the distance between them, so the particles behave as charged particles, although the charge is less than if there were no ionic shells. Colloidal particles with the same charge repel each other and are dispersed throughout the solvent. Then the colloid is in the sol state (from

Latin *solutus*—unbound, free), as is a suspension of clay. The finest clay particles stay suspended in water as long as they remain charged.

Soil colloids fall into three groups in respect of the charges they carry.

The first group, the acidoids, are colloids from which mainly cations are released. Humic acids and silicic acid are of this type.

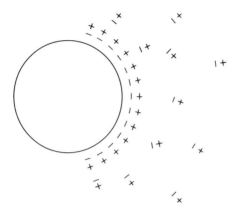

FIG 60. Diagram of a negatively charged colloidal particle, which releases (or attracts) cations, shown on only one side of the particle, which are bound by their attraction to the particle. There are free ions further out in the solution

+ Cations and hydrogen ions
− Anions and hydroxyl ions

The second group, the basoids, are colloids from which anions are released. Aluminium and iron hydroxides in strongly acid solutions are examples.

A third group, ampholytoids, are sources of both anions and cations. The charge on the colloidal particles may be either positive or negative, depending on the proportions of the types of ion released. It changes if the pH of the solution changes, since the proportions are affected by the hydrogen ion concentration of the system. At a particular pH value, the isoelectric point (IP), the ion shells contain equal numbers of positive and negative charges, and the ampholytoid particles are electrically neutral. If the pH of the solution is above that at the isoelectric point, the particles have an excess of hydrogen ions in relation to the solution, and therefore act as acidoids. If the pH is below the isoelectric point, the particles have an excess of hydroxyl ions, and have basic properties.

Because of their acidic and basic properties, the acidoids and basoids cannot generally exist free in the soil, but occur in the form of salts of some kind. Salts are formed between acidoids and mineral cations; between acidoids and basoids (forming ampholytoids);

and between ampholytoids and anions or cations from normal salts, acids or bases. The soil colloids are therefore usually complex colloids. The position of the isoelectric point for these colloids depends on the contents of basic and acidic components. The higher the content of acidic components, the lower is the iso-electric point, and vice versa.

Coagulation and dispersion

Like other reversible colloidal systems, the state of the soil colloids changes from sol to gel and from gel to sol. When coagulation occurs, the particles aggregate into complexes, the size of which varies in relation to the numbers of primary particles involved. In some conditions, these complexes later break up again and the particles disperse. Thus, the system dispersion (sol)—aggregation (gel) may sometimes be reversible. The state of the system, and any changes which take place, is determined by the charge of the colloidal particles, which depends on the properties of the ion shells. These are transitional stages between the two extremes: partially precipitated or partially dispersed phases, dense sols, loose gels, etc.

There are several causes of precipitation, of which the following are probably the most common:

1 When dispersed colloids with opposite charges are mixed, a mutual neutrality results, and double particles are formed, with an acidoid component such as humus or SiO_2 and a basoid component such as Al_2O_3 or FeO_3. The product is an ampholytoid, a clay-humus complex, for example.

2 An external supply of ions may increase or decrease the charge of the colloidal particles, by replacing some of the ions in the ion shells (see page 258 for discussion of ion exchange). Consequently, the state of precipitation or dispersion changes. If, for example, the added ions are more effective neutralisers than the ions they replace, there will be increased precipitation of the colloid. If colloidal clay in fresh water is carried out into salt water, it coagulates into secondary particles which are deposited and form sediments (Låg and Bergseth, 1957).

Differences in effectivity of neutralisation depend mainly on the following characteristics:

(a) The valency of the ions. Trivalent ions are more effective than divalent, which are, in turn, more effective than monovalent.

(b) The dissociation constants of the molecules from which the ions are derived. Ions which are more attracted by the colloid particles displace others from the ion shells, so getting nearer the particles and decreasing their net charge.

(c) The hydration of the ions. The ions attract a certain number of water molecules, which form shells around them and restrict their movements.

The water molecules closest to the ions are bound most strongly, and the attraction decreases with increasing distance from the ions. The degree of hydration seems to be related to the size of the ions, and increases with decreasing ionic radius. The smallest ions generally have the thickest shells of water molecules, and the thicker the shells, the greater the distance between the ions and the colloidal particles. This distance depends both on degree of hydration and on the electrostatic attraction between the particles. The capacity of ions to neutralise is therefore dependent on ion size. However, opinions about the exact nature of the relationship differ.

Among positively charged ions, hydrogen ions are especially effective in bringing about precipitation, probably mainly because they are small in size and relatively weakly hydrated, so that they lie in close proximity to any negatively charged colloidal particles and exert a relatively large neutralising effect.

Calcium is also effective, considerably more so than sodium. Mattson (1931) suggested that the difference between these two was because sodium was dissociated more strongly than calcium, therefore giving the parent particles a relatively greater negative charge. Besides, sodium ions are more hydrated and are pushed further away from the negatively charged particles, so that their neutralising effect is lessened. Hence the production of sodium ions results in a relatively large charge on the colloidal particles, which repel each other strongly so that the colloid is dispersed. Sodium therefore has a high dispersing effect, but a low precipitating effect; calcium is the opposite. Calcium dissociates less readily than sodium and so the particles acquire less charge. Hydration is less, so the hydration shell around the ions is thinner and the ions exert a large neutralising effect. The colloidal particles repel each other much less strongly and the colloid is less dispersed. If calcium ions are supplied, they replace other ions because of their strong electrostatic attraction and tend to bring about precipitation. In general, calcium causes aggregation and precipitation: sodium causes separation and dispersion. Irrigation with sea water, or other water containing large amounts of sodium ions, has unfavourable effects: the soil becomes sticky and viscous, and less permeable to water and to gases. The effect is considerable in semiarid regions, where salt may accumulate in the soil (Eaton and Horton, 1940). Addition of lime to soil rich in colloids causes some precipitation of the colloids and helps in the development of a crumbly, porous soil structure.

In general, cations which are strongly attracted (weakly dissociated) and therefore firmly bound in the ion shells, bring about precipitation and a porous soil structure; and those which are strongly dissociated bring about dispersion, so that the soil becomes dense, impermeable and badly aerated, sticky if wet, hard when dry. The effectivity of various ions in flocculating clay increases in the order, $Na^+ < K^+ < Mg^{++} < Ca^{++}$. This is the exact opposite of the degree of hydration, $Na^+ > K^+ > Mg^{++} > Ca^{++}$. However, for other ions the two sequences are not necessarily exactly opposite.

(d) The neutralising effect of ions also depends on the amounts present. The greater the concentration of any one type of ion, the greater its ability to displace others, even if these have a stronger electrostatic attraction. Calcium is present in the soil in considerably higher concentrations than the other cations, so that its importance in determining the state of the soil colloids is attributable to its high concentration as well as to its strong electrostatic attraction and low hydration (Lundblad, 1924).

The state of the soil colloids is affected by the plant community as well as by the mineral composition of the soil, because different plants take up minerals in different proportions, and there are different amounts of bases in the litter produced (see Table 17, page 85). Consequently, the mineral nutrients retained in the plants have an important effect on the colloidal state of the soil and on soil development.

3 The pH of the system governs the state of dispersion of the ampholytoids. At the isoelectric point, the particles have no net charge, and the colloids are in the gel state. Increase or decrease of pH may bring about dispersion, and the particles attain a positive or negative charge, respectively. If the pH changes back to the isoelectric point, the dispersed ampholytoids are precipitated. The pH at the isoelectric point is determined by the relative amounts of acid and basic components in the ampholytoid. The same pH may result in different states of dispersion in different ampholytoids, depending on their composition. Hence, information about the pH value alone gives no indication of the physical state of the soil.

4 Climatic conditions also affect the sol or gel state of the soil colloids, because they affect the salt concentration in the soil. Both freezing and drying out may cause coagulation. In each case, the colloids lose water, and the salt concentration and proportions of different ions change. When the soil thaws, or is rewetted, the colloids may disperse again, since the process is usually reversible. However, this reversibility is only partial: it decreases with time, especially after severe drying out. It is thought that brown coal consists mainly of precipitated humus colloids which have lost

their capacity for dispersion, mainly because they have been subjected to extreme drying out.

The significance of colloids to the soil structure

The sol–gel state of the soil colloids is important in determining the porosity and the coherence of the soil, and therefore permeability to oxygen, carbon dioxide and water, and mechanical resistance to penetration by roots or organisms. The coagulated particles form a network in which the spaces are filled by the hydration water of the particles, with the dissociated and dissolved substances contained in it. The water can diffuse in and out of the gel fairly freely, but it is bound by capillary and adsorptive forces and cannot be moved by the force of gravity, though it can be lost by evaporation. Running water cannot penetrate a layer of soil gels, such as fine clay or highly humified peat, and surface water may accumulate.

Since soil structure and the state of the soil colloids are related, all the ecologically important properties dependent on soil structure are affected by the conditions which affect the colloidal state: the amount and proportions of bases, the composition of the plant community, precipitation and evaporation, and frost. Climate has such a great effect on structure and other soil properties that it is possible to classify soil types on this basis (see Chapter 14).

The role of soil colloids as mineral nutrient stores

Ions and molecules are adsorbed by soil particles, i.e. bound to the surfaces of the particles by various forces. The amount of mineral nutrients which can be retained is related to the total surface area of the soil particles. The colloids with the largest surface area (see Table 44) have the highest adsorption capacity, and act as the main nutrient store in the soil. In a comparison of adsorptive capacity in brown earths and raw humus soils, Gorham (1953) found that increase in adsorptive capacity was related to increase in the content of organic matter, mainly humus colloids.

Ion exchange

If a solution of a salt, e.g. $CaCl_2$, percolates through a soil in which the colloidal particles have K^+ ions in the ion shells, the Ca^{++} may be retained in the soil and the K^+ be lost in the filtrate: the Ca^{++} ions replace the K^+ ions in the ion shells around the colloids.

$$\boxed{\begin{array}{c} - \ \text{K}^+ \\ \text{acidoid} \\ - \ \text{K}^+ \end{array}} + \text{CaCl}_2 \longrightarrow \boxed{\text{acidoid} = \text{Ca}^{++}} + \text{KCl}$$

In the same way, other ions around the colloids may be affected by those in solution, and various exchanges may take place.

Cation exchange has an important effect on the nutrient status of the soil and on its physical properties. Hydrogen ions differ considerably from other cations in their effects, and so two types of cation exchange, base exchange and hydrogen ion exchange, may be distinguished. Anion exchange is also important. Some of the phosphoric acid in the soil is probably bound by the colloids after exchange with hydroxyl ions.

The exchange processes are mainly based on the same properties of the ions which determine their capacity to neutralise the charged colloidal particles. Valency is important: divalent ions are generally attracted more strongly than monovalent ions. Hydration decreases mobility and consequently the tendency of the ions to replace other ions.

The cations can be arranged in order of their tendency to displace each other from the ion shells, as determined by their valency and degree of hydration. The sequence is broadly similar to that of neutralisation (see page 257) but there are some differences, e.g. hydrogen ions have a greater exchange capacity than calcium and magnesium ions. For the most common plant nutrients the sequence is: $H^+ > Ca^{++} > Mg^{++} > K^+ > Na^+$; and for monovalent ions it is: $H^+ > Cs^+ > Rb^+ > NH_4^+ > K^+ > Na^+ > Li^+$ (Jenny and Gieseking, 1936).

However, a particular ion may not always occupy the same place in the sequence, and the order may possibly be affected by the properties of the colloids.

The amounts of ions are also important. Using the same example of ion exchange: if the $CaCl_2$ is replaced by KCl, the process of ion exchange reverses, and potassium replaces calcium. The weaker exchange capacity of the potassium is compensated for by its greater concentration. Heavy fertiliser application with potassium may cause leaching of other cations, calcium or iron, for example, which are lost from the uppermost layers of the soil and accumulate deeper down (Lundblad, 1952).

The consequences of ion exchange have long been known to agriculture. Heavy and repeated applications of lime drive out not only potassium but also other nutrients from the colloids, even if these are bound so strongly that very little, or none, is taken up by plant roots. Because the calcium displaces these other ions, they become available to plants; some of the favourable effects of liming are attributable to this. However, the effect is only temporary, since it involves depletion of the nutrient reserves of the soil.

In the same way, repeated irrigation with sea water, or flooding of coastal areas, may cause loss of nutrients from the soil and replacement by the salts found in sea water. Sea water contains all the mineral nutrients required by plants, but the amounts and proportions are unfavourable to most land plants. There is too much sodium, and the irrigated or flooded soil has too high a concentration of sodium chloride to be tolerated by most mesophytic vegetation. If the soil is leached by rain water and then fertilised with calcium and cations of other bases, it reverts to its original state. These reversible changes can be represented in the form of a diagram showing the effect of the sodium chloride:

$$
\boxed{\text{acidoid}}
\begin{aligned}
&= \quad Ca^{++} \\
&- \quad K^{+} \\
&\equiv \quad Fe^{+++} \quad + NaCl \quad \rightleftharpoons \\
&= \quad Mg^{++}
\end{aligned}
$$

$$
\boxed{\text{acidoid}}
\begin{aligned}
&= \quad Na^{+}Na^{+} \\
&- \quad Na^{+} \\
&\equiv \quad Na^{+}Na^{+}Na^{+} \quad + Ca^{++}, Mg^{++}, Fe^{+++}, K^{+}, Cl^{-} \\
&= \quad Na^{+}Na^{+}
\end{aligned}
$$

To summarise: if a large quantity of a salt is added, or if small quantities are supplied repeatedly, the total exchange capacity of the colloid is used up, and the colloid is saturated. In general, the organic soil colloids have a higher exchange capacity than the inorganic ones, especially if exchange capacity is expressed per unit weight of soil. A colloid may be potassium-saturated, with a mixture of basic cations. If there are hydrogen ions in the ion shells, the colloid is only partially base-saturated, and if all the cations are hydrogen ions, the colloid is unsaturated. The degree and type of base-saturation determines the plant mineral nutrient economy.

The buffer capacity of the colloids

Partially base-saturated or unsaturated colloids, in which the ion shells contain mainly or wholly hydrogen ions, are acids and behave as acidoids, but they exert no immediate effect on the pH of the medium, since the hydrogen ions are electrostatically bound to the particles. If the hydrogen ions are displaced from the ion shells by the addition of a salt, the pH of the solution decreases, because the hydrogen ions, with the anions of the salt, form a free acid. This is called exchange acidity:

$$
\boxed{\text{acidoid}}\begin{array}{l} -\ H^+ \\ -\ H^+ \\ -\ H^+ \\ -\ H^+ \end{array} + CaCl_2 \;\rightleftharpoons\; \boxed{\text{acidoid}}\begin{array}{l} =\ Ca^{++} \\[1em] =\ Ca^{++} \end{array} + HCl
$$

If a base is the source of the cations, there is no change in pH:

$$
\boxed{\text{acidoid}}\begin{array}{l} -\ H^+ \\ -\ H^+ \\ -\ H^+ \\ -\ H^+ \end{array} + Ca(OH)_2 \quad \boxed{\text{acidoid}}\begin{array}{l} =\ Ca^{++} \\[1em] =\ Ca^{++} \end{array} + H_2O
$$

When Ca^{++} ions from calcium hydroxide replace hydrogen ions in the ion shells, and the hydrogen ions are then removed from the solution in some way, the acidoids release more hydrogen ions, so that it is possible for more calcium hydroxide to be added and used up. The acidoids are thereby acting as buffers towards the neutralising hydroxide. However, if the supply of hydroxide continues, the acidoids' reserves of hydrogen ions are finally exhausted: the whole of their exchange capacity for hydrogen ions is utilised by the calcium ions and they become base-saturated. The amount of exchangeable hydrogen can be determined by adding known amounts of a base and measuring the buffer capacity of the colloids towards the base supplied. However, the total exchange capacity of the colloids or the soil cannot be measured in this way, because the colloids may originally have been partly base-saturated. If the total base capacity or exchange capacity is to be determined, any bases must first be removed from the colloids by electrodialysis. This ensures replacement of all the cations and anions of the colloids by hydrogen and hydroxyl ions. Then a base of known concentration is added to the hydrogen-saturated colloids, and the pH of the soil suspension is measured at the same time. The figures can be treated in the same way as shown previously (in relation to measurements of the buffer capacity of litter, page 83 and Fig. 24). The basic buffers in the colloids can be determined similarly by titration with an acid. Figure 61 shows examples of such titration curves for three soil types. Peat soil had a greater buffer capacity (used more $Ca(OH)_2$, had a larger content of acidoids) than clay soil; and that of clay soil was greater than that of sandy brown earth. The base exchange capacity was therefore greatest in peat soils, next in clay soil, and least in sandy brown earth; i.e. to reach pH 7, the most time was required for peat, less for clay, and least for the brown earth.

In natural conditions, soils are not usually completely unsaturated, like the electrodialysed samples, but are base-saturated to some extent, so that their capacity to bind added bases is less than that of the samples. The magnitude of this capacity can again be determined by titration.

It is apparent from the foregoing discussion that soil pH is not a measure of base exchange capacity. If, for example, the degree of base saturation of the peat soil and the clay soil in natural conditions was such that both had a pH value of 5, attainment of complete base saturation (i.e. at pH 7) would require the addition of more bases to the peat than to the clay. Even if the peat originally had a pH somewhat higher than the clay, it might still need more bases than the clay to bring it to a state of base saturation.

The buffer capacities of soils towards acids can be compared similarly. In the example shown in Fig. 61, the brown earth was most effective as a buffer towards the added acid, showing that it had a high content of basoids. The clay and peat soils bound only a small amount of the added acid: their titration curves lie directly above that for the solvent itself.

A base-saturated soil is comparable with a salt of a strong base and a weak acid. It has an alkaline reaction, which is especially strong in soils containing large amounts of sodium (alkali soils). Soils that are only partially base-saturated behave like ampholytes; and soils that are wholly unsaturated behave like acids. Unsaturated soils are very poor in nutrients, and have other unfavourable characteristics: the high concentration of hydrogen ions is disadvantageous, or directly harmful, to physiological processes and to the biological processes which take place in the soil, and to soil structure.

As a result of the dynamic equilibrium between the ions in the ion shells around the colloidal particles and in the surrounding solution, the base-saturation of the soil colloids can be disturbed by external influences. The equilibrium is disturbed when cations are taken up by plant roots, from the soil solution or from the ion shells. The ions are taken up from the ion shells by direct exchange between the soil colloids and the plant cells, made possible by the intimate contact between root hairs and soil particles. Hydrogen ions from the plants replace the base cations from the soil colloids.

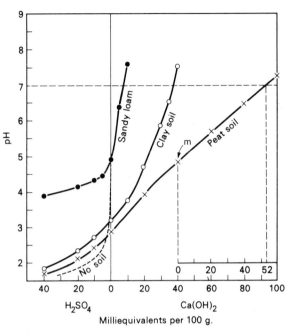

FIG 61. **Titration curves for three electrodialysed soils (saturated with hydrogen and hydroxyl ions), showing how the pH of the soil suspension (the ordinate) is changed by the addition of a base or an acid. The electrodialysed (unsaturated) soil exhibits the total base exchange capacity of the soil. If the soil was not electrodialysed, e.g. if the base content of the peat soil was m (see diagram), it would require only 52 milliequivalents of Ca(OH)$_2$ to raise the pH to 7 instead of the 93 milliequivalents required by the electrodialysed form. (After Mattson, 1946.)**

This is a reversible process: if a plant is planted in electrodialysed soil there may be an exchange between base cations from the plant and hydrogen ions from the soil colloids.

The materials taken up by plants from the soil return in the litter. However, if crops are harvested and removed, or grazed, and no replacements are supplied, the soil colloids slowly become impoverished and acid. As the ions are removed from the system, they are replaced by new ones, as long as the reserves last. Because of the buffer effect, fertiliser treatment with mineral salts may have a relatively small effect on soil acidity and on the composition of the ion shells. However, a rapid change in acidity may result if the added salt affects the exchange acidity. For example, when nitrogen is added in the form of ammonium sulphate, the ammonium ions are taken up and the sulphate ions mostly remain in solution and bind a corresponding number of hydrogen ions. Consequently, there is a relatively sharp rise in acidity, and ammonium sulphate is therefore regarded as an acid nutrient salt. If nitrogen is added as nitrate, the nitrate ions may be taken up more rapidly than the cations, in exchange for CO_3^{--} ions. Consequently, the content of hydroxyl ions in the solution increases and the acidity decreases. Nitrates are therefore basic nutrient salts.

The ionic equilibrium is also disturbed if the solution between the colloidal particles is leached away by rain, as is common in regions with a wet climate. The ions which have the least electrostatic attraction are lost most readily: K^+, Na^+, Mg^{++}, SO_4^{--}, NO_3^- and Cl^-. As a result of leaching, most soils in wet regions are not base-saturated and often have high concentrations of hydrogen ions. If a salt solution is filtered through such soils, the base cations are retained in the soil and the filtrate may contain the acids which have been released. In contrast, soils in arid zones have a high degree of base saturation. A salt solution filters through unchanged, or is changed to another salt as a result of ion exchange.

Soil absorption

The discovery of salt storage in the soil, and of exchange of elements, dates from the middle of the nineteenth century, when Thompson and Way found that soil retained ammonium if a solution containing the ammonium was filtered through it. It was soon shown that this retention was in some way associated with the exchange of equivalent amounts of another substance, and that the soil colloids were involved. The phenomenon was termed *absorption*. The terms *soil absorption* and *absorption capacity* are still used to apply generally to the retention of added plant nutrients and other substances by soil, so that they are protected against leaching. The term comprises not only retention brought about by ion exchange processes, but also that resulting from the formation of insoluble substances, such as calcium sulphate and calcium phosphate, or from the adsorption of volatile substances by soil particles or their solution in the soil water. Ammonia, hydrogen sulphide and other unpleasant-smelling gases produced as a result of fermentation and decay of organic materials are absorbed by the soil, provided that the processes by which they are produced take place at some distance below the soil surface. However, ion exchange by the soil colloids is by far the most important of these processes. Because of their large ion exchange capacity, the soil colloids are the site of the most important reserves of plant nutrients.

Water storage by the soil colloids

Another ecologically important property of the soil colloids is their capacity to take up and bind water. The soil colloids are hydrophilic or lyophilic, so that the individual particles adsorb water molecules. The binding forces are greatest nearest the particles. At some distance from the surface of the particles, the water molecules are free, provided that the colloid is water-saturated. Thus, like the ions of simpler compounds, the colloidal particles are hydrated and every particle is surrounded by a shell of water molecules bound to it.

In dry soil, the hydration forces are expressed as hygroscopy, and the soil absorbs water from the atmosphere, the particles are forced apart by the water, and the colloid swells.

Since the hydration water is bound, it cannot move under the influence of gravity nor be pressed out. However, it may be lost by evaporation. If the water supply is greater than the amount which can be adsorbed, the excess collects between the colloidal particles. The particles may then become dispersed, so the excess water becomes the liquid phase of the colloidal sol; or the particles may remain undispersed and the excess water may then be held by capillary forces. In soil colloids which form large aggregates, the least strongly held capillary water and the free water can be moved by gravity and may be lost.

As discussed in more detail in Chapter 3, there are various categories of soil water: free water, more or less strongly bound capillary water, and more or less strongly bound hydration water. Like capillarity, hydration is a surface process. The greater is the total surface area of the particles, the more water can be bound. Hence, water storage in the soil depends on the colloidal content, and the colloids are the site of the most important reserves of water for plants.

A highly-humified peat can absorb many times its own weight of water, and this water is all colloidally bound. Humus and clay are the soil materials which store most water and which have the greatest resistance to drying out. The water economy of a sandy soil can be improved by the addition of soil materials with a large colloid content. If equal parts of humified peat and sand are mixed, the resulting soil has a water-holding capacity eight times higher than that of the sand (Feustel and Byers, 1936).

13

The physical properties of the soil

Consistency, cohesion, adhesion, plasticity

The term 'soil consistency' is a comprehensive expression covering several rather different properties which are important ecologically or agriculturally: e.g., firmness or mobility, or any intermediate state; coherence and the ability to withstand pressure and tension; and all the characteristics which can be felt, such as softness, hardness, stickiness and crumbliness. These all depend on simpler properties, mainly the following:

1 The size and shape of the particles has an important effect on consistency: gravel and sand in soil confers a different hardness and coherence from sand and clay, while round particles make up a looser, more mobile soil than angular, oblong or discoid particles. The consistency changes as the amount of colloids increases and the soil attains some of the properties characteristic of colloids, i.e. softness and stickiness, with changes in these properties occurring as the sol–gel state changes.

2 Water also has an important effect on soil consistency. In firm soil, the water moves between the immobile particles, but if the water content increases beyond certain limits the particles come into suspension and become mobile. Then, if the right weight and pressure conditions prevail, there may be soil movement, for instance, the finest soil material, especially clay, moves under its own weight if the water supply is sufficient, whereas coarser material moves only if it is subjected to pressure, on slopes, for example, where soils or stones higher up the slope exert a pressure on the soils lower down.

The effect of water on consistency depends not only on the amount of water, but also on the way in which it is held around and between the particles. In the water shells around the colloidal particles, the water is firmly bound by polar forces. Further away from the particles, the molecules move more freely, but they are attracted to each other by cohesion. Baver (1956) points out that the larger particles attract water by adhesive forces and are therefore not in direct contact with each other, being separated by the water, but, at the same time, the water has a binding effect since the particles are held together by the cohesive forces between the water molecules; it is this cohesion that is responsible for the surface tension, which, in turn, if there is sufficient water to form menisci in the angles between the surfaces of adjacent particles, may cause soil particles to move nearer each other. As the menisci change, so does the soil consistency: if the water content decreases, the menisci become more concave, the particles are pulled closer together, and the system becomes firmer and less mobile. By the time all the water is gone, the particles are in contact with one another and held together by mutual attraction, and the soil is at its most solid. If the water content increases, the menisci flatten out and lose their capacity to hold the particles together, so that the soil particles become more mobile. If the water content is high enough, they may move because of their weight or in response to pressure, and finer soils, especially, tend to slip.

In both firm, intermediate and mobile types of soils, the degree of soil coherence brought about by a particular water content is reflected in the soil viscosity, in its toughness, and in its degree of plasticity.

Clay soils may occur in most of the possible states of soil consistency. In dry clay, the particles are in direct contact, with no water in between, and the soil has a high coherence and firm consistency, like concrete. In damp clay, the particles are surrounded by thin sheaths of water, and so they are held together not by mutual attraction, but by that of the water, and they are soft, sticky, tough and plastic, and move under pressure. The finer the clay (i.e. the higher the colloidal content), the narrower are the menisci between the particles, the greater is the force that holds the particles together, and the greater is the plasticity. Very wet clays are sticky and viscous and move in response to their own weight, and the higher the water content, the higher the mobility. Mobility is also related to the colloidal content, since the colloids act as lubricants between the larger particles.

The consistency of the soil is dependent not only on the water content, but also on the size and shape of the particles and on the proportions of colloidal and non-colloidal material. Smaller particles have narrower menisci between them and greater coherence, hence, the coherence of clay is greater than that of sand, and that of sand is greater than that of gravel: stiff clays are the most cohesive of all. Furthermore, since the consistency of the soil depends upon this cohesion and adhesion phenomena within the particle-water system, it is possible to make changes in it; for instance, the coherence of a sandy soil can be increased by the addition of clay or clay can be made less stiff by the addition of sand.

Swelling and shrinkage

Soil swelling due to an increase in the water content is seen as an increase in soil volume, or in pressure, if the volume is prevented from increasing. Drying-out of the soil brings about a corresponding decrease in volume and causes soil shrinkage which, in turn, causes the soil crust to split as a result of tension; it is usually divided into blocks of variable shape by a system of cracks. It is the soils with a high content of colloids, clay soils and peat soils, which are prone to changes of this type. The formation of cracks is harmful to the vegetation, since roots and other underground parts may be broken. However, the cracks have a favourable effect in increasing drainage and permeability, particularly in clay soils. Silt and silt-rich soils shrink less when they dry out, and there are few or no cracks. Such soils are more difficult to drain than clays (Tamm, 1940).

Soil structure

Simple structure, aggregate structure. The conditions for development of soil structure

The term 'soil structure' covers the size, shape and arrangement of the soil particles, the nature of the pore space between the particles, and the proportion of particles to space. Soils differ widely in these respects, and the subject has been given detailed study because of its importance in plant ecology and in agriculture.

Primary particles are the simple particles formed as a result of weathering or precipitation from a solution (colloidal particles): secondary particles are aggregates of two or more primary particles. Primary colloidal particles, which lose their electric charge, stick firmly to each other and form aggregates; other particles clump together with colloids as the binding agent.

Coarse soils, such as sands or gravels, containing only insignificant amounts of colloids, have a simple structure with only simple, primary particles. Finer clay or humus soils usually have a more or less pronounced aggregate structure, termed 'powdery', 'granular', etc., depending on the size of the aggregates. The nature of the structure, particularly the size, stability and arrangement of the aggregates, determines many of the properties of the soil important in plant ecology. The pore system is determined by the structure, and on this depends permeability to water, soil water capacity and aeration, and, hence, productivity. The hardness and resistance to ploughing of arable soil is also related to the aggregate structure. Soil structure may be improved or it may be worsened by cultivation and fertiliser application, depending on the circumstances, so that knowledge of the conditions for formation and stabilisation of aggregates of particles is essential for the solution of many ecological or agricultural problems.

Several edaphic, climatic and biotic conditions encourage aggregate formation. One of the most important is high colloid content, since colloids are the binding agents in the soil. The amount of inorganic colloids (clay) in a particular place is relatively constant and changes only slowly. However, the content of humus colloids may change rapidly in response to vegetational changes, particularly those brought about by man.

The humus colloids and other organic substances may influence aggregation in several ways:

1 The humus colloids themselves may form aggregates, containing calcium humate, for example.

2 The polysaccharide slime which is a product of microbial decomposition of stable humus acts as a cement to hold the particles together.

3 Because the organic matter contains nutrients, the soil is penetrated by fungal hyphae and actinomycetes, which also hold the particles together. This effect persists only as long as the organic matter is relatively unexploited; it is similar to the way in which loose soil particles are held together by plant roots and thus protected from erosion.

4 The most stable aggregates are soil particles held together by clay and humus complexes made up of organo-mineral gels (see also Laatsch, 1954).

The aggregate structure of a soil is therefore mainly dependent on the amounts and types of clay and humus colloids. Clay soils, with only sparse aggregate formation, can be improved by adding humus, because this has a favourable effect on the aggregation of the clay. The more humus produced by the vegetation, the better is the soil structure. Luxuriant plant communities, which produce large amounts of organic matter, usually grow on soils in which there are large numbers of highly specialised aggregates: this is particularly true of grass communities, which have high productivity and which form large quantities of humus colloids.

The productivity of plant communities also depends on the nutrient status of the soil, and this productivity affects both colloid formation and aggregation. Acid soils, poor in bases, are low in productivity and produce little aggregation: one of the causes of this is the leaching to which such soils are subject in wet climates, leading to loss of salts and sometimes of stable humus. Nutrient-rich soils, particularly those rich in lime, produce a large degree of aggregation, probably because of the effect of the base cations in causing precipitation. However, according to Baver (1956), the main cause of aggregation is the increased humus production resulting from the nutrient effect of the ions, in which case the effect of the mineral nutrients is mainly indirect.

The state of precipitation or dispersion of the colloids directly determines the formation or breakdown of the aggregates; any process that changes the status of the colloids in the soil changes the aggregate status. Some of the most important are:

1 Drying-out and rewetting of the soil. This promotes aggregation in several ways: electrolytes and colloids are concentrated by drying out, so that the effect of ions in bringing about precipitation is increased; the colloidal and other particles come into contact and form aggregates, the strongest binder being the clay particles (Baver, 1956). Drying-out causes the colloids to shrink and cracks to form in the soil, leading to the formation of large conglomerates or clods. If these are then wetted by rain, the colloids swell rapidly and unevenly and the tensions which result cause the clods to break up. Also, the water moving into the capillaries compresses the air it displaces so strongly that the clods tend to split. Finally, aggregate formation is promoted by the pressure exerted on the particles as a result of swelling, which tends to bring them into contact with each other.

Soils which never dry out have very little crumb structure. When they are drained for cultivation, aggregation is promoted by the subsequent alternate drying and wetting, and also by the increased oxygen supply, which favours the precipitation of the humus.

2 Alternate freezing and thawing has a similar effect. Ice cystals form as the soil freezes, and particles are pushed towards each other and aggregate. The large clods are broken apart by the growth of ice crystals. After thawing, gaps are left in place of the ice crystals, causing increased porosity. Thus, the loosening effect which autumn ploughing has on arable land is enhanced by frost action on the water in the upturned sods during the winter: if the soil is dry, the ice crystals are too small to exert much pressure; if the soil is wet, particularly if thawing occurs during a rainy period, the aggregates break up and the beneficial effect of freezing is lost. Ice formation also affects aggregate formation in the same way as drying out, in that water is removed from the colloids and other

particles, so that they are drawn closer together at the same time as the electrolyte concentration increases.

The humus content is very important to the stability of the aggregates. Soils poor in humus, some clays, for example, have unstable aggregates which break down as soon as the soil water content becomes high. Increase in the humus content slows the rate of breakdown, so the supply of organic manure to arable clays is important in improving and maintaining soil structure. The climatic factors can only exert a beneficial effect on soil structure if the humus content is sufficient.

3 Organisms affect soil structure in many respects. Roots, underground stems, etc. increase porosity by pushing tunnels through the soil, but at the same time the soil along the walls of the tunnels is compressed. Water uptake by roots, etc. causes local drying out, with consequent effects on aggregate formation. Dead roots form humus, sometimes deep down in the soil, so that aggregation is favoured there, while root secretions probably affect soil structure by changing the chemical environment. Bacteria and fungi also favour aggregate formation, but little is known of the mechanisms involved. Animals affect soil structure in various ways: their holes and tunnels increase aeration and thereby accelerate oxidation and precipitation of the humus colloids. The direct effect of animal digging is to compress some of the soil so that the particles aggregate. The ingestion of soil by worms, particularly earthworms, causes increased aggregation, and the high stability of the aggregates in the faeces is because the particles are cemented together by bacterial mucus (Kollmannsperger, 1951–52).

The effect of arable cultivation on soil structure

The size and form of the particles and the pores in the soil are dependent on the mechanical forces to which the soil is subjected, and are therefore different in arable soils and uncultivated soils. Arable soil is broken up mechanically into pieces, the shapes of which are determined by a three dimensional network: a system of breaks arises along a reticulate pattern of planes of weakness in which adhesive and cohesive forces are low (Andersson, 1954).

Arable soil is broken up in many ways in the course of cultivation. First, the natural vegetation cover is removed and with it one of the most effective promoters of the aggregate structure of the soil. Next, the natural pore system is destroyed and compressed by the plough and other implements, and the soil is laid bare and exposed to the direct effect of rain. Raindrops break up the aggregates on the exposed soil surface, the finer particles are carried down into the pores, and the surface becomes denser and forms a crust after drying-out. Artificial irrigation has similar but more marked effects. Having destroyed the soil structure in this way, the farmer must loosen up the soil time and time again with ploughs and harrows in order to maintain the soil's natural condition.

FIG 62. Stubble field after a harvest of wheat. The ground water is at 1·3 m, but there has been a capillary rise up to the bottom of the ploughed layer. Above this the soil is coarser and the capillary connection is broken. The wheat roots were in contact with the capillary water and despite a very dry summer (1955) the yield was satisfactory. Västergötland.

Cultivation of the soil is beneficial only if the soil water content is neither too high nor too low. For instance, the structure of wet soil may be destroyed because, during cultivation, the aggregates are crushed and the pores are blocked; a high humus content offsets this to a considerable extent. If the soil is too dry, damage results because the larger aggregates crumble.

Cultivation may also destroy soil structure by using up the humus reserves. The plant material which gives rise to humus is removed, and the rate of humification is increased. The structure degenerates and the size and number of aggregates decreases as the humus content declines: manures containing humus delay this, but are usually insufficient to prevent the overall decrease. Mineral fertilisers may also delay it if they bring about increased colloid precipitation, and if they result in an increased amount of roots and stem-bases left in the soil after harvest.

Porosity and permeability

The pore volume or porosity of the soil, i.e. the percentage of the total soil volume which is not taken up by solid particles, consists of pores varying in size from the submicroscopic and microscopic spaces between primary particles in clay and humus to the large pores and tunnels made by roots or burrowing animals.

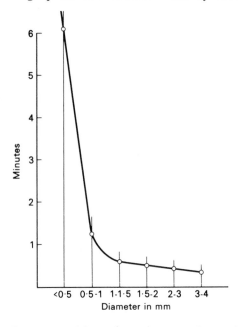

FIG 63. The relation between the size of the humus particles (abcissa) and infiltration time (ordinate). The time for infiltration of water increases rapidly as the particle diameter decreases below 0·5 mm. (After Kosmat, 1956.)

Pores resulting from aggregation of soil particles form a group intermediate in size. The numbers and sizes of the pores vary, because the numbers and sizes of the aggregates, and the state of swelling, depend on the cycles of wetting and drying or of freezing and thawing, on the colloidal status of the soil and its salt content, and, in general, on all the conditions affecting soil structure; this has important ecological effects. The permeability of soil to gases and to water is also determined by the porosity; and porosity is one of the conditions affecting soil compaction and mechanical resistance to penetration by roots and soil organisms.

The pore volume can be determined from:

$$p = 1 - \frac{a}{s} \quad \text{(Tamm, 1940)}$$

where a = the weight of a certain volume of soil, with the gases and water contained in it; and s = specific gravity of the soil particles.

The specific gravity of soil particles is similar for similar types of soil: *c.* 2·6 for clays; 2·7 for quartz sand; and 1·4 for humus (Baver, 1956). However, the volume to weight ratio varies according to the prevailing external conditions. Hence, the changes in pore volume are more or less parallel to changes in volume to weight. The pore volume is usually about 30 to 50 per cent. It is larger in sorted than in unsorted soils, since in unsorted soils the smaller particles tend to fill the spaces between the larger ones, e.g. the pore volume of tills is thus less than that of sands and gravels. However, the ecological importance of the pore volume depends not only on the size of the pores, but also on the type. Capillary pores are more important in storing water than non-capillary pores, but the non-capillary pores are more important in gas exchange. Tills have a smaller pore volume than sandy soils, but a larger proportion of the pores are capillary so the capacity for water retention is greater.

In soils with a good crumb structure, the pore volume is mainly determined by the degree of development of the aggregates. Table 51 shows how pore volume, capillary and non-capillary, changes with aggregation. Total porosity increases with increasing aggregation, but some of the capillary pores become so wide that they lose their capillary properties, and capillary porosity may decrease while non-capillary porosity increases. All the newly formed pores may come into the non-capillary category.

The proportion of capillary to non-capillary pores thus depends on the balance between aggregate formation and breakdown and is subject to the influence of environmental conditions. After heavy rain, bare, arable soil develops a dense, hard crust, because the raindrops break up the aggregates and the pores become blocked; trampling by animals has a similar effect. However, in natural communities the soil is protected by the litter and vegetation.

As aggregation increases and capillary pores become non-capillary, the soil permeability to water and gases increases, and so does biological activity; the water-holding capacity, however, decreases (Table 51). In comparison with sandy soils and other soils with a simple structure, clays and other colloid-rich soils have a larger capillary volume and, therefore, a larger water-holding capacity, but they also have a smaller volume of large, non-capillary pores and less good aeration. Sandy soils and cultivated peat soils have many non-capillary pores and are loose, porous and well-aerated, but they have only a small capillary volume. The practice of rolling arable soils of this type converts some of the non-capillary pores to capillary ones. In general, the porosity of arable soils and soils under pasture is less than that of forest soils. Trampling by animals, cultivation, changes in the humus complex, and other disturbances to the process of aggregate formation, prevent the normal development of porosity.

The pores in the soil contain air and other gases, and water. As the soil dries out, some of the water is replaced by air, and some

TABLE 51 The relation of porosity of air-dried soil to the size of the aggregates. (Krause, 1931, page 603.)

Soil properties	DIAMETER OF AGGREGATES (mm)				
	< 0.5	0.5 to 1.0	1.0 to 2.0	2.0 to 3.0	3.0 to 5.0
Porosity,					
total %	47.5	50.0	54.7	59.6	62.6
capillary %	44.8	25.5	25.1	24.5	23.9
non-capillary %	2.7	24.5	29.6	35.1	38.7
O_2 content in soil air %	5.4	18.6	19.3	19.4	—
Nitrate formation (mg N/kg soil)	9.0	19.1	—	34.0	45.8

by soil shrinkage. If water is added, the gases are displaced and the soil swells. In waterlogged soils almost all the air has been displaced. The soil therefore has a solid phase (the solid particles), a liquid phase (water with dissolved and dispersed substances), and a gas phase (air and other gases).

The liquid phase

Many dissolved and dispersed substances are contained in the liquid phase:

1 Substances dissolved in the rain, e.g. traces of sea salts, volcanic gases and gases produced as a result of humification processes or electrical discharges in the atmosphere.

2 Dissolved substances from the soil minerals, mainly the ions of salts.

3 Products of humification, such as mineral salts and the various organic compounds formed as a result of decomposition of cells. The soil water contains:

NH_4^+, K^+, Na^+, Ca^{++}, Mg^{++}, Zn^{++},

NO_2^-, NO_3^-, SO_3^{--}, SO_4^{--}, BO_3^{---}, and others.

Some ions may have two or more valencies:

Fe^{++}, Fe^{+++}; Mn^{++}, Mn^{+++}; Cu^{+}, Cu^{++}.

Others dissociate in stages, such as:

CO_3^{--}, HCO_3^{-}, PO_4^{---}, HPO_4^{--}, $H_2PO_4^{-}$.

4 Organic and inorganic compounds derived from living organisms, e.g. plant exudates, soluble constituents of animal excrement, etc.

These substances are usually present in low concentrations. Troedsson (1952, 1955) measured the following concentrations (mg/litre) of Na, K, Ca, and Mg in soils in central Sweden:

in surface water (6 moraine localities)   19·2 (5·6 to 43·4);
in gravitational water (4 moraine localities)   18·2 (17·2 to 19·5);
in ground water (4 moraine and swamp
   localities)   8·2 (5·4 to 11·0).

The total amount of Na, K, Ca, Mg, Mn, Fe, Al and SiO_2, in six moraine localities, was 19·5 mg/litre (14·2 to 28·2) and 115 mg/litre in one brown earth locality.

The soil water usually contains all the essential plant nutrients in balanced concentrations, so that the ions do not have the poisonous effects which some of them may have if they occur alone. Its composition is broadly similar to that of a balanced nutrient solution.

The gas phase

The gases present in the soil pores are those of the air, oxygen, nitrogen, carbon dioxide and water vapour, and others, more transitory, produced as a result of humification or of fire above or below the soil surface, such as ammonia, hydrogen sulphide and sulphur dioxide. The quantitative composition of the gas mixture below the soil surface is more or less the same as it is in the lower layers of the atmosphere, except for the higher concentration of carbon dioxide in the soil air. Russell and Appleyard (1915) give the following average percentages by volume:

Atmosphere: N_2 79·0, O_2 20·97, CO_2 0·03.
Soil air: N_2 79·2, O_2 20·6, CO_2 0·25.

Carbon dioxide is produced by the oxidation and breakdown of stable humus and by the respiration of the soil organisms. Some enters the gas phase, some the liquid phase, a small amount is assimilated by autotrophic organisms in the soil.

The soil water is relatively rich in carbon dioxide, some of which is given off if the water comes in contact with the atmosphere. A constant stream of carbon dioxide diffuses away from the soil (see Chapter 8, page 144). Carbon dioxide production decreases with depth, because there are fewer organisms, less organic material and less oxygen (Romell, 1932; Lundegårdh, 1949). Nevertheless, carbon dioxide concentration increases with depth and oxygen concentration decreases, because the diffusion paths are longer (Table 52).

Table 52 shows that the carbon dioxide concentration in the soil air is many times (eight to fifty-five) greater than in the atmosphere. The amounts are variable, depending on the season, and on climatic, edaphic and biotic conditions. The carbon dioxide concentration in fertilised arable soil may reach several per cent: the more luxuriant the vegetation, the more carbon dioxide is produced in the soil. The concentration is higher in warm, dry soils than in cold, wet ones, and in summer than in winter.

TABLE 52 The CO_2 content (volume %) of the soil air at different depths. (After Lundegårdh, 1949, page 302.)

	DEPTH (cm)			
	15	30	45	60
Pasture	1·46	—	1·64	—
Arable soil	0·34	—	0·45	—
Beechwood	—	0·33	—	0·39
Sandy soil	0·25	0·31	—	—

Carbon dioxide diffuses from the soil to the air because of the gradient in its partial pressure. Thus, temperature, wind or precipitation do not usually affect this movement to any great extent (Romell, 1922). Diffusion is also dependent on the area of free pore surface (pores unblocked by water), but is relatively independent of the capacity of the pore system, as long as it is above a certain minimum. The diffusion process goes on at about the same rate in soils of varying porosity, with the exception of clay soils, from which the rate is lower (Baver, 1956): a layer of clay in a soil may decrease diffusion to one-hundredth of what it would otherwise be (Romell, 1922).

The carbon dioxide content of the air is normally 0·03 per cent by volume, or 0·3 ml/litre (0·57 mg/litre). Solubility in water is high, many times greater than that of oxygen or nitrogen (Table 53). The high solubility compensates for the low concentration in air, and the amount of dissolved carbon dioxide in water is about the same as that in air, e.g. 0·3 ml CO_2 dissolves in one litre water at 15°C and 760 mm Hg (Table 53). The amount dissolved is dependent on temperature, and the solubility of CO_2 is affected by temperature more than that of oxygen and nitrogen. An aqueous solution contains the undissociated acids and the ions resulting from dissociation. The proportions vary according to the pH of the solution (see Fig. 33, page 140).

Considerably more carbon dioxide dissolves in the soil water than in water above the soil surface, partly because of the dissolved bases in the soil water, which bind the carbon dioxide, partly because of the higher carbon dioxide content in the gas phase in the soil (Table 52). Sea water at 15°C normally contains 150 times more carbon dioxide than air, although only 1 part in 150 is free carbon dioxide (Krogh, 1904). In fresh water containing lime, the content of CO_2 is less: in the Ljusna river in Sweden it is only 30

TABLE 53 The amounts of N_2, O_2 and CO_2 (in volume %, A, reduced to the volume at 0°C) which dissolve in water in contact with air at 760 mm pressure. Relative solubility, B = solubility relative to that of nitrogen. (Calculated from *Handbook of Chemistry and Physics*, 1947, page 1396.)

	VOL. % IN AIR	0°		10°		15°		20°		30°	
		A	B	A	B	A	B	A	B	A	B
N_2	79	1·86	1	1·47	1	1·32	1	1·22	1	1·06	1
O_2	21	1·02	2·1	0·80	2·1	0·73	2	0·65	2	0·55	2
CO_2	0·03	0·051	73	0·036	64	0·031	60	0·026	57	0·020	50

times more than in air; and in the Fyris river 130 times more (Hofman-Bang, 1905).

The high content of dissolved CO_2 is important to the photosynthesis of aquatic plants. If the total amount of CO_2 in water were as low as air, the photosynthesis of aquatic plants would be limited to a small fraction of what it is now, because CO_2 diffusion is about 10 000 times faster in air than in water.

The movement of oxygen in the soil is mainly in a direction opposite to that of carbon dioxide, since most of the oxygen is used in biological processes such as humification and respiration, which produce carbon dioxide. A small amount is used in other ways, e.g. oxidation of ferrous iron, manganese, sulphur, etc. Carbon dioxide production and oxygen uptake by soil, the so-called *soil respiration*, are therefore two opposite movements tending towards equilibrium, initiated by the biological and chemical processes in the soil on the one hand and by the atmosphere on the other. The system can be illustrated diagrammatically:

Atmosphere $\rightleftharpoons$ Gases in the soil air $\rightleftharpoons$ Gases in the soil liquid $\rightleftharpoons$ Chemical and biological processes in soil organisms and soil constituents

A supply of oxygen is essential for the aerobic metabolic processes in the soil. The supply is limited by resistance to diffusion, so, for the same reason that carbon dioxide accumulates in the soil, the oxygen concentration in the soil air is less than that in the atmosphere (Table 54), and decreases with depth. Water is the greatest hindrance to the exchange of oxygen, preventing the movement of gases by filling the pore system of the soil. Diffusion may

TABLE 54 The oxygen content of the soil air (volume %) in a cacao plantation at Rivers Estate, Trinidad. (After Russell, 1952, page 342.)

SOIL DEPTH (cm)	OXYGEN CONTENT %	
	WET SOIL OCT.–JAN.	DRY SOIL FEB.–MAY
10	13·7	20·6
25	12·7	19·8
45	12·2	18·8
90	7·6	17·3
120	7·8	16·4

be reduced 1000 times as a result (Romell, 1922). The finer the soil structure, the more easily the pore system is blocked. In soils where surface water and gravitational water can run away, e.g. tills, gravel and sandy soils, there is a good oxygen supply. Waterlogged soils suffer from lack of oxygen (Romell, 1922), and there is probably some oxygen deficiency in colloid-rich soils with insufficient aggregate formation, such as heavy clay soils, in which the water-holding capacity is high, but there is very little air. Since sand and gravel soils often have a water deficit because of their low water-holding capacity and high diffusion capacity, a satisfactory compromise for plants between water and oxygen economy can be achieved through mixing these soils with heavy clay.

Any tendency to oxygen deficiency is most marked in rainy periods or in autumn, when the soil air is replaced by water, and is least during dry periods and in spring (Table 54).

The properties of waterlogged soil are mostly a consequence of oxygen deficiency: drainage is a long-established practice to counteract the condition in agriculture, and has more recently been applied in forestry, too. The free water drains away, oxygen supply increases, and aerobic metabolic processes are favoured. Oxidation of stable humus leads to better aggregate structure, and then the composition of otherwise undisturbed vegetation on the drained land changes. Plants, such as the stunted spruces seen in bogs, which have grown very slowly because the water content was too high, begin to put out longer shoots, new species invade, and species which could tolerate the high water content gradually disappear. Even species, like birch and Scots pine, with a high tolerance to fluctuations in soil water content, grow better because of the increased nutrient mobilisation brought about by the better oxygen supply.

The oxygen content of the soil water, or of water in lakes or other bodies of water, depends on the relation between temperature and oxygen solubility (Table 53), on the content of organic material, and on the depth and rate of movement of the water. Because oxygen diffuses so much more slowly in water than in air, the uptake of oxygen by water is facilitated by movement of the latter. Still water, particularly if it is rich in organic matter, has very little or no oxygen in it; flowing water is rich in oxygen because there is a large surface area in contact with the air and there is good mixing: narrow, swift-flowing streams take up most oxygen. Troedsson (1955) made the following measurements of oxygen concentration in waters in central and south Sweden (the measurements are not strictly comparable because samples were not taken simultaneously):

Water on swampy ground	0·3 mg/l.
Water from well-drained land	0·5 mg/l.
Water running over a layer of humus	0·1 to 3·1 mg/l.
Water filtering through till with surface stones	0·8 to 4·5 mg/l.

Soil temperature

Soil temperature affects the rate of biological and physico-chemical processes in the soil. The soil temperature, and fluctuation in temperature, depends on the balance between the incoming radiation and other sources of heat supply, and on the conduction of heat and on the depth of its distribution.

Heat supply

Almost all the heat in the soil is derived directly from solar radiation; very little comes from other sources. Respiration of soil organisms, and other exothermic soil processes, is normally insufficient to affect soil temperature, for instance, a significant amount of heat is produced only when very large quantities of organic matter are humified (in compost heaps or heaps of farmyard manure). The annual heat supply from the centre of the earth outwards is also relatively small, *c*. 250 J/cm², equivalent to an increase in temperature of *c*. 1°C at 60 m depth in loose soil.

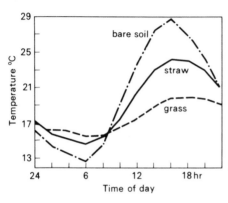

FIG 64. Temperature fluctuations in a quartz sand soil at 10 cm depth throughout one July day. Bare soil; bare soil covered with 0·5 cm chopped straw; soil with a cover of grass. (After Wollny, 1883.)

The heating effect of solar radiation depends on the angle of incidence of the radiation on the soil surface, and, hence, on the height of the sun, the slope of the ground and the latitude. Diurnal and annual rhythmic changes in height of the sun above the horizon cause similar rhythmic changes in heat supply to the soil (Fig. 64). The greatest fluctuations are close to the soil surface; variations decrease with depth and cease altogether at a certain depth, the latter depending on the intensity of radiation and the heat conductivity of the soil (Fig. 65).

The amount of radiation falling on a particular area of soil surface also depends on topography. In hilly areas, ground facing the sun is warmed fastest and to a higher temperature. The flora of south-facing mountain slopes has many special features (see Lundqvist, 1965) which, according to Selander (1955, page 341), can be explained as a result of the high soil temperatures and the mobility of the soils: the temperatures are several degrees higher than in the surrounding areas, and the vegetation period is longer. The ledges on such slopes may suffer from drought in the summer, so summer-flowering species are restricted, but, instead, there are numerous spring-flowering lowland species together with early-flowering mountain species (see Chapter 11, page 236).

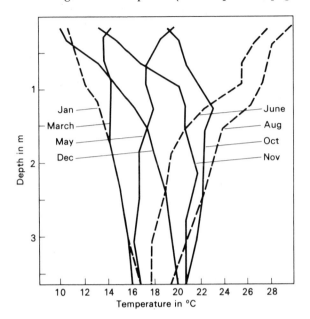

FIG 65. **Monthly means of soil temperature at different depths in bare soil at Davis, California. (After Smith, 1932.)**

Insulating media

There are various media in the atmosphere and on the ground which absorb both incoming and outgoing radiation (Fig. 64). Air absorbs almost all ultraviolet radiation, but most of the visible light and thermal radiation reaches the soil surface, provided that the air is dry and the sky cloudless, since cloud and mist absorb a large proportion of all radiation; water vapour absorbs thermal (longwave) radiation in particular. Hence, the proportion of solar radiation which reaches the earth's surface is greatest in arid climates.

The total annual incoming radiation at Rothamsted (England) is 320 000 J/cm^2 (Russell, 1952), of which *c*. 270 000 J/cm^2 is for the seven months from March to September, inclusive. The latent heat of vaporisation of water is *c*. 2100 J/g water, so that 270 000 J/cm^2 is sufficient for the evaporation of 1270 mm precipitation, or an average of 181 mm per month.

Clouds absorb not only incoming radiation during the day, but also outgoing thermal radiation during the night, so that the risk of night frosts is greater on clear nights than on cloudy ones; differences between day and night temperatures are greatest in clear weather. In autumn, cloud, mist and high humidity act as an insulating sheath towards the outgoing thermal radiation from the soil. Solid particles, such as smoke, also absorb thermal radiation; the practice of protecting fruit blossom against night frosts by creating smoke throughout an orchard is based on this.

Some of the incoming solar radiation is reflected, the rest is absorbed. Bare, black damp soil reflects *c*. 8 per cent of incoming solar radiation on a clear day, about twice as much as the corresponding dry soil. If the soil is covered by vegetation, there is much more reflection: vegetation wet with rain reflects *c*. 24 per cent before the rain has evaporated and *c*. 32 per cent after (Ångström, 1925). Some of the absorbed radiation is used in evaporating water; some is used in photosynthesis; the rest is stored temporarily in the soil, water and vegetation as heat, bringing about a rise in temperature and consequent increase in the outgoing thermal radiation to the atmosphere.

The vegetation exerts a large insulating effect between the atmosphere and the soil. Some of the radiation falling on the plants is reflected, some is absorbed and is reradiated to the atmosphere as thermal radiation, some is stored as heat in the vegetation and the air held within it, some is transferred by conduction and convection to the air passing over the vegetation, and some is utilised as latent heat of vaporisation, in evaporation and transpiration. The proportions of these categories depend on the height and foliage arrangement of the vegetation, and on the rate of transpiration. If transpiration is rapid, large amounts of water are evaporated and a correspondingly large proportion of total energy absorbed is used.

A luxuriant spruce forest in southern Sweden uses an average of 4000 to 7120 kJ/m² in transpiration, the equivalent of about 31 to 56 per cent of the total incident solar radiation (Table 55). If transpiration is rapid, the air is perceptibly cooler over vegetation than over bare ground. In dry periods, when transpiration is only a fraction of the maximum, the energy turnover involved is insignificant, and most of the absorbed energy returns to the atmosphere as outgoing thermal radiation and as sensible heat loss by convection from the vegetation, as the temperature difference between plant and air is large.

Because of the shading effect of the vegetation, the summer temperature in vegetation-covered soil is less than in bare soil, and least under luxuriant vegetation of several layers. Soil temperatures in clear-felled forest areas are higher, and heat penetrates to greater depths than in afforested areas (Homén, 1893; Tamm, 1931, 1940).

The vegetation has a similar effect on diurnal soil temperature changes (Fig. 64). Day temperatures of soil under vegetation are lower than air temperatures, night temperatures are higher. In bare soil the reverse occurs. The crown of a tree protects the vegetation under it from night frosts, especially on still nights (Geiger, 1965). The vegetation also insulates the soil from winter cold, so that penetration of frost into the soil is slower, and ice is formed at greater depths in bare soil and in soil under sparse vegetation than in soil under luxuriant vegetation. In general, the

TABLE 55 Latent heat used in transpiration in a forty-year-old, closed-canopy spruce forest in southern Sweden, calculated on the basis of the mean daily transpiration from May to August (Stålfelt, 1944) and compared with the incoming net radiation, assumed to be the same as at Rothamsted (Russell, 1952), i.e. 38 000 J/cm^2 per month during the vegetation period.

	DRY AREA	WET AREA
Daily transpiration from the forest, in g/cm^2 soil surface	0·19	0·34
Corresponding daily heat requirement (J)	400	712
Corresponding monthly heat requirement	12 000	21 360
Heat used in transpiration as a % of the net incoming radiation	31	56

vegetation, like the atmosphere, has a damping down effect on soil temperature: it lowers the maximum soil temperatures during the day and in summer, and raises the minimum temperatures during the night and in winter.

Litter has a similar effect (Fig. 64), depending on the depth of the litter layer: the effect of a thick layer of deciduous leaves, for example, may be considerable. A litter layer may protect young shoots and seedlings from too much heat in the summer and from too rapid thawing in spring.

Snow, particularly powdery snow containing a large amount of air, has a considerable insulatory effect. It prevents incident radiation reaching the soil surface, but on the other hand it slows down the penetration of frost into the soil, so that less ice is formed. A snow cover protects vegetation from the extreme changes in temperature which may occur in winter. Autumn-sown cereals generally survive the winter better under snow than if there is no snow cover, especially if the soil is frozen. The same is true of rosette plants and hemicryptophytes generally.

Water is a good insulator because of its high specific heat. Soils at the bottoms of lakes and other bodies of water warm up more slowly than terrestrial soils, but they are better protected against frost with the result that aquatic organisms experience smaller fluctuations in temperature than terrestrial ones. Water can also affect the climate because, with its high heat capacity, its buffering effect damps down the fluctuations in climate in surrounding areas of land, an effect which is greatest along sea coasts.

Heat capacity, thermal conductivity

Some of the solar radiation which penetrates to the soil surface is absorbed. Soils are heated at different rates, because of differences in their specific heat. Table 56 shows that the specific heat of mineral soil is less than of organic soil, which is less than that of

TABLE 56 Examples of the specific heat of soil materials, i.e. heat required to raise the temperature of 1 g (or 1 ml) material by 1°C. (According to Geiger, 1950, page 27.)

MATERIAL	SPECIFIC HEAT (calories)	
	CALCULATED BY WEIGHT	CALCULATED BY VOLUME
Water	1·0	1·0
Wet sand	0·3	0·4
Humus	0·44	0·57
Wet peat soil	0·8	0·7
Dry peat soil	0·44	0·1

water; water requires three times as much heat as the same volume of sand in order to bring about the same increase in temperature. Soil with a higher water content, therefore, has a higher specific heat, and more heat is required to raise its temperature. Loss of heat by evaporation of water also causes wet soil to warm up more slowly than dry.

Although sand has a lower specific heat than peat, for example, it warms up more rapidly (Table 57), since it conducts heat more rapidly. Thermal conductivity decreases rapidly as particle size decreases; it is greatest in bedrock, next in stones and boulders, less in gravel and sand, and least in clay and humus. A quartz boulder conducts heat about ten to twenty times more rapidly than mineral soil in general; if the boulder is crushed the conductivity is decreased by the air in the spaces between the particles. If the crushed material is saturated with water the conductivity increases, because the water conducts about thirty times faster than the air it has displaced, and two to three times faster than wet mineral soil (Table 57). Consequently, soil conductivity decreases as the content of air increases, and loose, granular soils conduct heat more slowly than compact soils. The conductivity of clay is increased by the addition

TABLE 57 Thermal conductivity of soil materials (calories flowing per sec across a centimetre cube of the substance, having a temperature difference of 1°C on opposite face). (Mainly from Tamm, 1940, page 185.)

Ice at 0°C	0·005
Water	0·001
Granite	0·011 (Geiger, 1950)
Fine sand (air dry)	0·0005
Fine sand with 10% water	0·0020
Fine sand with 20% water	0·0035
Clay with sand (air dry)	0·00045
Clay with sand (saturated)	0·0032
Clay (air dry)	0·00033
Clay (saturated)	0·0021
Humus	0·003 (Geiger, 1950)
Peat soil (air dry)	0·00027
Peat soil (saturated)	0·0011
Snow	0·00025–0·0008

of sand: addition of humus has the opposite effect. Heat is conducted relatively rapidly in stony soils because the conductivity of the stones themselves is high. The temperature at a depth of 10 cm may be several degrees higher in soils with a stony surface than in soils free of stones (Troedsson, 1956). Soils on south-facing boulder-covered slopes are warmer in spring than other; because of the combination of high insulation and high heat conductivity of the boulders.

Solid rock has a similar effect. A south-facing rock face stores heat because of its high heat conductivity, so that the air in the vicinity may be warmed for some time after sunset. Such a rock face is a warm habitat for plants, and branches of trees and bushes growing near it show earlier development of leaves and flowers in spring than branches further away.

Thermal conductivity is little affected by the type of mineral, and it is more or less the same in soils made up of different mineral materials: particle size, water and humus content are more important.

Warm and cold soils

Dry porous soils, soils with a good crumb structure and high content of air, which only occur in places with a low water table, become relatively warm at the surface, because of their low specific heat and their low conductivity, but heat penetrates only

FIG 66. Juniper roots exposed by frost heaving. Alvar community on Gotland.

slowly. The surface of such soils feels warm, giving them the name *warm soils*, and because of their low conductivity there is little thermal radiation loss from them, and consequently they do not freeze easily. Soil colour has some effect, because dark soils absorb more radiant energy than light ones.

Wet, compact soils, in which the water table is high, are called *cold soils*. Because of their high specific heat, they warm up slowly; because of their high conductivity, heat is rapidly conducted to the deeper layers, and the surface feels relatively cold. Evaporation of water from the surface also contributes to the low soil temperature. When the air temperature decreases, the heat in the soil is rapidly lost, ice formation occurs readily, and to greater depths than in warm soils (Tamm, 1940). Bog soils, and other humus-rich soils with a high water table, are colder than mineral soils and lose heat rapidly in frosty weather, especially from the surface, so they are subject to night frosts: bog and fen soils are well known for their proneness to frost. Ice in these soils tends to thaw only slowly, because of the high soil water content and the loss of heat involved in evaporation. Arable land prone to frost because it has a high water table and a compact soil can be improved in this respect by drainage and the addition of mineral soil.

Among other characteristics, the warm or cold properties of the soils affect the development of plant communities in various

FIG 67. Water meadow. The fence is the dividing line between mown meadow (the dryer part) and a tussocky grazing area (the wetter part). Uppland. (Photo. ATA, M. Sjöbäck.)

habitats, for instance, species intolerant of cold are mainly found on warm soils, least prone to frost. A suitable soil climate is necessary for many cultivated plants, especially species grown in gardens, but many garden species have their origin in warm climates and are cultivated outside their natural range of distribution, in conditions which are suboptimal or supraoptimal in many respects. They are more sensitive than native species to environmental conditions such as soil temperature, thus, to do well, they need all the help the gardener can give.

Temperature equilibration processes in the soil

Heat transport in the soil, like surface heating, depends on the specific heat of the soil, the thermal conductivity, and the temperature gradient. In daytime and in the summer, heat transport is mainly downwards, away from the surface: at night and in the winter it is mostly towards the surface. Some of the heat absorbed at the soil surface is conducted downwards, some is transferred to the air layer near the surface by conduction and convection, and a significant amount is used in the evaporation of soil water. When the water vapour condenses the latent heat utilised in evaporation is released again.

FIG 68. Polygons on alvar, Gotland. The material consists of stones up to the size of tennis balls together with smaller particles, even clay. When freezing occurs the finer fraction takes up water, expands and pushes the coarser fractions out to the sides.

The water vapour and warm air near the surface are transferred away by the wind, so that the heat is distributed along and above the soil surface. Hence, the temperature climate of any one place is partially dependent on the winds and the precipitation.

Similar transportation of heat, but across smaller distances, goes on in the soil itself. Some heat is used as latent heat of evaporation: in moist soils, the soil air is practically saturated with water vapour, which generally diffuses into the atmosphere. The direction of movement varies with time of day and season of the year in such a way as to reduce the temperature changes going on in the soil. The greater the incident radiation and soil surface temperature, the greater is the evaporation and heat loss from the surface. In the same way, a decrease in soil surface temperature is counteracted by condensation of water vapour there. If the surface temperature is less than that of deeper soil, water vapour diffuses upwards and condenses in the cold surface layers, so that latent heat is released. Because heat is transported in this way, temperature fluctuations in soil are less than in the air, and, hence, soil organisms, like aquatic organisms, live in a climate of temperature and humidity which is more stable than the aerial climate.

Freezing and thawing are other processes which tend to reduce temperature differences. As ice is formed, about 336 J/g is released, thus slowing down the rate of decrease in temperature. When thawing starts, the same amount of heat is taken up, and the rate of increase in temperature is slowed down.

Ice formation in soil subjects the vegetation to the process known as 'frost heaving'. Since ice occupies a larger volume than the equivalent amount of water, the soil volume increases as a result of freezing; the volume also increases because the frozen part of the soil accumulates water from deeper layers. Moreover, when a colloid freezes it loses water, and the total volume of the colloid and the water is larger than before. These factors cause the soil surface to be raised to a height which depends on the amounts of water and colloids involved. This movement of the soil, or 'frost heaving', causes root damage, and young and weakly rooted plants, in particular, may be lifted, the damage being mainly caused by the repeated freezing and thawing which may occur in late winter and early spring; autumn-sown cereals and seedlings in nurseries are often damaged in this way. Covering the seedlings may prevent damage, while in rosette plants the effects of frost heaving are counteracted to a certain extent by the contractile roots (see Chapter 11, page 240).

Frost heaving may be a contributory cause of the tussock formation characteristic of fens, water meadows and other wet ground. The dense masses of stems are mostly formed through branching and adventitious shoot formation of a single plant or a group of several plants. The frost lifts the plants each year, but any damage to the root system is made good by subsequent growth.

Stones in the soil are also lifted in places where the soil contains large amounts of water and colloids. When the ice thaws, the water which has accumulated as a result of freezing is released and the resulting wet soil may be subject to soil slip, particularly if it is silt or clay (see Chapter 11, page 239), and some movements of this kind form polygons and circles of stones. This happens when wet soil lying on a solid base, on rock or on a layer of ice, is subject to pressure if the surface freezes, because the soil between the frozen surface and the solid base expands. At weak points, the material is squeezed outwards, and the coarser particles tend to become arranged in bands, squares, etc., such as are often seen on mountain heaths, and plants being unable to take root there, the patches are bare.

However, ice formation has some advantageous consequences to plants. As the soil shrinks at the time of thawing, cracks often appear between soil and stones (C. O. Tamm, 1954), and rain runs down these cracks off the stones so that it penetrates relatively deeply. Troedsson (1955) found that the amount of precipitated water running off boulders which had their upper parts sticking up through the humus layer was greater than the amount of rain falling on an equivalent area of other parts of the surface. Roots are able to grow in the cracks without encountering any resistance to penetration and to come into contact with large amounts of well-protected water. Since soils with stony surfaces also warm up quickly, because of the high thermal conductivity of the stones, and as long as the material is not too coarse to retain a sufficient amount of water, conditions are favourable for species with deeply penetrating roots. There is often good forest growth on such soils (see Chapter 11, page 236).

14

The relation between vegetation and the climatic soil types

Soils are influenced both directly and indirectly by climate. The effect of any one factor is usually more or less dependent on the others. Rainfall, for example, determines the soil water content, and hence, together with soil temperature, also influences the development of soil organisms, humification, salt accumulation and depletion, etc. (see Fig. 1, page 17).

If precipitation exceeds evaporation, the run-off and the amount of water in the streams and rivers increases. Tamm (1956) expressed the water balance of a site as the humidity (H, in mm of water), defined as the difference between the mean annual precipitation (P, mm) and the mean annual evaporation (E, mm):

$$H = P - E$$

He used an empirical relationship containing only mean annual temperature (t, °C) to calculate annual evapotranspiration for coniferous forest on moraine:

$$E = 30 \cdot 4\, t + 220 \cdot 9 \text{ mm}$$

and obtained the following values of H for coniferous forest in different parts of Sweden:

Kalmar, $H = 47$ mm; Växjö, $H = 257$ mm;
Esmared (SE Halland), $H = 625$ mm.

Other, more refined methods can be used for calculating evapotranspiration from standard meteorological observations (see, for example, Penman, 1963) and for classifying the humidity of localities (Mohr and van Baren 1954, page 66), e.g. Birse and Dry (1970) derive potential water deficits for 183 stations in Scotland by deducting the monthly average rainfall from the potential evapotranspiration, and deriving a total for those months in which evaporation exceeded precipitation.

If precipitation is less than evaporation, there is no run-off, no watercourses develop, and practically all of the precipitation evaporates from the place on which it fell, even if some of it may sometimes filter down through the soil to become part of the ground water, or become part of lakes which have no outlet. Climates of this type are termed *arid*. Semihumid and semiarid climates may also be distinguished; these are intermediates in which evaporation and precipitation are equal, or precipitation is slightly in excess of evaporation, or slightly less, respectively.

Soil formation is more closely linked with humidity (as defined above) than with evaporation or precipitation alone, primarily because it is the humidity which determines the amount of leaching or accumulation of dissolved or dispersed substances in the soil. In humid localities there is some loss of these substances in the run-off water: in arid localities there is no such loss, but an accumulation as a result of weathering of the minerals. The balance between litter production and humification, or between production and decomposition of gyttja or dy, is also dependent on site humidity. In humid regions there is a large excess of humus, which becomes part of the soil: in arid regions this amount is only small.

Humid soils and their vegetation

There are humid areas in both cold and hot regions of the world. Hence, soil processes are subject to very varied conditions of temperature and precipitation, and proceed in various directions, so that many different types of soils are formed.

Arctic-alpine regions

In north Russia, Siberia, Greenland and arctic Canada there are large areas of tundra, an open tree-less landscape in which the soil is permanently frozen, except for the surface layer in the summer months, with a plant community consisting mainly of mosses, lichens, dwarf shrubs, and various grasses and herbs which are tolerant of cold and drought.

Vegetation. On level ground, mosses are dominant, but on hilly ground there are more lichens, and moss tundra and lichen tundra can be distinguished. There are bogs in moss tundra, with alternating areas of *Sphagnum* and open water, and dry areas covered by *Polytrichum*, for example. There are heathlike tundras, such as the *Grimmia* heaths on the volcanic lava on Iceland. Species of *Di-*

cranum, Hylocomium, Hypnum and other genera may be present. There are also dwarf shrubs, particularly *Empetrum, Vaccinium myrtillus, Rubus chamaemorus, Betula nana, Salix* and others. Arctic tundra is similar in some respects to Scandinavian mountain heath. In both, heath and bog grade into one another (Selander, 1955), and at the highest altitudes there are patches of vegetation alternating with patches of bare soil (raw soil, Kubiëna, 1953).

Lichen tundra is common in Lappland, Iceland, Greenland and Siberia in places where the bedrock is exposed or covered only by a thin layer of raw humus. Species of *Cladonia, Cetraria, Alectoria* and other lichen genera grow together with dwarf shrubs of the same types as in the moss tundra. There are also grasses such as *Nardus stricta, Deschampsia flexuosa* and *Festuca ovina*, various species of *Carex* and *Juncus*, and herbs like *Dryas octopetala, Chamaepericlymenum suecicum, Astragalus arcticus, A. frigidus*, etc. Lichen tundra is conspicuous at a distance as a yellow-grey expanse. There is a gradual transition from lichen tundra to mountain heath, the main difference between the two types of vegetation being the greater abundance of lichens in the lichen tundra.

Climate. The winter lasts for most of the year. The temperature may sink very low, to $-50°C$ in Siberia, for example, and the summer temperature is also low. The July mean in Siberia is scarcely over $+10°C$. The ice in the soil extends to a great depth, and only the surface thaws out, so the soil is permanently cold. The vegetation period is only 2 to $2\frac{1}{2}$ months, and resembles the spring of temperate regions rather than the summer.

There are strong winds, especially over the extensive areas of tundra in Siberia, up to 40 m/sec in winter. The wind dries the vegetation and causes mechanical damage, breaking, deforming and abrading the plants, by the action of the driving snow. In early spring the roots are still frozen, and the phanerogams' shoots must resist severe desiccation.

Precipitation is mostly in the summer, and is comparatively low, less than 300 mm per year. Snow cover is also small, 40 to 60 cm. The snow is blown away from exposed hummocks and ridges and collects in the hollows, which therefore have a smaller depth of frozen soil.

Relative humidity is high, and mist is frequent. Partly because of this, partly because of the low temperature and the short growing season, there is insufficient time for even the small amount of precipitation to evaporate, and excess water cannot run away if the surface is level or only slightly sloping, as on the large Siberian tundral plains, nor can it penetrate downwards through the frozen layers of soil, and hence, the thawed soil surface is always wet and soil slip is frequent.

If the ground is hilly, as it is in Greenland, the water runs off the slopes, and the soil dries out. Consequently, there is less tundra, and large areas of heaths occur.

Soil. Humification is slow, because of the high soil moisture content and low soil temperature, and probably also because of the low salt content in the vegetation. As a result, layers of peat several metres thick accumulate, especially in moss tundra. Only small quantities of litter are produced (*Salix* shoots grow only 1–5 mm per year), but the rate of humification is even slower. The snow is blown off the occasional hummocks in the moss cover so that the soil underneath is frozen to a greater depth. When the snow melts, the water collects in the hollows, forming pools. In the summer, the surface of the hummocks dries out and may crack open; the wind penetrates the cracks, causing erosion and levelling down. The cracks may also widen because of the weight of the mosses, especially if they are wet after rain, and then the soil may move outwards, leaving bare patches. As a result, the surface of tundra is a mosaic of moss hummocks, pools and bare patches.

The mineral soil has mostly been formed by mechanical weathering, brought about by the extreme temperature fluctuations, it therefore has a coarse structure and angular particles. Chemical weathering is very slow because of the low temperature. At the boundary of mineral soil and peat, there has been some mixing, giving rise to a blue-grey layer of soil, with yellow patches and streaks. This is called *gley*, following the Russian term, because of its characteristic appearance. The colours are those of iron compounds, present in the ferrous state in the ground water and under the peat, but oxidised in places where they are in contact with air. The yellow patches and streaks are due to ferric hydroxides.

When the soil surface freezes in autumn, the wet loose material between the surface and the deeper layers which are permanently frozen is compressed and oozes out through any areas of weakness. This results in some mixing of the peat and gley and gives rise to solifluction, and formation of polygons or circles of stones.

Because of this periodic soil movement, there is never an equilibrium between soil and plant community: new surfaces are frequently being exposed and the plant community remains in some kind of colonisation stage (Johansson, 1897; Sigafoos, 1952).

Cold and temperate regions

In these regions the soil is frozen for only part of the year: there is no permafrost. The water released when the ice and snow melt, sinks downwards through the soil, or runs off on the surface.

Because of this natural drainage, the soil surface dries out and becomes warmed, so that right from the beginning of the growing period the edaphic conditions are more favourable than they are in the tundra.

Vegetation and humus production. Forests are the dominant plant communities in the humid regions and cover the major part of the land surface. In the cold and temperate regions, evergreen coniferous forests predominate.

FIG 69. Good quality spruce forest. Småland, Sweden.

The forests may be pure stands of single species, spruce, pine, beech or oak, for example, or various mixed communities. There are most species in the warmest areas, where the forests are mostly deciduous; and least in the coldest, where conifers are most common.

Water consumption by forests is considerable, and probably

greater than by any other community, especially in the case of hydromorphic and tropomorphic forms of forest (see Rutter, 1968). Total leaf area of vegetation per unit area of land reaches a maximum in forests. Only in the humid regions is the water supply sufficient to support a closed forest and the associated understoreys of vegetation, and it is in such regions that forests have their natural areas of distribution and achieve their greatest luxuriance, density and productivity.

If the site humidity is not sufficient for an optimum water supply, then water becomes a limiting factor. In the monsoon area (southeast Asia, India, Africa, Brazil), where precipitation is *c.* 1800 mm per year, dense closed forests occur; but in other areas, with less precipitation, trees are more sparse and form part of the type of vegetation called *savanna*, an intermediate between forest and steppe. Savanna consists of grasses, herbs, and scattered trees and bushes. In drier areas it gives way to steppe. Savanna is most common in Africa, Australia, India and South America.

The zonation and development of forests is also affected by temperature. The warmer the climate, the more variation and the more species in the forests, provided that no other factors are strongly limiting. Tropical rain forest is the most luxuriant and species-rich.

Litter production is considerable, especially in closed forests. The humus which accumulates, dependent on the balance between litter production and decomposition, varies between the relatively small amounts in the soils of the poorest forest heaths and in savanna soils and the thick organic layers which accumulate in forest on fens and marshes. The type of humus also varies, from the acid raw humus in coniferous forest to the more or less neutral, or slightly acid humus formed in the deciduous forests in the less humid or semihumid regions. Raw humus production is particularly large in coniferous forest with a dense understorey of dwarf shrubs.

Scrub communities of bushes and small trees are common in temperate and cold regions, but are not usually very extensive. They are particularly characteristic of the belt just below the tree-line on mountains. They are also found in patches in a narrow belt along coastal beaches, between forests and meadows or other open land, and in places exposed to high winds. Productivity is probably rather low.

Dwarf shrub communities, perhaps with some grass or herbs, are most widespread in the most humid areas of the cold regions, generally forming heaths or bogs. At low altitudes, heaths have probably usually developed as a result of overexploitation of soil or vegetation by man. However, mountain and alpine heaths are

FIG 70. Snow patch on mountain heath. No vegetation has yet become established in the newly thawed zone, the mineral soil is bare, with no organic content. Bessheim, Norway.

FIG 71. Mountain heath at 1400 m altitude at Bessheim, Norway. The vegetation cover is incomplete, being broken by islands of gravel and stones.

probably natural communities. Productivity per unit area is probably small, but there are few precise measurements. Productivity of the above-ground parts of alpine tundra vegetation is 0·5 to 4·0 g dry weight per square meter annually (Billings and Mooney, 1968). Accumulation of peat or raw humus may be considerable in cold areas, because humification is slow.

Grass and herb communities also cover substantial areas of the humid regions, mainly forming meadows. Some are natural, such as alpine meadows, meadows in fens, and meadows along the coast, and serve as pastures for wild animals. Other meadows are the result of forest clearance by man, especially those on dry ground at low altitudes. Here, land has often been used in turn as pasture, arable land and meadow. Meadows are most common within the semihumid regions (a more detailed description is given in Chapter 18).

Productivity, and humus production, in grass and herb communities is usually high, and significant amounts of humus accumulate, especially in the wetter types of meadow in fens and around lakes.

Leaching, transport and precipitation of soil constituents. The nutrient reserves in the soil depend on the balance between supply and demand. Weathering and humification are sources of supply, dependent on temperature and on the types and amounts of bedrock and vegetation. Nutrients are removed by plants and by leaching and transport processes in the soil. These processes are affected by temperature and humidity, and by the type of vegetation and the course of humification.

Leaching and nutrient transport in the soil depend mainly on the amount of water. Water permeating downwards in the soil carries with it dissolved and dispersed ions, molecules and molecule complexes released in humification or in weathering of minerals. Some of the plant nutrients and nutrient-storing substances are lost from the soil in this way, particularly basic cations such as K^+, Na^+, Ca^{++}, Mg^{++}, etc., basoids like $Al_2O_3 . n\ H_2O$, $Fe_2O_3 . n\ H_2O$ and basic silicates and humates, anions SO_4^- and HPO_4^{--} etc., and acidoids such as $SiO_2 . n\ H_2O$, acid silicates and acidic stable humus.

There is often an excess of acids and acidoids remaining in the upper layers of soil, a change which is important to the vegetation. It takes place partly because basic ions are carried down to a greater extent than acidic ones, partly because there is an insufficient amount of bases in the medium (see Tables 47 to 50, pages 248 to 250), and partly because of insufficient mixing of mineral and organic materials (see Chapter 10). If the base content of the mineral soil and the litter is sufficient, and if the products of weathering mix with each other and with the humus, the humic acids and the inorganic colloids are neutralised and precipitated. Salts are bound by these precipitated colloids and are thereby retained in the solum. However, if the litter and mineral soil is poor in bases and there is no mixing, the humic acids remain in solution and are carried downwards through the soil in the water. Then there are insufficient precipitated colloids to retain the salts, which

FIG 72. **Pine-heather heath on podsolised sand. Gotland.**

are also leached away. The most soluble salts are lost most readily, i.e. salts of the alkali metals and the alkaline earths. The acids accelerate leaching because they slow down colloid precipitation and accelerate mineral weathering. The rate of leaching therefore depends on the balance of acids and bases in the soil, and hence primarily on the composition of the mineral soil (Tamm, 1931).

In the cold and temperate regions of the humid zones there are large areas of soils which are relatively poor in basic constituents, i.e. soils derived from rocks resistant to weathering or from minerals poor in bases; sand and gravel soils which have lost their finer and more easily weathered particles; and soils in which some of the products of weathering have been lost by leaching. Because these soils are poor in nutrients, the vegetation consists of species with relatively low nutrient demands, mostly coniferous forest with mosses, lichens and dwarf shrubs (Table 48). The wide differences in nutrient demands of different plants may be illustrated by the following examples (after Mork cited in O. Tamm, 1947, page 261):

Used per hectare and year (kg)	Potassium	Phosphate	Nitrogen
Spruce stand (south Norway)	15	10	40
Wheat	60	30	90
Turnips	187	57	118

Mineral salt utilisation by plants is also discussed in Chapter 18.

The forests of the coniferous forest circumpolar belt mostly grow on soils poor in nutrients. The vegetation of the under-storeys is mainly dwarf shrubs and mosses, producing litter poor in bases, like that of the trees themselves, so that there is a tendency towards mor formation. Acid substances in the mor humus accelerate weathering and leaching (Tamm, 1931).

The primary causes of leaching are therefore to be found in the humidity of the locality and the base-poverty of the mineral soil. The result is vegetation consisting of base-poor species, and these, in turn, promote raw humus formation and acid production and, consequently, weathering and leaching.

Because of leaching, the upper layers of the soil become poorer and poorer in colloids and readily soluble salts and the pH rises. Other changes result from the rise in acidity: the average size of simple particles increases, while aggregation is reduced; the humus fraction and the finer mineral fractions become greasy in consistency, and therefore denser and less permeable. When they freeze, they become more open and granular in structure, but the aggregates are not stable in water. Other changes depend on the composition of the particular soil. Usually the soil profile has some characteristic features resulting from the leaching out of some substances from the upper layers and redeposition lower down, or to the loss of large amounts of substances such as calcium, which have important effects on the soil properties. Various types of leached soils can be distinguished.

1 LEACHING OF THE COLLOID-FORMING SUBSTANCES IN THE SOLUM

(a) *Leaching in soils poor in calcium and in till soils*

When the upper layers of the mineral soil weather and are leached, the grains get covered by a white powder and the soil appears ash grey in colour, and is described as *bleached*. The whole soil profile is called a *podsol* (after the Russian *pod*—under, and *sola*—ash). Still more marked colour changes result from the deposition of the iron and humus leached from the bleached layer in the layer beneath the accumulation zone. If the dominant colour of the accumulation zone is rusty brown or red because of the accumulation of iron hydroxide, the profile is called an *iron podsol*. If the zone is black or dark brown because of precipitated humus, the profile is called a *humus podsol*.

The commonly accepted nomenclature is to denote the leached layers as A, the accumulation layers as B and the unchanged layers beneath as C, with subdivisions into A_1, A_2, etc., if different zones can be distinguished within each category. The organic complex is included in the A zone: A_{00} is the litter, and A_0 the raw humus, for example.

Iron podsol. An iron podsol may be of the following type (mainly according to Tamm, 1920, 1940; Hesselman, 1932).

A horizon

A_{00} litter, several cm deep.
A_0 raw humus (mor) 2–15 cm.
A_1 bleached soil 5–15 cm.

In the A_1 layer, particles are strongly weathered and covered with a white powder and with quartz projections, the quartz being unweathered. Felspar particles are covered by loose felspar. The layer is poor in iron, aluminium and bases (Tables 58, 59, 60), and in colloids, so that the water content is generally lower than in the layers beneath (Table 59).

B horizon (*accumulation zone*). The depth is variable, often 5 to 40 cm. The upper boundary with the bleached soil is sharp, but the lower boundary is irregular. The rust colour is darkest at the top, gets weaker downwards and fades out gradually. There are vertical and horizontal streaks, usually corresponding to cracks or passages formed by root penetration, where water movement has been especially rapid. The pH is higher than in the bleached soil (Table 62). Stones and gravel have a covering of hydrates of Fe_2O_3, Al_2O_3, SiO_2, and humus, i.e. of colloids which have been carried down from the bleached layer and precipitated (Table 60). Under this covering, the mineral particles are unweathered (Tamm, 1920).

The B horizon is denser than the other layers, because of the accumulation there. The mineral particles may be cemented together by the precipitated substances, forming sandstonelike clumps and layers known as *pans*. These are particularly common in raw humus podsols on west European *Calluna* heaths (Tamm, 1929). They are unfavourable to plant growth, as they hinder root penetration and natural drainage.

The accumulation layer sometimes also contains clay, carried down from the A horizon or formed *in situ* from the silica and aluminium which has been carried down.

TABLE 58 The loss of nutrients from the bleached layer of an iron podsol. (According to Tamm, 1920, page 107.)

	SUBSOIL %	BLEACHED LAYER % OF SUBSOIL	LOST FROM BLEACHED LAYER %	DEGREE OF WEATHERING = % OF ORIGINAL AMOUNTS
SiO_2	75·54	65·00	10·54	14
TiO_2	0·55	0·53	0·01	2
Al_2O_3	12·06	8·20	3·86	32
Sil. Fe_2O_3	3·27	1·13	2·14	65
Limonite Fe_2O_3	0·08	0·07	0·01	—
Mn_3O_4	0·05	0·03	0·02	40
CaO	2·04	1·51	0·53	26
MgO	1·30	0·52	0·78	60
Na_2O	2·14	1·42	0·72	34
K_2O	2·84	2·01	0·83	29
P_2O_5	0·12	0·02	0·10	83
Total	99·99	79·76	19·55	

Not all of the iron, aluminium and silica is precipitated in the accumulation zone, and some may move on into the ground water and the water courses (Table 63).

The C horizon is the unchanged mineral soil below.

Humus podsol. There is some humus precipitated in the B layer of an iron podsol (Tables 59, 60). The amount is variable, but generally increases with soil moisture content, and is greater if there is an excess of acid constituents. If there is a large humus component, the B horizon is brown or black, and the profile is termed a *humus podsol* (Table 62; Fig. 73). The difference between a humus podsol and an iron podsol is only quantitative, and there are all possible transitions between the two.

Humus podsols are common in wet places, for example, in forest with *Sphagnum* or in strongly leached, base-poor soils, such as those on mountain heaths. They are often associated with gley formation (see pages 296 and 310).

When rain water runs off bare rock onto a layer of till which has

TABLE 59 Example of the composition of the layers of a podsol profile (Västerbotten, N. Sweden). (According to Tamm, 1920, page 263.)

	A_1 LAYER 10–15 cm %	B LAYER 10–25 cm %	C LAYER 50 cm %
Humus	3·64	2·82	0·54
Water	1·30	3·20	1·93
SiO_2	76·93	68·95	73·30
TiO_2	0·64	0·52	0·56
Al_2O_3	9·68	12·31	11·59
Sil. Fe_2O_3	1·33	3·77	3·67
Limonite Fe_2O_3	0·09	0·93	0·38
Mn_3O_4	0·037	0·045	0·049
CaO	1·80	2·04	2·03
MgO	0·62	1·30	1·42
Na_2O	1·69	1·59	1·73
K_2O	2·38	2·65	2·49
P_2O_5	0·02	0·04	0·11
SO_3	0·024	0·029	0·029
Total	100·18	100·20	99·83

TABLE 60 Example of the composition of the layers of a podsol profile (Teindland State Forest, Morayshire, Scotland). (After Muir, 1934, page 44.)

	ANALYSIS AS % OF IGNITED FINE EARTH (< 2 mm)					
	A_1 5–7 cm	A_2 7–14 cm	B_1 14–16 cm	B_2 16–26 cm	B_3 26–50 cm	C 60–70 cm
SiO_2	89·43	91·05	82·49	78·86	76·40	83·12
TiO_2	0·14	0·12	0·24	0·21	0·21	0·11
Al_2O_3	6·11	5·81	10·10	10·30	11·08	8·56
Fe_2O_3	0·89	0·55	2·71	3·59	3·32	2·19
CaO	0·63	0·45	0·52	0·68	0·94	0·59
MgO	0·30	0·54	0·40	0·22	0·71	0·46
Undetermined residue	2·50	1·78	3·54	6·14	7·34	4·87
Total	100·00	100·00	100·00	100·00	100·00	100·00
Loss on ignition	38·73	1·77	7·87	5·66	3·92	1·63
Moisture (105°)	12·02	0·65	5·41	4·72	4·12	1·46

been sorted by wave action, lying over a layer of unsorted till, it penetrates into the lower layer carrying stable humus, which is precipitated to form a dark horizon (Troedsson, 1955). If the stable humus is not precipitated, it is carried further into water courses and lakes, where it is deposited as dy (see pages 220 and 310).

The physico-chemical basis of the processes of leaching and precipitation is not completely clear. According to Tamm (1931),

TABLE 61 Analysis of an iron podsol (Västerbotten, N. Sweden). (After Tamm, 1931, page 327.)

HORIZON	pH	COLLOIDS OF			
		HUMUS %	SiO_2 %	Al_2O_3 %	Fe_2O_3 %
A_2 (bleached layer)	4·3	1·35	0·03	0·18	0·43
B_1 (accumulation layer, upper part)	4·2	2·48	0·09	0·37	1·42
B_2 (accumulation layer, lower part)	4·8	1·68	0·33	1·31	0·70
C (subsoil)	5·6	0·57	0·13	0·61	0·41

TABLE 62 Analysis of a humus podsol (Västerbotten, N. Sweden). (After Tamm, 1931, page 334.)

	pH	COLLOIDS OF			
		HUMUS	SiO_2	Al_2O_3	Fe_2O_3
Bleached layer	4·5	6·37	0·03	0·37	0·04
Accumulation layer	4·8	2·06	2·02	0·31	0·09
Subsoil	4·5	0·60	0·07	0·26	0·09

the iron and aluminium are first transported as cations (Fe^{+++}, Al^{+++}), then the balance of basic ions and hydrogen ions changes and the pH rises (Table 61). The pH may change in the following way:

A_0 4·2; A_1 3·7; A_2 4·6;   B 5·8;   C 6·2.

Soon the point at which iron and aluminium are no longer stable as cations is reached, and they change to the hydroxides Fe_2O_3.

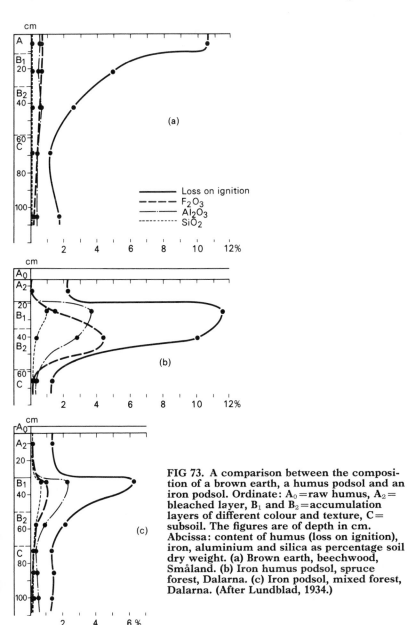

FIG 73. A comparison between the composition of a brown earth, a humus podsol and an iron podsol. Ordinate: A_0 = raw humus, A_2 = bleached layer, B_1 and B_2 = accumulation layers of different colour and texture, C = subsoil. The figures are of depth in cm. Abcissa: content of humus (loss on ignition), iron, aluminium and silica as percentage soil dry weight. (a) Brown earth, beechwood, Småland. (b) Iron humus podsol, spruce forest, Dalarna. (c) Iron podsol, mixed forest, Dalarna. (After Lundblad, 1934.)

TABLE 63 Composition of the ground water in two types of forest in central Sweden, differing in soil and vegetation. Weights in mg/litre. (After Troedsson, 1952, pages 14–15).

	GRENHOLMEN Gneiss-granite bedrock with till influenced by Cambro-Silurian rocks. Brown earth, herb-rich spruce forest with occasional deciduous trees (Mean of 45 analyses through the year)		BJURFORS Leptite-gneiss till. Iron podsol. Pine and spruce suitable for pulp (Mean of 55 analyses through the year)	
	MEAN	RANGE	MEAN	RANGE
SiO_2	15·5	8·5–24·7	12·0	5·1–21·5
Al	0·2	0–0·6	(0·3)	–
Fe^{++}	—	–	0·2	0–0·6
Fe^{+++}	—	–	0·6	0–1·0
Mn	2·9	0–4·3	0	–
Mg	4·2	2·5–10·8	1·3	0·8–1·6
Na	9·3	3·7–19·3	2·8	2·0–3·6
K	3·2	0·6–8·0	0·4	0·1–0·8
Ca	79·5	36·0–128·9	3·0	2·0–5·0
O_2	1·0	0·1–0·8	0·5	0·1–0·8
pH	7·1	6·3–8·1	6·6	6·2–6·9

$n\,H_2O$ and $Al_2O_3.n\,H_2O$, which at first are dispersed colloids. These are electrically charged, like SiO_2 or the humic acids. They are amphoteric electrolytes, usually either weakly acid or weakly basic, and are sources of both H^+ and OH^- ions, e.g.:

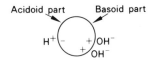

The proportions depend on the acidity of the soil water. At a certain pH, the isoelectric point (IP), the ampholytic colloids release equal quantities of H^+ and OH' ions, and the charge on the particles is zero. They therefore no longer repel each other, but coagulate and precipitate (Tamm, 1940).

According to Mattson and Gustafsson (1934), the colloids are deposited in a sequence depending on their isoelectric points, so that those with the lowest IP are deposited uppermost in the B horizon, where the pH is generally lowest, whereas the less acid colloids are deposited lower down. It is usually iron which is deposited first, and accumulates in the upper part of the B horizon (Lundblad, 1924).

However, the relationship is complicated by the fact that colloid precipitation can also occur in other ways. Colloids with opposite charges may cause each other to precipitate, forming a conglomerate with an ampholytic character.

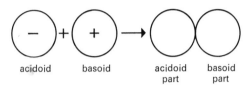

Cations in the solution may also bring about precipitation to a different extent for different colloids. Humic acids are relatively unaffected by such cations (Mattson, 1942). The cations can bind bases to a large extent, but the humic acids are not precipitated by the cations, but by the sesquioxides (Al_2O_3, Fe_2O_3) and other heavy metals in the A horizon, or in some cases by oxidation. Another complication arises because the humic acids, if present in large quantities, form protective colloids around the iron and aluminium colloids and so prevent their neutralisation.

The precipitation of iron, aluminium and humus can thus be regarded mainly as a colloidal precipitation process, which proceeds to the extent that humic acids and other acids are neutralised.

Colloids which are not precipitated are carried into the water courses, which therefore contain some silica, aluminium and iron (Table 63). These may retain their colloidal character or may become non-colloidal solutes. Some of the aluminium and silica colloids ($Al_2O_3 . n\ H_2O$, $SiO_2 . n\ H_2O$) are carried into lakes or the sea, where they form clay deposits, so that the clay in such places may be both allochthonous and autochthonous in origin.

Transport and conversion of iron is affected to a considerable extent by the stable humus, and thus by the plant community above and below the soil surface. Under humus podsols and other soils with a thick cover of moist, organic material such as peat, there is some ferrous iron in the ground water. In such places the conditions favour anaerobic processes and reductions, because there is a deficiency of oxygen. Ferrous iron is readily soluble, so it tends to be carried away in the ground water, emerging in springs of water rich in iron, or in rivers and lakes, where the oxygen supply is sufficient for oxidation to ferric hydroxide ($Fe_2O_3\ n\ H_2O$).

Formation of a gley soil is a consequence of the oxidation of iron. In marshy ground, or under peat, where there is an oxygen deficiency, the iron is in the ferrous state, but where the ground water rises high enough to contact cracks or root channels, the iron comes in contact with oxygen and is oxidised, so that ferric hydroxide is precipitated to form a brown horizon in the soil profile (the gley horizon). This horizon marks the boundary of the zone within which the ground water level fluctuates. Gley is common in temperate and cold regions, occurring in till, in sand, silt, and clay soils, wherever there is a high soil water content (Tamm, 1931, 1940).

The occurrence of dy is explained by the absence of suitable conditions for the precipitation of humic acids. Large quantities of acid stable humus are formed without being precipitated, and are carried into the water courses in the surface and ground water; water from peat bogs is often coloured brown by this humus. In lakes and flooded areas, like fens and water meadows, precipitation eventually occurs as the stable humus is oxidised, and the sediment formed in this way is the main constituent of the type of soil known as *dy*.

In the bleached soil layer, weathering and leaching go on, especially in the lowermost part; the most readily soluble substances have already been lost from the upper part. The degree of podsolisation varies, and may be distinguished by the thickness of the bleached layer (Tamm, 1920, 1921). It generally increases with greater humidity (H), and in Sweden the deepest podsol profiles are found in the north. There is also some relation between degree of podsolisation and amount of raw humus formed, so that the thicker the raw humus layer, the deeper the podsol. The two processes have a common cause, the lack of basic buffer substances in the vegetation (see Chapter 10).

Plant communities such as coniferous forests or heaths with mosses and dwarf shrubs produce large amounts of raw humus and usually grow on podsolised soil (Tamm, 1920, 1921, 1940).

Iron podsols occur generally throughout the coniferous forest region of the northern hemisphere and may be regarded as the typical soils of the region.

Rate of leaching. Leaching is a relatively rapid process. Tamm (1920) found a perceptibly leached layer in soil profiles only 40 years old. His measurements showed that a bleached layer 2 or 3 cm thick, and a 10 to 15 cm thick layer of rust-coloured soil could already be distinguished after 100 years, even though weathering and colloid accumulation are still relatively insignificant (a normal well-developed podsol profile takes 1000 to 1500 years to develop). The rate of leaching probably depends mainly on the type of plant community and on the humidity of the climate; the more water that filters through the soil, the more rapid the leaching. The concentration of minerals in the water which has filtered through is more or less constant, according to Troedsson (1952) (see Table 63, page 308), and independent of normal rates of movement of the water. Hence, in these conditions, leaching will be more or less proportional to the volume of water going through. Ionic equilibrium between the medium and the soil water is achieved rapidly, and is about the same in the surface water and ground water, and on sloping or level ground. There is some seasonal variation, but snow, rain, drought, frost, etc., have little effect (Troedsson, 1955).

(b) *Leaching in soils rich in calcium*

If the soil contains sufficient calcium, development is different in several respects from that described above. The pH of the soil water is high, humus colloids and inorganic colloids are precipitated, nutrient salts released in humification are bound by the colloids in relatively large amounts, the soil structure is granular, and there is good permeability. A more demanding type of vegetation grows on such soils, the litter is relatively rich in basic nutrients and humification results in neutral or only slightly acid humus, with no raw humus formation. In other words, the soil is a brown earth (described in more detail on page 324), which is a soil type typical of the slightly humid regions and occurring sporadically in the areas with mainly podsol soils. Brown earths may be formed even in humid localities, if precipitation does not greatly exceed evapotranspiration and if local conditions are favourable, for example, in warm sites where there is a good calcium supply and in other places where vegetation, soil composition, topography and microclimate are favourable. This does not mean that there is no leaching, but that leaching is not so intensive that podsolisation has yet begun. The soil may be young, or a brown earth-producing community may have remained undisturbed.

Calcium in soils may occur as a constituent of the bedrock, in limestone, dolomite, diabase and other greenstones, or as a constituent of the loose soil layers, of till, clay, etc. If the calcium content is high (up to 10 per cent CaO or more), the soil is called a *marl*. Various types of marl are distinguishable, depending on the main constituents of the soil. There are till marls, clay marls and marls consisting mainly of the remains of animals.

In Sweden, lime-containing clays and tills occur in southern and central areas of Skåne, Öland and Gotland, and in other important agricultural areas. Limestone is the bedrock in several parts of the country, including Dalarna and Lappland.

In the humid climatic regions, limestone, like any other mineral, is subject to weathering and leaching. Neutral calcium carbonate is converted to readily soluble acid bicarbonate by soil water containing carbon dioxide.

$$CaCO_3 + H_2CO_3 \longrightarrow Ca(HCO_3)_2$$

The bicarbonate dissolves in the ground water and is carried away, into rivers, lakes and springs, where some of the carbon dioxide may be released, so that the neutral carbonate is formed again and deposited in various forms, such as calcareous tufa.

$$Ca(HCO_3)_2 \longrightarrow CaCO_3 + H_2CO_3.$$

The rate of weathering and leaching is determined by the temperature and humidity. Scandinavian soils are young, and in places where the humidity is not particularly high, leaching has not proceeded far enough to affect the soil to any very great extent. Lime-containing soils in south Sweden are still of the brown earth type, with no raw humus formation or podsolisation. However, in more humid regions in north Sweden, leaching has already removed all the lime from the uppermost soil layers, down to a depth of as much as 70 to 100 cm (O. Tamm 1920, 1940).

Tamm (1920) gave an example of the rate of such changes: in 1796 in Jämtland the Indal river, which has a high lime content changed its course so that the Ragunda lake was drained. By 1914, the level of lime on the terraces which were exposed in 1796 was at a depth of about 0·5 m. Another example is cited by Crocker and Major (1955). They found that the lime content of soil exposed by glacier movement in Glacier Bay, Alaska, forty or fifty years ago had decreased from 5 per cent to an insignificant amount, and the pH had decreased from pH 8 to pH 5.

When calcium has been lost from the uppermost layers of soil, so that it is no longer accessible to plant roots, it no longer has any effect on the vegetation or on soil processes, which therefore proceed in the same way as in soil with very little or no calcium. As calcium leaching proceeds, the flora changes character, raw humus-forming plants become established and podsolisation begins.

2 LEACHING OF NUTRIENT SALTS

Potassium, calcium, magnesium and other base cations released by weathering and humification are present in solution in the soil water, or bound to the soil particles, especially the colloids. There are various ideas about the distribution of the anions. Probably these are distributed between the soil water and the soil particles in a similar way to the cations. Some are probably bound to amphoteric colloids, particularly in dry soil.

The proportions of free and bound ions depends partly on the proportions of soil water and colloids. The colloids possess a reserve of ions which can be released into the soil water. There is a tendency towards equilibrium between the two phases. The attainment and the position of the equilibrium depends not only on the exchange processes and solution processes, but also on the other processes affecting the amount of salts and ions present. Some

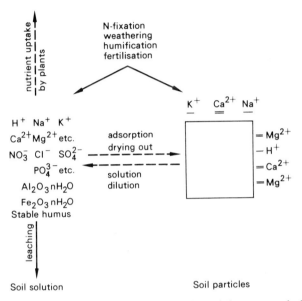

FIG 74. Diagrammatic representation of the accumulation and leaching processes in the soil (solid arrows) and the retention of the nutrients in equilibrium systems and cycles (dashed arrows).

of these processes increase the amount (weathering, humification, fertiliser application) and some decrease it (nutrient uptake by plants and leaching). An increase in water supply dilutes the ions so that more go into solution: drying out has the opposite effect. There is an equilibrium even at the low concentrations prevailing in the soil water. This is dependent on the type of soil, and the concentration is higher in the soil water of more fertile soils than in that of podsols (Table 63, page 308).

When leaching takes place the ions are lost in order, depending on the force with which they are bound to the surfaces of the particles; this, in turn, depends on the specific binding force and the distance between the ion and the particle. If large amounts of ions are added, particularly those which are attracted most strongly, such as hydrogen ions or calcium, they replace other ions, according to the rule mentioned in Chapter 12 ($H^+ > Ca^{++} > Mg^{++} > K^+ > Na^+$). The acids in acid soils therefore have an important effect on the leaching of base cations (Table 64). Potassium and sodium are lost first: calcium less readily. Nitrate and chloride are

TABLE 64 The composition of sea water (mean of 77 analyses) and river water (mean of 19 rivers). (Clarke, 1920.)

| | PERCENT OF TOTAL SALT CONTENT | | EQUIVALENTS IN TERMS OF $Mg = 1$ | |
	RIVERS	SEA	RIVERS	SEA
Cl	1·85	55·292	0·133	5·098
Br	—	0·188	—	0·153
NO_3	2·82	—	0·117	—
SO_4	8·22	7·692	0·445	0·523
CO_3	41·33	0·207	3·602	0·020
Na	3·47	30·593	0·395	4·346
K	1·33	1·106	0·089	0·092
Ca	20·46	1·197	2·670	0·196
Mg	4·65	3·725	1·000	1·000

the most readily lost anions. Carbonate and sulphate ions are retained to a greater extent, at least when in the form of their least soluble salts. Phosphate ions, which are important to plants but present in soils only in low concentrations, are most resistant to leaching. In podsols and other acid soils the uppermost layers are more or less free of mineral phosphate (Tables 58 and 59, pages 304 and 305), but phosphate ions are retained strongly further down the profile, generally at depths accessible to plant roots, so that the ions are taken up by the plants and returned in the litter to the surface layers. Only insignificant amounts of phosphate are carried down to the ground water.

Boron is widespread, but occurs only in small amounts. Soils in humid regions usually contain 0·1 to 2·5 ppm boron. Primary rocks contain about 0·001 per cent B_2O_3; marine sediments 0·03 to 0·05 per cent and sea water 0·0015 per cent (Landergren, 1945, according to Philipson, 1953, and Scheffer and Schactschabel, 1956).

3 THE EFFECT OF LEACHING ON THE HABITAT

The soil is impoverished by leaching, whether this takes the form of podsolisation or loss of material, mainly calcium, so that the composition of the vegetation changes to include less demanding species, which are generally the species which bring about more acid conditions and podsolisation. Because it produces acid humus, and this causes more rapid weathering, such vegetation obtains sufficient mineral salts despite the impoverished character of the soil. However, since this in turn increases the rate of leaching, a vicious circle is set up.

Plants also have other ways of adapting themselves to worsening soil conditions, in that the roots tend to avoid the bleached horizon and to prefer the raw humus and accumulation layers (Tamm, 1921; Mattson, 1938).

Trees blown down by storms tend to counteract the continued impoverishment of the soil. The bleached layers and accumulation layers are mixed together again, and mix with the deeper layers to some extent so that a new surface layer is built up. There is little information on how often this occurs. Möller (1957) calculated that the numbers of trees blown down by storms in the sixty-two year period from 1894 to 1956 was equivalent to about 3 per cent of the total number felled. For deciduous broad-leaved trees the figure was 0·3 per cent, and for conifers it was 9 per cent. During the period required for podsol formation (1000 to 1500 years in north Sweden, for example), a significant amount of soil would probably be mixed by such means. More soil is mixed by animal activity and by solifluction and frost heaving (Tamm, 1920). However, these processes have a more limited effect and are restricted to certain soil types.

During their passage to the sea (Tables 63, 64), the mineral salts pass through various types of habitats, the fertility of which they influence. If the movement is along the soil surface, as on sloping terrain, the salts act favourably on the vegetation of the low-lying places, where the soil becomes richer in nutrients, attains a better structure and higher pH than the soil higher up the slope, and consequently supports a more luxuriant vegetation. The pH of soil on mountain slopes generally decreases with increasing height. This effect is particularly marked if the salts transported include large quantities of calcium (Tamm, 1920, 1940).

The eutrophic character of fens is partly brought about by this kind of nutrient supply. Potassium is usually a limiting factor in well-drained soils, probably because it is bound by silica, which is produced in large quantities as a result of weathering and is lost by leaching because of its negative charge. The adsorbed potassium accumulates in places to which the silica is transported, such as fens. Troedsson (1955) found high potassium concentrations in ground water running off slopes and in water in marshy places (see page 382).

There are similar effects where the ground water emerges,

around springs and along streams: the luxuriance of the vegetation in such places does not depend merely upon the good water supply (Troedsson, 1955).

Tropical and sub-tropical regions

Vegetation. The humid soils in tropical and sub-tropical regions also support forest or have done so in the past.

1 TROPICAL RAIN FORESTS. Tropical rain forests occur in equatorial regions, particularly in a belt across Asia, from Ceylon and South-East Asia to New Guinea and Polynesia, and in equatorial Africa, southern Mexico and Brazil. The mean annual temperature is 24° to 30°C; no monthly mean is less than 18°C; rainfall is usually daily, at definite times, and in amounts up to 2000 to 4000 mm per year. Temperature and rainfall are among the environmental conditions which most often limit species distribution, but, since they do not do so to any great extent in rain forest, there are large number of species, life forms and types of growth. For example, there are more than 20 000 species of flowering plant on Java and Sumatra.

Rain forests include several layers of vegetation, up to five or six layers of crowns of trees, sometimes composed of over a hundred species, with the lowermost layer transitional to a bush layer (Richards, 1952). Vertical gradients in microclimate are large, and affect the mode of growth. In the uppermost layer, the light intensity is high and the tree leaves are leathery and xeromorphic; further down they are mesomorphic and in the lowermost layers they are hydromorphic. The same is true of the lianes, which, together with phanerogamic epiphytes, attain their greatest development in tropical rain forests. Algae and mosses live epiphytically on twigs, branches and leaves. The field layer consists of tall-growing bushlike herbs with perennial but non-woody stems (*Begonia* and *Scitamea* species, for example). The climate has no periodic variations and the trees bear leaves continuously. The leaves develop and are shed at all times of the year.

The rain forests outside the tropics, and on high ground within the tropics, in which temperature and precipitation are somewhat lower (precipitation 1000 to 1500 mm) and more variable, differ in several respects. There is less variety of form: the large-leaved and thin-leaved types of trees and cauliflorous types, all of which occur in tropical rain forest, disappear; there are fewer lianes and phanerogamic epiphytes, but more ferns, especially tree ferns and epiphytic ferns, and also more epiphytic lichens and mosses. In the wettest mountain rain forests the trees are covered by a sheath of mosses.

The productivity of rain forests is thought to be the highest in existence, probably in the range of 10 to 20 g/m² daily throughout most of the year (Billings, 1965). Water use is probably also very high.

2 MONSOON FORESTS. The word *monsoon*, from an arabic word meaning season, is applied to a seasonal wind, the direction of which is dependent on the season of the year. Monsoons are caused by the seasonal changes in temperature at the boundaries of sea and land. Because of the monsoon, the precipitation is also seasonal, and there is a rainy season and a dry season each year. The vegetation, including the forest vegetation, has become adapted to this periodicity and in the monsoon regions (South-East Asia, east Java, Africa, Brazil) the forest trees bear leaves in the rainy season and are leafless in the dry season.

Some of the trees in forests transitional between rain forests and monsoon forests are leafless in the dry season, the proportion increasing as the seasonal variation in the climate increases.

Monsoon forests have closed canopies if the precipitation is about 1800 mm per year or more. In areas where precipitation is less, there is insufficient water for forest growth and the forest is more open, tending towards savanna.

3 SAVANNA. The position of the rain belt along the Equator changes according to the time of year, so that the subtropical regions each side of it experience both rainy and dry seasons annually. In the rainy season the vegetation develops leaves and flowers: in the dry season the trees are generally leafless. As the annual precipitation decreases, the tree layer disappears and the vegetation becomes steppe, which is one of the semihumid or semiarid vegetation types (discussed further on page 336).

Savanna is intermediate between forest and steppe. It is a vegetation type developed in some subtropical regions with precipitation between 900 and 1500 mm, consisting of grassy plains with occasional trees and bushes. The field layer is made up of xerophytes and succulents, mostly tall grasses which are often tussocky, with hard stiff leaves, together with Cyperaceae and tussock-forming perennial herbs and bushes. The vegetation is not completely closed, and there are patches of bare soil here and there among the tussocks.

The development of savanna depends on the precipitation, but is also affected by edaphic and biotic factors. A lack of water-retaining soil constituents, because of the rapidity of breakdown of the stable humus, has the same physiological effect as low rainfall. Grazing animals with a preference for herbs and grasses tend to give a competitive advantage to the woody plants; and fires, which are common in savanna, have the opposite effect. Interference by man may cause savanna development, as exploitation of forests and consequent soil impoverishment has done on Java and Sumatra, for example.

The most important areas of savanna are in Africa, bordering the region of rain forests, Australia, South-East Asia and Brazil.

4 SCRUB. At the edge of the rain forest there is a belt of evergreen broad-leaved scrub. Similar types of vegetation occur in sub-tropical regions with winter rain, such as the Mediterranean countries, the Cape region of South Africa, California, parts of Chile, and west and south Australia. If rainfall is 500 to 900 mm, there is usually a scrub of dwarf trees and bushes together with climbing plants, and a field layer of various small bushes and bulbs. The trees and bushes grow to a height of 1 to 3 m, their leaves are xeromorphic, small, stiff, heavily cuticularised with a lot of sclerenchyma. Scrub of this type has been given various names: *macchia* (Italy), *maqui* (France), and chaparral (California). These communities are often the result of overexploitation of forests, or of fires, resulting in soil impoverishment and loss of water-holding capacity. Further impoverishment leads to development of bush-steppe and desert (see page 336).

Humus production is considerable, but little organic material accumulates because conditions of soil moisture and temperature are favourable for biological processes.

Weathering and leaching. Laterite. The soil processes in tropical and subtropical regions take place at high temperatures, and in the humid regions there is a good water supply. For instance, at latitudes between 20°N and 20°S, in tropical rain forest, for example, soil temperatures regularly reach 25°C and precipitation 2000 to 4000 mm. Consequently, weathering and humification are rapid; leaching is also rapid, so there is no accumulation of nutrients in litter or humus. It is only in regions like these which have the most luxuriant vegetation that so much litter is produced that the soil retains a thin humus layer or a small amount of humus mixed in the mineral soil: in spite of the abundant litter production, there is only a thin layer of leaves and other litter covering the soil (Richards, 1952) because humification is so rapid.

Laterite formation. Leaching is of a different type from that in cold and temperate regions. Cations of bases and silica are lost to a relatively greater extent than iron, aluminium and manganese, so that the aluminium and iron accumulate in the upper soil layers and the ratio SiO_2/Al_2O_3 decreases. At the same time, some of the properties of the soil change: the colour is red or yellow because of precipitated iron, the plasticity decreases, and capacity for swelling and cohesion also decrease. When wet, the soil is relatively soft and easy to cultivate, but when dry it is hard and bricklike, at least in extreme cases. The name *laterite* (from the latin *later* meaning brick) is given to such soils, which are used in the tropics as building materials. The hardness is stable, since it is brought about by irreversible processes.

Podsolisation. Laterites vary in acidity from weakly alkaline to strongly acid: podsolisation may occur in acid laterites. The processes involved are partly the reverse of those involved in laterite formation, in that iron and aluminium are lost in larger amounts than silica. It is particularly in the transitional areas between the tropical and temperate zones that laterites are exposed to such changes, and especially in tropical rain forest with high productivity and with the consequent humus production and infiltration of humus into the laterite. Hence, in tropical rain forests, as in other communities, podsolisation occurs in connection with accumulation of stable humus. The same is true of the monsoon forests, in places where there are stands of vegetation dense enough to produce an excess of humus. In the most acid localities, there may even be formation of raw humus. The lower the pH, the more rapid is podsolisation. The course of podsolisation seems to be similar to that in the cold and temperate regions (Wilde, 1946).

TABLE 65 The Si:Al ratio in colloidal clay in the A and B horizons from a series of soils, representing some of the large groups of the podsol soils of cold regions and the laterites of warm regions. (From Lutz and Chandler, 1946, page 407.)

	A HORIZON	B HORIZON
Podsol	3·41	0·81
Brown podsol	2·48	0·79
Grey-brown podsol	2·33	2·42
Red podsol	1·65	1·88
Laterite	1·01	1·78

There are many transitional forms between the podsols of warm and cold regions, yellow, brown and grey brown in colour. Yellow podsols occur in wetter areas, but otherwise the type of podsol and the Si/Al ratio seems to depend most on temperature. Table 65 shows how the Si/Al ratio in the A horizon decreases as temperature increases. The reason for this is the decrease in silica content of the laterite A horizon, and the accumulation of aluminium in the podsol A horizon and the brown earth B horizon, but not in the other soils.

Compared with the circumpolar soils, which mostly date from the last glaciation, laterite soils are very old, perhaps going back as far as the Tertiary period (Robinson, 1949), so weathering extends to a considerable depth, often twenty metres or more. The results of the many processes which have played a part during this long period of development have been modified by more recent processes and by these going on now, making research on laterite development a difficult task with many current hypotheses. The term *laterite* may be interpreted in different ways.

The relation between formation and vegetation is not well understood. However, laterite extends to greater depths than do plant roots, so its formation can be regarded as a geological process rather than one of solum formation.

Laterite is the forest soil of tropical and subtropical areas. Rain forests, monsoon forests and savanna all grow on various kinds of laterite, and so does some *macchia* vegetation. The difference in forest types may therefore derive to some extent from the difference in laterites, but the primary cause is undoubtedly climatic, mainly the humidity (H), with the effect of climatic differences accentuated by the lack of water-storing soil colloids. All the types of vegetation are xeromorphic to some degree.

Laterite generally forms dry, poor soil types, sometimes partially sterile in savanna areas, where the laterite has formed a hard crust, more like rock than soil (Wilde, 1946). It has little water-storage capacity and a very low content of nutrients and exchangeable bases, because of the lack of colloids and the high degree of leaching of base cations.

The nutrient reserves are mainly in the vegetation, at a stage of the cycle between plants and soil. Hence if the vegetation is destroyed the nutrient reserves decrease considerably and a new generation of forest will never attain the original luxuriance. If the soil is cultivated, the crops suffer from nutrient deficiency within a few years (Walter, 1951).

Even the most productive lateritic soils, such as those with tropical rain forest, have only limited nutrient reserves. If the forest is cleared and the soil cultivated, the reserves of nutrients and humus decrease rapidly. Humification is many times more rapid than in soils of temperate regions and so are weathering and leaching, because of the high temperature and precipitation. Hence the vegetation is more sensitive to interference than forest in temperate regions. The widespread distribution of scrub vegetation in areas which have been highly populated for a long time is probably mainly due to overexploitation of the vegetation, particularly the forests and pasture.

Semihumid and semiarid soils and their vegetation

At the boundaries of regions with humid or arid climates there are intermediate climates, and soil which are also intermediate in some respects between the humid and arid types. These soils are mainly mulls of various depths, i.e. brown or black mixtures of humus and mineral soil, called *brown earths* or *black earths*: the Russian chernozem or black earth is a characteristic example found in eastern Europe.

Brown earths

Vegetation and humus production. 1 FORESTS. Brown earths are first and foremost the soils typical of deciduous forests, which occur in the northern hemisphere in regions with a semihumid climate, with outlying more scattered occurrences in more humid areas to the north.

These forests may be pure stands or mixtures of species of *Betula*, *Fagus*, *Quercus*, *Tilia*, *Fracinus*, *Ulmus*, *Populus*, etc., forming one to three layers of vegetation, with a shrub layer of lianes such as *Hedera*, *Humulus*, *Clematis* and *Lonicera*, and a field layer of spring-flowering herbs and of grasses. Litter production is large and the soil flora and fauna is species-rich and luxuriant.

Some of these deciduous forests are of beech. Since beech is shade-tolerant, the trees form dense, pure stands with dense foliage and high litter production per unit area of soil surface. Beech leaves humify comparatively slowly, so the litter accumulates in thick layers representing several years production, and the field layer is sparse or non-existent. On better soils, brown earths are formed under beech woods, but on poorer soils there is podsol formation.

Oaks require more light and water than beech, so oak woods are more open, with space for other tree species. Central European oak woods may include many other species of deciduous trees, such as *Tilia*, *Acer*, *Ulmus* and *Fraxinus*, and a shrub layer of *Crataegus*, *Rhamnus*, *Euonymus*, *Viburnum*, *Lonicera*, etc. Typically there is a rich field layer of grasses and herbs.

Deciduous forests were formerly much more widespread than they are today, but they have been decimated or destroyed altogether by overexploitation; it is difficult to assess how long this exploitation has gone on. In Denmark there is evidence of exploitation by man in the late Iron Age (Iversen, 1941). Destruction of the solum often followed forest clearing, so that forest could no longer be supported and dwarf shrub communities developed, resulting in replacement of forests by heaths (see Chapter 18 for further discussion).

Deciduous forests were one of the most important natural resources, especially for prehistoric man. Timber, fuel, fibres, and building materials, materials for tools and animal fodder (leaves, acorns) mostly came from the forests (Sjöbeck, 1927). There were also edible fruits, from apple, rowan and *Prunus* trees, among others (see Chapter 18). The brown earth developed under the trees was even more important and its large nutrient reserves were

FIG 75. Deciduous woodland. Oak, ash, sycamore, hazel, ferns. Brown earth. Kinnekulle, Sweden.

utilised in creating meadows and arable land. The remaining areas of the original deciduous forest are scattered remnants distributed among the arable land on ground unsuitable for cultivation because it is stony or uneven (see Chapter 18) or small patches retained for their amenity value. At the boundaries of the coniferous forest region, and also within it, there are sporadic occurrences of deciduous woodland. In central and northern Sweden, it is found in nutrient-rich valleys, alone or mixed with conifers.

2 GRASS AND HERB COMMUNITIES. For several centuries, perhaps for thousands of years, meadows were one of the most important communities on brown earth soils. They were created as a result of forest clearance by man, followed by rapid development of the grasses and herbs of the field layer left to utilise the site without competition from the trees. These grasses and herbs were cut for

hay by man. Only remnants of these meadows are now left (see Chapter 18).

In the humid regions the meadows were mostly on brown earths or on soils which had previously been brown earths, and so are the present-day remnants. The flora owes its character mainly to the soil, although the management of the meadows also has a large effect. These meadows differ in flora and in origin from natural meadows such as Alpine meadows or some of the meadows along the edges of rivers or lakes.

The flora is made up of herbs and grasses with soft, juicy tissues, rich in nutrients and good for fodder. There are numerous species of grasses, some of which are tussock-forming. Most families are represented among the herbs, many with several species, for example, Ranunculaceae, Compositae and Papilionaceae. The flora includes all species named *pratensis*. Meadows have long been admired for their beauty and large numbers of flowers.

Most of the plants are geophytes and cryptophytes with creeping or orthotropic rhizomes. Bulbs and plants with runners are less common, and annuals are rare. The field layer is a closed community and the grasses form a dense turf. There are sometimes several mosses in the bottom layer, depending on the treatment of the meadow and on its productivity. A fertile meadow is more or less free of mosses, but mosses increase and productivity decreases as fertility is reduced by harvesting and grazing. Fertiliser application soon results in the decrease of mosses again.

3 SCRUB. At the junction of forest and open terrain (meadows, arable land, or water) there are often various shrubs of the genera *Rosa*, *Rubus*, *Prunus*, *Crataegus*, etc. The field layer is relatively rich in species, probably because the shrubs protect it from grazing and support animal life, especially birds, which favour soil development and seed dispersal. The microclimate and other environmental conditions in the scrub are different from the forest and the open terrain in many respects: wind is less than in open terrain, evaporation is less and soil temperature is higher, precipitation is less, because of the rain shadow effect of the forest, and water availability in the soil is less, especially if the roots of the forest trees penetrate the scrub belt, as they often do. There is a supply of litter from the forest to the scrub, and material from the forest is transferred by animals to the scrub, for nest building or as food. The conditions are thus rendered favourable for brown earth development by the plant and animal life in the scrub.

The occurrence of brown earths. Brown earths are common in temperate regions in which the climate is humid in some seasons of the year and arid at other times. They are widespread in north, south and west Europe, and sporadic in adjacent wetter areas. The south and central Swedish deciduous forest and meadow soils are generally of this type (Tamm, 1930).

The characteristics of brown earths. Litter is broken down rapidly and is therefore present only in thin layers. Under it is a layer of mull, of variable thickness, from a few centimetres to a metre or more. There is generally a high humus content, decreasing with depth. The higher the humus content, the less marked the morphological difference between the brown earth soil and the underlying layers of gravel, sand or clay. The lowermost part of the mull layer is brown in colour because of precipitated ferric hydroxide. The soil colour depends on the relative amounts of black and brown stable humus and the red precipitated iron. Thus, there are various layers in brown earths, but the boundaries between them are indistinct (Lundblad, 1924; Tamm, 1930).

Brown earths differ from podsols in several respects. They have no leached layer, except for soils from which so much calcium has been lost that it hardly occurs in the upper layers. Such leaching takes place in humid regions where brown earths develop because of favourable soil and vegetation, in spite of the high rainfall. Colloids derived from weathering, salts and mineral particles are fairly evenly distributed throughout the soil profile. Only the humus colloids show a decrease with depth (Table 66).

TABLE 66 Colloid content in a brown earth (beechwood, south Sweden). SiO_2, Fe_2O_3 and Al_2O_3 calculated as % of inorganic dry matter. Loss on ignition (mainly humus) calculated as % of the air-dried sample. (After Lundblad, 1924, page 18.)

		COLLOID CONTENT		
LAYER	LOSS ON IGNITION %	SiO_2 %	Fe_2O_3 %	Al_2O_3 %
Brown earth 3–13 cm	12·18	0·02	0·62	0·53
Brown earth 15–32 cm	5·91	0·06	0·62	0·42
Brown earth 32–57 cm	3·52	0·08	0·52	0·58
Till 60–80 cm	1·64	0·07	0·36	0·37
Till 100–115 cm	2·27	0·07	0·16	0·37

The salt content is high. Brown earths are base-saturated to a relatively great extent, whereas the base exchange capacity in raw humus is largely taken up by hydrogen ions. The content of humus colloids and clay colloids (SiO_2, Al_2O_3) is also high. Hence, there is a higher water-holding capacity and more evenly distributed soil moisture than in podsols. Brown earths hold more water than podsols in dry conditions, less in wet conditions.

Brown earths are neutral or slightly acid. Their aggregate structure is stable and not destroyed by rain. They are more fertile than podsols.

The origin of brown earths. Mineral soil, vegetation, climate and fauna must comply with certain conditions if brown earths are to develop.

The effect of the mineral soil depends mostly on the content of bases. A high base content promotes brown earth development. Calcium is especially important, so brown earths tend to develop on limestone and greenstone soils which makes them more stable than in places poor in bases.

The vegetation plays a part because it acts as a store of nutrients and a source of litter. Where the mineral soil is base-rich the vegetation is demanding and produces nutrient-rich litter. There is a reciprocal effect between plants and soil in that the vegetation gives rise to a large amount of nutrient-rich humus, which, in turn, retains mineral salts and increases productivity. In deciduous woodland consisting of the more demanding tree species, the brown earth soil is usually richer than in woods of the less demanding oak, beech or birch. However, in favourable conditions, brown earths are formed even under these less demanding trees, for example, in beechwoods (Lundblad, 1924).

Because of their high nutrient demands, species characteristic of brown earths hold relatively large quantities of mineral nutrients in circulation. The tree roots also make an important contribution, since they penetrate deep into the soil and are evenly distributed throughout the profile (Tamm, 1940) so that the nutrient reserves in the deeper layers of soil are also brought into circulation. In this way, nutrients may accumulate on the soil surface of deciduous woodland, which may even become calcium-rich as a result (Brenner, 1930).

Climate affects brown earth formation in several ways. The temperature must be high enough to allow rapid weathering and humification, but not so high that the humus fraction is broken down as rapidly as it is in tropical and subtropical regions. It must be high enough for humification in the lower layers so that peat does not accumulate, because nutrients would be held in peat and prevented from entering circulation. A suitable balance between production and breakdown of humus is such that the soil has a large amount of humus which takes part in nutrient circulation, even in the deepest layers.

Precipitation must be sufficient in relation to evaporation for occasional downward movement of water through the soil, so that salts do not accumulate in the surface layers, but not so great that large quantities of nutrients are lost. If leaching is slight, some nutrients are lost from the surface layers, but the greater part remains within the root zone, either because the water does not move any further downward or because the minerals are precipitated and retained on the way down. In semihumid and semiarid regions the climate is generally suitable in these respects.

Brown earth is a mull soil and its characteristic properties depend partly on its structure, so its occurrence is dependent on the processes which cause mixing of the mineral and humus fractions. The precipitation in relation to evaporation must be such that the stable humus moves downward in the soil and is precipitated without being lost altogether. Precipitation onto mineral particles, particularly the basic minerals, occurs most readily, so that the particles are cemented together. The intimate contact between the acid humus and the mineral particles promotes weathering throughout the whole brown earth profile (Hesselman, 1926; Tamm, 1930, 1941).

The soil animals are important in mixing, especially worms and various larvae, which are always present in brown earth, being favoured by the high content of suitable nutrients in the litter (see Chapter 6).

Unstable forms of brown earth. At the boundaries between humid and semihumid climatic regions, i.e. between the main areas of podsol and brown earth soils, there are soils intermediate between the two (Malmström, 1949). The more leaching there is, the more the soil resembles a podsol. There is no even gradation between areas of podsols and areas of brown earth because of the way in which brown earth development is affected by mineral status, soil moisture, vegetation, etc.

Since the solum is the result of interacting chemical, climatic and biotic processes, it is in a dynamic state. This is particularly true of the brown earths, with their relatively high humus content. One of the manifestations of this dynamic state is the way in which the mineral characteristics, climate, vegetation and fauna can replace each other to some extent in their effects on humus and mull formation. Thus a high content of bases and other nutrients in the mineral soil may partly compensate for low temperatures or excess precipitation. At the edge of the podsol region, where the climate is unfavourable for podsol formation, it is mainly the mineral characteristics of the soil which affect the dynamic relationship between brown earth and podsol (Tamm, 1930). If the mineral soil is variable, brown earths and podsols may develop in close proximity, e.g. an area of moraine may bear brown earth whereas nearby areas of sand and gravel are podsolised. The moraine has more fine material and weathers more rapidly, so may be sufficiently rich in nutrients, and moist enough, for brown earth formation although the sand and gravel are not (Tamm, 1921). Good nutrient supply has the same effect. In north Sweden, brown earths occur in valleys supplied with water rich in nutrients from places higher up, especially in areas rich in lime (Tamm, 1930). Temperature may be decisive in other places, and on south-facing slopes brown earth formation requires less good mineral soil than on colder slopes.

Because of this sensitivity to environmental conditions, brown earths in the transitional belt between the main areas of distribution of brown earths and podsols tend to be unstable and they may change to, or develop from, podsols. Brown earths in Sweden are generally in this category. The instability is greatest where the conditions for brown earth development are least favourable, and then the vegetation determines the direction of development. If spruce forest invades an area of deciduous forest on brown earth, or succeeds the deciduous forest after felling, the brown earth gradually changes to podsol (Tamm, 1921, 1929, 1930). Spruce litter contains few bases (Table 67), the soil becomes acid, leaching

TABLE 67 The acid/base balance in green leaves. Acidity $(H)=$ potential acidity, i.e. the capacity of the material to bind bases, determined by electrometric titration to pH 7. Bases $(B)=$ base excess in the ash, determined by titration. pH = actual acidity in a suspension of the material in water. (After Mattson and Koutler-Andersson, 1941, page 29–30.)

SOIL TYPE	TREE	ASH %	ACIDITY H ml 100/g dry matter	BASES B	ACIDS BASES $\frac{(H+B)}{B}$	pH
Podsol	Pine	2·52	50·5	33·5	2·51	3·87
	Spruce	4·14	43·3	46·2	1·97	3·97
	Beech	5·58	13·6	84·1	1·16	6·10
Brown earth	Spruce	4·65	27·0	71·5	1·38	4·41
	Beech	4·89	13·5	112·6	1·12	6·21

increases, podsolisation begins, and soil structure changes. Podsolisation is slower to start if the soil is rich in lime or other bases, and the spruce woods on limestone on the island of Gotland are on soils which have kept their brown earth properties and support a rich herb vegetation (Lundblad, 1924; Tamm, 1930). Podsolisation is also slower in coniferous forests in which there are some deciduous trees. The greater productivity of mixed forest may be partly attributable to this.

Podsols may change to brown earths if the conditions become more favourable, for example if plants with a high content of bases invade an area. Tamm (1921, 1930) has described such changes, following replacement of coniferous forest by deciduous trees. Dimbleby (1952) cites an example from Yorkshire, where raw humus on a pine-heather moor changed to mull in 60 to 100 years after invasion of birch, and the hardpan broke up. Grass allowed to grow undisturbed on a podsolised heath soil can bring about the same change (Mattson, Eriksson and Vahtras, 1948).

The lability of the soils in these intermediate regions is associated with corresponding variation in the plant communities, between deciduous, coniferous and mixed forests, for example.

Steppe soils and their vegetation

Vegetation. 1 GRASS AND HERB COMMUNITIES. Steppes, which are treeless grassy plains, cover large areas of Russia, Hungary, Central Asia, United States (the prairies) and South America (pampas), that is, about one-fifth of the total world land surface: grass and herb communities are at their most widespread in these areas. Perennial plants with bulbs and rhizomes, especially grasses, and annuals, constitute the major part of the vegetation. The moister steppes grade into meadows, the drier ones into bush-steppe and semidesert. The difference lies in the degree of closure of the plant community: steppe is more closed than bush-steppe, but not as completely closed as meadow. Steppe vegetation does not form a dense turflike carpet, but the plants grow alone or in groups and often form tussocks. They are often xeromorphic, so that steppe plants resemble those of bush-steppe more than those of meadows.

Steppes are treeless for several reasons. One is the lack of water. At the boundary of forest and steppe, trees grow in the moister places, where the water table is relatively high, but there is insufficient water for large numbers of trees. Forest trees planted in such places grow well at first, while they are small with low water requirements. But as they increase in size, lack of water limits their growth (Walter, 1943). Forest growth is also prevented by hay-making and grazing, and Walter (1943) states that in eastern Europe there are hardly any areas of steppe which are not utilised for these purposes. Hay-making changes the steppe vegetation in the same way as it does meadow vegetation: the field layer is protected from invasion by trees, and the sharp boundary between steppe and forest probably developed because of it. The development of steppe is also favoured by moderate grazing, such as that by wild animals. Grazing restricts growth of trees more than that of grasses; seed germination is favoured to some extent by animal trampling forming seedbeds. Numbers of wild animals on steppes are now insignificant in comparison with former herds of buffaloes, antelope and other large game animals. Osborn (1949) estimated that there were fifty million buffaloes on the North American prairies, but there are now only a few hundred, in reserves.

Frequent fires influence steppe vegetation, favouring grasses and herbs and preventing establishment of trees.

2 SCRUB. *Macchia* vegetation is common in the semihumid and semiarid regions of the subtropical zones. In dryer places the *macchia* is replaced by steppe: in wetter places it is more forestlike.

Soil. The reserves of mineral salts in steppe soils are smaller than in arid soils, but greater than in podsols; there is sufficient to support high production, but production is limited by other factors, mainly by lack of water. Steppe soils are variable in several respects, since they have developed over large areas under different conditions of climate and soil. Some properties can therefore develop differently, especially accumulation of lime, humus and other colloids, and there are several types of steppe soils, which can be grouped as prairie soils, chernozem soils and chestnut soils.

1 PRAIRIE SOILS. In the humid parts of the American prairies there is a soil which resembles the brown earth of deciduous forests, in its humus content, for example. Summer rainfall is sufficient to support a rich grass vegetation, with high humus production, since grasses give rise to especially large amounts of humus, probably more than other plants. Because of the large biomass per unit area, grasses give rise to more humus than herbs; and more than trees because, at the end of the growing season, all the material except for the underground parts becomes litter, whereas in trees and bushes some of the material produced is retained in the overwintering parts and growing shoots, and some supports respiration of these parts. Grasses also give rise to relatively large amounts of litter under the soil surface. Weaver and Zink (1946) found that the weight of roots, rhizomes and stem bases of three typical prairie grasses, three-year-old plants, were 12·2, 10·0 and 7·5 Mg/ha (dry weight at 32°C), and the ratio between the underground and above-ground parts was 0·57, 0·34 and 0·4, respectively. One-third to one-half of the material below the soil surface was in the top 10 cm of soil. Litter production by steppe grasses is considerable, even if the above-ground parts are completely removed, by grazing, for example. Production varies with species, and is greatest in perennials. Annuals, e.g. cereals, produce large quantities of root material, but this humifies rapidly and only a small proportion remains in the soil.

Prairie soils have a rich fauna which helps to mix the humus fraction with the mineral soil, resulting in a thick layer of mull, coloured grey brown to dark brown by the humus, redder in warmer areas. Moderate leaching contributes to the development of a soil profile. Winter rainfall is sufficient for some water to percolate through to the water table. Humus and clay colloids, and calcium and other bases are carried downwards. The clay is deposited at a particular depth, and calcium at a greater depth. As a result of this, and of leaching in general, the surface layers of the soil are weakly acid. However, base saturation is high, the soil is nutrient-rich, the crumb structure good and the soil porous. The deep-rooted grass vegetation keeps a large part of the nutrient reserves in circulation. Silicon turnover is probably also large, because grasses contain more silica than most plants, and more than the leaves of deciduous trees (Lutz and Chandler, 1946).

Prairie soils are found in the Mid-West of the United States, in a belt from north to south, corresponding with the maize belt. Productivity, especially of maize, is high. The prairies have been exploited and ploughed little by little, and now they are one of the foremost areas of cereal production.

2 BLACK EARTH (CHERNOZEM). (a) *Characteristics*. The forms of black earth found in the Ukraine can be regarded as typical. Chernozem is porous, and consists of a thick layer of mull with a well-developed aggregate structure. The depth may be 40 to 150 cm, but is usually between 70 and 100 cm; it is greatest in the northern, more humid areas of eastern Europe and least in the southern, more arid areas (Walter, 1951). The humus content is usually 8 to 10 per cent, it is highest at the surface and decreases with depth. The clay content is high, but not high enough to make the soil heavy. The soil is base-saturated, especially with calcium, and is neutral or weakly alkaline. There is some leaching, but less than in prairie soils and not enough to carry substances down to the ground water. Some carbonate and calcium phosphate are carried downwards, but no silica, no aluminium or iron hydroxides, and probably no clay. Materials carried down from the surface layers are redeposited lower down. At a particular depth (between 1 and 2 m), the calcium forms a definite horizon, and above this there are lens-shaped and mycelial-like deposits. Because temperature and moisture are suitable, the clay weathers rapidly. As a result, and because there is no loss by leaching, the soil is rich in nutrients and very resistant to deterioration (Walter, 1951). Its black colour is connected with the high base saturation and the way in which humification occurs in a neutral or alkaline medium and at relatively high temperatures (Robinson, 1949). The extent of chernozem development varies with the climate and mineral soil.

(b) *Distribution*. Chernozem covers large parts of most continents. In Europe and Asia it is found in parts of Rumania, Hungary, Czechoslovakia, Poland, Austria, Yugoslavia, Bulgaria and the Ukraine; there are also chernozemlike soils in some parts of Germany. These European areas can be regarded as the western parts of a belt stretching from the Ukraine to the Urals, through south Siberia and Central Asia (the Kirgis Steppe) to Mongolia and parts of South-East Asia. Chernozem is found in parts of Australia. In Africa there is one area from Senegal through the Sudan to Ethiopia, a second from Somalia to south Ethiopia, and others.

In America there is one belt through Canada, another from north to south through the central part of the US, and a third through South America.

(c) *Conditions for development of black earths*. The Russian and central Asian black earths have developed on a thick layer of mineral soil, finely divided and clayey, separated from moraines by the wind, or carried from surrounding deserts. This layer may be up to several tens of metres in depth. Chernozem can also develop on other types of mineral soil, provided that there is a sufficiently high content of clay and bases. The clay content is variable, e.g. 5 to 15 per cent in northern European areas of chernozem and 35 to 55 per cent in southern European areas (Walter, 1951).

The temperature climate of areas in which chernozem develops varies from that of the subarctic, wet and cold in winter (Ukraine, north central states in the US, and parts of Canada), to that of the subtropical, periodically dry savanna areas (South-East Asia, Africa). The feature in common is the long dry period to which the plants are subject, because of a dry season or a cold season. Russian and Asian chernozem develops in a climate with cold winters and hot summers. The mean January temperature varies from $-21°C$ (Siberia) to $-4°C$ (Ukraine) and the mean July temperature from $17°C$ (Siberia) to $23°C$ (south Russia). All chernozem areas are semiarid to some extent. Precipitation is less than in brown earth or prairie areas, varying between 300 mm and 500 mm per year in the Ukraine, for example. Sometimes the rain comes in such large amounts that the upper layers of the soil are leached, but there is insufficient to carry the leached substances into the ground water, so the colloids and nutrients remain in the rooting zone. In the cold districts the soil is frozen in winter to a considerable depth; some of the water resulting from thawing is lost as surface run-off.

Steppe vegetation of grasses and herbs, mainly tall, tussock-forming species of the genera *Stipa, Festuca, Poa, Bromus*, etc., forms the cover. *Stipa* species are particularly characteristic of Asian chernozem vegetation. There are other cryptophytes and hemicryptophytes, such as species of *Iris*, *Adonis*, *Salvia*, *Crambe*, *Statice*, etc. On bare patches of soil between the tussocks there are various therophytes: *Draba*, *Veronica*, *Myosotis*, etc. (Walter,1951).

Like prairie soil, chernozem is dependent for its development on the litter and humus produced by a rich grass flora. The climate is an important cause of humus accumulation. The growing period is short, but sufficient for the grasses to produce ripe seed. Then there is either a cold or a dry period, so that humification is slowed down and several years' production of humus accumulates.

Grasses also favour chernozem development because of their large root systems. The roots develop well in deep, porous soils like chernozems, and attain large numbers and great depths, so they can take up the nutrients leached from the uppermost layers. The extent and depth of the root systems also enables the grasses to survive drought, so that they are well-adapted to the dry steppe climate.

The fauna is also important, particularly the larger animals, worms, ants, and beetles and their larvae. Their activities improve the soil structure and increase its permeability and aeration so that roots can penetrate easily. The large number of animals and other organisms is probably one of the reasons for the high productivity which is characteristic of chernozem. The carbon/nitrogen ratio is usually low, about 10 (see page 185).

(d) *Exploitation*. Because of its high nutrient content and good structure, and consequent high fertility, chernozem has long been utilised for cultivation, and is now almost all arable land (Walter, 1951). Chernozem areas are the world's most important for wheat and rye production. They are less suitable for other crops, which generally need higher rainfall. Because of the large nutrient reserves, the effective weathering of the mineral soil and the way in which the nutrients remain in circulation, good harvests are possible even without added fertilisers. However, yields are less than in areas where agricultural methods are more intensive, such as in western Europe. Walter (1943) gives figures for yields in eastern Europe of 7 decitons per hectare, as compared with 21 decitons in Germany. Average harvests in Sweden from 1945 to 1950 were 22 decitons for winter wheat, and 18 decitons for spring wheat.

3 CHESTNUT SOILS (BUROZEM). At the edges of the arid zones in Russia and Asia, the soil is dryer and leaching is less than in the chernozem districts. Calcium carbonate accumulates near the surface, calcium sulphate is sometimes deposited under the calcium carbonate. The vegetation is still steppelike, but consists mainly of short-growing grasses. Rainfall is only sufficient to moisten the surface layers, productivity is less than that of chernozem, so

humus production, and depth of humus, is less; iron content of the surface layer is higher and the colour is therefore chestnut brown. 4 ALVAR SOILS (RENDZINA). The alvar areas of the Baltic islands of Öland and Gotland and in the Swedish province of Västergötland are covered by vegetation which is steppe-like in several respects (du Rietz, 1925; see page 328). The solum has a number of features characteristic of the type known internationally as *rendzina*. [The word *rendzina* is derived from a Polish word meaning rattle, because of the noise made by ploughing this stony

FIG 76. Alvar with juniper and pine bushes. Öland.

soil, Laatsch, 1954.] This is the stratified solum developed on limestone, dolomite or marl poor in quartz sand. In the course of time, a layer of loose soil develops on the surface as a result of humus production and weathering and splitting of the rock by root growth. The type of soil depends on the substratum. Hard limestones or dolomites with little clay give rise to shallow, very stony types of rendzina, poor in colloids and with low water-holding capacity. The lack of water is accentuated by exposure to wind and sun and the efficient drainage through cracks in the bedrock.

Rendzinas on lime and clay marls are deeper and richer in

colloids, and have higher water-holding capacity (Laatsch, 1954; Kubiëna, 1953). The wide variations in water economy, the high soil temperatures, the high lime content and the exposure are special conditions which affect the vegetation.

5 THE STABILITY OF STEPPE SOILS. At the boundaries of the humid and arid zones, the steppe soils are not very stable. Near the humid areas, where steppe grades into so-called *forest steppe*, the steppe tends to be invaded by forest trees, and this results in changes in the soil character. The calcium carbonate level sinks,

FIG 77. Alvar with fissures. Öland.

the humus fraction decreases and becomes more acid, the pore structure breaks up, iron and aluminium hydroxides are leached and precipitated lower down, and the black colour becomes grey. Chernozem which has changed in this way is called *grey forest soil*, or *degenerate chernozem*. It grades into the podsol soils, and in extreme cases it shows podsolisation. However, this process can also go in the opposite direction, for example, if a degenerate chernozen is invaded by grasses (Robinson, 1949; Walter, 1951).

Near the arid areas, changes are of the opposite kind. Increasing aridity causes decrease in humus content, browner or redder colour, and a thicker layer of calcium carbonate at a higher level.

Arid soils and their vegetation

Over large areas of the earth's surface, mostly in the hotter regions, evaporation exceeds precipitation and the ground dries rapidly after rain. The development of the vegetation is limited mainly by shortage of water and sometimes also by the salt which accumulates in the soil as a result of the lack of leaching.

The more arid the climate, the lower the humus content of the soil, partly because of the low productivity, partly because humification is very rapid. In relation to humus production above the ground, production below the soil surface is probably greater than in the plant communities in the humid regions, because the vegetation of arid soils has a relatively greater amount of plant material in the soil. In spite of the rapid mineralisation, the humus content is sufficient to colour the soil grey, so that arid soils of this type are called *grey soils* (serosem). A high salt content is asso-

FIG 78. Semidesert, with bushes, grasses, thistles. Fiuggi, central Italy.

ciated with aridity. Weathering is rapid because of the high temperatures, and there is no real leaching, only some movement of the salts in the soil. This movement depends on the amount of rainfall, the topography of the ground and the height above the water table. Because of differences in these conditions of the environment, there are several types of soil and vegetation.

FIG 79. Semidesert. Stones, gravel, and an insignificant amount of fine material. Fiuggi, central Italy.

Bush steppe

If there is enough rain to leach the surface layers of the soil to some extent, salts are reprecipitated further down; calcium is deposited just below the soil surface, partly as sulphate and partly as carbonate. The calcium level is high in the profile in arid soils. Steppe vegetation can develop because of the somewhat decreased salt content in the surface layers caused by leaching. However, the salt content is still high and, together with lack of water, this restricts the vegetation to strongly xerophytic grasses and bushes. The vegetation is sparse and open and is called *bush steppe*. It is widespread in Russia, Central Asia, south-western Africa, parts of Australia, and the western parts of the US, from west Texas to south Arizona. If the vegetation on the bush steppe is damaged or destroyed, by grazing, for example, wind erosion may remove the loose soil so that the hard deposits of calcium salts are exposed and the bush steppe becomes desert.

Bush steppe may also develop on hills and slopes from which salt has been washed away by rain sufficient in amount to allow some surface run-off.

Desert

In areas where rainfall is insufficient to bring about any leaching, or where leaching takes place, but evaporation brings the salts up to the soil surface again, the salt concentration at the surface is so high and the lack of water so extreme that the soil is completely or partially sterile and the area is desert, with very little or no vegetation.

Wind erosion increases the tendency towards desert formation because the finest particles, with highest water-holding capacity, are carried away. The whole of the loose soil may be blown away in some places, so that the bedrock or underlying hard layers are laid bare.

In the Arizona desert, the calcium layer (caliche) is only 10–20 cm below the soil surface (Lutz and Chandler, 1946; Walter, 1951) and there are similar desert soils in other places in the western parts of the United States.

Desert vegetation consists of grasses and bushes, the development of which is limited by the amount of rainfall. After rain, the vegetation develops, but then disappears until the next period of rain. Kassas (1952) gives a good example from the Egyptian desert of the relation between plant development and rainfall in desert areas. He distinguishes three life forms among desert plants: (a) *Perennials*, with root systems penetrating to the deeper layers where water collects. These are very resistant to drought. (b) *Ephemerals*, which only become apparent during the wetter seasons of the year, with superficial root systems not penetrating deeper than *c.* 20 cm. In dry years this vegetation does not develop to any extent, but in exceptionally wet years it may cover as much as 70 per cent of the soil surface in suitable places. (c) *Occasional desert elements* which develop in especially favourable conditions. Some of the oases (Kharga and Dakla oases) studied by Kassas have *c.* 10 mm rain every ten years. After rain, vegetation develops rapidly in the ravines where water collects off the plateaux, but disappears as rapidly when the water dries up, to reappear after the next rain several years later.

Salt steppe and salt desert

If the salt content of the soil is especially high, the terms *salt steppe*, *salt flora*, etc. are used. Salt may accumulate in valleys if it is washed down from higher up. Salt lakes form, leaving salt soils covered by a crust of salt after the water has dried up.

Salt accumulation can also occur in another way. Near marshy land and places subject to flooding, the water table rises to the soil surface and water evaporates, leaving the salts as a light grey crust. The salts are mainly chlorides, sulphates, carbonates and bicarbonates of sodium, potassium, calcium and magnesium, usually including large amounts of sodium chloride. The proportions of the various salts are variable, so that the pH of the soil varies from weakly to strongly alkaline. Such saline soils are known by various names, such as *solonchaks* by the Russians, or *white alkali soils*.

If saline soils are drained, naturally or by man, some of the salts are leached out, the neutral salts in relatively larger amounts than the alkali salts, so that the soil becomes strongly alkaline. Sodium carbonate is accumulated to a particularly large extent. These soils are called *solonetz*; they occur in Hungary, southern Russia, Central Asia and the US for example. The sodium carbonate disperses the humus in the soil, so that the soils become black or brown in colour. In America this has caused them to be known as *black alkali soils*.

Stocker (1929) interpreted leaching of the Hungarian solonetz soils as an upward leaching. Rain dissolves the salts, but there is no downward movement because of the low soil permeability. Instead, sodium carbonate moves up to the surface as the water evaporates. The humus and the iron, aluminium and clay colloids are dispersed by the sodium carbonate, and move up to the surface, where they are precipitated.

Salt steppe vegetation is markedly halophytic and xeromorphic. Most species are of the genera *Salicornia*, *Suaeda*, *Atriplex*, *Anabasis*, *Haloxylon*, *Halopeplis*, *Statice*, etc. characterised by a high salt tolerance. *Salicornia rubra*, for example, grows in places where the salt content is 0·5 to 6·5 per cent. The high ash content of the plants, which may be up to 25 per cent of the dry weight, is another characteristic feature. Stocker (1929) described the half xerophytic steppe vegetation of the Hungarian steppe as a mixture of life forms, best developed in the spring, when there is most water in the soil, but reduced to a species-poor bush steppe flora during the dry summer, with species of *Poa*, *Artemisia*, *Statice*, etc. This flora consists of species with roots that can penetrate the hard layer in the soil and reach the deeper, moister layers; *Statice* and *Gmelina*, for example, together with cryptogams which can withstand desiccation, such as *Nostoc commune*.

If the salt content is extremely high, salt deserts are formed, in or near areas of salt steppe. In some places, near the Dead Sea, for example, the salt content in the soil may be 14 to 27 per cent.

However, even this soil is not completely sterile, since some algae and possibly some microorganisms exist there (see Chapter 16).

Salt steppes and salt deserts occur in Spain, Hungary, southern Russia, inland Asia, Utah, Argentina and Australia.

The relation between vegetation, humidity and soil

There are podsol soils throughout practically the whole of Sweden, but mostly in the central and northern parts, where they are most highly developed. Brown earths are most common in the south of the country, and occur only sporadically further north. The development of both these soil types is related to the humidity (H).

Broadly, humidity changes with latitude and with the continental or maritime nature of the climate, but like other climatic factors it may vary considerably over a small distance particularly in mountainous or hilly districts. Humidity on a mountain slope differs from that in the valley below, and from that on an adjacent mountain slope facing in a different direction. Close study of climate and its effect on soil development requires reliable measurements of humidity and not only of precipitation, temperature and windspeed.

A simple expression of humidity is the amount of run-off. In the mountainous areas in north Sweden, run-off is 1350 to 1450 mm, equivalent to a mean annual precipitation of 1500 to 1600 mm, while in the south-east, and on Gotland and Öland, it is about 150 mm (R. Melin, 1957).

Hesselman made calculations of the humidity from the run-off of rivers in different parts of the country (Table 68), showing that humidity is lowest in the eastern and south-eastern areas, somewhat larger in the south-west and in the northern forest areas, and largest in the mountains.

The distribution of regions of differing humidity (H) in Sweden is shown in Fig. 82 and the regions are classified in Table 68. A detailed comparison between vegetation and humidity in Sweden was made by Hesselman (1932). In general, as humidity rises the temperature decreases, the occurrence of brown earths declines and raw humus formation and podsolisation become more widespread, with consequent changes in the vegetation.

In the least humid areas, especially on the islands of Gotland and Öland in the Baltic and coastal areas in the east and south-east, the soils are mostly brown earths or small areas of rendzinas with rapid humification and little or no podsolisation. A feature of the flora is the occurrence of several species of the Russian steppe, such as *Plantago tenuiflora, Ranunculus illyricus, Adonis vernalis, Carex*

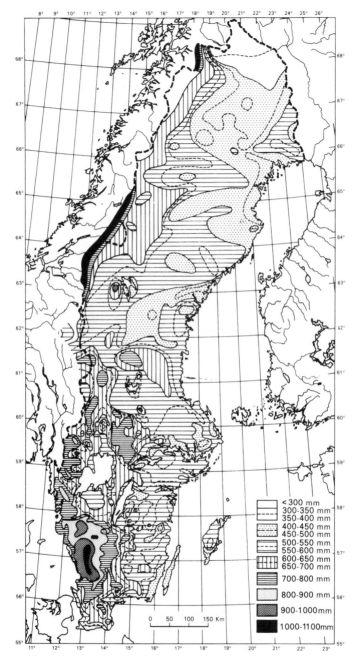

FIG 80. Precipitation map of Sweden. Means for annual precipitation 1911–20. (After Langlet, 1936.)

obtusata, Scabiosa suaveolens and *Minuartia viscosa*. In this region, too, there are large areas under commercial fruit production, and there is some tobacco cultivation.

In the weakly humid areas (Fig. 82) the soils are mostly brown earths, humification is rapid and podsolisation is weak. Bogs are mostly forest bogs at high altitudes. The flora includes several

FIG 81. Tobacco cultivation. East Skåne, Sweden (Photo. ATA, M. Sjöbäck.)

TABLE 68 The ratio precipitation: evaporation (= precipitation minus run-off) in the regions of some Swedish rivers. (After Hesselman, 1932, page 522.)

RIVER	PART OF SWEDEN	PRECIPITATION / EVAPORATION
Motala ström	SE	1·6
Mörrumsån	S	1·8
Lagan	SW	2·0
Dalälv	central (E)	2·3
Stora Luleälv	NE	3·0
Virijaure outflow stream	NW	4·0

continental species (Sterner, 1922) with their main areas of distribution in east and south-east Europe, e.g. *Pulsatilla pratensis, Cynanchum vincetoxicum, Draba muralis, Melica ciliata, Scutellaria hastifolia, Melampyrum nemorosum* and *Oxytropis pilosa*. Other continental species are common to this and the preceding region, e.g. *Trifolium montanum, Polygala comosa, Agrimonia eupatoria, Geranium sanguineum, Sedum album, Asperula tinctoria*, etc. Fruit growing is also widespread in this region.

The moderately humid region includes the greater part of north Sweden as well as parts in the south-west and central districts. In all the areas humification is slow and podsolisation strong. Large areas are covered by bogs and coniferous heaths. Only in the south-west is the flora remarkable for the inclusion of several species characteristic of Atlantic coastal areas, e.g. *Erica tetralix, Juncus squarrosus, Gentiana pneumonanthe, Narthecium ossifragum, Sagina subulata, Galium saxatile* and *Genista pilosa*.

In the highly humid regions, especially in the mountains in the north and west, podsolisation is strong and brown earths occur only sporadically in warm, flushed areas, for example. In general the soil is covered by a thick layer of raw humus and the dominant vegetation is forest heath and bog. Peat formation is also widespread and its development is carried to further extremes, even on level, well-drained ground in the extremely humid regions in the mountains and in Halland in the south.

The relation between humidity, soil and vegetation supports the idea that there is a causal connection between the distribution of certain species and the humidity, and that humidity, directly or indirectly, has an effect on the development of the plants. Humidity is a function of temperature, precipitation and wind, the first two determining the production in the soil of salts (by weathering), fine particles (by weathering and humification), and organic substances, while the degree of humidity affects accumulation or loss of fine particles. It is therefore humidity that mainly determines the fertility of the soil. Plant species are, of course, also affected by other climatic factors, such as day length, light intensity, temperature and duration of particular temperatures, etc., but not as much as by humidity. They may be differentially affected by these factors, so that the distribution of some species may correspond with some temperature factor or with height above sea level, for example. There is a clear relation only when one of the ecological factors has a dominant effect: the distribution maps in Fig. 83 illustrate this point. They are of plant species with clearly limited areas of distribution: the line for *Erica tetralix* separates the Atlantic and Continental species; the limit of alpine vegetation follows a particular altitude in a way that suggests control by temperature, and

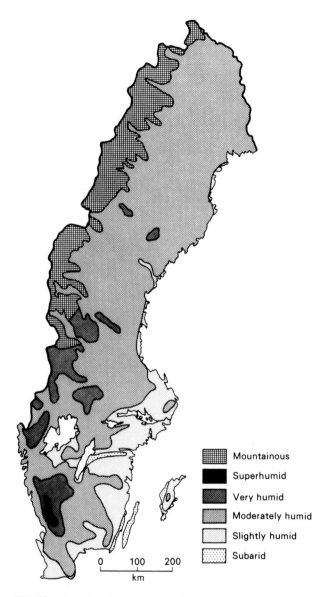

FIG 82. Map showing the humidity regions of Sweden. Mountainous region: *H* values (i.e. the part of the precipitation which does not evaporate) generally over 400 mm and very variable. Superhumid region: *H* > 600 mm. Very humid region: *H* 400–600 mm. Moderately humid region: *H* 200–400 mm. Slightly humid region: *H* 100–200 mm. Subarid region: *H* < 100 mm. (After Tamm, 1959.)

TABLE 69 Survey of the most important groups of soils.

CLIMATE		WEATHERING LEACHING ACCUMULATION	SOILS	VEGETATION
Humid	Cold	Gley formation	Tundra soils Peat soils	Tundra
Humid	Temperate	Podsolisation	Podsols	Bogs, Heaths, Coniferous forest
Humid	Warm	Laterite formation	Laterite soils	Savannas, Monsoon forests, Rain forests
Semihumid Semiarid	Subarctic to subtropical	Calcareous horizon lies deep or shallow	Brown soils Prairie soils Chernosem Burosem	Deciduous forests Meadows Arable soils Steppe {tall grass, arable {short grass, pasture
Arid	Cold to warm	Salt accumulation	Serosem Saline soils Soda soils	Scrub steppe, desert scrub Salt steppe, salt desert

possibly by light. It is more difficult to define the climatic or edaphic factors which are related to the northern limit of ash or the southern limit of spruce, for example.

Survey of soil types and their relation with climate

Table 69 shows the main features of the relation between climate, vegetation and soil in diagrammatic form: humidity and temperature are the only climatic factors considered, so the picture is necessarily incomplete. Nor does the table show the changes in vegetation and soil which result from interference by man.

The relation between climate, vegetation and soil holds generally across the continents from the Poles to the Equator. It is most clear in areas where the climate changes more or less regularly with latitude or altitude. A belt from north to south through European Russia and west Asia may be taken as an example. Several of the soil and vegetation types listed in Table 69 follow one another in sequence:

1 Tundra, on the coast bordering the Arctic ocean.

2 The podsol region, a broad belt of coniferous forest stretching transversely across the continent.

3 The brown earth region, with an associated belt of deciduous forest in central Russia.

4 The black earth region, extending across south Russia eastwards towards Siberia, originally steppe but now arable land or pasture.

5 Salt soils, forming steppes and deserts in south and southeastern areas, including the Kirgis steppe, Persia and Arabia.

A north–south belt from north Scandinavia, southwards through Europe and Africa, shows the same zonation. There is, in turn, tundra, coniferous forest, deciduous forest, *macchia* vegetation, desert, bush steppe, savanna and, around the Equator, rain forest.

On the slopes of high mountains, there are climate-soil-vegetation belts in a more compressed form. In Scandinavia, there is cultivated land in the valleys which originally bore deciduous forest or water meadow, then a belt of coniferous forest with raw humus and podsols, then scrubby forest with more and more peat over strongly podsolised mineral soil, then heath with thick peat and soil affected by frost, and, lastly, bare ground with no vegetation and very marked frost effects.

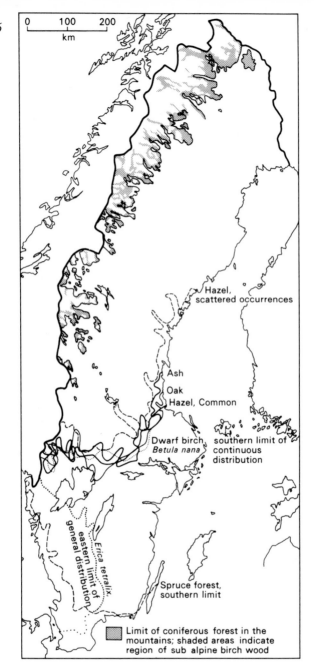

0 100 200
km

Hazel,
scattered occurrences

Ash

Oak

Hazel, Common

Dwarf birch, southern limit of
Betula nana continuous
distribution

Erica tetralix, eastern limit of
general distribution

Spruce forest,
southern limit

Limit of coniferous forest in the
mountains; shaded areas indicate
region of sub alpine birch wood

**FIG 83. Vegetation regions of Sweden. (After du
Rietz, 1952.)**

15

Plant nutrients and their physiological effects

Plant nutrients, especially mineral salts, have been one of the main subjects of plant physiological research since the middle of the nineteenth century, when the important discoveries about the nature of plant nutrients were made. It has been found that the exchange of substances between plants and the soil has many consequences, not only on the development of the plants and on their competitive ability and distribution, but also on various branches of agriculture and the appearance of the countryside. Plant nutrient problems are complicated and have many facets, and new effects are continually being discovered.

The nutrient economy of plants is a system of biotic, physico-chemical, climatic and geological processes, to which different plants react in different ways. Hoagland and Arnon (1948) rightly described it as the most complicated branch of biology yet studied. The nutrient system is itself a result of interaction between three separate systems:

> the plant, with all the features of a living organism, and with the special characteristics connected with photosynthesis; the soil, the properties of which are based on geological, mineralogical, colloid-chemical and biological processes; and the climate, with all its various combinations of light, temperature, carbon dioxide, wind and moisture. Water in its various forms is a common constituent running through the system.

Plants have long been classified as autotrophs or heterotrophs. Autotrophs use small, inorganic molecules poor in energy and relatively simple in construction. These molecules are assimilated by the plants and built up into energy-rich organic molecules, which are usually large and complicated. The heterotrophs (parasites and saprophytes) use these large molecules and break them down to the original small molecules. Heterotrophs also obtain from autotrophs various physiologically active compounds, including vitamins and hormones. Root culture on artificial media has shown that various vitamins in particular proportions are required for plant development (Addison and Whaley, 1953). Autotrophs produce these for their own use; then they are utilised by heterotrophs, including animals and man.

Autotrophic nutrition

The autotrophs take up all the various kinds of inorganic substances present in the soil and water. Uptake depends on concentration: the higher the concentration of a substance, the more is taken up. However, uptake is also selective, so that cells take up relatively more of some substances than of others. For example, in the cell sap of the marine alga *Valonia* there is less NaCl but more KCl than in the surrounding sea water (see Table 11, page 78), and somewhat less Mg, Ca and SO_4. Higher plants also show some selectivity, quantitatively and qualitatively. Root metabolism seems to control the selection, but the precise mechanism is unknown. As a result of the differential uptake, the proportions of mineral salts in plants are different from those in the soil, although all the mineral constituents of the soil are represented, including those which have no function in plant metabolism, since no substance is excluded completely. This imperfect selective ability has some significance, since a substance with no function in a particular plant may be of nutritive value to heterotrophs using the plant for food. Cobalt, for example, is possibly of no value to plants, but is important to animals; if cobalt is absent from a soil, sheep and cattle may suffer deficiency symptoms. Animals may possibly take up cobalt as a component of vitamin B_{12}, which is synthesised by microorganisms and occurs in soils and in higher plants (see page 359). Other substances important in animal metabolism, such as NaCl, are present in all plants to some extent, but seem to be of no importance to them: even halophytes, which normally contain large amounts of chloride, can grow on chloride-free substrates. It is the inability to exclude completely any of the mineral salts in the soil which is responsible for the comprehensive mineral composition of vegetation and the fulfillment of the special requirements of animals and man. Another consequence is that harmful amounts of poisonous substances may sometimes be taken up, such as selenium, copper, lead or other heavy metals (Lundblad, 1946; Svanberg, 1949).

Most of the inorganic substances taken up by autotrophs from the soil, air or water do take some part in metabolism, although the way in which they do so is not yet always fully understood. Some substances are building materials, i.e. are constituents of the molecules from which the organic material is synthesised. Another group act as tools in building and maintenance processes, for example, the substances which form part of enzyme complexes, or which act as 'activators' changing the charge on protein molecules (McElroy and Nason, 1954). Another group may inhibit or promote the effect of other substances, a phenomenon known as *antagonism*.

But this division into building materials, tools and antagonisms is artificial, and some substances may come into two or more categories, although their main effect may only be in one.

The elements of construction

Oxygen, carbon and hydrogen, in that order, are quantitatively the most important nutrients for the carbon-assimilating plant. Oxygen content varies between *c.* 12 per cent in dry seeds and 88 per cent in some algae. In leaves of herbaceous plants, it is generally 70 to 80 per cent. There is about 30 to 40 per cent oxygen in the dry matter: 40 to 50 per cent carbon, and 5 to 6 per cent hydrogen. More than 90 per cent of the dry matter is made up of these three elements, and they are constituents of all organic substances. Proteins also contain nitrogen, sulphur and sometimes phosphorus. Phosphorus is a constituent of lecithin and other phosphatides in the cytoplasm, including the nucleotides in the nucleus. Magnesium is a constituent of chlorophyll. Plants obtain these substances from carbon dioxide, water and salts, especially nitrates, ammonium salts, phosphates, sulphates, and magnesium salts, which can therefore be called *plant nutrients*: the organic matter is built up almost entirely from these.

Some of these substances have other functions as well as being nutrients. The oxidation and reduction systems are based on hydrogen and oxygen, and these elements play a part in almost all metabolic processes. Sulphur deficiency may cause chlorosis and slow growth, especially of roots. Nitrogen deficiency tends to have the opposite effect, in that growth in length may increase in grass and legumes, for example (see Pirschle, 1940; Hewitt, 1963). Because sulphur and nitrogen have such different effects on growth it is unlikely that the cause lies in the cell construction processes. However, the cause is not yet known, and to analyse the processes involved is very much more difficult than to establish the final result on plant growth. In most cases, the effect of various nutrients is only known through the final results, and the intermediate stages of the reaction chains involved are unknown.

Phosphorus plays a part in synthesis of nucleotides, which are important constituents of the cell nucleus and also prosthetic groups of some enzymes. The change from adenosine diphosphate to adenosine triphosphate, and the reverse, is an essential for energy transfer in the cell. Phosphorus also occurs in some intermediates in carbohydrate metabolism (phosphorylation), and so is important in both hereditary and metabolic processes. The rate of cell division in roots, and uptake of nitrogen and sulphur compounds is promoted by phosphorus. Phosphorus is generally present in larger amounts in young tissues than in older ones.

Magnesium also has some function, in addition to being a constituent of chlorophyll, because it is required in larger amounts than would be needed for chlorophyll synthesis. It may be a constituent of some enzymes, but its other functions are unknown.

Silicon is an element with a particular function in construction. It is an important constituent of the silicious coats of diatoms and it occurs in horsetails (*Equisetum*), grasses and other plants, in which it is deposited in the cells as so-called *silica bodies*. The whole cell contents may harden into a transparent mass, such as in the awns of barley, making the tips hard and prickly; it may also have other functions. Complete absence of silica may inhibit growth of higher plants and cause slower rate of division of diatoms. Pirson (1946), Burström (1949), Millar (1955, page 244), and Bollard and Butler (1966), give information about the nutritive effects.

Nutrients with a catalytic effect, and with effects of unknown types

There are several plant nutrients which are not part of the cell walls or the cytoplasm, but which affect plant development and metabolism by stimulating or inhibiting certain processes (see Nason and McElroy, 1963). These effects may be catalytic, in that the nutrients activate or inhibit enzymes, by acting as a prosthetic group, for example. Enolase, an enzyme involved in respiration, consists of a protein which has no effect except in the presence of magnesium, zinc or manganese. These metals have a similar effect in conjunction with phosphatases and peptidases, which also use copper. Magnesium works in conjunction with carboxylase; zinc with carbon dioxide anhydrase and zymohexase; copper with polyphenol oxidase (or tyrosinase), ascorbic acid oxidase and uricase (Arnon, 1950a and b).

1 Elements taken up in relatively large amounts. POTASSIUM. Large amounts of potassium occur in plant cells, especially in the meristem and in young tissues, and potassium is probably involved in several physiological processes (see Nason and McElroy, 1963). K_2O often constitutes 30 to 50 per cent of the plant ash. Potassium deficiency inhibits protein synthesis, polymerisation processes, substance production in general and consequently growth. The decreased growth rate is a result of the effect of potassium deficiency on cell division and cell elongation: internodes are shorter, leaves smaller and root production less. Leaves may become curly and more succulent because cell elongation is disturbed. Chlorophyll synthesis is inhibited and anthocyanin synthesis is increased; water uptake, and therefore transpiration, may be reduced, and respiration may also be affected. Potassium seems to have an important function in maintaining physiological balance in the cytoplasm, and this explains why potassium deficiency decreases the length of life of some organs, such as shoot tips and leaves, which wither prematurely.

Potassium-deficient plants are often less resistant to disease, e.g. potato to *Phytophora* infection.

Potassium can be replaced in some of its functions, to some extent, by sodium or rubidium.

CALCIUM. Calcium is similar in some respects to potassium. It affects permeability, respiration and the state of the cytoplasm. Plants subject to calcium deficiency age prematurely. One of the physiological effects of calcium is on ascorbic acid synthesis. Relatively large amounts are required (10 to 30 per cent CaO in the ash). However, in some fungi and algae (*Chlorella*, for example) there is a very small requirement and a concentration as low as 0·001 mg/litre is sufficient to give optimum growth rate. In such cases, calcium probably has a different type of function.

Several other elements, the trace elements or micronutrients, have their optimum effects at low concentrations. Calcium acts both as a macronutrient and a micronutrient (for literature, see Pirson, 1946; Burström, 1949; Pirson and Seydel, 1950).

The direct nutrient effect of calcium should be distinguished from its indirect effects on humification, soil structure, nutrient accumulation, neutralisation of acids, etc. (see Chapter 16). Large amounts are needed to have such effects (in agriculture several hundred kilograms per hectare are used for liming the soil), whereas very small amounts are needed to act as a trace element.

ALUMINIUM. This is the most common metal in the soil, constituting 7·3 per cent of the earth's crust. It seems that ferns are the only plants which require aluminium for normal growth, and they need very little, e.g. 0·16 mg/litre in experiments in which ferns are grown in nutrient solution. Some *Lycopodium* species contain particularly large amounts of aluminium, up to 30 per cent of the ash, but the function is unknown. A favourable effect of aluminium on higher plants has not often been established, but it occurs generally in the cells, since like other elements in the soil it cannot be entirely excluded. The amount in the cells depends on the amount in the soil. Faber (1925) gives an example of plants growing on volcanic soils in Java in which the content of Al_2O_3 is up to 14 per cent of the dry weight, compared with 0·1 to 0·2 per cent in the same species growing on other soils. Plants which invade these soils must therefore experience a tenfold increase in aluminium content.

If the aluminium in the soil is in the form of metal cations, large amounts may have deleterious effects on plant roots (Pirschle, 1941; Lundblad, 1946; Faber 1925, Bollard and Butler, 1966).

2 Trace elements. Plants need only small quantities of trace elements to produce the maximum effect. Hence, research has been difficult, and this aspect of plant mineral nutrition was long unknown. Normal methods of chemical analysis are often not sensitive enough to detect and measure physiological amounts of these elements. Optimum concentrations of some trace elements may be present as impurities in other chemicals or in the substances which dissolve out of the walls of experimental vessels, or the total requirements of a plant may have been contained in the seed from which it developed. The fact that the total amounts required are small tells us nothing about the concentrations at which these elements are active within the cell. Mothes (1939) pointed this out, and suggested that the architecture and differentiation in the cell would allow differences in concentration between different places. A mass analysis of cells and issues gives no information about such different concentrations.

The way in which these elements work is only partly known. Some are activating or inhibiting agents in enzyme reactions (see Brenchley, 1947; Pirson, 1946, Nason and McElroy, 1963), e.g. zinc, copper and manganese, but probably have other functions as well. The so-called deficiency diseases are caused by lack of the micronutrients, with abnormalities which are often characteristic of the particular element. Deficiency sometimes causes retarded development of some part of the plant, such as the root system, and then the above-ground parts of the plant readily develop a water deficit and a low transpiration rate (Schmied, 1953). The plant symptoms can sometimes be used as indicators that an element is absent from the soil (see Lundblad, 1946, for example). The importance of the trace elements in metabolism is a fairly recent subject for research and is now receiving a great deal of attention. One of the first pieces of work was published in 1914, by Magé, showing a relationship between boron deficiency and heart rot in beet. By 1946 (Lundblad, 1946) over 10 000 papers on effects of trace elements had appeared (see reviews by Pirson, 1946;

Lundblad, 1946; Burström, 1949, Baumeister, 1954; Scharrer, 1955; Hewitt, 1963). The symptoms, distribution and economic importance of several deficiency diseases of cultivated plants are now known, but the physiological mechanisms involved are only partly understood.

Arable land is fairly often deficient in trace elements. Like other nutrients, they are removed in every harvest. The annual yield from one hectare may contain only 20 or 30 g of the element, but if the soil content is low this may be enough to bring about deficiency. Additions of trace elements in fertilisers are now considered essential if good harvests are to be maintained.

Natural vegetation is also affected by trace elements, which are subject to leaching or concentration in the soil, like all salts, and which are required in different amounts by different plants. Beans require more boron than barley, maize requires more zinc than lucerne, for example (Arnon, 1950b).

Molybdenum is a trace element; and iron may also be considered as such, because only small quantities are required for normal plant development.

Some of the effects of the trace elements are discussed below.

Iron is required for chlorophyll synthesis, and iron-deficient leaves are chlorotic (pale yellow instead of green). Iron deficiency also causes accumulation of nitrate and sulphate in plants, and so plays a part in nitrogen and sulphur metabolism. It is also essential for nitrogen fixation by *Azotobacter*. Table 70 shows the iron content of various plants. About five-sixths of the iron is bound in the nucleotides and about one-sixth takes part in enzyme reactions. Although iron is very common in the soil—3·5 per cent of the earth's crust is iron—and plant requirements are small, iron deficiency is well known in agriculture. In Sweden, it is mostly *Brassica* species in which iron deficiency symptoms have been found, but other species are commonly affected elsewhere. Tomatoes, strawberries, raspberries, cherries, black and red currants, apples, plums, etc. are particularly susceptible. The leaves are chlorotic and often curled or deformed in some other way. Differences in susceptibility between one species and another probably depend on differences in ability to obtain the iron from the soil (literature in Pirson, 1955; Lundblad, 1946; Scharrer, 1955; Hewitt, 1963).

Little is known about boron deficiency in natural vegetation, but it is common in crop plants. Sugarbeet is affected by heart rot. The innermost, youngest leaves in the rosette blacken and die, small rosettes of shrunken leaves develop on the upper part of the beet, then the parenchyma is destroyed and the rot spreads downwards in the root. *Brassica* species often have boron deficiency symptoms, again it is the root parenchyma which is particularly affected. A section of the root has a marbled appearance, with some parts infiltrated with water, others spongy with large spaces. In celery, there are also brown streaks on the leaves. Cereals have other symptoms: internode elongation decreases, leaves become dark green and upright, and little seed is set.

Table 70 The iron content of some plants (above-ground parts unless otherwise indicated). (From Lundblad, 1946, page 449; *cf.* Güttler-Tetschen-Liebwerd, 1941.)

PLANT MATERIAL	IRON CONTENT (mg/kg dry matter)
Apples	29
Plum (eatable part)	44–55
Raspberry, fruits	62
Oats, at flowering stage	50
Oats, at milk-ripe stage	154
Turnips, root	59–92
Turnips, tops	618
Cabbage	147–208
Timothy grass, flowering	154–287
Red clover	162–302
Lettuce	690–1500
Spinach leaves	1–319

Boron deficiency also causes other abnormalities, such as decreased flowering in *Nasturtium*, *Gloxinia*, etc; decreased fruit set in grape vines, with some seedless fruit; slow growth in legumes; chlorosis in hemp; and morphological defects, such as inrolled leaf edges, in potatoes and hemp.

Most of these deficiency symptoms are related to chlorophyll formation, the function of the meristem, and cell elongation. The cause of the symptoms probably lies in disturbance of metabolism, since abnormally large amounts of carbohydrate accumulate and protein synthesis is inhibited. The youngest tissues are probably affected first, because they normally contain more boron than the other parts of the plant. The optimum amount is very small. One part of boron in five million may have a positive effect in laboratory experiments. Differences in sensitivity between plants may be due to different boron requirements or to different rates of uptake.

The boron content of a plant cannot be taken as a measure of the boron requirement. As with other elements, there is some luxury

TABLE 71 The boron content of various plants grown on the same type of soil. (From Lundblad, 1946, page 444.)

PLANT MATERIAL	BORON (mg/kg dry wt.)	PLANT MATERIAL	BORON (mg/kg dry wt.)
Barley	2	Mustard	22
Rye	3	Tobacco	25
Wheat	3	Red clover	36
Maize	5	Cabbage	37
Spinach	10	Turnips	49
Celery	18	Radishes	65
Peas	22	Fodder beet	76

TABLE 72 The effect of additions of boron in an experiment with sugarbeet (After Oswald, 1935.)

TREATMENT	BEET YIELD	SUGAR YIELD	SUGAR CONTENT %
Control	100	100	15·4
5 kg/ha borax	113	109	16·4
10 ,,	125	144	17·7
20 ,,	132	151	17·6
40 ,,	130	148	17·5

uptake. Boron which might have been lost by leaching may be retained in this way.

The boron content of soil is generally 10 to 30 mg/kg. Some is lost in harvests and by grazing so that sooner or later a deficiency arises, in the same way as for nitrogen, phosphorus and other elements present in small amounts. Agerberg (1958) found numerous instances of boron deficiency in arable land in north Sweden. The amount removed in harvested crops depends on the crop and on the yield (literature in Brenchley, 1947; Pirson, 1946; Lundblad, 1946; Philipson, 1953; Scharrer, 1955; Hewitt, 1963).

Copper is part of several enzymes and plays an important part in metabolism, particularly in photosynthesis and in nitrogen metabolism. Cereals and other grasses sometimes show copper deficiency symptoms: the leaf tips yellow, dry out and wither prematurely (German, *Urbarmachungskrankheit*), formation of ears is delayed or fails altogether and copious green shoots are formed instead. The better pasture grasses are replaced by less demanding, less valuable species. In tomatoes, there are chlorophyll defects, curled leaves, undeveloped internodes and poor flower development. The fruits of plums, apples and pears are spotted. Copper deficiency in plants causes copper deficiency diseases in cattle and other animals.

Plant copper requirements are small. Development of one tomato plant requires 0·002 mg (Arnon, 1950b). In a nutrient solution, 0·02 ppm is sufficient, and 2·0 ppm is poisonous. In general, 0·1 to 0·2 ppm in a culture solution is enough to prevent chlorosis and rosette formation (see Pirschle, 1941; Lundblad, 1946).

TABLE 73 Copper content of crops from Swedish soils. (After Svanberg and Nydahl, 1941.)

	COPPER (mg/kg)
Hay from normal area	
Red clover	9·8
Lucerne	8·7
Timothy	5·8
Mixed grass	6·0
Hay from copper-deficient area	
Legumes	5·2
Grasses	3·8
Cereals from normal area	
Oats, grain	3·7
Barley	5·5
Autumn wheat	5·5
Spring wheat	5·0
Rye	4·0
Peas	7·3
Oats, straw	3·0
Wheat, straw	1·7–2·0

In spite of this low requirement, 25 to 50 g/ha copper is removed in common crops harvested. Copper is relatively easily lost by leaching, and so copper deficiency diseases occur on bog soils, podsols and heaths in north-west Europe. Copper deficiency is common on arable soils in north Sweden (Agerberg, 1958).

Some plants have a characteristic high copper optimum or high tolerance to copper. Some uncommon mosses, such as *Dryoptodon atratus* and *Mielichhoferia elongata*, tolerate copper concentrations one hundred times greater than in normal soil (Mårtensson and Berggren, 1954).

Manganese deficiency is responsible for the well-known grey speck disease of oats and other cereals and grasses. The leaves have chlorotic spots which spread out over the surface, the tissues break down, the edges roll in and the leaf blades split at the base. There is decreased assimilation and weak development of roots, fruits and other organs. Many other crops also suffer from manganese deficiency, e.g. potatoes (the top leaves roll inwards on the upper side and have black spots), *Brassica* species (marbled, chlorotic leaves), peas (spots on the cotyledons), fruit trees (chlorosis), sugarbeet and mangolds (the leaves yellow, except for the veins, the edges roll inwards, and the leaves wither). In general, the meristem degenerates. Yeast also requires manganese for normal development.

TABLE 74 The effect of fertilising a Swedish bog soil (Småland) with copper. (After Lundblad, 1936.)

	STRAW (dt/ha)	GRAIN (dt/ha)	RELATIVE FIGURES	
			STRAW	GRAIN
Untreated soil	$71 \cdot 5 \pm 5 \cdot 5$	$6 \cdot 1 \pm 1 \cdot 6$	100	100
$CuSO_4$ added	$79 \cdot 4 \pm 1 \cdot 2$	$19 \cdot 8 \pm 0 \cdot 8$	112	325
$ZnSO_4$ added	$63 \cdot 8 \pm 2 \cdot 5$	$8 \cdot 6 \pm 2 \cdot 2$	90	141
$MgSO_4$ added	$70 \cdot 2 \pm 2 \cdot 1$	$7 \cdot 9 \pm 1 \cdot 4$	99	130

Most of these deficiency symptoms probably arise as a result of the enzymatic effect of manganese (see Hewitt, 1963). Manganese is thought to affect the oxidation process of ascorbic acid formation, starch formation and storage, and nitrate assimilation. The optimum concentration is again low and one part in fifty million stimulates growth of clover and beans, for example.

The manganese content of plants is very variable, both within a species and between different species (Table 75). Potato tubers contain 20 ppm, oat straw contains 700 ppm. Some of this is certainly an excess, and takes no part in metabolism. However, this accumulation prevents leaching and the manganese accumulates in the litter and humus, so that there is less tendency to manganese deficiency.

A high manganese content in plants has a bad effect on animals since it may cause anaemia, for example, in horses (literature in Burström, 1949; Lundblad, 1946; Pirson, 1946, 1955; Scharrer, 1955).

Zinc is an essential nutrient for both higher and lower plants. It promotes the activity of nitrifying bacteria, algae (*Chlorella*) and fungi (*Marasmius scorodarius*) (Lindeberg, 1944). Zinc deficiency in higher plants causes chlorophyll defects, disturbances in carbohydrate metabolism, lower auxin content, dwarfness in the leaves or in the plants as a whole, small leaves in fruit trees, absence of flowering and fruiting, leaf and flower abscission in beans, etc. Thus, the symptoms are similar in several respects to those caused by lack of manganese or copper, and it has been shown that the symptoms can sometimes be relieved by supply of these elements, so that Zn, Mn and Cu can replace each other to some extent, probably because they form prosthetic groups in the same type of enzyme.

TABLE 75 Manganese content of various crop plants. (From Lundblad, 1946.)

	MANGANESE (mg/kg dry weight)
Oats, grain	51
Oats, straw	733
Wheat, grain	49
Wheat, straw	60
Timothy, hay	86
Turnips, tops	185
Turnips, roots	19
Potato, tubers	19
Spinach, leaves	101
Red clover, hay	121
Lucerne, hay	60

Very small quantities are required, as for other microelements. A tomato plant needs only 0·002 mg (literature in Arnon, 1950b; Burström, 1949; Pirson, 1946, 1955; Lindeberg, 1944; Hewitt, 1963).

Molybdenum is a catalyst in the process of nitrogen fixation by *Azotobacter* and *Clostridium*. It can be replaced in this function by vanadium. Other fungi, such as *Aspergillus niger*, require molybdenum for normal growth. It also has a favourable effect on the growth of higher plants and is thought to be essential (Hewitt, 1963). The optimum concentration is extremely low. Tomatoes, mustard, lettuce, oats, etc., require only 0·01 ppm (Arnon, 1950b). The optimum soil molybdenum content for legumes is 0·4 ppm; more than 5 ppm is poisonous. Brenchley (1947) lists contents of between 4·3 and 69 ppm in some French soils, so that molybdenum does occur in harmful amounts. The seeds of some plants (e.g. legumes) may contain up to ten times as much molybdenum as is needed for normal development of the plant (Meagher, Johnson and Stout, 1952).

Vanadium is commonly found in plants. It is required for growth by *Scenedesmus obliquus* (Arnon and Wessel, 1953) but has not so far been shown to have any essential effect on higher plants.

Arsenic is a strong cell poison, but in very small amounts it increases the yield of some crop plants (Lundblad, 1946). It is doubtful whether it should be included in a list of essential nutrients.

Cobalt is a constituent of vitamin B_{12}, which is synthesised by a number of microorganisms, including algae, but not by higher plants or by animals. When this vitamin is present in higher plants, it has probably been taken up from the soil, where it has been released from microorganisms. The function of the vitamin in plants is not known. Cobalt has some secondary effects: an abundant supply of cobalt may bring about iron deficiency symptoms (chlorosis) and may also inhibit phosphorus uptake (Nicholas and Thomas, 1953).

Selenium is present in such large amounts in soils in some parts of North and South America that the plants there contain measurable quantities. No positive effect on plant metabolism is known, but in some species it is poisonous, and it is very poisonous to animals. Hence, soil containing selenium—and there are millions of hectares in America—is unsuitable for cereal or fodder production.

McElroy and Nason (1954) made a survey of the catalytic effects of microelements.

Balance, antagonism

An element can have a direct effect on physiological processes as a nutrient or as a catalyst. It can also have an indirect effect, known as *antagonism*, by inhibiting or promoting the effect of another element. It may be favourable or unfavourable to the plant, depending on the concentration of the first element. If this was present in supraoptimal concentration then an inhibitory effect by another element will have a favourable effect on the plant: if the first element was present in suboptimal concentration then antagonism by another element will have an unfavourable effect. In the same way, a promotory effect by a second element will be favourable if the first element is present in suboptimal concentration; and unfavourable if in supraoptimal concentration (see Fig. 7, page 23).

A physiologically favourable nutrient source must therefore contain all the required nutrients, in such proportions that the largest possible number of favourable interactions is achieved. The soil or medium is then said to be in a *balanced nutritional state*. The soil water is generally a balanced nutrient solution, at least in fertile soils, and so is sea water and complete synthetic nutrient solutions. Cell sap and other liquids in living organisms are also all balanced in their own ways.

The effect of various substances on each other depends on their physical and chemical properties, and on the amount and form in which they occur in the soil or in water. The antagonistic effect is sometimes caused by the substrate, so the antagonism is wholly or partly ecological in nature. The term *antagonism* must therefore include certain edaphic factors. The extent to which the antagonistic processes are understood has allowed some analysis of the balance mechanism.

Various types of antagonism. Antagonism may arise through competition for space, and one of the simplest examples is the restriction of oxygen supply by water in waterlogged soil. There is similar antagonism between ions in soil, if one substance is present in such large amounts as to exclude others. If, for example, high soil calcium content makes uptake of potassium and magnesium more difficult, it may be because of competition for space on the particles and adsorbing cell surfaces. It may also be because the diffusion paths through the cell walls have been changed in such a way that the passage of other ions is hindered (Waris, 1939). The addition of calcium causes release of potassium, which therefore becomes more available to plants. In the same way, supply of a large quantity of potassium can displace calcium.

Leaching may also be the cause of antagonism. A substance which has been displaced by another into the soil solution is liable to be lost by leaching. Supply of calcium causes increased leaching of potassium, and this effect is enhanced by magnesium, so that the risk of potassium deficiency is intensified. On the other hand, the supply of potassium to soil with a low calcium content may cause leaching of calcium. Ehrenberg's potassium-calcium rule, applied in agriculture, that the relation of concentration of potassium to that of calcium should be maintained within a certain range, is based on knowledge of these antagonisms—competition for space, displacement, leaching—between the two elements. The value of an element as a nutrient depends on its absolute concentration, and on the relation between its concentration and that of other elements.

The solubility of elements like phosphorus, which occur in the soil in small amounts, may cause antagonistic effects. Insoluble phosphates are formed if there is a good supply of calcium and iron, especially in very acid soils, in which phosphate is precipitated in the form of iron and aluminium salts. Phosphorus deficiency may result, even if quite large quantities of the element are present in the soil.

Other forms of antagonism are more complex, for example, that between calcium and boron. Calcium can cause decreased uptake of boron, can decrease the effect of boron deficiency, and can counteract the effect of a poisonous excess of boron. The mechanisms involved are not fully understood. In turn, boron has an effect on the calcium economy of plants. It seems that the soluble calcium content of a plant is more closely related to its boron content than to the total calcium content. There is therefore reciprocal antagonism between calcium and boron, depending on the relative proportions of the two elements: good growth of sugarbeet, for example, requires a ratio 1:100, approximately, of boron to calcium; and tobacco requires 1:1200 (Brenchley, 1947).

Plants sometimes suffer chlorosis on soils with a high calcium content. This is sometimes related to iron deficiency caused by the excess amount of calcium (literature in Hewitt, 1963).

The manganese economy is affected by complex antagonistic effects. Manganese deficiency is brought about by excess calcium in the soil, and occurs most in areas of chalk and limestone. It can also occur in conditions of calcium deficiency linked with potassium excess, or of iron deficiency or iron excess. The proportions between the elements, not the absolute amounts, are important here. Good growth of soya beans requires about twice as much iron as manganese, but whether there is 0·002 ppm manganese or 2 ppm seems to be of little importance. Deficiency symptoms develop if the ratio changes significantly from 2:1, and so are caused both by manganese excess and iron deficiency and by manganese deficiency and iron excess. The effects of a harmful excess of manganese can similarly be counteracted by liming (see page 424).

Zinc is antagonistic to iron and copper. Harmful effects of excess iron and copper are counteracted by the right amount of zinc. However, excess zinc is also poisonous (Lundblad, 1946; Pirson, 1946).

Cells cannot exclude an element completely, but can exert some selection between elements closely related chemically. Hence such elements may be antagonistic to each other—phosphorus and arsenic, for example. Plant tolerance to arsenic is increased by a good phosphorus supply, but decreased by phosphorus deficiency. This has been called *mass antagonism* (Lundblad, 1946; Brenchley, 1947).

Hydrogen ions and other cations are antagonistic in several ways. The importance of hydrogen ions in the processes of weathering and leaching has been discussed earlier. Acidity also effects the solubility of various substances in the soil. Colloidal iron hydroxide is least soluble at pH 7·1, and at this pH iron is precipitated and is relatively unavailable to plants. Supply of lime or other bases, or any other treatment leading to this pH therefore causes precipitation of iron and, if the soil iron content is low, iron deficiency may result. Deficiency symptoms resulting in this way are common on soils with a high calcium content. Copper and manganese deficiency may develop in the same way. Copper is least available at pH 5·5 to 6·5: manganese at pH 6·5 to 7·5.

Soil pH also affects uptake of calcium, molybdenum and nitrogen. Hence, all the conditions which affect pH (weathering, leaching, supply of basic or acid nutrients such as nitrate, nitro-chalk or ammonium nitrate fertilisers) also affect the plant nutrient balance with respect to these nutrients and to any others whose availability and uptake is sensitive to pH changes (literature in Lundblad, 1946; Burström, 1949; Scharrer, 1955).

Antagonism between anions is less common: van Goor (1951) showed that in nutrient uptake by *Larix leptolepis* there was antagonism between nitrate and phosphate: each reduced the uptake of the other, excess nitrate may therefore cause phosphate deficiency and consequent slow growth. An example of antagonism between phosphate and bicarbonate ions is given in Chapter 16.

The effect of hydrogen ion concentration is probably generally indirect. Slow growth of a plant at pH 4 to 5 is probably caused by deficiency of some nutrient, such as calcium; slow growth at pH 7 or over is probably caused by deficiency of iron or manganese. Supply of the appropriate nutrient counteracts the unfavourable effect of the hydrogen ion concentration (see also Chapter 16). Liming of acid arable land increases production not as a direct result of the change in hydrogen ion concentration, but as an indirect result, and because the calcium itself increases availability of potassium and other elements, increases humification and improves soil structure (Åslander, 1932a and b).

The occurrence of the elements in plants and in the soil

Besides the nineteen elements which have already been considered, all except arsenic and selenium essential to plants, sodium and chloride occur in large amounts in sea water and have various ecological effects, particularly osmotic effects, and fluoride, bromide and iodide are present in both plants and animals. All the essential elements are in the lighter part of the periodic system (Fig. 84) and in the main they lie along a line from carbon, in its central position, to argon. Only hydrogen and molybdenum are not on this line, called the *nutrient line* by Frey Wyssling (1935, 1945).

The four most important building elements (carbon, hydrogen, oxygen, nitrogen) come from the atmosphere and hydrosphere: the others come from the lithosphere. The elements are present in plants in different proportions from those obtaining in soils (Table 76). Amounts of iron in plants may vary as much as between 40 and 2000 ppm of the dry matter; and manganese between 10 and 2000 ppm, for example (Svanberg, 1949).

No fundamental differences between the lithosphere elements and the other nutrients would be expected. It is true of all the elements that they are often present in suboptimal concentrations; some elements may be present in optimal or supraoptimal

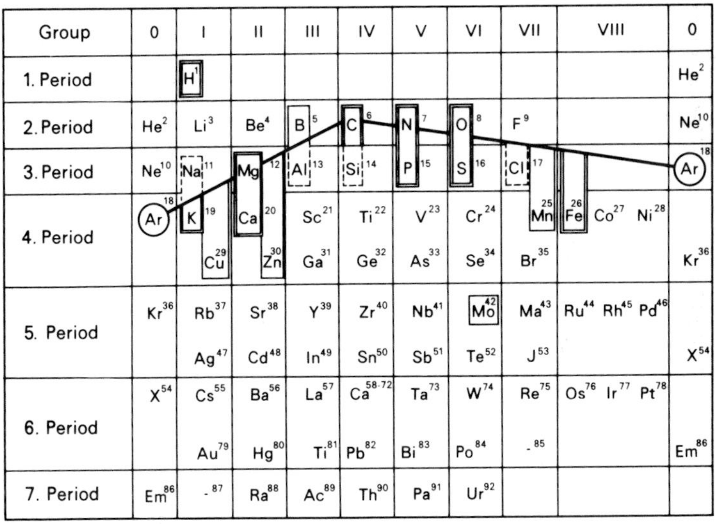

Group	0	I	II	III	IV	V	VI	VII	VIII			0
1. Period		H^1										He^2
2. Period	He^2	Li^3	Be^4	B^5	C^6	N^7	O^8	F^9				Ne^{10}
3. Period	Ne^{10}	Na^{11}	Mg^{12}	Al^{13}	Si^{14}	P^{15}	S^{16}	Cl^{17}				Ar^{18}
4. Period	Ar^{18}	K^{19}	Ca^{20}	Sc^{21}	Ti^{22}	V^{23}	Cr^{24}	Mn^{25}	Fe^{26}	Co^{27}	Ni^{28}	Kr^{36}
		Cu^{29}	Zn^{30}	Ga^{31}	Ge^{32}	As^{33}	Se^{34}	Br^{35}				
5. Period	Kr^{36}	Rb^{37}	Sr^{38}	Y^{39}	Zr^{40}	Nb^{41}	Mo^{42}	Ma^{43}	Ru^{44} Rh^{45} Pd^{46}			X^{54}
		Ag^{47}	Cd^{48}	In^{49}	Sn^{50}	Sb^{51}	Te^{52}	J^{53}				
6. Period	X^{54}	Cs^{55}	Ba^{56}	La^{57}	Ca^{58-72}	Ta^{73}	W^{74}	Re^{75}	Os^{76} Ir^{77} Pt^{78}			Em^{86}
		Au^{79}	Hg^{80}	Ti^{81}	Pb^{82}	Bi^{83}	Po^{84}	$-^{85}$				
7. Period	Em^{86}	$-^{87}$	Ra^{88}	Ac^{89}	Th^{90}	Pa^{91}	Ur^{92}					

FIG 84. **The position of the nutrient elements in the periodic system, showing their grouping along the so-called *nutrient line*. (After Frey-Wyssling, 1945.)**

concentrations. Concentrations of carbon and fixed nitrogen are generally suboptimal; oxygen deficiency occurs where anaerobic processes are going on; water (hydrogen) is often deficient in arid regions. Plants use only small amounts of those elements which are most common in the lithosphere (Si, Al, Fe), whereas some of the least common ones are essential. They must all be derived from weathering and can be lost by weathering or by precipitation in the soil in some form. A single element therefore often becomes deficient and limits the growth and range of distribution of a species.

The nutrient, catalytic, antagonistic and inhibitory effects of the mineral nutrients together form an ecological-physiological reaction system with innumerable combinations, giving rise to innumerable habitat differences. These combinations play a part in the selection which governs the distribution of plant species in different habitats. The particular nutrient requirements of individual wild species are mostly unknown—only certain extreme differences have been studied—but studies of crop plants have shown that even closely related types, the various cereals, for example, may have different requirements for nutrients and different rates of uptake; the nutrient balance in a habitat may therefore suit one species but not another, a species can only succeed in competition with others if the particular conditions obtaining are suitable.

Heterotrophic nutrition

Carnivorous plants, parasites, saprophytes

Where there are autotrophic plants there are also parasites and saprophytes, which obtain their food directly from living plants or from the litter-humus complex. The carnivorous plants obtain their food from animals.

The term *carnivorous* is usually applied only to higher plants, although there are many bacteria and fungi which attack living animals. There are parasites and saprophytes among many groups of plants, from bacteria to the flowering plants. The degree of heterotrophy is variable; it may be with regard to a group of nutrients or to a single element, compensating for one missing link in an autotrophic pattern.

The heterotrophic mode of nutrition of carnivorous plants is mostly of secondary importance, since the plants contain chlorophyll and obtain mineral salts like autotrophs do. However, the nitrogen economy, in particular, is strengthened, so that better growth may result.

Parasites are usually divided into full parasites and hemiparasites, although there is no sharp distinction between them, and various degrees of heterotrophy occur. *Euphrasia* is a hemiparasite, with chlorophyll, roots and root hairs, and it can exist autotrophically. However, the weak root system favours the development of parasitism. *Rhinanthus* has chlorophyll and roots, but no root hairs; it is more dependent on parasitism than *Euphrasia*, and by penetrating the host root system with haustoria it can probably obtain both salts and water. *Cuscuta* has haustoria and usually has no chlorophyll, but it can develop chlorophyll if necessary. *Lathraea* and *Orobanche* are wholly heterotrophic for carbon and salts, since they have no chlorophyll and only rudimentary leaves. The same is true of *Helosis* (Balanophoraceae), which has a shoot system reduced to a leafless, rudimentary stem. Finally, *Rafflesia* has no stem except for the base of the inflorescence. Besides being carbon and salt heterotrophs, these plants are probably also heterotrophic for some organic catalysts.

There is a similar series of transitional forms among the higher plant saprophytes, from partial saprophytism to full heterotrophy,

TABLE 76 Comparison between approximate amounts of various elements found in plant material and in bedrock (means of many analyses). (Mainly from Boresch, 1931, page 297; Svanberg, 1949; Holleman, 1952.)

	PLANT DRY WEIGHT (mg/kg)	BEDROCK (mg/kg)
K	15 000	24 000
Ca	9 000	34 000
Mg	2 000	19 000
Na	1 500	26 000
Fe	400	47 000
Al	200	75 000
Mn	50	800
Zn	20	1–10
Cu	6	100
Mb	1	0·1–1·0
Co	0·1	2–100
As	0·1	1–10
H	8 000–55 000	–
O	130 000–900 000	–
C	400 000–500 000	–
N	1 300–50 000	–
Cl	5 000	1 900
Si	4 000	250 000
P	2 000	1 200
S	2 000	600
B	20	10
Sn	10	6
F	2	100–1 000
I	0·2	2 (sea water)
Se	(0)	trace

and among flowering plants which utilise humus nutrients by developing mycorrhiza. *Orthilia secunda* has a stem and well-developed green leaves, but lives symbiotically with fungi, which possibly allow the uptake of carbon and other nutrients in an organic form. *Goodyera repens* is pale green and probably unable to assimilate sufficient carbon itself. *Epipogium aphyllum* has neither leaves nor chlorophyll; and *Thismia* (Burmaniaceae) is even more reduced, with vegetative parts consisting of the mycorrhiza and a rudimentary stem.

The degree of heterotrophy for carbon, salts and other nutrients, and the requirement for them, is variable. The heterotrophic mode of nutrition is secondary in the phanerogams, since they have all evolved from autotrophic ancestors.

Most heterotrophs, by far, are cryptogams, including most of the saprophytes playing a part in the humification and mineralisation of the litter (see Chapter 6). There are also large numbers of plant and animal parasites among the cryptogams, and there are many partial heterotrophs, so it is probable that heterotrophy has again usually been derived from autotrophy.

Saprophyte nutrition is broadly similar to the old idea about the origin or organic materials first put forward by Aristotle and still accepted, as the humus theory, up to the beginning of the nineteenth century. Now that autotrophy, saprophytism and parasitism are understood, the application of the humus theory is restricted to the saprophytes, a seemingly insignificant group, but including large numbers of individuals and species.

The nutrients

Like autotrophs, heterotrophs need building materials and catalysts. However, these need not be synthesised from inorganic low energy compounds, but can be derived from energy-rich organic compounds.

The raw material used by heterotrophs is complex and variable compared with the simple, constant materials used by autotrophs, and the heterotrophs must be specially adapted in order to utilise it. A saprophyte or parasite can attack a particular kind of litter or part of the litter because it possesses suitable enzyme systems or the capacity to develop suitable systems, or because the type of litter suits the physiological character of the saprophyte in some other way. Similarly, a parasite can only attack a particular kind of organism. In a symbiotic relationship, each organism is dependent on the other. It is the requirement for specialisation which results in the restriction of a parasite to a definite host plant species and the restriction of a saprophyte to a definite type of litter. As a consequence, the breakdown of litter is dependent on the activity of several types of saprophyte. One species may attack certain cells, tissues or parts of cells, but utilise them only partially. The remaining parts, which may comprise both structural and enzymatic components, are subsequently utilised by other species.

The chemical basis for all these processes is only partly known. In the mycorrhizal symbiosis, the fungus can take up nitrogen, phosphorus and calcium from the external medium and transfer them to the host plant, and, in return, obtains growth-promoting substances from the host (Melin, 1952; Melin and Nilsson, 1952, 1954, 1957).

Considerable research on fungal nutrition has been carried out, because of the importance of fungi in soil processes, in the food industry and in medicine. In 1934, the Swiss plant physiologist, Schopfer, showed that aneurin (thiamin, vitamin B_1), which is important in the developmental physiology of animals and man, is also an important growth factor for some fungi (*Phycomyces*). Earlier, the Belgian, Wildiers (1901) had discovered that yeast developed in a sterile medium only if the small amounts of certain organic substances, which he called *bios*, were present. Subsequent research has shown that fungal growth requires the presence of various vitamins, and sometimes also auxinlike hormones. Fungal growth therefore requires the same kinds of growth substances (vitamins and hormones) as that of higher plants, and probably also involves the same kinds of enzymes and catalysts. Schopfer (1943) drew up the following scheme of the complex of catalysts which regulate fungal growth:

Biocatalysts
{ Inorganic, e.g. Mn, Zn, Mb, etc., with oligo-dynamic effects.
Organic (enzymes, active *in vitro*)

Ergons (growth factors)
{ Vitamins (coenzymes, active only *in vivo*)
Hormones (having morphological effects, among others)

The development of heterotrophy

In contrast to the inorganic catalysts, which are taken up from the external medium, the organic catalysts are synthesised by plants themselves. However, the ability to carry out such syntheses is variable. Green plants can usually synthesise all the organic biocatalysts they require, but the parasites and saprophytes have partly lost this ability. Fungi and other non-green plants lack the ability to a considerable extent. The culture of fungi on artificial media has shown that sugar and mineral salts alone are usually insufficient for normal growth: certain growth substances are also required. One species may require vitamin B_1, another vitamin H, a third both, and so on.

Table 77 lists some of the growth substances required by fungi. The amount and types required depends on the degree of heterotrophy. It is thought that fungal heterotrophy is also secondary, the heterotrophic fungi having been derived from autotrophic ancestors. The capacity to synthesise growth substances may be wholly or partially lost, through a series of consecutive mutations, for example.

Like many other catalysts, the growth substances are ogliodynamic. Aneurin, which is the coenzyme for the enzyme carboxylase (Schopfer, 1943) is effective in concentrations as low as 1 to

TABLE 77 Fungal growth factors. (Fries, 1943, 1948; Norkrans, 1950.)

Growth Factors	Essential for
Aneurin (thiamin, vitamin B_1)	many phyco-, asco- and basidiomycetes
Biotin (vitamin H)	many ascomycetes, some basidiomycetes
Pyridoxin (adermin, vitamin B_6)	some ascomycetes
Pantothenic acid	yeast, some *Tricholoma* sp.
Inositol	some ascomycetes
p-amino-benzoic acid (vitamin H)	*Rhodotorula aurantiaca,* *Tricholoma brevipes*
Hypoxanthin (factor Z_1)	*Phycomyces*
Oleic acid	*Pityrosporum ovale*
Heteroauxin (B-indoleacetic acid)	*Pyronema confluens? Tilletia caries?*

10^{11} (Fries, 1948). However, the effect depends more on the amount taken up than on the external concentration, as is demonstrated by determining the economic coefficient, i.e. the relationship between the amount of fungal material produced and the corresponding amount of catalyst used. Schopfer (1943) gives the following examples of the economic coefficient for aneurin:

Phycomyces blakesleeanus	1 to 200 000
Rhodotorula rubra	1 to 200 000
Polyporus adustus	1 to 380 000
Cercosporella herpotrichoides	1 to 433 000
Ustilago violacea	1 to 2 500 000

The effect of growth substances is manifested as a stimulation of growth. The means by which this is brought about is largely unknown. Some of the growth substances inhibit growth if they are present in too high concentrations. Melin and Norkrans (1942) found, for example, that growth of a mycorrhizal, aneurin-autotrophic species was inhibited by additions of aneurin.

Aneurin is often required by fungi, e.g. wood-decomposers (Fries, 1938; Lindeberg, 1946) and several mycorrhizal fungi (Melin and co-workers, 1939, 1940, 1942). Some are only partially aneurin-heterotrophic and synthesise a part of the substance themselves. Aneurin has two components, pyrimidine and thiazole, and some fungi can synthesise one of these but not the other.

There are also fungi which cannot synthesise either component, but which retain the ability to put them together if they are supplied. Kögl and Fries (1937, according to Fries, 1948) set up an artificial fungal symbiosis by cultivating thiazole-synthesising and pyrimidine-synthesising species together in an artificial medium. Since the two species exchanged their products, the system was aneurin-autotrophic. Symbiosis between other fungi can develop in the same way, if the autotrophy of one compensates for the heterotrophy of another. Fries (1943) cultured eight *Ophiostoma* species, of which four were aneurin-heterotrophic, three were adermin-heterotrophic and one was aneurin-adermin heterotrophic. Out of twenty-eight combinations, symbiosis developed in twelve, i.e. in all cases where one component was aneurin-heterotrophic and adermin-autotrophic and the other was adermin-heterotrophic and auxin-autotrophic.

Some fungi require the addition of certain amino acids or purines, or grow better if these substances are present in the medium (Fries, 1950). Relatively large amounts are required, so they act as nutrients, and not oligodynamically. It seems that these fungi have become so heterotrophic that they have even begun to lose the ability to synthesise amino acids.

Heterotrophy is very variable in type and extent of development. Within a single species, there are mutant forms which differ from each other and from the parent in type of heterotrophy ('physiological mutations', Wikberg and Fries, 1952; Fries, 1953). Heterotrophy is therefore only one of many characteristics which distinguish species (Fries, 1948).

In natural conditions, fungi obtain growth substances, amino acids, etc., from the litter, since such substances are synthesised by green plants. Fries (1952) calculated that the annual supply to the soil in a beechwood, in the beech leaves, was 400 mg/m^2 nucleic acid and about 13 mg/m^2 adenine. Some of the substances are supplied in other ways. Some, perhaps the majority, pass through animals without being broken down, and are supplied in the animal excrement. The good effect of farmyard manure on the soil is probably partly due to its nutrient supply to soil microphytes, stimulating their development and hence humification and nutrient mobilisation.

Fungi and bacteria probably obtain growth factors and other

nutrients from plant roots, leaves and other organs, which normally secrete small amounts of various organic substances. More than 30 different substances, including sugars, organic acids and amides, are secreted from the roots of *Pinus strobus* growing in nutrient solution (Slankis *et al.*, 1964. See also Lundegårdh and Stenlid, 1944; Melin and Rama, 1954; Gyllenberg, 1956; literature in Slankis *et al.*, 1964). The rhizosphere is considered to be rich in growth substances. Experiments by Melin and Norkrans (1942) indicated that mycorrhizal fungi obtain aneurin or its components direct from tree roots. Another rhizosphere effect is the inhibition of some bacteria and stimulation of others by the root systems of some species (Gyllenberg, 1956). Fungi can also secrete growth substances to the environment, and so can other microphytes in the soil (Schopfer, 1943). However, the litter makes the greatest contribution and the surface layer of the soil contains more aneurin than the deeper layers. Soil extract has a favourable effect on cultures of fungi (Schopfer, 1943), bacteria (Nilsson *et al.*, 1938) and algae (Levring, 1945) because of the various growth substances present in the surface layers of the soil.

The movements of growth substances from one organism to another sometimes form part of some sort of cycle, e.g. from an autotrophic plant to fungi, via animals and the soil, and from fungi to phanerogams, saprophytes and soil animals. Marine algae, like other CO_2-assimilating plants, produce carotene. This is converted to vitamin A by planktonic animals, and this vitamin A accumulates in the liver of the fish which eat the plankton. It is extracted from the liver for medicinal use.

Apart from growth substances, the litter probably contains other, as yet unidentified, substances which are important for the growth and development of heterotrophs. Melin (1941, 1946) found that extracts of various kinds of leaves stimulated (or inhibited) fungal growth, and that the stimulation was not caused by growth substances hitherto known.

Extracts of leaves and other plant organs contain substances which, in oligodynamic concentrations, affect the state of cytoplasm, including its viscosity. They have a destructive effect at higher concentrations. These substances are also released as a result of humification and are present in soil, in soil water, rivers and the sea (Stålfelt, 1948). Thus, as well as nutrient and catalytic effects of the litter, there are effects on cytoplasmic structure.

A particular species of fungus is restricted by its special nutrient requirements to a particular kind of litter and biotic environment. The development of heterotrophy has brought about this increased dependence on biotic environment and consequent restriction to some extent to habitats where certain autotrophs are present.

There are some heterotrophic species of algae, e.g. species of *Euglena*, *Chlamydomonas* and *Ulotrix*, unable to synthesise certain growth substances. Like fungi, there are flagellate algae which have lost the ability to synthesise aneurin, but which can put pyrimidine and thiazole together if these two components are supplied. Ascorbic acid is another substance which these plants must obtain from the environment: it plays a part in their cell division. It can partly be replaced by aneurin, in the heterotrophic forms (Pirson, 1946).

These examples show that there are various degrees of heterotrophy, with a range of intermediates between the plants which can themselves synthesise the organic compounds and those which use only organic substances as food (certain fungi, and wholly parasitic or saprophytic higher plants). It is therefore difficult to draw a hard and fast line between autotrophs and heterotrophs. The difficulty is increased because partial heterotrophy cannot usually be detected from the easily observed, external characteristics of a plant. In a relationship between higher plants and bacteria, for example, the higher plants are often heterotrophic to some extent. The possibility cannot then be excluded that these plants utilise some of the numerous growth factors and organic compounds which are present in the soil, to compensate for a lack of balance in their own production and consumption of such substances. There are phanerogams which seem to be heterotrophic to some extent, though they have normal root systems. For example, *Camellia japonica* grown in sand requires a supply of vitamin B_1 (Schopfer, 1943); *Bougainvillea glabra* or *Arbutus unedo*, or others, will grow if no vitamins are supplied, but grow more rapidly if aneurin is added. They seem to have insufficient ability to synthesise aneurin (Schopfer, 1943; Melin, 1941). Even if uptake of the vitamin from the external medium is not essential, it can take place if the opportunity arises and can therefore affect metabolism and development. Von Witsch and Flügel (1953) found that seedlings of wheat, *Helianthus* and *Sinapis* grown in nutrient solution containing aneurin took up the vitamin. This affected transpiration and CO_2-assimilation in such a way that the seedlings developed shade-plant characteristics: transpiration increased, but resistance to exposure to high light intensity decreased; CO_2-assimilation became more sensitive to light, and so on.

16

The ecological effects of plant nutrients

The relation between the salt content of the soil and plant growth

The amount of plant nutrient in the soil has a major influence on its productive capacity, affecting vegetation both quantitatively and qualitatively. It has both direct and indirect effects on productivity. Among the processes affected indirectly is the formation of the organic soil component which, in turn, influences the accumulation of nutrients in the soil. The direct effects, expressed as the luxuriance or poverty, and the composition, of the vegetation, will be discussed in this chapter.

The optimum effect

The total effect of a single nutrient is the sum of its effects on several primary physiological processes. Calcium, for example, has several direct primary effects: it acts as a catalyst, neutralises acids and alters pH values, affects ion balance, permeability, ion uptake, water transport, etc.; in addition, it has edaphic-ecological effects because it affects humification, soil formation and soil structure. The development of a species, i.e. the number and size of individuals, their hardiness, reproductive capacity and competitive ability, etc., depends upon the overall influence of all these effects.

The separate processes, as well as plant growth as a whole, attain a maximum rate and efficiency at a particular concentration of a nutrient, the *optimum concentration*. The optimum can be measured experimentally if all other conditions, of light, water, temperature, etc., are held constant. The optimum concentration varies, however, depending on the strength and combinations of other factors. This modifying effect cannot usually be eliminated

by having all other factors present in excess, since an excess sometimes has an inhibitory effect. Hence, the optimum concentration
changes with the experimental conditions, within a certain range.
The wider this range, the broader the optimum. Furthermore,
there is a broad optimum if the individual processes are not particularly sensitive to excess or deficiency of the nutrient (see Fig. 7,
page 23).

The optimum concentration of a nutrient is different for different processes (see Fig. 90). For example, the optimum calcium concentration in respect of permeability to a particular substance may
be different from that which allows the best root development;
and the optimum concentration for humification may not be
optimal for phosphate uptake, etc. Because the optimum for
various processes is different, and because it can shift, the optimum
for the total effect is also variable (see Fig. 94, page 402). The
total effect is expressed as size and number of individuals, reproductive capacity, etc., and since these characters are not
necessarily very closely related, their response curves to a particular factor are not necessarily coincidental. For example, as the
number of individuals increases, their size may decrease (see Fig.
94). Since the various processes have different requirements for a
nutrient, a measure of its importance to one process cannot be
assumed to be a measure of its importance to the total. However,
by analysing the requirements and sensitivity of several processes,
it is possible to appreciate the total effect.

The total effect of a salt can be determined directly, by measuring yield, total amount of leaves or dry matter, etc., but the result
depends not only on the absolute amount of the nutrient present,
but also on the state in which it exists in the system and on the prevailing conditions.

The amount of a salt which is available to a plant in natural conditions, determining its effect as a nutrient, is not constant, but depends on the following factors:

1 THE CONCENTRATION IN THE SOIL WATER. The salt content
in the soil water is usually small, about 0·05 to 0·2 per cent in
arable soil. It changes continuously, partly because of variations in
consumption and supply, partly because of change in soil water
content, depending on rainfall, evaporation and run-off. However,
every change is followed by a shift back towards the original value.
There is an equilibrium between dissolved and adsorbed salts
which is established so rapidly that run-off water and water which
has percolated through the soil has a more or less constant salt content (Troedsson, 1952). This equilibrium is part of the balanced
system of consumption and production of salts (Fig. 85).

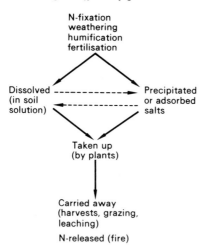

FIG 85. Schematic representation of the salt-equilibrium in the soil and its dependence on consumption and production.

2 THE MOVEMENT OF SOIL WATER. The concentration in the immediate vicinity of the plant roots is more important than the average soil water concentration; and concentration decreases sharply close to the roots. Hence rate of salt uptake is determined by the rate of supply of the salt from the environment. Ions can move along the surfaces of soil particles and from one particle to another. This is a slow movement, to the extent that it depends on diffusion, but mass flow in moving water is rapid, and can quickly replace the salts which have been taken up at the root surfaces. The luxuriance of vegetation where there is moving water is probably partly attributable to the good salt supply by mass flow. Some laboratory experiments support this idea. In still water, the layer of liquid in contact with the absorbing root surface is soon depleted of salts, and the rate of uptake decreases. The higher the original salt concentration, the more slowly this occurs, and there is no such effect if the solution is kept moving, e.g., by a stream of air bubbles (Fig. 86) (Olsen, 1950, 1953b). Roots take up as much salts from the flowing, dilute solution as from the more concentrated still solution, so that increase in concentration has little effect, except at very low concentrations. The effect of movement of the solution is certainly largely because of the supply of nutrient ions to the roots, but may also be related to increased oxygen supply, because the region nearest the root is also depleted of oxygen in a still solution, and to the maintenance of a constant pH. The pH changes because of differences in rate of uptake of anions and cations and because the roots produce some CO_2 (see page 412 for a discussion of the effect of pH on salt uptake).

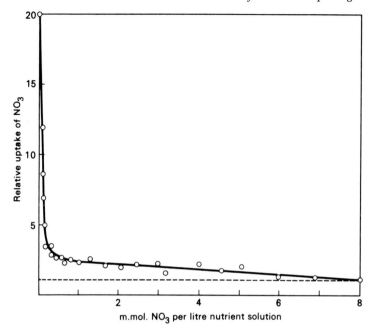

FIG 86. **The effect of movement by a stream of air bubbles of nutrient solutions of various concentrations (abcissa) on salt uptake by roots of rye plants (ordinate). Dotted line: uptake of nitrate from still solutions (taken as 1). Solid line: uptake from continuously moving solutions. (After Olsen, 1953.)**

3 ION BALANCE. The uptake of a particular ion is affected by the ion balance. Increased supply of a cation results in increased uptake of that cation, but also in decreased uptake of other cations to a similar extent (see Fig. 22, page 77). However, this antagonism only arises in the most dilute solutions (0·000 3 me/litre or less, Olsen, 1950).

In extreme cases, the relation between ions may be such that the roots secrete salts into the soil. A clay soil saturated with sodium depletes the plant roots of potassium and calcium, because of ion exchange between roots and soil.

4 CONTACT EXCHANGE. Salt uptake also depends on direct contact between the roots and the soil particles. At the surface of contact the ion concentrations are different from those in the soil water generally. There is also a high concentration of carbon dioxide and other acids produced by the root cells. The root hairs exchange the hydrogen ions of these acids directly with the cations from the particles, both adsorbed cations and cations from the crystal structure of the mineral itself. However, even this type of ion uptake is

probably mediated by water, in the film which surrounds the root hairs, since only dissolved ions can be taken up.

Thus, the optimum amount of an ion or combination of ions is dependent on the interaction of several factors, including the amount of undissolved salts and salts in solution, the proportions of the various ions and the mobility of the soil water. In other words, a particular ion can be optimal at different concentrations not only for different species, but also for the same species in a range of environmental conditions. In agriculture, numerous experiments to determine the optimum amount of a nutrient have been carried out, and fertilisation practices have been worked out on the basis of the results obtained, which, of course, depend on all the various conditions mentioned above.

Natural soils are usually complicated systems which cannot be precisely defined. This makes experimental attempts to determine the optimum amount of nutrients required in such soils somewhat difficult, so artificial, definable systems like quartz sand or nutrient solutions are often employed. Tank cultures are one way of using nutrient solutions. The plants are grown in containers which are regularly lowered into tanks of nutrient solutions. There are several advantages of this method: the concentrations of nutrients in the solution are known and can be controlled, the solution surrounding the roots is replaced every time the container is lowered into the tank, and the roots are well aerated.

The optimum amounts of nutrients required by wild species are less well known; there has been comparatively little research on this subject. The variation is probably very large, for instance, low optima would be expected in some groups, such as mosses, high optima in halophytes.

Limiting effects

Nutrient deficiency is a well-known occurrence in agriculture and horticulture, and various fertilisation practices to counteract it have long been used. Crop yield may be limited by deficiency of one or more nutrients such as nitrogen, phosphorus, potassium, or micronutrients (see Chapter 15), if application of fertilisers is insufficient.

Because of the deficiency, metabolism and growth is slow, various deficiency symptoms appear, production is low, and weeds invade. On old pasture, which is grazed but not fertilised, less-demanding grasses and herbs gradually become established, to be followed by the least demanding species of all, the mosses. If nutrients are supplied, grasses and herbs return and the mosses disappear. A nutrient supply which is insufficient for one species is optimal for another.

The extent to which the growth of a particular individual or species is limited by nutrient deficiency is known only in a few cases. Tamm (1953) showed that the development of forest mosses was limited by nutrient deficiency in the glades between the trees, and in other places outside the areas directly under the crowns, where there was no nutrient supply in drips from the crowns. Mosses grow in such situations despite the insufficient nutrient supply, possibly because they form pure stands and a closed moss turf, thereby preventing invasion by other species.

A species with high nutrient demands may give way to another with a lower optimum requirement, depending on the nutrient status of the habitat, and the requirements of the species. If the nutrient supply in a locality is low, the plants there may be existing under suboptimal conditions. Heath is poorer in nutrients than meadow, but probably offers a nutrient supply nearer the optimum for many heath species, which have low nutrient requirements and the ability to take up nutrients which are unavailable to other species. Only if the heath plants are stunted, like the dwarf trees on heath on cliffs, can it be assumed that nutrient conditions are limiting.

Nutrient deficiency may limit the growth of forests. Forest trees can grow on drained peat in northern Sweden, provided that nutrients are supplied (Malmström, 1952; see Chapter 10, Fig. 48, page 216). Tree growth is limited just as much by nutrient deficiency as by excess water, in these localities. On peat in south Sweden, where the nutrient supply is better, forest can grow on drained peat even without the addition of fertilisers.

Different forest tree species have different soil requirements: this has long been known. The most demanding species, elm, ash, lime, sycamore, etc., do well on soils with a good nutrient supply; beech, larch, spruce, aspen and alder need less; and pine needs least (Carbonnier, 1945–46). At the boundary between the demanding and less-demanding tree species, the nutrient supply is more or less suboptimal for the former. The competitive ability of the tree changes, in such a situation, with the changes in soil nutrient status. Limiting effects of nutrient deficiency have also been demonstrated in water plants (Fig. 87).

Inhibitory effects

Several of the anions and cations of nutrient salts have unfavourable effects if they are present in supraoptimal amounts. If there is an excess of soluble salts, the osmotic value of the soil water rises, so that water becomes less available to the roots. The soil is then sometimes said to be physiologically dry, since an excess of salts may have the same effect as a shortage of water. The yield of crops on salt-rich soils may be decreased as a result. A total salt content of 0·25 per cent is too high to allow cultivation of crop plants. Artificial irrigation can reduce the salt content, so that the soil can be used (Repp, 1951).

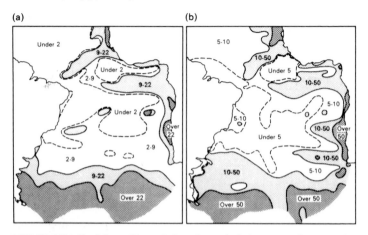

FIG 87. **The limiting effect of phosphate deficiency on the growth of planktonic organisms. (a) Phosphate concentration (mg/m³). (b) No. of organisms per ml, in the surface water (0–50 m) in the south Atlantic. (After Hentschel and Wattenberg, from Steeman Nielson, 1944.)**

In the soils of arid regions, the carbonates, sulphates and chlorides of magnesium, potassium, sodium and calcium accumulate to the greatest extent, so that there is an excess of alkali salts, the soil becomes alkaline and its physical state deteriorates. The crumb structure breaks down, permeability decreases and plant

TABLE 78 The effect of salt content (sodium sulphate) on yield from a sand culture. (After Eaton, 1942.)

Osmotic value of solution (bars)	0·7	1·8	3·5	5·1
Dry weight yield (kg per unit area of 18 square feet)	11·4	8·9	6·3	3·7
Total transpiration (1000 litre)	6·4	4·8	3·5	2·4

physiological processes are disturbed. Insufficient salt content makes cultivation difficult in wetter regions: excess salt content makes cultivation difficult in arid regions, so that only crops sufficiently salt-tolerant are worth growing. Salt tolerance varies considerably in different species, and even in different races of the same species. In artificially irrigated arable land in arid regions, the salts in the irrigation water accumulate in the soil as the water evaporates; water must then be applied occasionally merely to leach out the salts. This technique requires that the surface is level, otherwise salts accumulate in the hollows.

An excess of salts has both direct and indirect harmful effects. One of the indirect effects is the disturbance of the physiological ion equilibrium. Excess calcium, for example, can cause deficiency of potassium and magnesium, and possibly of some trace elements.

Some ions have so-called *poisonous inhibitory effects* which are not fully understood. Sodium and chloride are examples of ions with poisonous effects at relatively high concentrations; and manganese, boron, copper, etc., are poisonous even at relatively low concentrations (see Chapter 15).

The inhibitory effect of supraoptimal concentrations has often been utilised by man, in the battle against harmful organisms, for example. Salting food preserves it against attack by fungi or bacteria; copper salts are used to protect fruit trees and other plants from fungus diseases; and iron sulphate, sodium chlorate, etc., are used as weedkillers.

Poisonous effects and disturbances of the ionic balance are usually only localised phenomena, but saline soils which cause low growth rates because of their high osmotic value and high alkalinity are widespread in arid regions. Excess salt in arid soil types is the equivalent of salt deficiency in soils of humid regions: the salt content is supraoptimal for most plants in the first case, suboptimal in the second. In the same way that the number of species in heaths and coniferous forests is limited because of low content of salts, the number of species in salt deserts and salt steppes is limited by the opposite condition and only species which are sufficiently salt-tolerant can survive these.

The effect of mineral nutrients in selection

The mineral nutrients which are most important to plants are not those which occur most abundantly in nature, such as aluminium, sulphur and silicon. If this had been so, nutrient deficiency as a factor limiting plant growth and distribution would hardly exist. Nor would these elements be lost by leaching to any extent, since their salts are relatively insoluble. They have very little effect on soil pH and are not usually poisonous. However, the important nutrients are elements which occur in very much smaller quantities (phosphorus, manganese, boron, etc.), potassium, which is present in relatively large amounts in minerals, but which is easily leached, and nitrogen, which in its gaseous form is unavailable to higher plants and must be fixed and retained through a number of processes. The supply of plant nutrients is usually so small that it limits plant growth. In addition, most nutrients are readily soluble, so that they are transported in moving water, and this accentuates the shortage of nutrients in soils of humid regions and the excess in soils of arid regions.

This variation in salt supply has important ecological effects. The differences in the amounts and concentrations of salts ensure that somewhere on the earth's surface there are nutrient conditions suitable to meet the requirements of any plant species or new variety. This variability in nutrient conditions tends to bring about variability in plant characteristics.

Differences in salt supply cause competition, directly or indirectly. Suboptimal supply of salts acts directly, and also, like supraoptimal supplies, indirectly: if a species is subjected to stress because of deficiency or excess it is more sensitive to competition for other nutrients and site requirements. The interaction follows Mitscherlich's law, in that the ability to utilise a factor which is deficient is decreased if the species is also suffering from deficiency of another factor.

TABLE 79 The relation between salt content and the distribution of some submerged phanerogams in the water around the coasts of Denmark. (After Steeman Nielsen, 1944.)

SALT CONTENT (‰)	30	18	10	2	0·2	
AREA	SEA WATER	POLY-HALINE	MESO-HALINE β	MESO-HALINE α	OLIGO-HALINE	FRESH WATER
Zostera marina	———	———	———			
Z. noltii	———	———				
Ruppia maritima		———	———			
R. spiralis		———	———			
Najas marina				———		
Zannichellia palustris			———	———	———	———
Potamogeton pectinata			———	———	———	———
P. filiformis				———	———	———
P. perfoliatus					———	———
Ranunculus baudotii					———	———

A species grows best in places where the nutrient conditions are optimal, but it can tolerate a certain amount of deviation from the optimum, especially if its competitors are similarly handicapped. However, if the conditions suit one species better than another, the more closely adapted species tends to spread at the other's expense. The sensitivity of a species to nutrient deficiency is accentuated by competition. The great majority of dispersal units of various species which fall in a habitat fail to become established because of competition from neighbouring plants. Of the few which do become established, the majority are eliminated after a time by their better-adapted neighbours. As a result, each habitat is occupied by those species which are best suited to it. The greater the competition, because of lack or excess of nutrients or other requirements, and because of a copious supply of dispersal units, the more effectively selection operates. There is rarely a stable final stage, because the edaphic conditions change with time. Instead, an equilibrium between soil and vegetation is attained. Forests are certainly stable in comparison to many other plant communities, and may even be regarded as a 'climax' in community development, particularly as this development tends to move in the direction of forest, but they are not completely stable. Forests may be destroyed or damaged by natural phenomena (fires or storms), or by other organisms (disease or man), and be succeeded by other types of community.

Because of selection, and changes in the edaphic conditions, the composition of a plant community is unstable; there is a succession from species to species, with associated changes in the soil.

The effect of selection is often expressed as obvious differences in the appearance of the vegetation. If the nutrient status of the soil is extreme in some respect, the associated plant community is often made up of species characterised by some immediately obvious features, so that they have been given a special name, such as heath plants, salt-marsh plants, etc. The object of giving such special names may have been ecological, plant geographical, or more practical, so different principles have been applied in different cases. Particular features of the soil nutrient composition may have been considered (chalk plants, nitrate plants, etc.); or nutrient concentration or plant morphology (halophytes); or productivity (eutrophs, oligotrophs). Some of these types of vegetation are discussed below.

Eutrophy and oligotrophy

Eutrophic habitats are highly productive, nutrient-rich and characterised by species-rich, luxuriant vegetation made up of plants with high nutrient demands. The term *eutrophic* is also applied to plants and to plant communities. In the same way, the term *oligotrophic* applies to localities poor in nutrients, and to the plants which grow there. Optimal nutrient concentrations are lower for oligotrophic than for eutrophic species.

Terrestrial eutrophic localities are generally those where the soil has a high content of humus and clay. The vegetation is luxuriant and species-rich, and the species have high nutrient demands. Many deciduous trees, herbs and grasses are eutrophic species; and deciduous woods, fens, and brown-earth communities generally, are eutrophic communities.

Terrestrial oligotrophic communities include heaths, coniferous forest, and podsol communities generally. There are relatively small numbers of species and the soil has a low content of fine materials and of nutrients in a soluble or adsorbed form. On heaths, for example, nitrification is slow, and the nitrogen is bound in an organic form; there is also a poor supply of phosphates and potassium (Olsen, 1921). Oligotrophy is even more marked on lichen heaths; lichens and mosses which comprise the ground layer are among the least-demanding plant species. Moss communities are usually oligotrophic. Species of *Sphagnum* are one of the main constituents of these communities, and they are characterised by a very low optimal nutrient concentration (see pages 214 and 424).

There are also various degrees of trophy distinguishable in aquatic communities. The water has three main sources of nutrients: the bottom sediments, precipitation, and run-off from the surrounding land. The latter is the most important, and largely determines the degree of trophy. When the water comes from cultivated land and populated areas, nutrients accumulate in the water courses or in lakes, where rich vegetation and animal life develops. If the water comes from peat, unpopulated mountain areas or areas of poor moraine, it is poor in nutrients; such water may be coloured brown by stable humus from the peat and vegetation is sparse and so is animal life (see Lundh, 1951, page 122).

Halophytes

Localities with a high salt content support plants of a special kind, morphologically and physiologically distinct from other plants; these are known as *halophytes* (greek *hals* genetive *halos*—salt, *phytón*—plant). Marine plants, mainly brown and red algae, form the largest group. Phanerogamic halophytes mostly grow in terrestrial situations—sea shores, salt deserts and salt steppes. The term *glycophyte* (greek *glykýs*—sweet), analogous to the term *halophyte* can be applied to plants which occur in localities of low salt content. The vegetation of peat bogs, especially *Sphagnum* mosses, consists of extreme glycophytes.

Saline habitats. Plants can be categorised as halophytes and glycophytes; water as salt water or fresh water; and terrestrial habitats similarly as salt-rich or halic, and salt-poor or glycic. There are intermediates between the two sorts of habitat, and between the two sorts of plants (Stocker, 1928), so the distinction is only quantitative.

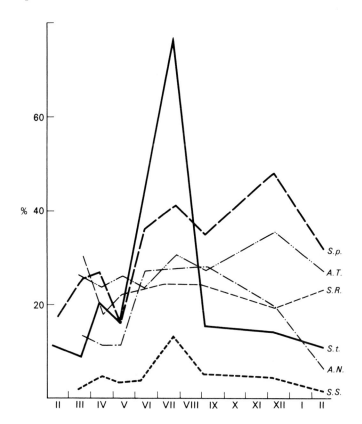

FIG 88. Annual changes in salt concentration of the soil water in various plant communities along the coast of the Dead Sea and in towards the desert. The values are percentages of dissolved salts in the soil water. *S.p.*: Suaedetum palestinae; *A.T.*: *Arthrocnemum-Tamarix* association; *S.R.*: Salsoletum-Rosmarinii; *S.t.*: Salsoletum tetrandae; *A.N.*: *Anabasis-Notoceras* association; *S.S.*: *Salsola tetranda–Stipa tortilis* association. (After Zohary and Orshansky, 1949.)

Oltmanns (1892) found that several marine algae grew at salt concentrations down to 0·2 to 0·3 per cent, and that the upper limit of salt concentration for many freshwater phanerogams was about 1 per cent. The transition between salt conditions suitable for marine and freshwater plants seems therefore to be about 0·5 per cent. Stocker (1928), who made an intensive study of halophyte ecology, took this into consideration and applied the term *halophytes* to plants which tolerated higher concentrations than freshwater vegetation generally, i.e. *c.* 0·5 per cent or more. Such a distinction is partly subjective, but is a useful aid to the general survey and classification of a mass of information; and it has some biological basis, because typical halophytes are morphologically, physiologically and ecologically different from glycophytes.

There are several types of saline habitat: aquatic (the sea, salt springs, salt lakes in arid regions, in the centre of continents, or in depressions just above the shoreline near the coast (Warming, 1918; Berger-Landefeldt, 1957)); aero-halic (areas to which salt is carried in the wind (Stocker, 1928)); or terrestrial (seashores, salt steppes, salt deserts).

1 AQUATIC HABITATS. The fresh water which enters the sea and lakes in large amounts influences the salt content, which is therefore higher in arid than in humid regions (Table 80).

The types of salt, and their relative proportions, vary with the position, climate and vegetation. Table 81 is a survey of the most important.

TABLE 80 Salt content (%) of surface water. (Mainly after Warming, 1918; Uphof, 1941.)

Dead Sea	20–26
Great Salt Lake, Utah, U.S.	16–22 (bays and marshland 1–12)
Red Sea	4
Persian Gulf	4
Mediterranean Sea	3·5–3·9
The oceans	3·5
Skagerack	3
Kattegatt	1·5–3
Baltic Sea	0·1–0·5
Gulf of Finland	0·3–0·7
Inland lakes	0·006–0·03

Salt concentration increases with aridity, and may be very high in some salt lakes. There is a layer of salt up to several metres thick on the bottom, with a saturated salt solution above it. In spring and summer, when evaporation is high and there is little addition of ground water, even the surface layers may have a salt content as high as 38 per cent (Table 82). The salt content decreases towards the surface in relation to the amount of fresh water entering the lake.

In Great Salt Lake, Utah, the salt content is 16 to 22 per cent at the surface and about 27 per cent at the bottom (Uphof, 1941). A Norwegian fiord (Romereinsfiord) had the following salt concentrations (Nordgaard, 1903): 1·06 per cent at the surface, 2·39 per cent at 1 m, 2·98 per cent at 2 m, 3·06 per cent at 5 m, and 3·42 per cent at 20 m.

The salt content is not constant, but is changed by the supply of water from precipitation, rivers, ground water and tidal water.

TABLE 81 The composition of salts in sea water and in Lake Geneva. (From Stocker, 1928; Steeman Nielsen, 1944.)

Salts in the Oceans			Salts in Lake Geneva	
Per cent of total salt content		mg / kg solution (mean of 77 analyses)	(mg/litre)	
Cl	55·29	NaCl	27 213	$CaCO_3$	73·9
Br	0·19	$MgCl_2$	3 807	$CaSO_4$	47·9
SO_4	7·69	$MgSO_4$	1 658	$NaCl + KCl$	18·8
CO_3 (carbonate)	0·21	$CaSO_4$	1 260	Na_2SO_4	15·0
Na	30·59	K_2SO_4	863	SiO_2	3·7
K	1·11	$CaCO_3$	123	$Al_2O_3 + Fe_2O_3$	1·9
Ca	1·20	$MgBr_2$	76	$Ca(NO_3)_2$	1·0
Mg	3·72			$(NH_4)_2SO_4$	trace
		Total	35 000	Organic matter, and losses	11·9
		or 3·5%			
		P_2O_5	0·1		174·1
		NO_3	0·25		
		or 0·0174%			

Plants on the seashore may be exposed to salt concentrations which are higher than that of normal sea water, and which are also very variable. This is particularly true of plants in the intertidal zone, which dries out at low tide, so that the salt content in the remaining water may rise to 6 per cent. Dilution by rain may lower the salt concentration to below that of sea water (Stocker, 1928). Indian mangroves are probably subject to the greatest changes: at low tide the salt content of the water rises to 8 to 12 per cent NaCl, because of evaporation (Faber, 1923).

2 AEROHALIC HABITATS. Vegetation at the coasts is affected by salt even if the plants are never in direct contact with the sea. Spray and salt crystals are carried inland by the wind and some are deposited on plants. The salt crystals dissolve and form concentrated salt solutions. If large amounts are involved, the plants may

wilt. Damage is also caused because of the increased transpiration and deformation of leaves and other organs brought about by the wind itself.

3 TERRESTRIAL HABITATS. The salt content of terrestrial habitats is more variable than in aquatic ones. Precipitation, evaporation and movement of water above and below the soil surface bring about changes in salt distribution, some of which are temporary, some seasonal, in both the horizontal and vertical directions in the soil. In soils in humid regions, the changes are mostly temporary, and are most marked in the surface layers of the soil (Stocker, 1928; Pompe, 1940). In arid regions, it is the seasonal changes which are most pronounced (Berger-Landefeldt, 1957) (see Fig. 88).

TABLE 82 The content of sodium salts in some lakes in Wadi Natrun (a hollow in the desert north-west of Cairo). Samples taken February–March. (After Stocker, 1928.)

LAKE	SPECIFIC GRAVITY	SALT CONTENT (g per 100 ml water)			TOTAL
		Na_2CO_3	NaCl	Na_2SO_4	
Abu Mamar	1·261	7·13	23·15	7·57	37·85
Zugm	1·229	4·34	24·10	5·82	34·26
Gaar	1·198	0·36	23·35	6·42	30·13
Fazda	1·165	3·78	10·95	7·62	22·35
Abu Gebara	1·124	4·63	9·50	2·33	16·46
Zawia	1·073	3·30	3·45	1·32	8·07

Soil solutions may vary from salt-saturated to almost salt-free.

Bog peat has a low salt content. Malmström (1923) analysed water from 37 habitats (fens, forests, sedge and dwarf shrub bogs) in northern Sweden (Västerbotten) and found that the total amount of inorganic substances was 9 to 45 mg/litre (0·000 9 to 0·004 5 per cent). Troedsson (1952) found total amounts of 14 to 115 mg/litre (0·001 4 to 0·011 5 per cent) SiO_2, Al, Fe, Mn, Na, K and Ca, in the ground water of various soils. The salt content of the soil water of arable soils is somewhat higher: 0·05 to 0·2 per cent; in seashore soils it may reach a few per cent; and in soils of salt steppes or salt deserts it is 10 or 20 per cent or more (Zohary and Orsansky, 1949). The higher the salt concentration, the greater the variation caused by evaporation and precipitation.

Salt tolerance. 1 PLANTS GENERALLY. Plants vary in ability to tolerate high salt concentrations. Glycophytes have some degree of tolerance even though they grow in places where the soil solution contains only small amounts of salt, usually less than the amounts generally present in arable soils, i.e. 0·05 to 0·2 per cent, corresponding to an osmotic value of 0·5 to 1·5 bars (Table 87, page 398). This concentration is lower than that in the cell sap, which is between 5 and 10 bars for most mesophytes.

Glycophytes grow best at low salt concentrations. Increased salt concentration in the external medium leads first to increased salt uptake, so that the concentration in the plasma and the cell sap rises (Fig. 89). If the salt concentration is below the optimum, for

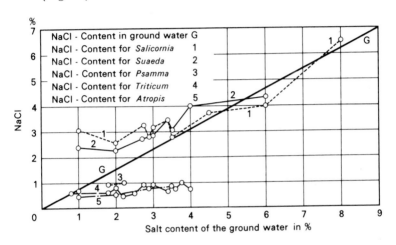

FIG 89. The relation between salt content in the plant and that in the surrounding medium. Ordinate: Mean NaCl – content of the plant material expressed as a percentage of the water content. Abcissa: Mean total salt content of the ground water in the habitat. (After Arnold, 1936.)

separate processes or for growth as a whole, this increase in concentration is favourable: if it is already above the optimum, the consequences depend on the salt tolerance of the particular species (Fig. 90).

The distribution of glycophytes may be limited by their salt tolerance, if they are unable to spread into habitats with higher salt contents. A species is checked within an intermediate zone, separating areas of high and low salt content, in which it may spread and regress in turn. Uptake of salt increases, particularly of sodium chloride, in areas of high salt content. Some species also accumulate considerable amounts of calcium salts, e.g. *Soja* and *Pisum sativum* (Repp, 1951).

Several morphological and physiological characters change in relation to salt accumulation (Table 83). The osmotic value of the cell sap rises, so that water uptake is facilitated. The difference in osmotic value between plant and soil may even be maintained, despite the larger amount of salt in the soil, as long as this does not become too extreme. This enables the plant to continue to take up water from the soil.

However, the increase in cell sap concentration is generally insufficient to counteract the physiological dryness of the soil fully, and water uptake and transpiration by plants on soils with high

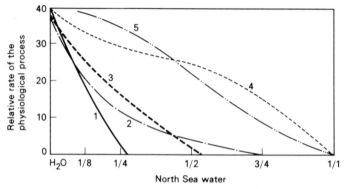

FIG 90. The effect of salt on various physiological processes in glycophytes: 1. Starch synthesis resulting from carbon dioxide assimilation by leaves of *Syringa* (ordinate; relative starch content). 2. Breakdown of starch in starch-containing *Syringa* leaves (ordinate; relative starch content). 3. Root growth in young seedlings of *Lepidium* (ordinate; relative root length). 4. Percentage germination of *Lepidium* seeds. 5. Root growth in young seedlings of *Aster tripolium* (ordinate; relative root length). N.B.: 3/4 North Sea water = 3 parts sea water + 1 part tap water. (After Montfort, 1927.)

salt content may be less (Table 78, page 378) (Repp, 1951; see Magistad, 1945, and Slatyer, 1967, Chapter 9).

Table 83 shows that salt accumulation in cells does not always keep pace with the external salt content. For example, the salt content (ash) of *Pisum sativum* increased from 16 to 18·1 per cent in response to an increase in the medium from 0·13 to 0·33 per cent. It has already (Chapter 5) been pointed out that plants do not take up salts in the same proportions as they are present in the soil solution. To some extent, plants can prevent the entry of some salts (Table 83), so that salt accumulation is determined not only by the salt concentration of the external medium and the plant water economy, but also by physiological processes (Schratz, 1936; Repp, 1951, 1958). Even closely related species differ in their ability to prevent salt uptake: *Ranunculus aquatilis* rapidly takes up salt and is consequently poisoned; but *R. baudotii* does not (Gessner, 1940).

The physiological processes involved in regulating salt uptake in this way are not understood. The permeability and organisation involved in normal salt uptake presumably play a part. Salts are transported into the cells and through the various parts of the cell and are then actively secreted into the vacuoles (Arisz, 1953). This transport is against the concentration gradient (Magistad, 1945). Both uptake and transport involve active physiological processes and various types of resistance, including diffusion resistance and

TABLE 83 A comparison of certain physiological and morphological characteristics of plants cultivated in normal soil (N) and in an area (S) bounding on a saline area, in Hungary. (After Repp, 1951, page 283.)

SPECIES	SOIL	SOIL SALT CONTENT IN ROOT REGION (%)	CELL OSMOTIC VALUE (bars)	ASH % DRY WEIGHT	DEGREE OF SUCCU- LENCE	NO. OF STOMATA (per mm^2)
Pisum	N	0·13	11·6	16·0	1·51	85
sativum	S	0·33	16·0	18·1	1·69	92
Helianthe-	N	0·13	14·2	17·7	1·23	184
mum annuus	S	0·33	15·5	19·6	1·46	224
Zea mays	N	0·19	14·2	12·1	—	73
	S	0·33	15·5	13·0	—	92
Trifolium	N	0·19	19·4	9·1	0·82	207
repens	S	0·35	20·8	13·7	0·98	286

electrostatic resistance. Regulation of salt uptake can therefore be brought about by change in the transport processes or in the resistance to transport.

Increased salt uptake not only causes higher osmotic values of the cells, but also changes certain cytoplasmic functions. Degree of hydration, permeability, ion exchange, and the state of the cytoplasm in general, are all affected, and metabolism is consequently affected indirectly as well as directly (Stocker, 1928). Some of the consequences are visible in morphological changes. The leaves are smaller, there are more stomata per unit area of surface, and the leaves are thicker because of an increase in cell size. The ratio of surface area to fresh weight is smaller, and the leaves are succulent (Table 83; Uphof, 1941). The resulting increased water content in the plant, and dilution of the cell sap, counteracts the effects of the high salt concentration in the external medium to some extent (Schratz, 1936; Steiner, 1939; Repp, 1958).

The root to shoot ratio also increases (Repp, 1951). The increased size of the root system again compensates somewhat for the decreased effectivity caused by the salt.

All these changes—higher salt content, higher osmotic value, increased succulence, larger root system, smaller leaf area and decreased transpiration—are phenotypic modifications, which do not occur in progeny growing in places where salt concentration is low (Repp, 1951).

However, reduced uptake of some salts, large root systems, low water potentials and low transpiration rates are insufficient to counteract the harmful effects of supraoptimal salt concentrations. Such harmful effects may result from disturbance to permeability, hydration or other processes affected by the state of the cytoplasm. The relation between salt and various physiological processes has been studied in a few cases. Boyer (1965) obtained evidence of a slow decline in net photosynthesis per unit leaf area in whole cotton plants grown in saline culture solutions with concentrations up to the equivalent of 8·5 bars. The stomata were open, so the effect could be attributed to salinity. Transpiration was not reduced.

The cultivation of crops on salt land is difficult or impossible because of the harmful physiological effects of excess salt; the alkaline pH, which is often a consequence of salt accumulation, adds to the difficulties. The availability of phosphate, iron and magnesium is reduced as a result of the high pH. Salt also adversely affects the physical condition of the soil through colloid dispersion and reduced permeability to water and gases (literature in Magistad, 1945). The yield is affected when the salt concentration reaches 0·25 per cent, equivalent to about 1·7 bars (Repp, 1951); it becomes very small when the salt concentration is equivalent to about 10 bars; and no crops at all can be obtained at about 47 bars (Magistad, 1945) (see Table 78, page 378; and Table 87, page 398). However, experiments in Israel indicate that many crop species can be grown by irrigation with salty water, even up to a content of 4 per cent dissolved solids, if they are in sandy or gravelly soil (Boyko, 1967).

2 HALOPHYTES. So far, research into the so-called *halophyte problem* (Stocker, 1928), i.e., how plants survive high salt concentrations in the soil, has shown several means by which the unfavourable effect of excess salt are mitigated or counteracted altogether. A particular species often uses more than one.

Apparent halophytes. These species are not really halophytes, although they grow on salt steppes and in other saline habitats, since they develop at times of the year when the salt content is temporarily very low. Keller (1925b) found that the surface layers of the soil of salt steppes in Turkistan are leached during the rainy season, so that species which then develop rapidly, with a superficial root system, are more or less unaffected by the salt. Stocker (1928) cites similar examples from Thüringen: at the time of seed germination in the spring the soil salt content was at its lowest. Keller considers that the annuals, succulents and spring ephemerals, which develop in saline habitats under these conditions, are only apparently halophytic species; they are glycophytes which make use of the salt areas when they are temporarily leached.

True halophytes germinate and develop at relatively high salt concentrations. They react to the salt in the same way as glycophytes, initially at least. Salt is accumulated and the osmotic value of the cell sap rises, salt uptake is hindered to some extent, succulence develops, etc. (Table 84). The difference is that the effects are

TABLE 84 The relation between salt content of the plant and of the surrounding medium. Experimental cultivation of *Salicornia europaea* in nutrient solutions with increasing NaCl content. (From Keller, 1925a, page 234.)

NaCl (g per flask)	PLANT ASH CONTENT (g per fresh weight)	PLANT NaCl CONTENT (g per fresh weight)	NaCl REQUIRED TO BRING ABOUT PLASMOLYSIS (g per 100 ml solution)
0	3·0	0·13	3
1	2·4	1·03	4
10	2·7	1·4	4
30	3·5	2·7	5
50	5 0	4·3	7–8

more marked in halophytes: halophytes have a greater ability to control salt uptake, and the cytoplasm is less sensitive to high salt concentrations. As a result, and possibly as a result of other special characteristics, the halophytes have a greater salt tolerance than glycophytes. The special characteristics are:

(a) *Osmotic value.* As in glycophytes, salt uptake in halophytes is dependent on the salt concentration in the external medium: the greater the concentration, the more salt is taken up (Table 84). Halophytes therefore generally contain more salt than glycophytes, and their cells and cytoplasm have higher osmotic values. In ten mangrove species investigated by Faber (1913), the osmotic values were between 24 and 72 bars. Even higher values are obtained if salt is added to culture solutions in which the plants are grown, e.g. up to 205 bars in *Avicennia* (Faber, 1923).

Desert halophytes also have high values: e.g., over 50 bars in *Limoniastrum*, up to 60 bars in *Aeluropus* and up to 70 bars in *Halocnemon* (Killian and Faurel, 1933). Walter (1936), Walter and

Steiner (1936) and Steiner (1939) give other examples. There is no hard and fast distinction between glycophytes and halophytes in amounts of accumulated salts. Some glycophytes may contain as much salt (the equivalent of 10 to 15 bars or more) as those halophytes which contain least salt.

Most of these salts are chlorides, particularly sodium chloride. Only relatively small amounts of other salts, such as sulphates, are taken up (Steiner, 1939).

The requirement for salt, and the ability to store it, like other physiological characteristics, are genotypically determined, and therefore vary from species to species. It is the group of species known as *halophytes* which accumulate relatively large amounts of salts, even in habitats with a low salt content (Steiner, 1939; Takada, 1951). In glycophytes, the amount in such habitats is generally insignificant (Steiner, 1939).

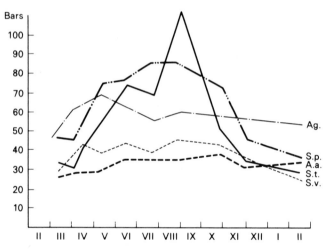

FIG 91. **Annual variations in the osmotic value of the cell sap in some halophytes from the coast of the Dead Sea and the nearby desert.** *A.g.=Arthrocnemum glaucum*; *S.p.=Suaeda palestina*; *A.a.=Anabasis articulata*; *S.t.=Salsola tetranda*; *S.v.=Salsola vilosa.* **(After Zohary and Orshansky, 1949.)**

Again like other physiological characteristics, salt accumulation is a character subject to modification. It is affected by the salt content of the external medium, by climate and by the season of the year (Figs. 91, 92). The more salt in the external medium, the more is accumulated. If the concentration changes, the plants adjust accordingly. In the cells of mangrove species, the highest salt content is reached at low tide, and the lowest at high tide

(Faber, 1923). As a result of this adjustment, the salt content of different species growing in the same place is often about the same (Schratz, 1936; Zohary and Orsansky, 1949). However, the salt tolerance of different species is different, and is determined by the prevailing salt content of the environment and other external conditions, as well as by the genotype (Stocker, 1928).

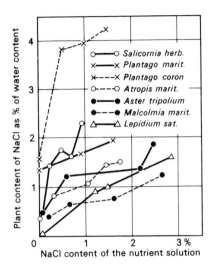

FIG 92. The relation between plant salt content, as a percentage of the water content, and the salt content of the external medium (nutrient solution). (According to Steiner, 1939, after data from Schratz, 1936.)

TABLE 85 Examples of frequently occurring osmotic values in cells of various groups of halophytes. (After Stocker, 1928, page 308.)

	BARS
North Sea halophytes, leaves	17–36
Indian mangroves, *Avicennia* leaves	max. 163
,, ,, ,, roots	max. 96
Rocky desert at Biskra, Algeria, leaves	15–100
,, ,, ,, ,, mean of 46 species	50
Egyptian rocky desert, leaves	19–91
,, ,, ,, mean of 5 species	41
,, ,, ,, roots	35–48

By means of salt accumulation, terrestrial halophytes achieve the high osmotic values and high suction force required to regulate the turgor pressure of the cells and to counteract the resistance of the external medium to water uptake. *Salicornia*, for example, generally accumulates so much sodium chloride that the concentration of the cell sap is greater than that of the soil solution (Schratz, 1936). As long as salt accumulation is maintained in proportion to the external salt content, the difference in osmotic value between plant and soil is about the same as in vegetation on non-saline soils. In these circumstances, the physiological dryness of the soil is counteracted and the plant water balance is not disturbed (Steiner, 1939). However, water uptake by halophytes is probably more complicated than this interpretation would suggest. Collander (1941) found that in *Atriplex hortensis* there was salt accumulation in the stems and leaves, while the salt content of the root system was about the same as in glycophytes. This implies that the parenchyma cells of the roots are subject to large water losses before they attain the high suction force needed to take up water from the surrounding soil.

However, even the most salt-tolerant species have only a limited ability to accumulate salt, so that accumulation does not keep pace with increase in salt concentration in the soil at the highest salt concentrations, but lags more and more behind, so that the suction force of the plants fails to increase in relation to that of the soil (Table 84, page 391). Eventually, the salt concentration in the plant cells becomes the same as that in the soil, and then only the increase in suction force due to other osmotically active substances, such as sugars and organic acids, can provide a gradient of suction force between plant and soil. Even this reserve may be exhausted if the external salt concentration increases still further (Steiner, 1939), and in many saline soils the concentration is at or above this critical level. The osmotic value of the soil solution is up to 200 bars in some soils (literature in Magistad, 1945).

The ability of a species to grow in saline situations therefore depends on its capacity for salt accumulation and its cytoplasmic salt tolerance (Steiner, 1939). The wider the range between the maximum and minimum salt concentrations at which a species can grow, the more situations it can invade: the differences between halophytes and glycophytes lie in the differences in position and width of this tolerated range.

The submerged halophytes also accumulate salt: there is no equivalent of the transpiration stream, but uptake of salts maintains the osmotic balance of the cells and their environment so that normal turgor pressures obtain.

(b) *Succulence.* Many halophytes, particularly seashore species, are characterised by succulence. The hypodermal parenchyma is developed as a water tissue with small intercellular spaces and low chlorophyll content; there are virtually no hairs or mechanical tissue (Keller, 1925a). At least in some cases, succulence is a direct consequence of high salt content, as it is in glycophytes. However, in halophytes it is more extreme, and subject to some modification (Fig. 92). Out of eighty-one halophytes investigated (Lesage, cited by Uphof, 1941), fifty-four developed succulence when grown in a saline environment, but did not in a non-saline one; other species reacted less strongly. In some mangrove species, a tendency for stomata to be sunken was associated with succulence (Faber, 1923; Stocker, 1928). Faber interprets succulence as a consequence of cell hypertrophy resulting from high osmotic values which bring about very high turgidity when the roots are surrounded by osmotically weaker solutions (tidal water, brackish water). However, the reasons for this are not clearly known. In any case, it is probable that succulence of halophytes is brought about somewhat differently from xerophyte succulence.

It can be assumed that the water-holding tissues in halophytes serve as a reservoir of water, as in xerophytes (Steiner, 1939). These reserves of water serve to delay or prevent wilting and associated damage. Other properties of succulents operate in the same way: the shape and position of the thick and generally short leaves is maintained even after the loss of large amounts of water.

(c) *Transpiration.* Transpiration of halophytes is about the same as that of glycophytes, if expressed per unit area of surface of the transpiring organs, but usually lower if expressed on a weight basis (Table 86). This is because the ratio of surface area to fresh weight is less in halophytes than in glycophytes, particularly in the succulent types.

Montfort (1926) made the interesting observation that the guard cells of halophyte stomata have a greater salt tolerance than the mesophyll cells. Even if the mesophyll cells have ceased to function because of the salt, the guard cells continue to regulate the stomatal aperture.

(d) *Mechanical features.* In dry, saline places there are often halophytes with a large proportion of mechanical tissue. This enables the organs to maintain their shape and position, even if the cells lose their turgidity, and also delays wilting, and the damage caused by deformation and tearing of the cells and tissues (Keller, 1925a).

(e) *Water deficit.* It is not known whether halophytes tolerate higher water deficits than other plants. It is possible that their drought resistance may even be lower. If the cell salt contents are supraoptimal, the salt would be expected to make the plant more sensitive to higher deficits, since the salt concentration in the cells increases as the deficit increases.

(f) *Salt regulators.* (i) Salt barriers. The ability to prevent salt uptake is greater in halophytes than in glycophytes. For example, if *Salicornia europaea* is grown in salt solutions of increasing concentrations, salt is accumulated in the cells, but the salt content rises more slowly in the plants than in the solutions (as shown in Table 84, page 391). When the concentration in the medium is fifty times the original value, the concentration in the plants is only about five times the original value. When the salt concentration of the external solution is low, the plants take up relatively large

TABLE 86 Mean transpiration rate of plants from the beach, sand dunes and salt-free soil at Borkum, Germany. North Sea coast. (After Schratz, 1937, pages 626–7.)

| | | TRANSPIRATION | |
		mg/g/min	mg/dm²/min
	Phragmites communis	17·3	17·7
	Helianthus annuus	16·1	24·8
	Fragaria	12·9	14·5
H	*Artemisia maritima*	11·1	12·9
	Erica tetralix	11·0	24·5
H	*Triticum junceum*	10·5	20·6
H	*Elymus arenarius*	6·9	15·6
H	*Limonium vulgare*	7·4	16·1
H	*Glaux maritima*	6·4	15·2
H	*Aster tripolium*	5·0	22·7
H	*Cakile maritima*	4·1	17·0
H	*Plantago maritima*	2·9	18·1
H	*Salsola kali*	2·1	14·0
H	*Salicornia europaea*	1·5	14·6

H = halophyte.

amounts of salt, but they take up relatively less as the concentration increases—there is thus a 'salt barrier' (Stocker, 1928). Mangrove species, which grow in situations where the salt concentration in the water is a few per cent (26 to 41 bars), accumulate salt up to a certain concentration in the cells, but then a salt barrier comes into operation and the osmotic value remains constant. The salt content of sea water is 3·5 per cent, equivalent to 25·3 bars of which 23·6 is due to chlorides, mainly NaCl, and 1·7 to sulphates, taken as Na_2SO_4 (Walter and Steiner, 1936). If these plants took up the external solution as such, the salt content of the cells would more than double in a single day, since the cell water content is lost in transpiration and replaced between once and twice a day. When the cells have reached their physiologically determined salt content, further uptake is probably prevented altogether, so that the water in the conducting systems is free of salt (Walter, 1936;

Steiner, 1939; Collander, 1941). The barrier mechanism seems to be best developed in *Suaeda*, the cytoplasm of which is almost completely impermeable to sodium chloride (Höfler and Weixl-Hofmann, 1939).

In some succulent halophytes the water content of the organs increases with age, so that the cell sap is diluted and the effect of salt accumulation is counteracted. This is probably the result of osmomorphosis, changes in the cells brought about by the salt (Schratz, 1936; Steiner, 1939). Since succulents have a rapid water turnover (Table 86), the effect of such an increase in water content would be small unless a barrier operated to prevent the entry of salt during water uptake. However, an increase in water content can compensate for an imperfect barrier mechanism, acting as a fine control for the adjustment of the effects of the barrier.

(ii) Salt secretion. Some halophytes have epidermal glands which secrete salt. Ruhland (1915) studied this phenomenon in *Statice gmelini*, and found that leaves placed in a solution of sodium chloride took up salt through their cut edges, but secreted it through epidermal glands so that there was no salt accumulation in the cells. Intact plants take up salts through the roots. Salt-secreting species therefore also have two regulatory mechanisms: the salt barrier in the roots and the salt-secreting epidermal glands. The glands can again be regarded as a fine regulator, or compensation for an imperfect barrier mechanism. Salt secretion is oxygen-dependent and is linked to cell metabolism; it seems similar to salt secretion by root parenchyma into the vessels (Arisz *et al.*, 1955).

These salt-secreting plants also have high osmotic values and limited salt accumulation. When the limit is reached, excess salt is secreted by the glands. The salt content rises if the external concentration is increased, and falls if the external medium is diluted, by rain, for example (Pompe, 1940). Through this equilibrium system, the salt content of the plant changes so that a suitable suction force is maintained.

Salt secretion takes place in succulent and in non-succulent halophytes (Zohary and Orshansky, 1949). It occurs in succulents mostly in those species in which there is no osmomorphosis and so no associated regulatory mechanism. Faber (1923) found osmomorphosis only in species which do not secrete salt; and Stocker (1928) points out that the most succulent forms of halophytes, e.g. *Salicornia* and *Mesembryanthemum*, do not secrete salt.

In general, the salt-secreting halophytes are found in places with a good water supply, on sea beaches and in salt marshes, where there is plenty of water to sustain the glandular activity. Examples are *Limonium vulgare*, *Armeria maritima* and *Glaux maritima* along north temperate costs; *Avicennia aegiciros* and *Acanthos ilicifolius* in Indian mangrove swamps (Steiner, 1939; Stocker, 1928); and *Aegialitis annulata*, a mangrove in the intertidal region in Queensland, Australia (Atkinson *et al.*, 1967).

There are also salt-secreting halophytes on steppes in south Europe and in North African deserts, e.g. *Tamarix, Frankenia, Statice*, etc. (Stocker, 1928). The relatively large amounts of water required for the activity of the glands are provided by the ground

TABLE 87 Osmotic values of NaCl solutions, at the freezing point of the solutions (the osmotic values are somewhat higher at 20°C—generally 8 to 10%). Molecular weight = 58·45. (Mainly after Walter, 1936, page 336.)

$\dfrac{N}{100}$	PERCENTAGE SALT IN SOLUTION	bars
1	0·06	0·44
10	0·6	4·20
20	1·1	8·32
30	1·7	12·43
40	2·3	16·52
50	2·8	20·61
55	3·2	23·6
60	3·5	25·7
75	4·3	31·8
100	5·8	42·8
125	7·3	53·7
150	8·8	64·2
175	10·0	73·4

water, and the growth habit of these plants must be such that the roots reach this water. *Statice* roots penetrate to a depth of 2 m or more, to soil layers which have more water and less salt than at shallower depths. *Tamarix* roots have been found to penetrate as far as 30 m in the region of the Suez canal (Stocker, 1928).

(g) *Regulator weak or absent.* There are some halophytes which accumulate salt through the whole of the vegetation period, and so can be regarded as having no regulator, or only an ineffective regulatory mechanism. Steiner (1934 and 1939) called these plants 'accumulation types': *Juncus gerardii*, which grows on marshland with a moderate salt content, is an example. The osmotic value of the cells rises from the middle of May to the end of August from *c.* 20 to *c.* 40 bars; this type of plant must have a large salt tolerance.

(*h*) *Summary.* The ability of the halophytes to tolerate salt depends on many interacting factors and properties. The salt regulator is foremost, with the most important mechanism being the ability of the roots to prevent salt uptake to some extent. Together with this, either osmomorphosis or salt secretion by epidermal glands act as fine regulatory mechanisms. Regulation systems are developed to very different extents: mangrove species can effectively prevent salt uptake, whereas other halophytes allow salt to enter in more or less the same concentration as in the external medium. The latter must rely completely on high salt tolerance, and then resemble the most salt-tolerant glycophytes in this respect.

Because of this salt regulatory mechanism, halophytes can accumulate the amount of salt necessary to maintain a sufficiently high suction force for water uptake to be possible. Other characters affecting the water economy are relatively low transpiration rates and anatomical and morphological features which prevent wilting resulting from loss of turgor. High cytoplasmic salt tolerance is also essential. This is another character developed to varying degrees: there are some plants which tolerate high salt concentrations in the cells, whereas others tolerate only as much as the more salt-tolerant glycophytes. Sometimes halophytism seems to be characteristic of a whole genus: there are many examples of this among genera in the Chenopodiaceae. In other families, such as the Ericaceae, halophytes are rare (Stocker, 1928). The halophytic characters can be modified and adapted according to the requirements of the environment. The range of possible modification can be seen from the shape of the curve of response of growth to salt concentration. The broader the optimum, the greater the adaptability; some species are so adaptable that they can live either on saline or normal soils, e.g. some mangrove halophytes (Faber, 1923).

Even the greatest degree of salt tolerance is insufficient for survival in the most saline environments. On high plateaux in the Egyptian desert, Stocker (1928) found salt concentrations of 190 to 430 per cent in the soil solutions. Since there was undissolved salt in the soil, presumably the soil solution remains saturated even after rain. There is no vegetation in such places, but there were plants in nearby valleys where the salt concentration in the soil solution varied between 8 and 51 per cent. Algae are probably the most salt-tolerant plants. Woronin (1926, cited by Stocker, 1928) observed luxuriant algal growth at a salt concentration of 11·5 per cent, at the Great Tambucus Lake in the north Caucasus. Some *Chlamydomonas* species may be able to tolerate concentrations up to 25 per cent (Oltmanns, 1892). However, organic life is probably no longer possible when the salt content reaches 20 per cent, as in the northern part of the Dead Sea (19·5 per cent) and in salt lakes in Wadi Natrum (see Table 82, page 386). Stocker (1928) found no macroscopic vegetation in Wadi Natrum.

Bernstein (1962) has summarised the literature on saline soils and their vegetation.

3 THE SELECTIVE EFFECT OF SALINE HABITATS. The most extreme environmental conditions, those which differ most from the optima for the majority of species, are most important in determining the variety of species growing in a particular place. In saline habitats, only salt-tolerant species can survive. If the relation between salt concentration and growth, for example, is expressed in the form of an optimum curve, the salt requirement is shown by the suboptimal concentration range, the salt tolerance by the supraoptimal range, and the ecological range by the curve as a whole (Fig. 93; Stocker, 1928). Moreover, the salt-tolerant species is

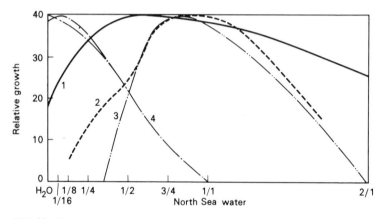

FIG 93. The relation between growth and salt content in marine algae and terrestrial halophytes. 1. *Salicornia europaea* (terrestrial halophyte). 2. *Fucus serratus* (marine alga). 3. *Nitschia putrida* (marine alga). 4. *Aster tripolium* (terrestrial halophyte). 3/4 North Sea water = 3 parts sea water to one part tap water. (Montfort, 1927.)

relatively insensitive to changes in the region of the optimum: the greater its ability to regulate the internal salt content, the broader the optimum and the wider the ecological range.

As the salt content in the external medium increases, selection of species becomes more rigorous and the chances of germinating seeds and young seedlings surviving the more critical stages in their life cycle diminish. Once the roots have penetrated the surface layer of soil, probably the most saline, a successful plant water balance is more likely to be achieved. In the Egyptian desert, Stocker (1928) found *Phragmites stenophylla* growing in soil covered by a crust of salt, but with the roots in deeper less saline layers. At a depth of 70 cm the salt content in the ground water was only 1·7 per cent.

As the soil salt content increases, the numbers of species and of individuals per unit of surface area decrease, the vegetation becomes more and more sparse and desertlike (Fig. 94).

There are generally few annuals in saline situations, although the open vegetation would otherwise be particularly favourable for this type of plant. The difficulties encountered by germinating seeds and young seedlings are too great (Stocker, 1928). The vegetation consists of perennial herbs and grasses, together with low bushes. These are less dependent on seed germination than annuals, and the root systems and underground stems can exploit large volumes of soil.

Knowledge of the way in which selection is affected by salt optima, salt requirement, salt tolerance, etc. is still only fragmentary. Kolkwitz (1919 and earlier) found that *Salicornia europaea* grew in places with up to 14 per cent Na Cl, *Halimione pedunculata* at 4·2 per cent, and *Antennaria maritima* at 3·2 per cent. In *Salicornia-Atropis* dunes on North Sea coasts, the distribution of some halophytes is related to various upper limits of salt concentration (Schratz, 1936; Fig. 94): *Triticum junceum* < 1·5 per cent; *Atropis maritima* < 2·5 per cent; *Suaeda maritima* 1 to 5 per cent; and *Salicornia europaea* 10 per cent. At concentrations higher than 10 per cent, there are only small scattered individuals of *Salicornia*.

Salt has the same effect on the distribution of glycophytes, among which there are tolerant and more sensitive species. There is no sharp distinction between glycophytes and halophytes: plants vary very greatly in salt tolerance and in salt requirements. Environments also range from those with very little salt, such as peat bogs, to those with a large amount. The plants which grow in low salt habitats, the glycophytes, have relatively low salt tolerance, but many are able to adapt themselves to survive somewhat higher salt concentrations, by modification of their osmotic values and other properties so as to prevent the harmful effects of the salt. Glycophytes are excluded from saline situations by competition from halophytes; and halophytes are excluded from non-saline situations by competition from glycophytes. The salt requirement of many halophytes is so low that they could otherwise grow in non-saline situations. Some even have very low optimal salt concentrations, but insufficient ability to survive in competition with glycophytes in situations where their optimal salt concentration obtains (Stocker, 1928).

In general, the range of salt tolerance seems to be narrower in aquatic than in terrestrial halophytes (Montfort, 1927; see Fig. 93). Extreme aquatic halophytes cannot tolerate non-saline situations (Stocker, 1928): *Salicornia olivieri* does not flower if the salt content is lower than 3·5 to 5 per cent (Uphof, 1941). The optimum salt concentration for *Zostera* lies within a narrow range, corresponding to that of sea water, and adaptation to the low salt concentrations

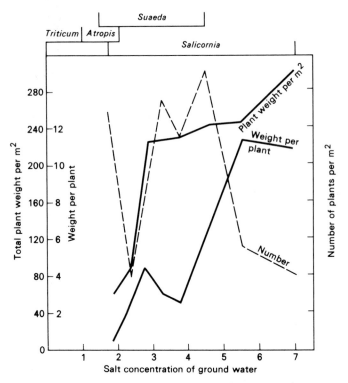

FIG 94. The relation between the salt content of the ground water, numbers and weights of individual *Salicornia* plants, and total weight of *Salicornia* per unit area. The ecological ranges for *Triticum junceum*, *Atropis maritimus*, *Suaeda meritima* and *Salicornia* are also shown. (After Schratz, 1936.)

of fresh water is not possible. On the other hand, *Ruppia* has a broad optimum and some adaptability. *Myriophyllum* has a low salt optimum and only limited adaptability, and can be regarded as typical of brackish water vegetation (Montfort, 1927; Stocker, 1928).

The relatively narrower range of salt optima of aquatic halophytes as compared with terrestrial species suggests that they are more highly adapted for their halophytic mode of life, so that in some cases they have lost the capacity for survival in fresh water. Perhaps the aquatic halophytes have developed over a longer period, or perhaps the difference is a consequence of the greater stability of the medium in which they live, in which they are not subjected to such large changes in salt concentration.

Among aquatic halophytes which grow on the seashore, species occurring higher up the shore tolerate a wider range of salt concentrations than those growing lower down. Deep sea algae tolerate salt concentrations equivalent to 0·5 to 1·3 times normal sea water; algae at low tide level 0·3 to 2·2; and algae between high and low water levels 0·1 to 3·0 (Biebl, 1937, 1949). The vegetation is zoned partly as a result of the salt gradient. Not only the aquatic algae, but also the lichens on sea cliffs and the phanerogams above high water level are zoned, so that there is a series of belts from deep water inwards to vegetation which is reached only by salt crystals from the spray.

reached only by salt crystals from the spray.

Sernander (1917) has described the vegetational belts on the shore, starting with the region between the tide marks, which he called *littoral* (from Latin *litus*—shore). He distinguished the following:

The littoral belt is sometimes covered by water and sometimes exposed. Along tidal coasts, it is covered at high tide and exposed at low tide. *Salicornia*, *Laminaria* and *Fucus* are characteristic species. In non-tidal areas, it is exposed at less regular intervals as a result of changes in water level brought about by storms, currents, waves, etc.

The sublittoral belt is always under water, but still with a complete cover of vegetation. It has its lower limit at the depth where the vegetation gets more sparse because of decreasing light penetration and the sea bottom becomes bare of plants, or communities of red algae begin to dominate.

The supralittoral belt is immediately above the littoral. It is occasionally covered by waves and spray, or may even be flooded. The lower supralittoral belt is covered by waves and is flooded on occasions (storms or spring tides). On rocky shores it is covered by closed vegetation, mainly lichens, with large numbers of *Verrucaria maura* making it brown-black in colour, so that it is often distinguishable at a distance and may be called the *black lichen belt*. On less stable substrata, sand or pebbles, there is no vegetation, or very little, e.g. *Aster tripolium*. The upper supralittoral belt is affected by splashing, by flooding on rare occasions, and by the sand, gravel and drift material which then accumulates. Washed-up

Fucus and other algae may form thick, nutrient-rich heaps which support a luxuriant vegetation. This belt is often rich in *Festuca rubra* and members of the Chenopodiaceae.

FIG 95. A shallow bay (Östersjövik) on Gotland. Background: littoral belt with partly exposed stones. Foreground: upper supralittoral belt with deep deposits of nutrient-rich drift material.

The epilittoral belt, which is above the supralittoral, is affected by salt carried from the sea by wind, either as spray or as crystals. In this belt, and including on the sand dunes, there may be several types of plant communities, with or without a desertlike vegetation: thorny shrub and coastal woodland, showing characteristic wind pruning and salt impregnation (wind-pruned woodland and dunes may also develop in places where there is no salt, e.g. around Lake Vänern in Sweden), and lichen communities on rocks. The epilittoral belt often extends for a long distance inland, so it has no well-defined inner limit. Fine salt crystals can be carried many miles in the wind (see Chapter 17).

The coastal belts of vegetation are not developed solely because of the influence of salt, but also because of effects of water and wind, mineral soil and humus, and light and temperature (see Chapman, 1964): special studies must be made if the effects of one factor are to be separated from those of the others, and there have been few so far. Hence, the effect of salt in each belt cannot be separated from the other environmental conditions, although there

competition between potassium and hydrogen ions. The hydrogen ion concentrations at these low pHs were large enough to inhibit potassium uptake from a dilute solution. If the potassium concentration was increased, the roots took up more potassium, so that the inhibitory effect of low pH was less. The uptake of Ca^{++} ions is also decreased by strongly acid solutions (pH 4 and 5) (Arnon *et al.*, 1942).

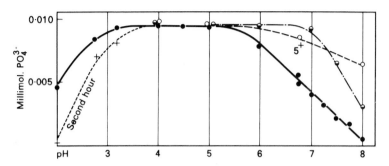

FIG 100. The effect of hydrogen ion concentration on the uptake of potassium phosphate by roots of rye. Ordinate: salt uptake per hour and litre of solution (initial concentration 0·3 millimol/litre). At pH 2 to 4, there is more uptake in the first hour (solid line) than in the next hour (dotted line). At pH 5·5 to 8·0, the dashed line represents the uptake when the solution is five times more concentrated. Alternate dots and dashes represent phosphate uptake by plants which were first saturated with bicarbonate, so that bicarbonate was no longer competing with phosphate. (After Olsen, 1953c.)

At pH values below 3·5, the hydrogen ions had an effect which increased with time, so that salt uptake decreased from hour to hour (dotted line on the left of Fig. 100). However, even at pH 2, a rye seedling is able to take up phosphate at first, but not after an hour: it has already been suggested that the loss of function is due to loss of charge by the cell membrane, so that the acid diffuses in and the cytoplasm is killed. Hydroxyl ions have similar harmful effects at pH values above 10·5. The two critical points, pH 3·5 and pH 10·5, are the same distance from the neutral point, pH 7·0 (Olsen, 1953c).

The experiments show that hydrogen ion concentration has a twofold effect on salt uptake: direct, by displacing cations, and indirect, by cytoplasmic damage or by favouring the bicarbonate ions which compete with phosphate ions. Both these effects occur outside the optimum pH range: within the optimum range there seems to be no effect of change of pH.

These experiments also show that the pH range of the rye seedlings is decreased by factors, chemical or biotic, which reduce anion and cation uptake. Competition between species or individuals has a similar effect: competition for nutrients, water, light or any other requirement narrows the pH range tolerated by the seedlings. If nutrients, for example, are supplied in amounts sufficient for all the plants in a locality, there is no longer any competition for nutrients, and the selective effect of pH is diminished. This is at least a partial explanation of the differences in opinion with regard to indicator species. Åslander (1929, 1952) found that *Chenopodium album* and *Stellaria media*, and other species which were previously considered to be characteristic of neutral or alkaline soil, grew on acid soils if the nutrient content was high. In the same way, indicators of acid soil, such as *Rumex acetosella*, grow on slightly acid or alkaline soils if there is no competition for nutrients. Growth of barley grown in nutrient solutions is poor at low pH (3·7), but only if the solution is very dilute. In more concentrated solutions at the same pH growth is good.

In these examples, the effect of acidity is mediated through the consequent impairment of nutrient uptake and hence plant growth. Acidity can also affect plant growth in other ways, some of which were discussed in Chapter 14 and will therefore be mentioned only briefly here.

5 *Edaphic effects*. Podsolisation and leaching are accelerated at low pH, which therefore contributes to the onset of soil oligotrophy and consequent increased competition for nutrients; this, in turn, causes development of oligotrophic, species-poor plant communities. Oligotrophy can, of course, arise for other reasons: for example, removal of calcium and other bases by acid soil colloids, as in some peat soils (Höfler, 1952, page 396).

On the other hand, high pH is a contributory factor in the development of eutrophic soils and species-rich communities (Table 90). However, all soils with high pH are not eutrophic: if there is a large excess of calcium causing the neutral or basic reaction, there may be a deficiency of phosphate or other cations (Chapter 8).

TABLE 90 The relation between species richness and soil pH in a number of Danish soils. (After Olsen, 1923, page 93.)

pH	Relative no. of species per unit area	pH	Relative no. of species per unit area
3·6	3	5·2	19
3·9	7	6·2	20
4·2	12	7·5	39
4·7	13	7·7	44

There is also a definite relationship between the degree of trophy and the hydrogen ion concentration in water. In the Swedish lakes studied by Lohammar (1938), the content of electrolytes was related to pH; and in Denmark there were about four times as many species in lakes with alkaline or neutral water as in lakes with a low pH (Iversen, 1929). The content of salts and hydrogen ions depends mainly on the way in which the surrounding country is cultivated and inhabited. The richer the environment, the greater the nutrient supply to the lakes and the larger the number of species in them (Weimark, 1950). Sjörs (1950) found this sort of relation between the composition of the flora and the acidity and electrical conductivity of bog water.

The aluminium concentration in soil water increases with acidity, because the minerals dissolve more quickly. The concentration is sometimes so high that several workers have suggested that it is a cause of the deterioration of soil condition. Aluminium is precipitated inside plant cells and is thought, for example, to inhibit phosphate transport (Wright, 1943). This might be one of the causes of phosphate deficiency in acid conditions (see page 171).

High acidity may bring about such a high concentration of dissolved manganese salts that they accumulate in cells and disturb metabolism (Hale and Heintze, 1946; Neeman, 1955).

The physical state of the soil is affected by acidity, because of its influence on colloidal flocculation or dispersion and, as a consequence, on several physical properties of soil, such as swelling, coherence, porosity, water-holding capacity and aeration. The more a species is affected by any of these, the more it is affected by soil acidity.

Microorganisms are often tolerant of a wider range of pH than are higher plants. Bacteria are particularly tolerant of hydroxyl ions: nitrate bacteria grow in culture between pH 5·3 and 10·3, with an optimum at 8·4 to 9·3. Fungi have different requirements; they tolerate acid media: the saprophytic *Rhizopus nigricans*, and the parasitic *Candida albicans* can grow at pH 2·2 to 9·6. Starkey and Waksman (1943) found species which could tolerate even more acid media. In contrast, actinomycetes prefer alkaline media and generally have a pH range of 5 to 9, with an optimum of pH 7 to 8 (Henrici, 1947).

This is important in plant protection and plant pathology, since parasites and phytopathogenic species also differ in their pH preferences. Some do best in neutral soil, e.g. some *Fusarium* species; others do best in acid soil, e.g. *Rhizoctonia violacea* and *Plasmodiophora brassicae*. *P. brassicae* causes club root in cabbages, turnips, etc.; the activity of its zoospores is restricted by high pH (*c.* 6 or more), and, hence, the addition of lime to the soil helps to control the disease (literature in Wolf and Wolf, 1947). However, as yet little is known of the relation between soil microorganisms and environmental conditions.

Humification of litter is dependent on soil microorganisms, as is nutrient mobilisation and other soil processes. If some fraction of the community of soil microorganisms is favoured or inhibited, a change in some of the soil processes will result. For example, raw humus is usually the product of litter decomposition in coniferous woods and dwarf shrub communities, probably because the high acidity favours fungi more than other soil organisms (see Chapter 14).

Selection. The actual and potential acidity in a habitat has many effects, so that to distinguish its importance to the various species in a community, and its selective effect on different species, is a complicated and difficult task. For example, species frequency and soil pH represent the summation of a complex of ecological factors, including nutrition and biological and physico-chemical soil processes. The physico-chemical soil factors are themselves a summation of properties such as ion balance and permeability, each of which can be changed by hydrogen ions. The direction and extent of change is affected by water balance, temperature, light, and neighbouring plant competitors. Because of the large numbers of possible combinations of these factors, visible effects of acidity can be expected only if acidity dominates over all these other factors, i.e. when acidity is very low or very high. As other factors become suboptimal or supraoptimal to a greater extent than acidity, the visible effect of acidity gets less and less until it disappears completely. Acidity can, of course, still have an effect, but it is masked by that of other factors, and is extremely difficult to measure. Other factors may be correlated with hydrogen ion concentration, and their effect may be put down to pH alone, particularly if only pH is measured—it is one of the easiest soil factors to measure. If other factors operate in the opposite direction to pH, they may mask the pH effect completely. This may occur in the optimum pH range, so that the pH dependence of the phenomenon under study may appear to have twin optima (see Table 89, page 409, *Molinia* and the experiment with *Calluna* referred to on page 411). Measurement of soil pH alone, or of any other single factor, is insufficient to give adequate information about the plant environment.

The effect of acidity on selection of species in a habitat depends on the interference between pH and other factors. The potential survival of a species depends on the magnitude of the various environmental factors and the way in which the species is influenced by them and in competition with other species. The effect of acidity, as of other factors, depends on the position of the optimum and the width of the range of tolerance. Competition is more acute if other factors are suboptimal or supraoptimal, so the tolerated pH range gets narrower and the species is restricted to habitats with optimal acidity. Plants with low pH optima are therefore found on acid soils: plants with high optima occur on circumneutral or alkaline soils. The competitive ability of the first group is greatest on acid soils and that of the second group on alkaline soils; in other places their survival depends on the degree of competition from other species and on the environment in general.

The requirements of cultivated plants are met as far as possible by growing the plants on soils with approximately the pH which is optimal for the crop or by changing the soil pH to make it as near optimal as possible. The reaction of crop plants to acidity is known to some extent: e.g. rye generally grows best at pH 5 to 6, wheat at 6 to 7 and lucerne at 7 to 8. However, sensitivity to pH can vary considerably among different varieties of the same species (Neeman, 1955). It is a complicated task to satisfy the pH requirements of a species, because soil acidity is subject to rapid changes. Relatively more basic mineral nutrients than acid are removed in harvests; roots exchange basic cations and hydrogen ions; plants themselves produce acid organic substances of various kinds (organic acids, humic acids); and fertilisers may cause increased soil acidity. To counteract these changes, or to change soil pH in general, basic fertilisers must be supplied, such as lime, nitrochalk, saltpetre, etc., or acidic fertilisers such as ammonium sulphate.

In the silage method of fodder conservation, a pH which is suboptimal for development of microorganisms is employed. Addition of a suitable acid (hydrochloric acid, sulphuric acid or formic acid) reduces the pH to about 3·7. This reduces bacterial activity sufficiently to allow the fodder to be kept. Low pH can also be achieved by allowing the fodder to ferment so that the resultant acids, such as lactic acid, prevent development of microorganisms. Addition of sugar-containing substances, like molasses, accelerates acid production.

Calcicole and calcifuge plants

The occurrence and flora of calcareous areas. Botanists have long observed that limestone regions in Scandinavia, and in the humid parts of the cold and temperate regions generally, are characterised by a flora which is particularly rich in species and often particularly luxuriant.

In Scandinavia there are many areas rich in limestone (calcium

418

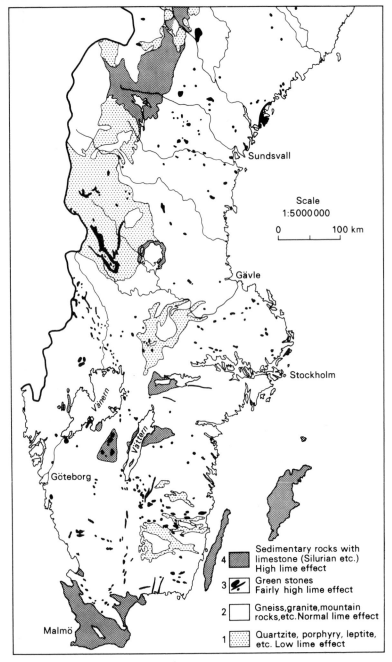

Scale
1:5 000 000

0   100 km

Sundsvall

Gävle

Stockholm

Vänern

Vättern

Göteborg

Malmö

4 Sedimentary rocks with
   limestone (Silurian etc.)
   High lime effect

3 Green stones
   Fairly high lime effect

2 Gneiss,granite,mountain
   rocks,etc.Normal lime effect

1 Quartzite, porphyry, leptite,
   etc. Low lime effect

FIG 101. Types of rock in Sweden. (Mainly from 'Geologisk över-
siktskarta över Sveriges berggrund', from Tamm, 1921.)

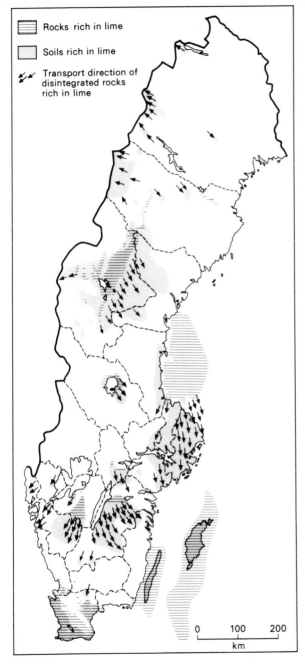

FIG 102. Swedish limestone outcrops and their
movements. (After Lundqvist in Magnusson, Lundqvist and Granlund, 1957.)

carbonate). This may be Silurian, Cretaceous, or older, as in Bergslagen and Södermanland, or derived from greenstone or banks of shells, or it may have been transported and precipitated in water. Tamm (1921) has summarised the effect of limestone on the vegetation, particularly on the forests in these areas. Because of the large number of species, including rare species, the Silurian areas in central Sweden, for example, contrast markedly with the surrounding areas, particularly those in which the bedrock is of nutrient-poor minerals. The same is true of heath on the island of Gotland, where there is a large number of species despite the shallowness of the soil (Hesselman, 1908). In east Uppland, where the inland ice has brought Silurian calcareous material from the sea bottom outside Gävle, there are rich meadows and patches of deciduous woodland.

Outside these areas, soils are generally deficient in calcium as a result of weathering and leaching. A calcium-rich component in the soil or bedrock is therefore often sufficient to allow a richer-looking vegetation to develop. The greenstones are a group of dark-coloured, basic igneous rocks, including diorites, gabbros, hyperites, diabases (trap), amphibolites, and basalts, which are all richer in calcium and magnesium than are the granite-gneiss group (see Chapter 11). There are many large or small outcrops of greenstones in Sweden, e.g. in a streak through the south Swedish granite gneiss region. This is marked by patches of deciduous wood with rich vegetation, particularly where the minerals are easily weathered (Hård af Segerstad, 1924; Weimarck, 1939).

In other places in Scandinavia there are scattered occurrences of greenstones. The hyperites in Värmland are well known for their productive spruce forest, rich in herbs, and for patches of deciduous wood with typical brown earths (Tamm, 1921). The oligotrophic porphyry and sandstone regions in Dalarna include some diabase occurrences on which are rich patches of spruce woodland with mull soil and herbs (Tamm, 1931). The greenstone regions, together with regions of chalk and limestone, are the special localities for spruce in the northern part of Sweden and for patches of deciduous wood in the south. In contrast, pine is found distributed over the oligotrophic rocks (Eklund, 1943; Hesselman, 1916–17; Tamm, 1921). On flat ground the calcium is deeper in the soil as a result of leaching, so that its effect on the vegetation gets less and less as time goes on. On sloping ground, the calcium moves sideways, benefiting the vegetation where it accumulates or through which it passes. In northern Sweden, for example, in the Jämtland mountains, the calcium moves downwards to the lowest-lying places, allowing the development of brown earth and a rich vegetation.

Wind can also play a part in the transportation of calcium, by blowing dry deciduous leaves rich in calcium away from the places where they were shed, often in significant quantities (Halden, 1950).

It is apparent that calcium affects the number of species growing

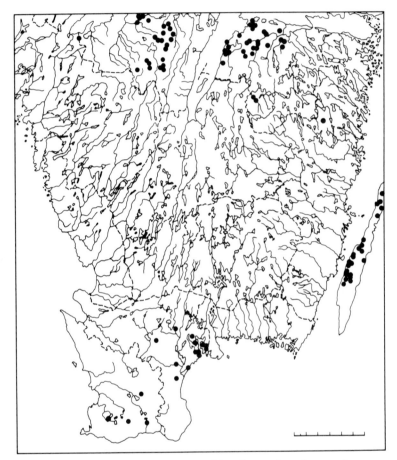

FIG 103. Distribution of *Ophrys muscifera* in south and central Sweden. (After Hård af Segerstad, 1924.)

in a locality, since the flora in calcareous regions is particularly species-rich. The way in which this is brought about is an age-old problem. It has long been observed that some species prefer places rich in calcium, the so-called *calcicoles*, whereas others, *calcifuges*, avoid such places. As long ago as the early nineteenth century there were experiments to find out more about such habitat preferences, and since then this question has been given a considerable amount of attention.

Indicator species. Because of the supposed relationship between certain plants and the calcium content of the soil, there have been attempts to use plants as indicators of the calcium status of the soil (see, e.g., Mevius, 1931). This is an old-established practice in agriculture: weeds such as *Rumex acetosella* and *Spergula arvensis* are regarded as indicators of calcium deficiency, and thus of a soil condition which can be improved by additions of marl. Detailed investigations have also demonstrated that there is a relatively close relationship between the distribution of some species of plants and soil calcium content. For example, Witting (1947)

TABLE 91 Plants considered to be restricted to calcareous soils (A) or avoiding calcareous soils (B). (After Walter, 1947; Du Rietz, 1949. For other lists see Mevius, 1931; Steele, 1955.)

A	B
Zerna erecta	*Nardus stricta*
Sesleria albicans	*Deschampsia flexuosa*
Many orchids	*Rumex acetosella*
Vincetoxicum officinale	*Scleranthus annuus*
Carlina acaulis	*Spergula arvensis*
Trifolium montanum	*Trifolium spadiceum*
Coronilla emerus	*Calluna vulgaris*
Melampyrum arvense	*Vaccinium myrtillus*
Schoenus ferrugineus	*Gnaphalium arenarium*
Carex lepidocarpa	*Jasione montana*

found that two fen species, *Carex lepidocarpa* and *Schoenus ferrugineus*, grew only where the calcium content in the water was more than 18 mg/litre. He therefore used these species as indicators of a high degree of eutrophy in fens. In the same way, Waldheim and Weimarck (1943) classified bogs and fens in south Sweden according to their degree of eutrophy, as shown by indicator species. 'Rich fens' were those with a high proportion of herbs and mosses; 'poor fens' were those with a high proportion of grasses and *Sphagnum* species; oligotrophic bogs had large amounts of dwarf shrubs, *Sphagnum* species, and lichens.

The presence or absence of indicator species is probably often a fairly reliable indication of the calcium status of the soil. However, both field and laboratory experiments have shown that the relation between species distribution and soil calcium content is not as close as was first thought. Calcicoles are much less reliable indicators than are plants favouring acid or basic soils in general, as is shown by the work of Brenner (1930), who examined the soil in various places in Finland where the 'calcicoles' *Aconitum septentrionale, Orchis sambucina, Listera ovata*, etc. occurred, but he found that the calcium content was very low. *Ophioglossum vulgatum, Isatis tinctoria* and other species, which were restricted in inland areas to soils with a high calcium content, were found at the coast on soils poor in calcium. In addition, some species thought to avoid lime, e.g. *Rumex acetosella*, grew in soils with a high calcium content. Many such cases are now known, and the number is increasing as information is collected and compared: species which are calcicoles in some areas are less demanding in others. The classification of species as calcicoles is therefore very difficult, and the number so classified decreases as investigations are carried out in larger areas. Furthermore, experimental cultivation of plants thought to be restricted to soils with a high calcium content has shown that these plants can grow without additions of calcium.

It seems therefore that plants behave as calcicole or calcifuge only within limited areas and under particular environmental conditions. Over wider areas, or in conditions tested by experiment, the differences are diffuse or indistinguishable. Like the dependence of plants of acid soils on hydrogen ion concentration, the cause of the dependence of calcicoles on calcium probably lies in the indirect effects. Calcium has an effect on several ecological and physiological processes and, through these, on the distribution of species. However, since these same processes are also affected by many other conditions, no hard and fast relation between calcium and the presence or absence of a particular species can be expected.

The ecological effects of calcium. The many physiological effects of calcium have been discussed in Chapter 15. Calcium plays a part in metabolism as a nutrient and as an antagonist; it promotes root hair formation and affects transport. In these cases the calcium has a direct effect on physiological processes, and it can be assumed that the most favourable effect is exerted when the calcium is present in a particular, optimum, concentration.

The optimum is probably different in different plants: plants with a particularly high optimum can probably justifiably be called calcicole. However, it is not known whether any such plants exist. Iljin (1938, 1940) considers that a physiological distinction between calcicoles and other species is that the calcicoles have calcium dissolved in the cell sap, e.g. as calcium malate, whereas in other plants it is precipitated as organic salts such as calcium oxalate.

There are many indirect ecological effects of calcium, which often cause it to have a dominant effect upon the vegetation, particularly in the following instances:

1 THE EFFECT OF CALCIUM ON SOIL ACIDITY. Since calcium

is the most common and most abundant base cation, its presence or absence in the soil is important in determining soil acidity and all the secondary effects of acidity. Calcium modifies the soil pH, so that weakly acid, neutral or alkaline soils develop, counteracting the harmful effects of hydrogen ions. In such circumstances, calcium is affecting plants only indirectly, by way of its effect on hydrogen or hydroxyl ions. Basiphilic plants may therefore appear to be calcicoles and acidophilic plants to be calcifuges.

There are many known examples of this apparent calcium-dependence. In humid regions, or in the mountains, there is often a lack of bases in the soil, and leaching causes it to be acid. Hence species which require neutral or basic conditions cannot survive and are found only in places where the soil has a sufficiently high content of bases, usually where there is lime, as neutral soils without lime are rarely found in humid regions (Olsen, 1923). Plants which require a high pH therefore seem to be calcicoles in such areas.

For the same reason, plants with low pH optima seem to be calcifuges (Olsen, 1923). *Sphagnum* species are particularly sensitive to calcium, but this is not because of their sensitivity to calcium ions as such (Mevius, 1921)—they tolerate calcium chloride or calcium nitrate—but because of the unfavourable effect of a high concentration of hydroxyl ions. Mevius grew *Pinus pinaster* and *Sarothamnus scoparius* in nutrient solutions and found that they required a certain amount of calcium for development, but that if calcium was added in larger amounts, as calcium carbonate, the hydroxyl ion concentration became supraoptimal and had unfavourable effects in damaging roots, destroying the shoot apices and lowering growth rate. Sodium and calcium have the same effects (Olsen, 1923b).

Again, the effect of calcium is indirect, operating through the hydrogen ion concentration. Plants seeming to be calcicoles are, in fact, acidophobic; plants seeming to be calcifuges are acidophilic.

As a result of these relationships, the number of apparent calcifuges increases northwards. Northern mountain regions are often rich in such plants, particularly mosses and lichens (Brenner, 1930; Aaltonen, 1948). However, in arid and semiarid regions, the majority of plants are indifferent to calcium because there is usually a large amount of calcium and the soils are mostly neutral or alkaline. The increased competition brought about by high acidity does not occur and plants can tolerate wide ranges of variation of other environmental conditions.

One of the ways in which differential plant sensitivity to calcium is brought about is through the effects of manganese. Some manganese is essential (see Chapter 15), but excess accumulates in the

cells, causing chlorosis, necrotic patches on the leaves, etc.
(Wallace *et al.*, 1945). Very large reductions in crop yield sometimes
result (Hale and Heintze, 1946). Calcium affects plant-manganese
relationships because the soluble manganese content of soil is
dependent on the soil pH; more manganese dissolves in acid soil,
in which manganese content readily becomes supraoptimal. However, the position of the optimum, and the ability to tolerate excess
manganese, is different in different species. According to Olsen
(1936), *Lemna polyrrhiza* and *Senecio vulgaris* are particularly
sensitive. As low a concentration as 2 mg/litre of manganese in
culture solution is poisonous, and 5 mg/litre is lethal. *Hordeum
distichon* and *Sinapis alba* tolerate 11 mg/litre, and are killed only if
the concentration is increased to 250 mg/litre. Growth of *Zea mays*
is promoted by addition of manganese up to a concentration of
50 mg/litre, and inhibited by concentrations greater than 250 mg/
litre. The optimum concentration for *Deschampsia flexuosa* is 11 to
250 mg/litre. Similar differences may exist between different races
of the same species (Neeman, 1955).

If soil calcium content is high, pH is high and a high soil manganese content is not harmful. Plants which have a low requirement for manganese are therefore generally associated with such
conditions, not because they require a high soil calcium content,
but because they are avoiding manganese. Species like *Deschampsia
flexuosa*, which tolerate large amounts of manganese, are more or
less independent of calcium and occur on acid soils.

2 SOIL NUTRIENT STATUS. The occurrence of raw humus and
brown earth, and the balance between these two soil types—one
can succeed the other—is dependent to an important extent upon
soil calcium content (see Chapter 14). Brown earths tend to develop
where calcium is present; podsols develop where there is a deficiency in calcium. The degree of trophy of soils, and consequently
the vegetation, is determined by calcium content operating in
several different ways.

However, there is a clear relation between species distribution
and soil calcium content only if the deficiency or excess of calcium
reaches extreme values. In either case, nutrient deficiency may be
a consequence. In humid regions, lack of calcium is accompanied
by leaching, podsolisation and the formation of poor acid soils
bearing an oligotrophic and acidophilic flora: excess calcium
brings about deficiency of iron, magnesium and trace elements.
Calcifuge species, such as lupins, growing on soil rich in calcium,
are sometimes chlorotic because of iron deficiency. Supply of an
iron salt may counteract the chlorosis. This may be because there
is too little iron available among the nutrients in the soil solution,
so that plants are dependent on the rate of weathering of the

minerals to meet their requirements. Weathering is dependent on pH, and therefore, in turn, on soil calcium content, so a high soil calcium content can cause slow weathering and consequent iron deficiency (Olsen, 1923; Mothes, 1939).

Excess calcium can also cause nutrient deficiencies by upsetting the balance between the various ions taken up by plants. The greater the demand for particular cations, the greater is the unfavourable effect of excess calcium. Some calcifuges are plants which require a good supply of potassium, e.g. the reason why the chestnut (*Castanea sativa*) does not do well on calcareous soils may be potassium deficiency caused by the calcium (Steiner, 1939). Other effects of calcium are accentuated by chloride, probably because this increases the amount of calcium which is in solution and easily transported, so that the K:Ca ratio decreases. For this reason, a high NaCl content may cause potassium deficiency. This explains the observation that sensitivity to calcium and sensitivity to chloride may be parallel in some instances, and calcifuges are sometimes strikingly sensitive to chloride (see Pirschle, 1939).

Finally, high calcium content may cause phosphoric acid to be unavailable. Plants with high phosphate requirements are therefore absent from calcareous soils (see Chapter 15).

Calcium probably has similar effects on the mineral economy of aquatic plants: the flora in lakes rich or poor in calcium is very different in quantity and in quality (Waldheim, 1947).

3 COMPETITION. That calcium may affect competition has already been suggested. Competition for nutrients is intense in poor, acid soils which are deficient in calcium, and species which tolerate high hydrogen ion concentrations have high competitive ability and are therefore widely distributed in such soils. This is particularly true of species with low competitive ability for space or for light. According to older ideas (see Olsen, 1923), this type of competition explains why species such as *Rumex acetosella* grow best on acid soils. These species would actually grow better where there was more calcium, but in such situations they are crowded out by species with higher competitive ability.

4 SOIL STRUCTURE AND SOIL TEMPERATURE. A suitable amount of calcium brings about a porous soil structure with all the associated advantages, a relatively high oxygen content and low water content, contributing to raised soil temperature. Plants which require a high soil temperature tend to be associated with calcareous soils, because of this effect. They can grow equally well on non-calcareous soils, if these are warm enough and have a sufficiently high oxygen content, etc. *Zerna erecta* is found on soils poor in calcium in Mediterranean countries, but in central Europe it occurs only on calcareous soils, which are the only soil

types sufficiently warm for it (Steiner, 1939). For the same reason, many species at the northern limits of their range act as calcicoles, and it is not improbable that the presence of an area of calcareous soil allows a species to extend its range northwards (Brenner, 1930; Hård af Segerstad, 1924). Calcareous soils on south-facing mountain slopes are particularly effective, since the greater radiation and the rapid rate of weathering add to the favourable effect of the soil itself. The soil climate is improved by the calcium, the greater radiation and the topography, which provides shelter from north winds. The flora of south-facing slopes (Andersson and Birger, 1912; Selander, 1950; du Rietz, 1954; Lundqvist, 1965) is particularly luxuriant and species-rich, including thermophilous elements. This is especially true of south-facing slopes which include deposits of calcium (Halden, 1950).

These examples of the ecological effects of calcium show that some of the species in certain areas prefer calcareous soil, or avoid it, because the calcium affects soil pH, the amount and types of nutrients, the structure and temperature of the soil, etc. These species are also found in other areas with similar soil conditions, but in the absence of calcium, to which they are indifferent. Calciphily and calciphoby are thus not an expression of a direct relation between plant and calcium, but of plant dependence on factors of the environment influenced by calcium. Apparent calciphily may be acidophoby, eutrophy, thermophily, or merely an expression of low competitive ability in nutrient-poor conditions. In the same way, calciphoby may, in fact, be acidophily or oligotrophy, or the expression of sensitivity to certain elements (K, Fe, Mg, P, trace elements), or of high competitive ability in nutrient-poor conditions.

Because of the many different effects of calcium on soils and because of the wide variations in its amount and distribution, it is often of overriding importance in determining soil type, particularly in humid regions. The flora in such regions is particularly sensitive to calcium and numerous species are more or less restricted to calcareous or to non-calcareous soils.

In agriculture, the sensitivity of some species to calcium is utilised, for example, in controlling certain fungal diseases caused by calciphobic fungi by additions of lime. Mosses and various weeds can be controlled in the same way.

Nitrogen specialists

The availability of fixed nitrogen is usually low, and plants normally react strongly to addition of available nitrogen, the leaves increase in size and become darker green, shoots increase in size, but flowering and fruiting may decrease, effects which are well known in horticulture and agriculture. A large amount of nitrogen results in cereal crops with high leaf production and long straw, but with little sclerenchyma and therefore insufficiently strong straw; potatoes produce well-developed shoots and large leaves, but decreased tuber production; fruit trees produce strong shoots, but less flowers, etc. Such effects are favourable for the production of leaf crops, for grazing for example, but not always for root crops, fruit or cereals.

The results of nitrogen deficiency are in many respects the opposite. In some plants, e.g. *Tradescantia viridis* and *Impatiens balsamina*, the leaf parenchyma and leaf epidermis are affected and the organs tend to become succulent (Gessner and Schumann-Peterson, 1948). Other species become more xeromorphic. Mothes (1932) found that in tobacco, nitrogen deficiency caused a thicker cuticle, smaller epidermal cells, more stomata per unit of leaf surface, denser venation, etc. These observations led him to express the opinion that the xeromorphic nature of the vegetation of raised bogs was a result of nitrogen deficiency. Subsequent investigation supported this hypothesis. Muller-Stoll (1947) showed that nitrogen deficiency accelerates cell development and differentiation, walls are thicker, cells mature earlier, and the leaves are small and woody. There are some similarities between these effects and those of water. Simonis (1948) grew a common raised bog species, *Andromeda polifolia*, in dry and moist nitrogen-deficient conditions. The nitrogen deficiency caused xeromorphy, which, contrary to expectation, was accentuated by a good water supply. This agrees with earlier observations by Firbas (1931) on xeromorphic development in raised bog species and its dependence on soil moisture. He found a greater degree of xeromorphy in the wetter *Sphagnum* localities than in the drier *Calluna* localities, higher up and with more humification. Some of Muller-Stoll's fertilisation experiments (1947) agreed with these results, and showed that increased nitrogen supply caused decreased xeromorphy (softer leaves, larger leaf area, higher water content, intercellular volume and epidermal cell size, etc.) in raised bog species. According to Greb (1957), the reaction to nitrogen also depends on the temperature of the root medium. Species probably react differently to nitrogen. Lundqvist (1956) grew *Nicotiana rustica* and *Helianthus annuus* under conditions of nitrogen deficiency and found, like Mothes, that there were changes in the direction of increased xeromorphy. However, the effect tended to be the opposite in *Urtica dioica* and *Chenopodium album*.

Excess nitrogen, as well as deficiency, brings about anatomical

corresponding to a concentration of about 0·000 3 N (Karlsson, 1952).

The various forms of potassium are in equilibrium:

$$\text{adsorbed} \rightleftarrows \text{dissolved} \rightleftarrows \begin{cases} \text{fixed potassium} \\ \text{primary crystal potassium} \end{cases}$$

As a result, potassium which is not taken up by plants can be stored by adsorption or fixation. When needed, this again becomes available, though only in small amounts because the fixed form is firmly bound. In summer, the soluble fraction decreases, because plants use more than is released from the other fractions. In autumn and winter, the opposite obtains (Table 92).

However, leaching is only partly prevented by storage. The ground water in Swedish soils contains potassium equivalent to a concentration of c. 0·000 1 N. About 3 to 30 kg/ha K is lost annually in the ground water and in streams and rivers, the least from clay soils and the most from sand, silt and other soils poor in colloids. The potassium is bound mostly by colloids, which are therefore important in determining how much is lost. The amount of run-off is also important. Swedish arable soils in areas with an annual precipitation of 550 mm lose c. 7·5 kg/ha annually (Karlsson, 1952).

Calcium. Calcium is present in soils and minerals mostly as carbonate or silicate. The content varies very much, from c. 0·1 per cent in soils poorest in calcium, to soils which are almost pure calcium carbonate. Normal Swedish arable soils contain 0·1 to 0·5 per cent. Storage, mobilisation and turnover is similar to that of potassium. The mobile calcium reserves are found dissolved in soil water, adsorbed on colloids, particularly on humus, or in less soluble forms. The adsorbed and easily exchangeable fraction is several times larger for calcium than for potassium, and the same is generally true of the other two fractions.

Nitrogen. The nitrogen content of plants generally varies between about 0·5 and 2·0 per cent of the dry weight (Tables 12, 14, pages 78, 81). Various tree leaves contain 0·5 to 1·25 per cent (Lutz and Chandler, 1946). Nitrogen has a special place in the nutrient balance, since it is the only element which plants themselves must convert into a chemically bound form. The amount of nitrogen in the mobile nutrient fraction is therefore usually suboptimal. Nitrogen deficiency is frequent, an effect accentuated by the fact that nitrogen compounds tend to be unstable and easily lost (see Chapter 8).

The nitrogen in soils is of organic origin. It makes up part of the mobile nutrient reserves and occurs mainly in litter, humus and living soil organisms.

The nitrogen content of a particular type of soil depends mainly on its content of organic material. The nitrogen content of English arable soils is 0·1 to 0·2 per cent (Russell, 1952), and slightly higher in arable soils in Sweden (0·1 to 0·4 per cent), where a normal arable soil contains 2000 to 3000 kg/ha nitrogen. There is still more nitrogen in peat soils, up to 1 to 4 per cent. Fen peat may contain up to 15 000 kg/ha nitrogen. In contrast, forest soils in northern Sweden are poor in nitrogen, e.g. Hesselman (1937) found 320 to 770 kg/ha.

TABLE 92 Changes in the salt content of the soil water during the vegetation period. (After Burd and Martin, 1924, page 161.) Salt content in g per ton soil water.

| | DATE | | |
	30 April 1923	4 Sept. 1923	28 April 1924
pH	7·4	7·6	7·6
NO_3	149	58	252
HCO_3	83	155	142
SO_4	561	432	699
PO_4	1·1	0·6	0·6
Ca	242	193	336
Mg	91	47	76
Na	42	40	59
K	21	9	12
Total solutes	1190	935	1527

The accumulation of organic material and of nitrogen in soils is determined mainly by climate and vegetation, and by the balance between litter production and humification. If humification is slow compared to litter production, the content of nitrogen in the soil is high. In humid regions, the nitrogen content increases with distance from the Equator, i.e. with decreasing temperature; a mean temperature decrease of *c.* 10°C corresponds approximately with a doubled soil nitrogen content. The reason for this is not only the direct effect of temperature on soil organisms and on humification processes, but also the indirect effects on soil acidity and permeability and, therefore, on leaching.

The greater part of the mobile nitrogen reserves, that in living organisms, litter and humus, is unavailable to higher plants and can be utilised only by parasites and saprophytes. It is through their activity in the processes of humification and mineralisation that nitrogen is converted to a water-soluble form available to higher plants. This is rapidly used up. Any remaining is adsorbed by colloids or fixed by crystals. According to Karlsson (1952) ammonium ions can be bound by crystals in the same way as potassium. The nitrate bacteria convert this ammonium into nitrate, which dissolves in the soil solution and is rapidly taken up. In humid climates, some is lost by leaching; up to 30 to 40 kg/ha may be lost from Swedish arable soils annually.

A relatively large amount of nitrogen is removed in harvested crops (Table 102, page **496**): a crop of autumn wheat contains about 90 kg/ha nitrogen. Since this nitrogen, and that lost by leaching, comes from the soil water and the colloids, the nitrogen balance of the soil depends on the capacity of these sources and their content of nitrogen. However, this content is usually sub-optimal, and addition of nitrogen to most soils leads to increased productivity. Most types of intensive cultivation of plants are dependent on nitrogen fertilisation.

Phosphorus. Phosphorus is a constituent of all cytoplasmic material, though in considerably smaller amounts than nitrogen. The phosphorus content of leaves of various trees (both evergreen and deciduous) is only *c.* 0·04 to 0·3 per cent of the dry weight (Lutz and Chandler, 1946). In the soil, the content is about the same, 0·03 to 0·3 per cent. In harvested crops, 10 to 25 kg/ha are removed (Table 102). For the phosphorus cycle as a whole see Chapter 8.

The primary sources of phosphorus for plants are the minerals apatite and phosphorite present in many types of rocks, although only in small amounts.

The phosphorus content (P_2O_5) of some common rocks (Laatsch, 1954) is: limestone 0·04 per cent; granite 0·07 per cent; sandstone 0·08 per cent; clay shale 0·17 per cent; volcanic rocks 0·3 per cent.

Phosphorus is also divided into the four fractions: it is bound in organic material (living organisms, litter, humus); dissolved in soil water in concentrations of 1 to 3 parts per million; adsorbed on colloids and solid particles; and fixed in more or less insoluble inorganic compounds in the soil (see Chapter 8, page 171). Like nitrogen, some of the dissolved phosphate is consumed rapidly by bacteria and fungi; some of the remainder is bound to colloids and is readily released later; and some is converted into insoluble compounds, the nature of which varies according to the chemical composition of the medium and the acidity (see Chapter 8, page 171). Because of this system of stepwise increase in firmness of fixation, phosphorus is present throughout the soil, but only in small quantities.

Sulphur. In humid regions the sulphur content of well-drained soils is *c.* 0·05 per cent, and is considerably higher in soils in wet places (see Chapter 8). The other parts of the lithosphere contain an average of *c*. 0·05 per cent, and sea water 0·088 per cent. Sulphur is present in plants in significant amounts; some a result of luxury consumption. For example, the sulphur content is higher in plants in and around industrial areas. Thomas *et al.* (1950, cited by Wiklander, 1957) give a figure of 0·11 per cent for leaves of coniferous trees in non-industrial areas of United States, 0·34 per cent for deciduous trees; 0·23 per cent for bushes; 1·5 per cent for halophytes; and 0·16 per cent for the straw and grain of wheat and oats.

The sulphate ion is retained by adsorption by the basic soil components. However, it is loosely held and the loss from leaching is significant. Drainage water contains 4 to 16 mg/litre sulphur (Wiklander, 1957). Adsorption increases as pH decreases, so that soil acidity is very important in determining the proportions of sulphur retained or lost by leaching. The amount of colloids is also important, so adsorption is related to nitrogen content to some extent (Fig. 105).

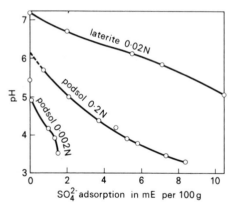

FIG 105. **The relation between sulphate adsorption and pH in the colloidal fraction from a laterite (solution: 0·02 N (NH₄)₂SO₄) and a podsol (solution: 0·2 N and 0·002 N (NH₄)₂SO₄). (After Wiklander, 1957.)**

Copper. According to Schlichting (1955), copper is bound to a large extent by stable humus, but remains more available to plants than the copper in minerals. Leaching is prevented, and instead there is accumulation in the A horizon, so that although there are usually only small amounts, the vegetation has a supply available even in soils originally very poor in copper. Despite this, there is some loss by leaching, since some humus, and with it some copper, is carried away in drainage water.

The dependence of the nutrient reserves on mineral soil, climate, and vegetation

The mineral soil

It is mainly the mineral soil that determines the amount and proportions of nutrient salts present. The clay content is of primary importance, together with the content of other fine-graded material and of colloid-precipitating bases, such as calcium, potassium and magnesium. A high clay content means that salts produced by chemical weathering are adsorbed and retained in the soil. Colloid flocculation depends on the amount of bases, not the absolute amount, but the excess of bases over acidic substances. In humid regions it is often a deficiency of bases which is the cause of the disappearance of colloids. The presence of sufficient calcium is important in bringing about colloid flocculation and retention of salts. Even if the vegetation consists of conifers and other species which produce litter which is acid and poor in bases, a rich type of soil, such as a brown earth, can develop if there is sufficient calcium.

Climate

In arid regions, weathering of minerals and humification is rapid and there is little or no leaching. Consequently, plant growth may be limited by excess of salts in the soil, where salt concentrations are high and often supraoptimal.

In semiarid regions, the soil is rich in organic and inorganic colloids, the salt content is about optimal for most plants, and weathering is relatively rapid and would lead to accumulation of supraoptimal concentrations of salts if this was not prevented by leaching. However, leaching is insufficient to cause nutrient deficiencies. Humification is also favourable to plants: it is slow enough for many intermediate products, such as humus, to remain in the soil, yet not so slow that peat accumulates. Semiarid soils can therefore usually retain large amounts of water and salts—having large amounts of humus and clay—and have a good supply of available nutrients.

In humid regions, soils are often deficient in salts as a result of leaching, particularly if weathering is slow because of low temperatures. Since humification is then also slow, large amounts of organic colloids accumulate, but often store no salts because of the leaching and slow weathering. As long as plant roots can grow down through these organic layers to the mineral soil below, salts are continuously brought up to the surface. Some mixing of material may take place, so that mull develops. However, at the other extreme, there may be so much accumulation of organic material that wholly organic soils, like peat or dy, are formed. Eventually the organic layer may become so deep that plant roots can no longer reach mineral soil and must rely on the nutrients available in the organic layer and those carried in wind and water. There are usually only suboptimal concentrations of available nutrients present in these organic soils.

The vegetation

The vegetation contributes quantitatively and qualitatively to the building-up of nutrient reserves. A luxuriant and demanding vegetation leads to the development of soils such as brown earths, rich in humus colloids and salts. The soil under deciduous woodlands and meadows is rich in nutrients, and has a large capacity to retain nutrients. An undemanding type of vegetation produces litter poor in bases and conducive to podsolisation, and a soil poor in colloids and nutrients. Dwarf shrub heaths on sand and gravel are an example. The nutrient reserves of pine-heaths in north Sweden are low; this is a contributory cause of the low timber production in these forests (see, e.g. Romell and Malmström, 1945). Partly as a result of the soil-plants relationship, a change in composition of the vegetation can lead to a change in soil nutrient status. In Chapter 14 mention was made of the possible change to podsol from brown earth resulting from planting coniferous trees in the transitional areas between humid and semihumid regions; and from podsol to brown earth if coniferous forest is replaced by deciduous forest or a grass-herb community. Vegetation poor in bases favours podsolisation and soil impoverishment, whereas vegetation rich in bases causes development of a soil rich in nutrients and colloids. The amount of nutrient reserves, their composition, and storage are thus dependent on the reciprocal relationship between the vegetation and the environment.

Storage of nutrient salts

During this cycle, nutrient salts remain stored for short or long periods of time at particular stages. The following are the most important:

1 Nutrient salts stored in an organically bound form, i.e. in living plants and animals; litter, underwater litter, humus, raw humus, peat, gyttja, dy; coal and fossils. The salts are released from this store by humification and mineralisation.

2 Nutrient salts stored in an inorganic form in:

(a) Soil water. The salts are in solution and immediately available to plants.

(b) Soil particles. Partly colloids and colloidal fragments ($SiO_2 . n\ H_2O$, $Al_2O_3 . n\ H_2O$, $Fe_2O_3 . n\ H_2O$, stable humus, humates and humic acid); partly coarser particles (clay, silt etc.). The substances adsorbed onto the surfaces of the particles are relatively available, and can be taken up by plant cells by ion exchange, either directly or after solution in the soil water.

(c) Precipitated insoluble forms of nutrient salts (e.g. calcium carbonate, iron phosphate). These dissolve very slowly, sometimes only under particular conditions of acidity, CO_2 content, etc.

The type of storage has an important effect on the plant nutrient economy. The nutrient requirement is met to some extent from fresh nutrient reserves, i.e. salts released by primary weathering and produced by assimilation of gaseous nitrogen by microorganisms. Because these are variable in type and in their capacity to bind salts, and particularly because of the shifting equilibrium between the various fractions, the supply is relatively constant. The system acts as a buffer against forces and processes which would tend to change the concentration of nutrients. The stability of the nutrient supply depends on the total amount of stored nutrients and the proportions of the various fractions. If a large proportion of a salt is firmly bound, then there is only a low concentration in an available form. The system is of most importance in relation to elements which frequently occur only in suboptimal amounts, such as phosphorus and nitrogen. The fact that nutrients are often stored in an organic form, in a fraction built up by plants themselves, is also of ecological significance. In the course of time, plant activity has built up the nutrient reserves on which the organic world now lives and, in the same way, plants have been largely responsible for building up the fractions in which the nutrients are now retained.

The buffer effect of the nutrient reserves means that only part of the commercial fertilisers supplied to a crop are of any benefit to it, part is held in store and used in subsequent years if fertiliser application is not repeated. If the stores are wholly or partly filled, a harvest can be obtained even if no fertiliser is applied in some years, particularly if the soil consists of easily weathered, fine-grained material. The harvest of one or several satisfactory crops without fertiliser application does not therefore mean that addition of mineral salts is unnecessary.

Distribution of the nutrient salts

How easily a plant community is disturbed by outside interference depends partly on the distribution of the nutrient reserves between the vegetation, litter, humus colloids and mineral colloids and on the particular substances involved (peat, dy, insoluble inorganic compounds, etc.). The analysis of the distribution of nutrients in a plant community is often an important ecological aim.

In a stable community the distribution of the mobile nutrients between living vegetation, litter and active humus (i.e. that in which decomposition is going on) is probably such that utilisation is approximately equivalent to mobilisation. A particular quantity of vegetation then corresponds to a particular amount of litter and active humus, e.g. both amounts are small in heath on rocky terrain, but larger in deciduous woodland.

FIG 106. **Deciduous woodland, Kinnekulle, Sweden.
The mobile nutrient reserves are mainly in the brown
earth.**

The distribution of nutrients is affected by climate, because of its effect on soil development. In semihumid regions there may be so much mull that the greater proportion of the nutrients are contained in it. The magnitude of these reserves and the effect of weathering of fine material explain the continuous productivity of soils such as chernozems.

The organic component increases with increase in humidity of the climate and it becomes the most important site of the nutrient reserves in temperate and cold latitudes. Raw humus, peat and dy contain large quantities of nutrients, of which only a small proportion are mobile. However, this mobile proportion is probably usually a larger source of nutrients than direct weathering and nitrogen assimilation together. Hence the vegetation is exploiting the nutrient storage capacity that it has itself built up.

Large nutrient reserves are not, of course, always related to eutrophy. Peat bogs are oligotrophic, despite the large nutrient reserves in the peat; these nutrients are not immediately available, but they may be used by the vegetation in the future. A change in

FIG 107. Pine-lichen heath on bedrock with a thin layer of moraine, Södermanland, Sweden. The mobile nutrient reserves are probably mainly in the plants.

climate may cause drying-out of peat, and some of the fixed nitro-gen then reenters the cycle. Increased oxygen supply has the same effect, through stimulating the activity of microorganisms. Ploughing and draining fens and bogs improves the oxygen supply. However, as the peat in these communities is exploited, the nutrient reserves and the organic material are used up and after a decade or so all the peat may have gone.

Even thin layers of humus or raw humus may have the largest nutrient storage capacity in the soil profile. Clear-felled forest heath is an example: in these habitats any large-scale interference with turnover in the soil carries with it the risk of damaging or des-troying the plant community: C. O. Tamm (1947) described a situation in south Sweden where forest tree seedlings on sandy soils died in patches, with clear symptoms of nutrient deficiency. If breakdown of the humus cover was accelerated by burning or soil cultivation, the risk of nutrient deficiency increased further.

In arid soils, the organic component is a much less important store of nutrients; most of the nutrients are adsorbed on clay, or precipitated between the mineral particles. Salt steppes and salt deserts often contain so many salts that they form a crust on the surface or a layer deeper in the soil (Berger-Landefeldt, 1957.)

There are probably also plant communities in which the greater part of the mobile nutrients is stored within the vegetation itself. Tropical rain forest (Richards, 1952) is a likely example, together with old coniferous forest growing on heaths in cold regions and, for some nutrient elements, peat bogs. Malmer and Sjörs (1955) showed that a considerable part of the reserves of potassium in a peat bog was in the living vegetation, and that some species were conspicuous as accumulators of potassium, others of manganese. The small salt supply to such communities is complemented by the ability of some of the plants to retain some of the important nutrients. If these plant communities are left undisturbed, more and more of the mobile nutrient reserves are accumulated in the vegetation. Knowledge of the proportions of the total amounts of nutrients which are in the vegetation is important in forecasting the results of exploitation of a plant community.

Movement of nutrient salts

Nutrient mobility is expressed in several ways. Minerals and nitrogen which have been taken up by the plant move within the plant body as well as outside it. Movements of nutrients during and after humification, and from one community to another, are of considerable ecological importance. Some of the movements form part of the cycle between plant and soil (Chapter 8); some have special features, since, to some extent, every element has its own pathway. Otherwise there is a mass transport, independent of the nature of the nutrients, mediated by plant roots and by wind, water, animals and man.

Root transport

The nutrients move to the above-ground organs in the roots and stems, and return to the soil in the litter. The soil surface is thereby a place of accumulation for nutrients from all the layers of soil to which plant roots penetrate. Despite this upward movement, soils in humid and semihumid regions have no accumulation of salts in the uppermost layers, because of the counteracting downward movement of leaching, and because of uptake by the roots, most of which are in the uppermost layers. Below these layers, the frequency of roots decreases fairly evenly with depth (Kivenheimo, 1947). The deepest roots, and species with the deepest root systems, achieve contact with parts of the soil little exploited previously. This is important to the nutrient economy, because fresh reserves are tapped, replacing the salts lost from the community by leaching in harvested crops, etc. Deep roots may also take up some of the nutrients in the drainage water, which would otherwise be lost. Potassium, for example, is present in larger amounts in the drainage water than in the surface water (Troedsson, 1955). In such cases, the possibility of luxury[1] consumption is an advantage to the salt economy of the community.

The depth of plant root systems varies considerably. The roots of aquatic plants usually remain in the 5 to 10 cm layer; roots of bog plants in the 10 to 15 cm layer (min. 5 cm, max. 50 cm, in various localities in Finland, Metsävianio, 1931; Murén 1934); and in better drained soil, roots penetrate deeper. Kivenheimo (1947) found that the root depths of ninety-three species of grasses, herbs and dwarf shrubs in forests in Finland varied from 3 cm (*Goodyera repens*) to 180 cm (*Orthilia secunda*). Table 93 illustrates the root distribution in these forests, which were of various types, such as mixed forest (deciduous and coniferous trees) rich in herbs, moss-rich coniferous forest (*Oxalis-Vaccinium myrtillus type*), and heath-forest (*Vaccinium-Calluna-Cladonia* type). Table 94 shows that root depth is greatest in the third type and least in the first. In general, root depth increases with decreasing water availability and probably also with decreasing availability of mineral nutrients,

---

[1] Luxury consumption is uptake of salts not needed by the plant for its metabolism, taken up because uptake is partly determined by concentration.

i.e. increasing oligotrophy. (For further information on root development in different species see Walter, 1951.)

Kivenheimo's investigation also showed that the roots of forest trees are mostly distributed in the same soil layers as the roots of field layer species, so that there is competition for water and nutrients. Some of the tree roots do penetrate deeper (3 m for pine; 2·8 m for birch; Laitakari, 1927, 1935), so that forest trees are better equipped than herbs to obtain water and nutrients from the deeper layers of the soil. The activity of deciduous trees is particularly significant because they take up relatively large amounts of basic nutrients. For this reason, deciduous woods are able to accumulate more salts than other communities, and to bring about the formation of mull. An easily observable illustration of this is the mull formation which takes place on large rocks and boulders in deciduous woods. Halden (1941) and Sörlin (1943) have described examples of this even in situations where there is a deficiency of calcium and other basic elements in the rocks and soil. The leaf fall, and the uptake of nutrients from deep soil layers, either directly or from ground water, explains this. In contrast, raw humus forms on large boulders in coniferous woods (Fig. 108).

Thus, in addition to its physiological importance, salt uptake by roots is important ecologically in counteracting loss of salts from the soil by leaching.

TABLE 93 Maximum root depth of grasses, herbs and dwarf shrubs in coniferous forests in Finland. (After Kivenheimo, 1947, page 129–30.)

ROOT DEPTH (cm)	NO. OF SPECIES	
	ABSOLUTE	PER CENT OF TOTAL
0–5	3	3
5–15	27	29
15–50	45	48
50–100	12	13
>100	6	7
Total	93	100

Guttation and other forms of salt excretion

Some of the salts taken up by the roots are returned to the soil by guttation or by salt excretion from the leaves. To a certain extent, salts can be both taken up and excreted through the epidermis of the shoot system. The cuticle is not completely impermeable: salts and organic molecules, e.g. herbicides sprayed on leaves, enter the cells (see Kaindl, 1953). The amount of salts lost from leaves by guttation and returned to the soil is not known. The phenomenon is restricted to certain species, particularly grasses and herbs: *Saxifraga* species secrete lime, and some halophytes secrete sodium chloride, especially members of the Frankeniaceae and Plumbaginaceae. It is sometimes characteristic of certain phases

TABLE 94 The relation between root depth and the quality class of the site. (After Kivenheimo, 1947, page 131.)

	HERB-RICH MIXED FOREST		MOSS-RICH CONIFEROUS FOREST		HEATH FOREST (MAINLY PINE)	
Root depth (cm)	0–15	15–>100	0–15	15–>100	0–15	15–>100
Percent of total no. of species	52·5	47·5	34	66	14	86

of development, such as young leaves, and since it often requires full turgidity of the organ and is therefore dependent on suitable weather conditions, it does not represent a large-scale contribution to nutrient turnover. However, guttation salts are in a free state and are immediately available for further uptake by roots or other organs. Treatment of trace element deficiency symptoms by foliar sprays containing the appropriate element in solution is effective because of salt uptake through the epidermis.

Wind and water

The wind blows away soil particles from wind-eroded soil and ash and smoke from volcanoes and from burnt ground. It also blows away hydrochloric and sulphuric acid produced when fossil fuels are burned, and in certain industrial processes such as smelting ores; ammonia produced in humification and burning of fossil fuels; oxides of nitrogen from denitrification and from lightning; neutral salts in spray from the sea; and organic material such as pollen, spores, seeds, dust, etc.

The amount of windborne material has probably increased in the last few hundred years. Mattson and Koutler-Andersson (1954) found more nitrogen and mineral salts in the surface layers of bog peat than in deeper layers. They consider that this is the result of increased wind erosion of arable soil and increased combustion of fossil fuels.

The amount of windborne material in the air varies from place to place (Emanuelsson *et al.*, 1954) (Table 95), and the quantity of salts in the rain also varies (Table 96). The rainborne supply of ammonium-nitrogen varied between 0·9 and 3·7 kg/ha annually in south Sweden (Egnér 1954). The equivalent amounts of calcium were 2·6 to 13·9 kg/ha. In Great Britain an annual supply of 280 kg/ha of soluble solid substances has been measured. This included 40 kg of chlorides, 60 kg sulphate and 14 kg calcium (Meetham, 1952).

FIG 108. **Boulder in coniferous forest, one year after felling. The surface is covered by a layer of raw humus *c*. 5 cm thick. On this, ferns, mosses and nine species of phanerogams, including a tree, are growing.**

The amount of ammonium- and nitrate-nitrogen in the air is greatest in the tropics, decreasing towards the poles. Both forms are dissolved in rain water, which therefore contains most nitrogen at the beginning of a rainy period. Short showers therefore supply relatively more fixed nitrogen than do longer periods of rain (Ångström and Högberg, 1952).

The salt content of the air varies because there are various sources of the salts, supplying different amounts, in different places. The exact contributions of these sources are not known. The sea is certainly a large source, perhaps the largest of all, and the salt content of the air is related to wind direction and distance from the coast. At the coast of New England, the chloride content of rain is 6 mg/litre, but in the vicinity of the Great Lakes it is only 0·2 to 0·6 mg/litre. A few kilometres inland from the Norwegian coast it is 3·5 to 57·5 mg/litre (see Mattson, Sandberg and Terning, 1944).

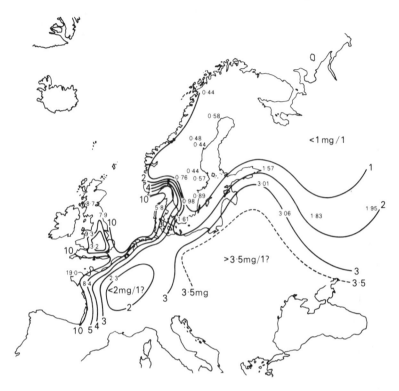

FIG 109. The chloride content in rain in Europe. (After Rossby and Egnér, 1955.)

The quantity of nutrients supplied in rain is certainly small (Table 96; Figs. 109, 110), but even such small amounts are important to plant communities living in suboptimal nutrient conditions, e.g. on raised bogs and sometimes in coastal areas. Witting (1948) found a relationship between the distance of a bog from the coast, its salt content and the richness of the vegetation. Coastal species may be dependent on a rainborne nitrogen supply. Steeman Nielsen (1944) calculated that the annual supply of nitrogen in rain is about 3 kg/ha corresponding to 200 mg/m^3 of water, or 1/3000 of the total amount of nitrogen in the sea; Carlisle, Brown and White (1966) measured 9·5 kg/ha in northern England. In comparison, the nitrogen brought annually to the sea in rivers is about 1/50 000. It seems probable that plants obtain certain rare nutrients, particularly micronutrients, as atmospheric salts. Elements which occur in isolated widely separated places in the earth's crust are weathered and leached and carried to the sea; there they become mixed and distributed over wide areas before being carried inland by the wind and precipitated in the rain, though in very small amounts. It is thought that the human requirement for iodine is met by windborne salts from the sea (Boyce, 1951). The amount of

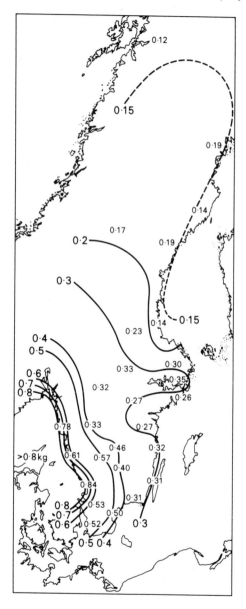

FIG 110. The supply of airborne nitrate in Sweden (in kg/ha/year). (After Emanuelsson, Eriksson and Egnér, 1954.)

boron supplied in the annual precipitation (at Uppsala, Sweden) is of the same order of magnitude as that in hay harvested from the same area of land (Philipson, 1953).

The proportions of the airborne salts reaching the vegetation and the soil depends to some extent on the nature of the vegetation. As the rain passes through the vegetation, the composition of the salts it contains may change quantitatively and qualitatively as some of the salts are retained on the plants and others remain in the water. Tamm (1953) found that rain which had passed through

TABLE 95 The content of plant nutrients in the air ($\mu g\ m^{-3}$). (After Egnér, 1954, pages 5, 45.)

PLACE	$NH_3=N$	$NO_3=N$	Ca	Na	K	Mg	Cl	S
(Sweden)								
Ultuna	2·9	0·8	4·3	10·0	4·0	2·4	1·1	10·5
Bjärka-Säby	4·4	0·7	6·0	6·9	1·9	1·3	1·4	2·4
Flahult	2·3	0·3	6·3	6·2	2·0	1·4	2·6	5·1
Bräkne-Hoby	3·2	0·4	5·0	5·2	1·0	1·3	1·5	5·3
Tomelilla	13·8	1·2	19·3	3·7	2·2	2·3	5·0	14·3
Hilleshög	3·3	0·8	3·4	3·3	0·5	1·2	2·4	11·3
Plönninge	8·6	3·5	15·3	30·0	3·6	12·7	8·3	9·1
Skara	4·8	1·0	13·8	9·0	2·0	2·6	3·5	7·0

TABLE 96 Plant nutrients in rain (kg/ha/year). (Mean of observations made from 1/9/51 to 31/8/53, Egnér, 1954, page 46; Carlisle, Brown and White, 1966.)

PLACE	$NH_3=N$	$NO_3=N$	Ca	Na	K	Mg	Cl	S	P
(Sweden)									
Bräkne-Hoby	2·6	1·1	4·0	6·1	2·5	1·3	7·7	7·0	
Plönninge	3·7	1·5	5·3	16·3	2·6	2·6	25·0	8·7	
Flahult	0·9	0·6	4·7	5·1	2·8	1·2	7·3	5·5	
Skara	1·3	0·8	5·5	5·5	2·2	1·2	8·9	5·5	
Bjärka-Säby	0·9	0·5	13·9	3·5	3·7	0·9	7·0	4·0	
Ultuna	1·1	0·6	2·6	1·6	0·6	0·6	2·8	3·6	
(England)									
Grisedale forest	9·5		7·3	35·0	3·0	4·6			0·43

spruce crowns was enriched with phosphate, potassium, sodium and calcium (Table 97), but had lost some of its nitrogen. The phosphate content in water dripping from the trees was sometimes higher than in the water of many lakes, and about the same as in soil water (Tamm, 1951a). The amount of potassium, sodium and calcium supplied in the rain to soil under the trees was 2·2 to 3·7 kg/ha of each in six weeks (Table 98). As a comparison, Tamm cites figures of 40 kg Ca and 8 kg K, per hectare and year, supplied from the litter of a productive spruce forest in south Norway (Morks, 1942).

The salts in rain dropping from trees are not derived wholly from rain water, but include some derived from the dust, smoke, soot and other deposits on the tree crowns. These are not only of

TABLE 97 Mineral content in rain (6·9 mm) collected in the open and under spruce during two July days and one September day. (After Tamm, 1953, page 88.)

	PARTS PER MILLION				
	ASH	P	K	Na	Ca
Open	2–5	≤0·1	0·0–0·3	0·2–0·9	0·1–0·3
Under spruce crown					
peripheral	1–17	0·01–0·12	1·0–5·3	0·8–1·0	0·5–2·1
,, ,, ,, central	6–36	0·06–3·5	3·1–14·5	1·0–4·3	0·7–5·4

TABLE 98 Nutrients supplied in the rain to soil under crowns of trees and in the open. Rainfall 101 to 104 mm in the open, sampled 19/10/50 to 30/11/50. (After Tamm, 1951a, page 187.)

	NUTRIENTS SUPPLIED TO THE SOIL IN THE RAIN FROM 19/10/50 TO 30/11/50 (kg/ha)		
	Ca	K	Na
Open	0·4	0·2	0·6
Under spruce crowns (*Picea abies*)	>2·2	>2·4	>3·7
Under pine crowns (*Pinus sylvestris*)	2·5	2·7	3·2

windborne origin, but also include animal litter, excrement, epiphyte litter and the bark-, leaf- and twig-litter of the trees themselves. Leaching of this litter begins before it falls, and appreciable amounts of salts may be leached from the surfaces of living leaves and shoots, especially if salt is secreted as in cotton or saltbush.

Water transports material horizontally as well as vertically. Vertical transport (leaching, humus infiltration) has already been described (Chapter 14). Water moves horizontally both above and below ground. Its salt content depends on weathering, humification and the prevailing climate. The salts are carried to the sea in the rivers. For example, the Nile carries 20 million tons of dissolved substances per year (Fox, 1950) and German rivers 34 million tons per year (Boresch, 1931). Table 99 is a survey of water transport of salts: some of the salts are carried back to land in the wind (Tables 95, 96), by sea birds, or by man's activities (fishing, seaweed fertilisers, sea salt).

TABLE 99 Survey of transport of soil and nutrients by water.

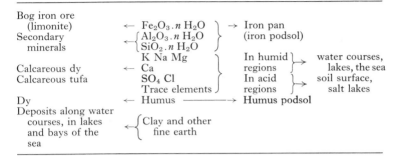

Bog iron ore (limonite)	$\leftarrow$ $Fe_2O_3 . n\,H_2O$	$\rightarrow$ Iron pan (iron podsol)	
Secondary minerals	$\leftarrow \begin{cases} Al_2O_3 . n\,H_2O \\ SiO_2 . n\,H_2O \end{cases}$		
Calcareous dy	$\leftarrow$ K Na Mg Ca	In humid regions	water courses, lakes, the sea
Calcareous tufa	SO_4 Cl Trace elements	In acid regions	soil surface, salt lakes
Dy	$\leftarrow$ Humus $\longrightarrow$	Humus podsol	
Deposits along water courses, in lakes and bays of the sea	$\leftarrow \begin{cases} \text{Clay and other} \\ \text{fine earth} \end{cases}$		

Plants, animals, men

Plants build up their own nitrogen reserves and make an essential contribution in retaining other nutrients so that they accumulate in the surface layers of soil and take part in the nutrient cycles. However, plants may also have the opposite effect: the raw humus flora causes or accelerates podsolisation (see Chapter 14); denitrification bacteria break down nitrogen compounds (see Chapter 8); and in large accumulations of organic nutrients microorganisms may develop so quickly that the ammonia formed is lost to the atmosphere before it is used by soil or vegetation (see Chapter 8).

Animals are usually responsible for the movement of only relatively small amounts of plant nutrients, but their effect may be locally intensive. Sea birds, herds of mammals, or communal insects cause local accumulations of organic material which may be large enough to affect the vegetation. The tips of rocks on or near sea cliffs, or other places where sea birds gather, are covered in guano, rich in plant nutrients, especially nitrogen and phosphorus; the local accumulation of nutrients usually brings about a zonation of the flora, e.g. of lichens. In the immediate vicinity of ant hills plants are often more luxuriant than they are further away; nutrient accumulation (excrement, deposits of nuts and other fruits) occurs around the dwelling places of mammals; banks of shells, and the associated calcareous soils, favour molluscs again. Such organic accumulations deposited long ago have been the source of fossil nutrient layers such as Chile saltpetre and Silurian chalk.

Man makes a significant contribution to the movement of plant nutrients through agriculture, forestry, the rearing of animals and the construction of town and villages. As a result, plant nutrients are moved from forests to arable land, from arable land to farmyards and gardens, from country to town, and, via sewage systems, to the sea or lakes. By burning organic materials, salts are converted to ash, accelerating leaching. Even fossil forms of plant nutrients, coal or peat, have been returned to circulation by various means. Recently, salts have been made industrially from minerals and the air, bringing about an opposite movement. The significance of these contributions is discussed further in Chapter 18.

Loss and replacement of the mobile reserves

Some of the salts lost from one area become available to plants in another, with no net loss to the plant world as a whole. There is a net loss, however, if the salts break down or are fixed so that they become unavailable. Leaching (of substances dissolved in the soil water), evaporation (of ammonia and the oxides of nitrogen), sedimentation of organic material, fossil formation (coal, peat, shells, bones, etc.) and fire are processes which bring about loss.

Some of the material which is not permanently lost is recirculated relatively rapidly. This is probably true of the ammonia and oxides of nitrogen, and the leached salts. Some of the leached salts are taken up by deep-rooted plants and so return to the vegetation, but a significant proportion is converted so that it remains unavailable for a long period, e.g. some of the salts which accumulate in the sea. In the Baltic Sea, most of the salts are in the bottommost layers, covered by layers of fresh water (Gessner, 1933). Salts in deep water in the oceans are similarly unavailable to plants. Nutrients may sometimes be lost in stages. Some of the salts carried from dry land to the sea are taken up by planktonic algae, but then sink to the bottom when the algae become litter. Pettersson (1950) gives examples of marine sediments with humus-coloured constituents at a depth of 12 m. The amount of nutrients lost in this way over the years is probably very large. Material deposited in other forms of sedimentation and fossil formation also disappears from circulation for long periods.

Fire is responsible for considerable losses, especially of nitrogen, which is released and returns to the atmosphere. Other elements become converted into soluble forms, mainly oxides, which are lost by leaching. Fire also destroys other nutrients, such as vitamins and other growth factors essential to the soil microphytes.

To balance these processes of loss, there are a few sources of fresh nutrient reserves: mineral weathering, nitrogen fixation, and industrial production of plant nutrients.

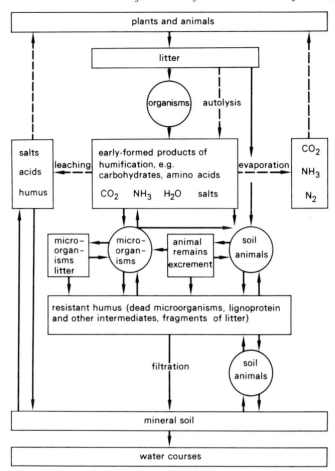

FIG 111. The balance between processes of loss and of gain involved in the turnover of the mobile nutrient reserves on dry land. Rectangles denote the places where the salts and soil materials occur; circles represent the organisms responsible; arrows represent the process by which the salts and fine materials are moved; the dashed arrows are the shortest routes.

Loss and formation of colloids and other fine soil particles

At the same time as the movement and loss of the mobile soil nutrients, there is an equivalent loss of the colloids and fine particles by which the nutrients are retained in the soil. The forces operating are again wind, fire, water and man.

Of the fine soil particles, it is really only the burned stable humus which is totally lost; the other constituents are only moved.

Wind erosion

Where soil has a close cover of vegetation, the humus, clay and other fine material is protected from the wind; if the vegetation is removed and the soil laid bare, it is exposed to the effect of external forces. The finest particles can then be blown upwards and carried away by strong winds, sometimes to very great heights, and they

FIG 112. Wind-eroded bottom of calcareous deposits. Mästermyr, Gotland.

are transported across seas and continents. The greater part is finally deposited in the sea; coarser particles are carried only short distances; leaves and other litter are blown away by wind and accumulate in piles or in holes; sand is transported in a way depending on the particle size—sand dunes are built up from the coarser particles. Like other deposits of wind-sorted material, these are characterised by even-sized particles. Desert sand dunes and the thick layers of fine earth which comprise loess soils are the result of previous transport by wind. The phenomenon occurs on a smaller scale on sandy coasts and newly burned surfaces where sand and ash, respectively, are blown about.

The wind generally causes surface erosion, removing material fairly evenly from all over the surface, but hollows and ravines may also be formed. If the vegetation cover on aeolic soil is destroyed, the wind may blow the exposed sand on top of the remaining vegetation, which is killed as a result, so that the soil underneath it may eventually be blown away in turn.

All forms of wind erosion are dependent on the dryness of the soil. The soil particles separate easily only if the soil is very dry, so erosion increases as soil dries and as wind velocity increases.

Water erosion

Raindrops falling on bare ground break up the crumb structure and carry away loose particles. If there is any surface run-off, fine particles are carried away in the water. The effect is accentuated as the soil pores get blocked up, infiltration is slowed, and surface run-off increases progressively. The water in ditches through arable land contains solid particles of various sizes. The heavier the rain, the coarser the material which is moved.

Raindrops cause erosion over the whole surface of the land. On sloping ground the water collects in furrows along which material is loosened and carried away. The furrows deepen into channels

FIG 113. An eroded slope at the edge of Lake Våttern.

and even ravines. The water has the greatest effect on steep slopes and on fine soils, since fine material is most readily transported away: streams and rivers flowing through areas with deep layers of clay or other fine material cut downwards so that winding ravines and gorges are formed. In contrast, water courses in areas of moraine are generally shallow, since the coarser material accumulates on the bottom and prevents erosion.

Fine material carried by water is deposited in hollows, in lakes and estuaries, and at all places where the rate of flow is slow. The soil in fens, water meadows, lake bottoms and deltas is usually built up from fluvial deposits (Chapter 10). In delta formation, the riverbed becomes raised, causing flooding and transport of material to the sides. If the supply of material is sufficient, the river channel comes to lie higher than the surrounding country, increasing the risk of flooding. The Yellow river (Hwang-Ho) in China transports about 2·5 thousand million tons of solid particles annually, enough to load a goods train sixteen times as long as the Equator (Hjulström, 1940). The water in the river and in the sea in the vicinity is coloured yellow by this material.

Erosion by agricultural implements

Sloping arable land is subject to erosion by agricultural implements. Ploughs and harrows, and other machines that lift up the soil, move the soil particles, bringing about a net movement downwards so that eventually all the soil comes to lie at the bottom of the slope, and stones and bedrock are exposed higher up. On steep slopes, which are particularly subject to this type of erosion, construction of terraces has sought to counteract it. There is often a river in the valley bottom between such slopes, and this carries away much of the eroded material. Moreover, in the course of time, the stream or river gradually may change its course and the loose material which has come down from higher up the slopes, and is often used for arable cultivation along its banks, is washed away.

Most of the material eroded by wind or by water is transported to the sea. According to Pettersson (1950), the layer of red clay at the bottom of the Atlantic increases in depth by 5 to 10 mm in 1000 years; sedimentation in the Pacific is less. Between the Atlantic ridge and Martinique, a depth of 1300 m of sedimented material has been measured, and between the Atlantic ridge and Madeira the sedimented layers are between 1550 and 4000 m thick. In the Pacific this depth is only 150 to 500 m.

Erosion has many ecological effects. Because the fine particles are carried away, the remaining soil tends to be dry and poor, favouring the development of heaths in the humid regions and bush steppes and deserts in arid regions. Podsolisation is a consequence of heath development. Thus, in the places from which the material is lost by erosion, there is a change in the direction of oligotrophy, while, in the places in which it is deposited, there is a change in the direction of eutrophy.

FIG 114. Desert scrub, where the fine earth has been lost by erosion by wind, water and agricultural implements. Fiuggi, central Italy.

Severe erosion may cause devastation of the landscape. In extreme cases, all the loose soil is removed and bare rock is exposed; the land is broken up by ravines and rendered inaccessible; and ponds, streams and river basins are blocked so that the lower reaches or rivers become marshland (Fig. 140, page 530).

Erosion occurs with varying intensity, depending on climatic and edaphic conditions and on exploitation by man. It is worst in areas where there are deep layers of fine soil, strong winds and low rainfall, or long, dry periods, and where there has been continuous cultivation for centuries. It is bad in areas of loess, and in tropical and subtropical regions of high and very heavy rainfall. For example, southern Chile has an annual rainfall of about 3000 mm, and some parts of India about 12 000 mm. Such values are common in the tropics. In comparison, the tendency for erosion to take place in western and northern Europe is insignificant. Precipitation in Sweden is generally 500 to 800 mm, distributed over 125 to 225 days. The British Isles have precipitation of about

TABLE 100 Conditions bringing about or affecting movement of nutrient reserves.

OPERATIVE FORCES AND CONDITIONS		PROCESSES OF MOVEMENT OF NUTRIENT RESERVES	ECOLOGICAL EFFECTS OF MOVEMENT OF NUTRIENT RESERVES	
Primary	Bedrock	Mineral composition and weathering	Soil formation and chemical-physical-biological condition	Fertility
	Humidity temp., CO_2	Weathering		
	Topography	Humification		
		Radiation, temperature		
		Transport of soil and water		Selective effect
	Wind water	Leaching	Movement of material	
		Accumulation		
		Movement of fine earth		
	Plants	Litter content of acid and basic substances		
	Animals	Grazing, excrement		
Secondary	Man	Hay-making		Composition and productivity of the plant community
		Grazing		
		Arable cultivation		
		Forestry		
		Drainage		
		Fertilisation		
		Soil cultivation		
		Change of vegetation		
		Harvest		
		Transport of materials		
	Fire	Fire		

600 mm, on about 150 days. Cloudbursts do occur, but are local and only rare.

Environmental conditions which bring about movements of nutrient reserves

Several of the conditions which lead to accumulation of nutrient reserves also cause their movement (Table 100). However, accumulation and movement are affected differently. Conditions which are the primary cause of the formation of soil material and which determine the properties of the soil, i.e. certain geological, climatic and topographic conditions, usually have very little effect on movement. Table 100 is a diagrammatic representation of these interactions, in which geology, climate and topography are considered to be primary factors. During short periods they bring about no significant movement of nutrients. However, over long periods, changes in climate and topography affect radiation and humidity and therefore influence the effect of humification, water and wind. Because of these changes, soils have probably changed in many respects during post glacial time.

The secondary factors are considered to be those which affect the soil after its formation; they exert an immediate effect. Soil properties can be changed by wind, water or fire within several days or even hours. Man may bring about equally rapid change, and his interference has the most varied effects of any of the secondary factors. It influences processes of synthesis and of breakdown, movement of material and of vegetation; it extends over practically all types of soil and plant community. This is discussed in the next chapter.

18

The human influence

The balance between soil and plants

Soil is important to plants because it provides space and a place of attachment; it acts as a source of water and nutrients, and it stores water and nutrients so that they are available to plants.

On the other hand, plants themselves influence the formation of soil and its properties.

1 Autotrophic plants are a source of nutrients to soil organisms, which fix nitrogen, they remove plant and animal remains from the soil, and produce organic constituents which are particularly important to the physical properties of the soil.

2 Plants contribute to soil development by supplying material derived from the environment. In the course of time, plants have converted some of the large reserves of carbon, hydrogen, oxygen and nitrogen in the sea and the air into a solid state, and so into soil constituents. The amount depends on the prevailing balance between synthetic and destructive processes in a particular place. There may be so much accumulation that the soil is wholly organic.

The supply of mineral particles to the soil is also assisted by plants because some of the products of their secretion and decomposition accelerate the solution and weathering of minerals.

3 Soil chemical properties are also determined to a large extent by plants, depending on the size of the mobile nutrient reserves. The larger these are, the larger the effect of plants on turnover. The availability of nitrogen is entirely dependent on plants. The amount of other salts depends also on the capacity of the organic colloids and soil particles which retain the salts. The effect of plants is particularly important in the accumulation of phosphorus.

4 The physical properties of the soil—the crumb structure, coherence, elasticity, porosity, permeability, and storage capacity

for gases, water and heat—depend on the nature and amount of stable humus, especially humus colloids.

5 Plants have a strong influence on soil development by providing material and by affecting the chemical and physical properties of the soil. The vegetation determines, for example, whether soil development is in the direction of raw humus and podsols or of mull and brown earths. The soil changes in response to qualitative or quantitative changes in the plant cover.

6 Finally, the vegetation affects the soil microclimate, through light interception, minimising temperature fluctuations and increasing accumulation of CO_2. Plants protect the soil from erosion by water or wind. The wind speed is greatly reduced, especially near the soil, and rain is intercepted, so that some evaporates directly and some passes so slowly downwards through the soil that the structure is not disturbed.

All these processes contribute to a series of balanced interactions between soil and plants. When plants or plant communities are exploited by man, this balance is disturbed, and the direct effects of disturbance are followed by indirect effects. For example, if the vegetation is wholly or partly removed, the soil and nutrient economy is changed, which, in turn, affects the capacity for survival of the plants and animals. Extensive changes have been brought about during the course of time by cultivation of the land, causing changes in plant and animal life and hence in conditions affecting the future of man himself.

To obtain food and a place to live, man exploits nature in more or less the same way as animals do. However, man's interference is more consistent and calculated and therefore more extensive and more likely to cause change. When the human population was smaller, and the numbers of large animals were greater, the ecological effects of the two were comparable in importance. Now that the human population has become so large and exploitation of natural resources so much more extensive, the ecological effect of man has increased enormously, while the effect of the larger mammals is much less. Man has determined and regulated the influence of animals and other conditions of the environment more and more. He has brought fire, water and soil into his service, and used them to make use of the plant and animal worlds. Man is now more important than most other environmental factors in bringing about ecological change. The movement of nutrient reserves, i.e. accumulation and loss of salts, and increase or decrease in proportions of colloids and fine soil particles, is now mainly determined by man. The ways in which the plant world and the reserves represented by the soil are exploited are of primary importance to soil productivity.

Nature as it is today has to a large extent been shaped by the influence of man exerted over several thousand years, and the study of this influence is largely a study of the history of the settlements and the cultivation of the area in question.

The oldest traces of cultivation of plants in Sweden date from the early Stone Age, but man was already present much earlier. The climate was probably warmer than it is now and the land was covered by luxuriant vegetation and was rich in game. This virgin forest probably covered large parts of northern Europe during the Stone Age. It was deciduous, mainly oak, growing on brown earths which developed readily in the warm and semiarid climate. Fens and bogs were other dominant plant communities in which organic and inorganic material accumulated during the course of time, building up reserves of various plant nutrients. As the original hunters and fishermen settled and began to keep livestock, the exploitation of vegetation and soil reserves which had built up over centuries began. These reserves were large enough to support the population and its culture until fairly recent times. The partial utilisation or loss of humus and mobile nutrient reserves was an inevitable result of the primitive agricultural methods and of the increasing population. Presentday arable fields in areas of brown earth or clay soils are mainly on land which itself was, or still is, brown earth or clay. The original deciduous forest soils remain only in places unsuitable for cultivation because of their stoniness or topography, e.g. the stony, but nutrient-rich moraines in south and central Sweden, which still bear deciduous forests rich in herbaceous species.

Land has been exploited in various distinct ways, namely, shifting cultivation involving burning, meadows, grazing, arable land, or forests. In Europe, the oldest practices, i.e. shifting cultivation and use of meadows, now retain very little of their original importance. However, they are of considerable interest from a historical point of view, as they had an important effect on the development of the soil and plants in the presentday landscape. In some areas, particularly in south Sweden, these practices were probably the most important influence on the landscape in their time.

Shifting cultivation, involving burning

History

It seems to be generally accepted that the men who first moved into northern Europe at the beginning of the post-glacial period were hunters and fishermen, largely dependent on animals as a source of food. Their successors, who kept livestock, must also have found animal food much easier to obtain than vegetable food, as it still is today for people living in Arctic regions. The dividing line between an animal and plant-based food economy has moved northwards in the course of time. In the middle of the eighteenth century it seems to have been as far south as central Sweden. Linnaeus (1734) wrote that people in the western parishes obtained their meat mostly from the forest, as fish, birds, reindeer, elk and several other wild animals.

Conditions in the late Stone Age, considered to be the period when fixed settlements began, were favourable for such an economy. There was probably plenty of game, including the aurochs (extinct wild ox), deer and wild boar. Conditions for keeping the first livestock, probably oxen, pigs and sheep, were also good (Klindt-Jensen, 1956).

As long as the climate was mild enough for winter grazing, no barns were needed for the animals; the only requirement was the pasture for grazing. There was plenty of this along the banks of lakes and rivers, where there were large areas of grass communities, as there often are today. Even better grazing was available if the forest could be cleared, and in the absence of suitable tools, this was done by burning. Subsequent burning procedures probably date back to these early practices, the original aim of which was probably to obtain more land for grazing. Grasses and herbs grew in the places where the fire had penetrated the forests, so it was only necessary to keep the area clear in the future, a relatively easy task, because of the activity of the grazing animals themselves. A form of this primitive economy has survived until recently in central and northern Scandinavian summer pastures (Sjöbeck, 1953).

There was land in plenty for these early people and their livestock, and in Scandinavia and the whole of northern Europe there were probably large herds of domestic animals. Tacitus wrote of the German peoples that they were rich in livestock and that the numbers they owned were their only wealth, very valuable to them. Caesar wrote in the *Gallic Wars* that some of the Germanic tribes ate very little cereals, but used mostly milk and meat of domestic animals and also devoted themselves to hunting.[1]

Cultivation of cereals and other crops began early, but it was much later before it achieved any great importance. In pieces of clay pottery found in south Sweden, dating from 2500–2000 BC (Early Neolithic), there are impressions of barley (*Hordeum polystichum*), emmer wheat (*Triticum dicoccum*) and Einkorn (*Triticum*

---

[1] The same word is used for livestock and wealth in the old Scandinavian language. In Latin, too, the word *pecunia* (money) is derived from *pecus* (pig).

monococcum) (Hjelmquist, 1952). Among the 4000-year-old finds from Dagsmosse in Östergötland, Sweden, are large quantities of six-rowed barley together with wheat and about sixty weeds of arable land (Berggren, 1956), so cereal cultivation in Sweden probably started in the late Stone Age. Oats and millet (*Panicum miliaceum*) were grown in Denmark in the Bronze Age, and rye in the Iron Age (Iversen, 1941). Cultivation of rye in Västergötland probably began early in the first century AD (M. Fries, 1958).

How the first cereal cultivation was carried out, and how extensive it was, is not known. For practical reasons, it is probable that it was only on a small scale at first, as in gardens today. The diet was predominantly animal, and all types of vegetable food were an enjoyable and beneficial addition rather than a major constituent. Even in its most primitive form, cultivation of cereals and utilisation of the grain requires some tools and considerable knowledge. However, agriculture on a large scale probably started with the help of burning, the only feasible method of preparing a large area of ground for sowing, since there were at first no tools suitable for clearing forest and breaking up ground held together by roots and rhizomes. The loose surface layer left after burning could be broken up with the simplest of tools. At the same time, the nutrient reserves were mobilised: in addition to the ash, nutrients were released by humification of the soil organic material (raw humus, litter, roots, rhizomes).

Burning was probably carried out without previous clear-felling, since in the late Stone Age there were no implements for large-scale forest clearance, although the land was largely covered with deciduous forest, which is not easily set on fire. Nevertheless, at that time the climate in north Europe was warmer and dryer than it is now, and the vegetation would therefore have had a lower water content, particularly on the areas of moraine first used for agriculture. Also, the primeval forest would have included many dead trees and other dry material, which would have helped spread the fire sufficiently to destroy the undergrowth and kill the trees. Complete removal of the trees was unnecessary. In Finland, up until the last century, land on which the largest trees were ring-barked but left standing was burnt over. There were up to 300 large trees (diameter at breast height up to 40 cm) per hectare (Heikenheimo, 1919).

Implements

Several implements were required for cultivation following burning, particularly rakes, pick-axes and sickles, and preferably also axes and saws. Without these it would scarcely have been possible to bring enough land under cultivation to be of any significance. Rakes could have been made of wood, if necessary, but the other tools required metal. A type of sickle was already in use in Sweden in the Bronze Age, but this became an effective tool only during the Iron Age, when the scythe was also developed (Malmström, 1951). According to Brögger (cited by Sjöbeck, 1932), forest clearance first got under way in Norway, in Valdres and other valleys, in 200–400 AD, when effective chopping implements were first made. This probably occurred later on the island of Öland, in the sixth century AD (Sjöbeck, 1933). In more southern countries, where iron was utilised much earlier, forest felling was also much earlier, 800–700 BC in Italy, for example (Hyams, 1952); the Egyptians already used copper axes and saws in 2600 BC.

From the Iron Age onwards, it should therefore have been feasible to practise shifting cultivation following burning, in Scandinavia. From then on, burning became the most important means of cultivation and remained so over a long period, in many places until recent times.

Permanent arable land developed alongside shifting cultivation, but for a long time it was of much less importance. Even when burning became less common relatively recently, it was still carried out as a complement to permanent arable land, especially in less fertile areas. The burned land generally gave a better crop yield than the exhausted arable land. In Sweden, as in Scandinavia generally, burning occurred up until recently in many places, e.g. in central Skåne at the beginning of the nineteenth century (Sjöbeck, 1932) and in northern Skåne at the beginning of the twentieth century (Weimarck, 1952) (Fig. 128, page 492). In Finland 50 000 ha of forest land were burned in 1870; 17 300 in 1890; 7700 in 1900; and 3800 in 1910 (Heikenheimo, 1919). In some provinces in Russia in 1900, burned land accounted for 43 per cent of all cultivated land, including meadows.

The advent of mineral fertilisers, bringing about increased productivity of arable land, caused the practice of burning to die out rapidly.

Mode of operation

It is not known how burning was carried out in prehistoric times. In historical times, up to the beginning of the twentieth century, the trees and bushes were cut down and allowed to dry out and then the land was burned over at a suitable time in the spring (Linnaeus, 1749).

Picks and hoes were the implements used to break up the soil surface and make it suitable for sowing. Small boulders, roots and other rubbish loosened by the picks were collected into heaps. Each time the area was used more stones were removed and added to the heaps, since otherwise they would have hindered the work and covered some of the productive area. Removal of the stones was also essential to make the surface even enough for scything cereal crops. The humification and resultant shrinkage of tussocks and other litter also contributed to producing an even surface. Land which has previously been burned therefore has a more even surface than other moraine areas (Weimarck, 1952). A root crop was planted on the burned area the first year; this was formerly turnips, but more recently has been potatoes. After the first root crop, barley or rye was sown (Linnaeus, 1749, 1751), or oats or wheat in some areas (Wibeck, 1911). The sequence was the same at the beginning of the twentieth century. Because of the rapid mobilisation of the soil nutrient reserves and the absence of the worst weeds, the yield was high and of excellent quality; at the end of the nineteenth century, burned land still gave higher yields, of better quality, than permanent arable land. For example, barley and rye yielded sixteen to twenty times the weight of grain sown (Weimarck, 1952); there was also a good yield of turnips and potatoes and they were of first class shape and flavour, as mentioned by Linnaeus (1751). In addition, the crops were usually strikingly free from disease, as the soil was not infected by disease organisms.

However, the nutrients in the ash and the organic soil complex were soon used up, because of the high nutrient requirements of the crop plants. Mork (1946) analysed ash from burned land originally bearing oligotrophic plant communities and found that it contained 25 to 56 kg/ha K_2O, 11 to 28 kg/ha P_2O_5, and 42 to 68 kg/ha CaO. A potato crop uses about 110 kg/ha K, 4 kg/ha P, and 3 kg/ha Ca; and wheat about 49, 17 and 14 kg/ha, respectively (Table 102, page 496). Some of the more soluble ash constituents are lost by leaching (for examples, see Uggla, 1958). Burned land does not therefore remain fertile for long. After a few years, only one or two on the poorer land, the yield decreased steeply so that cultivation ceased to be worthwhile. The subsequent use of the area depended on the situation, requirements for grazing, for meadows or for arable land. These, in turn, depended on the stage of development of the population and the extent to which land had already been utilised.

The subsequent use of the burned areas

Fire was an excellent agent for clearing areas for new dwelling places and farms, and it was used for this purpose in Sweden up to the end of the nineteenth century. First, the area around the farm-house was burned and used for growing root crops and cereals; after some years, the nutrient reserves were almost exhausted and the productivity was too low for continued cultivation, but high enough to support grass and other pasture plants which could be cut for hay. Hay-making on these areas was possible because the period of cultivation had resulted in a more even surface, and the removal of loose boulders, half-burned tree trunks, branches and other large litter, and the hay-making itself helped to make the surface more even. The amount of hay decreased each year, because of the annual removal of nutrients in the crop, but although the hay meadows became exhausted they were not generally fertilised: the available fertiliser was reserved for the permanent arable land.

Not all the burned ground nearest the farmhouse became hay meadow. The most level areas, and the areas of the best soil, were kept cultivated and were fertilised, so becoming permanent arable land: stable manure was used on these. This represented an attempt to intensify and concentrate the cultivation of crop plants. How-ever, for many centuries the productive capacity of permanent arable land was less than that of the burned land, because of the primitive methods employed. Nevertheless, the arable land had several important advantages: it lay near the farmhouse and was easier to work, since the surface was already cleared and draught animals could be used.

Forest-covered land provided new land for burning, and the nearest and most accessible areas were naturally selected first. After a few harvests, these areas became hay meadows or arable land, if more of either was needed, or else they provided grazing. Unless they were kept cleared, these pastures were invaded by trees again, at first mainly deciduous trees and bushes. Deciduous tree seedlings tend to be grazed, particularly in spring and early summer for the young leaves, and in autumn and winter for the buds; coniferous tree seedlings and juniper bushes are usually avoided, so they tend to become more frequent if grazing pressure is severe. Sheep and goats are most effective at preventing regrowth of forest: they stop colonisation by deciduous trees completely. Spring grazing also holds back spruce growth, and spruce seed-lings may remain in a shrub stage for many years, whereas juniper bushes increase in numbers. On better soils, or with less grazing, deciduous woodland develops first, and then spruce grows up underneath. Depending on the grazing pressure, mixed, coni-ferous or, occasionally, deciduous forest might establish itself. Mixed forest tended to develop in inland areas, where there was relatively more land and less severe grazing (see Sjörs, 1954).

In other respects, development was governed by population density. As long as the population was small, there were always new areas available for burning. Up to the Iron Age, the availability of land would not have limited the practice of burning, and the burned areas were probably then used only for the first one or two years when yields were high: both Caesar and Tacitus, at about the time of the birth of Christ, wrote that the Germanic tribes made only one sowing on their arable land. Caesar (*Gallic Wars*, **4**, 1 and **6**, 22) wrote that these people made little use of arable cultivation and that their food was mostly cheese, milk and meat. They owned no particular areas of land but families were allocated land for a year by the leaders and were then moved to a different place.

The coastal regions, which were colonised first and had the densest populations, would have been the first to run out of new land. Then it would have been necessary to reuse areas which had already been burned, and this practice would have become more common as the people built permanent dwellings, and as new forest areas became more difficult of access. Burned areas which no longer gave satisfactory yields could be burned again if necessary. The litter, tussocks, raw humus, and any remains from the previous burning, were collected and burned, but it was not a very rewarding practice and was only used as a last resort: according to Heikenheimo (1919), reburning took place in Germany and Austria, as well as in Scandinavia, if lack of new land made it unavoidable. If possible, forest was allowed to grow on the previously burned areas for some years before clearing and reburning, and when it was eventually done not only the dwarf shrubs and other litter were burned, but also some of the humus that had been disturbed and mixed with litter. This last occurred in patches, and was not intentional: it happened accidentally on hot, windy days (Johnsson, 1921). This type of burning brought about relatively good fertilisation by the ash and yields increased again to a good or satisfactory level. Burning was repeated every three to five years in some places, but was much less frequent in others (twenty to thirty years according to Linnaeus 1741, 1749), the frequency depending on the density of population: in the less populated inland areas, the forest might even have been left to mature before it became necessary to burn again. During this period, salts from weathering accumulated in the soil, so the effect was the same as letting arable land lie fallow.

Heather moor

As a result of repeated burning the humus and nutrient reserves in such areas were used up and the soil eventually became unsuitable for cultivation; they were not, however, abandoned completely, but continued to be used for grazing. The vegetation was heather and other dwarf shrubs, and, where there were sufficient nutrients to support them, scattered grasses and herbs. Grazing impoverished the soil still further and sooner or later heather became dominant, so that heather moor succeeded the original forest.

Heather has some value as emergency fodder: the young shoots are eaten by animals if nothing better is available, but after a few years the shoots become unsuitably coarse and woody. Spruce and juniper invade the moor, if it can support them, and the field layer

FIG 115. A Calluna heath reserve (Mästocka) in Halland, Sweden. Part of the heath has been burned, and grasses, herbs, dwarf shrubs and tree seedlings have grown up after burning, e.g. *Deschampsia flexuosa, Avena pratensis, Luzula campestris, Festuca rubra, Potentilla erecta, Lathyrus montanus, Lotus corniculatus, Scorzonera humilis, Rumex acetosella, Vaccinium vitis-idaea, Calluna vulgaris, Betula, Salix* and *Sorbus aucuparia.*

is suppressed. In these circumstances, grazing could be improved by burning, since, if the heather is burned in early spring a crop of new shoots grows up from the undamaged underground stems (see Uggla, 1958). In this way, heather moor became permanent grazing land in the poorest communities. Heather burning to improve grazing was carried out up to the beginning of this century in much of south Sweden; the province of Halland in the south-west was practically treeless by the end of the seventeenth century (Malmström, 1939).

Heather burning has been practised in historical times in the British Isles, Germany, countries around the Baltic, and elsewhere (Wibeck, 1911). There were about 6000 km² of heather moor in Denmark around 1840 (Romell, 1952), and about the same in Holland, but this has decreased more recently because of afforestation, either by planting or by natural regeneration.

If heather had not been a member of the flora of west Europe, large areas of the land which degenerated to heather moor might well have become desert. On the poorest parts of the heather moors even the heather may not be able to form a closed canopy and bare sand may remain exposed. Heather can grow on the impoverished soil because it has especially low requirements and wide tolerance in relation to other phanerogams. It is found in dry and in wet places; and tolerates a wide range of pH. In the absence of competition it can grow on soils of any pH, although if there is competition it has a preference for acid soils (see Chapter 16). It probably requires only small amounts of salts, or it may be able to accelerate the rate of mineral weathering and fulfil its salt requirements in this way. Thus heather is able to colonise bare, impoverished areas of soil and to begin to build up new soil and new nutrient reserves; it is a particularly effective agent for soil formation because of its high rate of litter production (Table 9, page 75).

The Atlantic heath complex

The Swedish coastal heaths form part of the so-called *Atlantic heath complex*, including the west coast of Norway, Jutland (Denmark), western parts of Germany, parts of Holland, Belgium and France, and the British Isles. The primary reason for the development of heath in these areas is the length of time that man has lived there. Relatively dense populations built up early and exploitation of vegetation and soil has therefore gone on longer than in the inland areas which were colonised later. There are large numbers of Bronze Age burial mounds along the coasts in the Atlantic heath belt—evidence of the relatively large population during that early period. The good communications provided by the sea, and the mild climate, made these areas inviting for settlers. Animals could graze outside during the mild winters: Romell (1952) emphasises the importance of winter grazing for early colonisation, since barns and stables were unnecessary, at least in the warmer periods; and

FIG 116. Mosaic of *Calluna* heath (dark) and cotton grass bog (light). Mörkhult, Halland.

it was not necessary to collect winter fodder. There were, in any case, no suitable tools to do so. Grazing was therefore heavy and the vegetation and soil exploited to such an extent that the heather had to be burned to provide more grazing. Burning tended to be more severe at the coast than it was inland, as the strong winds often caused the humus to burn, particularly in high areas, and even if the burning itself was done on calm days, strong winds during the next day or two might cause the humus to burn. It is this fact that is probably one of the causes of the widespread distribution of heather along the coasts.

Another cause for burning, but operating later, was the lack of material for fencing. Sjöbeck (1933) considers that the forests had been cleared so thoroughly that there was insufficient timber to make fences to enclose arable land and meadows. This led to a lack of fodder, so that the cattle had to subsist on grazing heather even in the winter, and heather burning became essential.

The more severe inland climate limited the time available for grazing, so hay meadows and barns therefore became necessary. Hay-making required good tools and consequently could scarcely have taken place before the Iron Age. Inland, too, following exploitation came the heather moor, but it has had less time to develop there than it has had at the coasts.

Soil

As a result of the repeated burning and removal of grazed material, and of the consequent change in the vegetation, the brown earths, together with the greater part of the nutrient reserves, became exhausted. The soil processes changed in the direction of podsolisation, even in areas where this had not previously taken place. Leaching also contributed to the continued soil impoverishment.

Meadows

History

Like hunting and fishing, the oldest methods, the use of meadows was a means of obtaining animal food: land for grazing was the only requirement for the most primitive way of keeping cattle. However, in areas where winters were severe, winter fodder was essential and barns had to be built. While the population remained low, grazing land was plentiful. Winter fodder became necessary only after increase in the population led to migration of people northward and inland, or after worsening of the climate. Meadows were used, and deciduous foliage was gathered, to supply winter fodder. Use of deciduous foliage probably preceded hay-making, because hay-making required good tools. Natural meadows, such as water meadows, were probably exploited first. Late Stone Age dwelling places in South Sweden, dating from a time when stock was kept, have generally been found along coasts and lake shores, where there would have been natural meadows (Sjöbeck, 1931). Forest land was cleared to make meadows only when the natural meadows were insufficient or unsatisfactory because they were too wet. The time when the dryer meadows became common is uncertain. In Sweden they were probably widely distributed relatively early, particularly along the shores of the Baltic as far north as Uppland. Iron Age remains have most often been found in or near remnants of meadows or former meadows (Selander, 1955, page 315).

Sjöbeck (1932) considers that meadows in Sweden have been the material basis of the survival and growth of the population from the beginning of the first century AD until recent centuries. Meadows and grazing land supported the livestock on which the people depended for their main source of food.[1]

---

[1] There is evidence that culture of meadows is an ancient practice from the west European languages, which have words for meadow derived from a common root (related to Latin *metere*—to mow) e.g. German *Matte* and *Madh*, from *mähen*, English *meadow*, Swedish *mad* (for large lake or river meadows).

FIG 117. A farm surrounded by arable land which was formerly meadow and water meadow. Outside this area, which is bounded by stone walls, is enclosed pasture and forest. North Skåne, Sweden.

The origin of meadows

Dry meadows originated on land cleared by burning in the nineteenth century and earlier. It was only by burning, followed by the use of picks and rakes, that the litter, roots, surface stones and other obstructions could be removed to leave an even surface for hay-making.

The original areas burned were in the vicinity of the dwelling places, and it was here that the earliest meadows and arable land were established. As new areas were settled, up to and during the nineteenth century in Sweden, they were first burned, then houses were built. Near the houses, arable land was established in the flattest and best places, only in small areas at first. The meadows played much the most important part in agriculture (Sjöbeck, 1927). As tools developed it became possible to produce more winter fodder. This meant that there were more animals and more manure, so allowing cultivation and maintenance of larger arable areas. As a consequence, many meadows were converted to arable land.

Meadows and arable land together were enclosed, in contrast to the surrounding forest. Sjöbeck (1932) points out that Swedish laws dating from the early Middle Ages refer to fences around meadows and arable land and reports that eighteenth century maps also show these types of land within a common enclosure: Linnaeus made the same observation (1741, 1749), and this practice was maintained until the beginning of the twentieth century. The aim was to keep animals off the hay meadows before they were cut: grazing was to take place in the forests, on old burned land or clear-felled areas, or on the heath lands. In autumn, when the grass in the forest was coarse and old, the animals were moved to the meadows, where new grass had grown since the hay had been carted.

Selection of plants

The special character of the meadows developed as a result of hay-making and grazing. Late-flowering species were put at a disadvantage by hay-cutting, which was formerly carried out up until August. A strong, resistant type of hay, which was economical and occupied the animals in chewing it, was obtained by waiting until the grass had matured, preferably flowered and set fruit (Gustafsson, 1954). The important thing was to enable the animals to sur-

FIG 118. Meadow with pollarded ash trees. The fence forms the boundary between arable land and former meadow, now enclosed pasture. Småland, Sweden. (Photo ATA, M. Sjöbäck.)

vive the winter—their yield was of secondary importance since this depended on conditions in the summer rather than the winter. Hay-making also changed the flora by favouring species such as geophytes and cryptophytes which could develop adventitious shoots and grow from dormant reserve buds. Perhaps the most important effect was the continuous elimination of trees and bushes. However, the meadows were not completely without trees: those growing between stones, or along the fences, or in other places inaccessible to a scythe were left to grow into the characteristic single trees or groups of trees. Groups of trees also grew up from the piles of rubbish which collected where twigs, bushes, etc., were piled, protecting the seedlings from the scythe (Sjörs, 1954). As was realised very early (Krook, 1765), the deciduous trees in the meadows contributed to the maintenance of the meadow species because their roots penetrated to nutrients deep down in the soil and not otherwise available. The deep root systems and copious humus production of the grasses were also important in this respect (see pages 329 and 332).

Water meadows were treated in the same way as the drier meadows, except that originally they were generally treeless and were thus natural meadows. Trampling by the animals, and frost, favoured the development of tussocks, so that the proportion of tussock-building grasses and sedges increased (see page 289, Fig. 67; Linnaeus, 1749).

Deciduous foliage

Deciduous foliage was an important component of animal fodder, particularly for sheep and goats. It was probably used long before meadows were widely exploited. In a book on agriculture by Cato (died 149 BC) the use of leaves for fodder is advised. 'Give cattle leaves of alder, poplar, beech and fig; and give sheep leaves as long as they are available', he wrote. At that time, iron tools were available for collecting the leaves. However, leaves would have been used even earlier; they can be gathered without tools since the branches of some trees, e.g. beech and aspen, can be broken off as easily as they can be cut; even so, a knife for cutting branches would have been a significant improvement. According to Klindt-Jensen (1956), knives with flint blades were used for this purpose in the early Stone Age.

The trees in the meadow were the first to be used, since this cleared the land. If the foliage was particularly valuable as fodder, only part of the crown of the tree was cut, so that the lowermost branches could continue to produce new shoots for use in subsequent years; the trunks of such trees often attained great ages and large diameters, but were usually hollow. At the beginning of the century there were still many such trunks of lime, ash and even birch in south Sweden (see Sjöbeck, 1946). These pollarded deciduous trees, single or in groups, were characteristic features in the physiognomy of the meadows. Some of these groups still stand in presentday arable land which was once meadow (Fig. 122, page 486).

Trees with leaves not used as fodder, but which were useful as timber, or as sources of tannins, etc., such as oak, beech and sycamore, were left when the meadows were cleared, but coniferous trees, and bushes, were felled (see Linnaeus, 1741).

Remnants of deciduous woodland were able to persist, in spite of the grazing animals, because of the fences, and in Sweden today tend to be in the vicinity of farms and villages, where there were enclosed meadows.

Clearing twigs, etc.

The deciduous trees in the dry meadows made removal of branches and other litter from the ground necessary. These would otherwise have obstructed scything and become mixed with the hay. Early in spring, before the grass and herbs started to grow, leaves, twigs and other litter were raked into heaps and removed. This left bare soil in many places, encouraging seed germination. Trampling by animals also resulted in areas suitable for germination. Nevertheless, there were not many annual plant species in the flora of the meadows.

The disappearance of the meadows

Meadows were most widespread in Sweden in the Middle Ages and were gradually replaced by arable land. Arable land produced not only animal food, but also cereals and other vegetable food; it lay close to the farm houses and could be cultivated using draught animals; moreover, the arable land was fertilised with material derived from the meadows: 'Meadows are the mother of arable land' was a well-known saying. Continuous use of the same area could only go on if the material removed was replaced, and the available stable manure was therefore primarily required for this purpose. Leaves and other litter from the forest were used in the barns for bedding, or composted, and then put on the arable land, which was therefore dependent on the exploitation of the meadows and the forest, from which there was a continuous outgoing stream of nutrient salts and humus. The forests were impoverished by this as well as by burning; so, too, were the meadows.

The water meadows were the least rapidly affected, since their reserves of salts and humus were replaced by regular annual flooding. In many places, dams were constructed to increase the extent of the flooded areas. These were still used at the beginning of the twentieth century in northern Skåne, for example (see Selander, 1955, page 317).

FIG 119. Use of deciduous foliage for fodder. (Photo. ATA, M. Sjöbäck.)

Like the delta regions of big rivers, water meadows have been exploited for centuries with no decrease in yield and no added fertilisers. In prehistoric times, such meadows would have been a never failing source of animal food. They are less used now, not because of decreased yield, but because of presentday requirements for hay with a higher food value by more demanding types of animals, and because their use is no longer economic.

The dryer meadows had relatively high productivity at first, because large salt and humus reserves had built up when the land bore forest and had been only partly depleted by burning for a few seasons. However, removal of the hay crop involved new losses only partly compensated by weathering of the soil minerals and by nitrogen fixation. In areas where there were small populations and relatively large areas of meadow, or where settlement was relatively recent, the meadows may still retain their productivity. Sjörs (1954) gives examples of hay meadows in Dalarna, still in use in 1944, which have been used for at least a century, and probably for two or three. The type of mineral soil, the water supply and other factors would affect the duration of good productivity: fine soil would weather more rapidly; a water supply from adjacent areas might bring with it nutrient salts and oxygen. The plants themselves would have an effect, since deep-rooted species exploit the soil more efficiently than others; and species producing litter which is rapidly humified allow more rapid turnover in the soil (see Sjörs, 1954).

As the population density increased so did exploitation of the dryer meadows and their productivity went down. Complaints about this were common in eighteenth century Sweden (Sjöbeck, 1932). Only five loads of hay were obtained from areas which had previously yielded fifty to sixty (Tiburtius, 1761, cited by Lindell, 1953). Experience in the nineteenth and twentieth centuries was the same. After a meadow had been in use for several decades, it was no longer worth cutting the hay, and the flora became more oligotrophic: less demanding species replaced the more demanding; the proportion of mosses and dwarf shrubs increased, and the dwarf shrubs were avoided during scything, because they blunted the scythes, and so tended to increase still more.

From ancient times up to the nineteenth century, the cause of these changes was thought to be 'ageing' of the soil, and the mosses and dwarf shrubs were 'the soil's grey hair' (Olsson, 1955).

Because of loss of nutrient salts, particularly basic cations, the soil became more acid and podsolisation started or increased. In old hay meadows in Västergötland, Mattson *et al.* (1950) found that the uppermost layer of soil was more acid and had a lower phosphate content than the layer beneath. They suggest that leaching, hay-making and grazing over a period of centuries, with no compensating fertilisation, is the cause of this.

As far as possible, new meadows were made further away from the farm, on productive burned land, so there came to be isolated fenced areas of meadow and, subsequently, arable land in the midst of the forests.

When mineral fertilisers came into use in the latter half of the nineteenth century, the total area of arable land increased rapidly (Fig. 121), and so did crop yields, with the result that hay production was transferred to arable land. Timothy, clover, lucerne and other suitable fodder species were sown from time to time. Thus, hay production was intensified and yield increased. Production was also cheaper, since machines could be used (Table 101). After these

TABLE 101 Hay production from meadows, in Sweden 1866–1945. (From *Statiska årsbok*, 1946, pages 93–5; and 1956, pages 76–7.)

	AREA OF MEADOW USED FOR HAY (ha)	ANNUAL YIELD OF HAY FROM MEADOWS (Mg)	YIELD PER (ha) MEADOW (Mg)	AREA OF MEADOW (ha) USED FOR GRAZING AND SILAGE	AREA OF (SOWN) PASTURE USED FOR HAY (ha)	ANNUAL YIELD OF HAY FROM SOWN PASTURE (Mg)	YIELD PER HECTARE PASTURE (Mg)
1866/70		1 900 000				1 480 166	
1886/90		1 600 000	0·9			2 090 511	2·5
1911/15	631 572	810 000	1·2	691 476	1 218 019	4 062 402	3·3
1941/45	316 759	362 831	1·2	686 877	1 325 219	4 047 475	3·0
1946/50	240 000	304 310	1·27	700 000	1 318 140	4 486 640	3·4
1955	130 000			590 000	1 186 700		

changes, the remaining dry meadows rapidly disappeared, being recolonised by forest, or used for grazing or arable cultivation.

Change to forest

Since the dry meadows were created by man's activity, they could only survive if the same activity was maintained (Sjörs, 1954). If

FIG 120. **Overgrown meadow. Gotland.**

they were neglected, bushes and trees grew up and after a few decades the area reverted to forest. However, this occurred only rarely, because the meadow land was too valuable to be given up.

Change to arable land

The flattest and best meadow land was converted to arable use, a change which occurred in Sweden in the eighteenth century

(Sjöbeck, 1932) and was still going on at the beginning of the twentieth century (see Figs. 121, 123). It probably began much earlier.

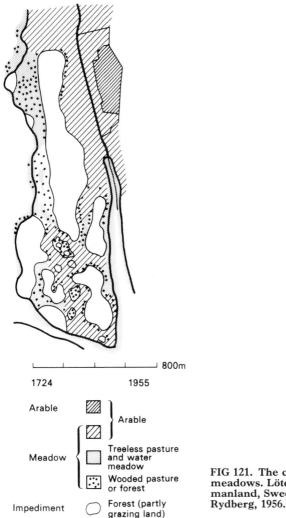

FIG 121. The change in use of meadows. Löten farm, Södermanland, Sweden. (After M. Rydberg, 1956.)

Water meadows have also been cultivated. Some of the clay plains which are now important arable land were probably formerly water meadows or forest marshes (Sjöbeck, 1932). New cultivation of such areas was still going on on a large scale during the

FIG 122. Arable land which was formerly meadow. Some of the groups of trees remain. North Skåne, Sweden.

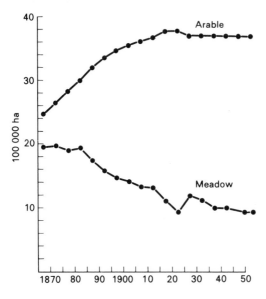

FIG 123. Areas of arable land and natural meadow (hay, grazing, green fodder) in Sweden from 1866 to 1953. (From Statiskisk Årsbok, 1946 and 1953.)

first half of the twentieth century. Drainage has often brought about a fall in water level, or complete disappearance of lakes and other bodies of water, and consequent fall in the ground water level. Over large areas a dryer microclimate has resulted. All such changes have affected the vegetation and landscape.

Change to grazing

Meadows which were too wet or too stony to cultivate, but had a good cover of grasses and herbs, have continued to be used, at least partly, for grazing. At first it was mainly water meadows which were used in this way (see 'Grazing' below).

Secondary effects of the disappearance of the meadows

One of the most characteristic and interesting-looking components of the landscape was lost as the dryer meadows disappeared: the meadows were famed for their succession of plants in flower. The numbers of many species of insect, such as humble bees and bees, have been decimated. The decrease in beekeeping in south Sweden over the last few decades (described by Cederholm, 1946) is probably a consequence of the disappearance of the meadows, and, more recently, of the use of chemicals for the control of insect pests. Between 1931 and 1940 honey production was an average of only ten kilogrammes per hive per year, whereas in Norrland it was twice as much. Disappearance of meadows occurred much later in the north than in the south, and there are still many more flowers in the northern landscapes.

The hollow, pollarded tree trunks, often centuries old, also disappeared as the meadows were ploughed, and with them the nesting places of many birds, especially owls. Bird life was depleted, and the balance between birds, insects and rodents upset.

Finally, the light character of the landscape given by the scattered deciduous trees was replaced by the darker appearance of coniferous trees colonising the open areas, even in the vicinity of the farms (Carl Fries, 1954). Selander (1955) has given a comprehensive description of meadows and their disappearance.

Grazing

The use of land for grazing cattle is probably the oldest form of agriculture, and the cheapest and most rewarding. Cato the older (died 149 BC) was the greatest agricultural expert of his time. It is said that when asked by a friend what he thought was the most reliable source of income he replied, 'Good grazing land,' and when asked what the next best was he said, 'Fairly good grazing land'. In presentday farming, grazing is the most economic way of raising some kinds of animals, especially in climates in which the animals can be outside all the year round. In more northern winters, the animals must generally be fed inside, although pigs formerly fed on acorns outside.

Before the development of suitable machinery, winter fodder was very difficult to get, while it was relatively easy to find good summer grazing. The result was that there were usually more animals than could be fed adequately in winter, when they often starved. Winter grazing was sometimes the only alternative (see Linnaeus, 1741; G. Weimarck, 1952). It was only in the summer that the animals grew and were a source of income to the owner. Like the meadows, grazing and its development has had a great effect on plant communities, but detailed examination of its effect is only possible for the last century or so, generally corresponding with the period of decline of the meadows. Some of the latter started to be used for grazing, at first perhaps alternately with hay production. If the meadows were used solely for grazing, they were fenced in, and the vegetation changed considerably. Species intolerant of grazing and trampling disappeared, while tolerant species flourished. Tolerant species include grasses, and plants with well-developed underground stems and capacity to form adventitious shoots, and also the seedlings of coniferous trees, juniper, thorny bushes, *Salix* species, and some Ranunculaceae and thistles, which are avoided by the grazing animals. Other changes are caused by trampling and frost: together with the presence of species which readily form adventitious shoots, they bring about tussock formation in wet or soft ground. The tussocks affect the distribution of species: grasses, sedges and certain tree species prefer the surface of the tussocks, whereas the ground between the tussocks is often covered by mosses. But the more resistant species may disappear if the grazing is so intense that the shoots are bitten off before they can assimilate enough to maintain the root system, and bare patches may therefore develop. Management (removal of bushes and tree seedlings), climate, soil moisture, structure and mineral content, and type of grazing animal also have their effect. On dry ground in south Sweden, birch, rowan and aspen, and various dwarf shrubs, tend to spread, while grasses and herbs disappear. Dry ground grazed by sheep and goats tends to become covered by juniper.

The same sort of development takes place in more southern latitudes. In south-western Africa, the savanna is destroyed by overgrazing. This is accentuated by the animals avoiding the bushes, which become more and more frequent, particularly species of *Acacia* (Walter, 1954).

FIG 124. Rough pasture in the mountains (Gendesheim, Norway). The pasture – separated from the forest by a fence – is a dwarf shrub heath. There are grasses, sedges and herbs in the forest.

FIG 125. Enclosed pasture with mixed forest.

In the past, grazing land was often allowed to revert to forest, the fences broke down and the animals grazed in the forest. It became too expensive to build fences, partly because of the rise in timber prices. More recently, forest grazing has ceased, because the grazing is poor and widely dispersed, because of the damage it does, and because of the time needed to find and bring in the animals every day. To provide more intensive grazing, the productivity of smaller areas has been increased, by clearing bushes, by drainage, soil preparation, fertilisation and the introduction of species such as white clover and more suitable grasses.

Arable cultivation

History

In its most primitive form, arable cultivation could have been carried out using picks or hoes, like the shifting cultivation following burning, but draught animals and special implements of the little plough and harrow type may have been used. Little ploughs (Latin *aratrum*) were used in Jutland and probably in

FIG 126. Improved enclosed pasture. The area on the right is unfertilised, that on the left has been fertilised with nitrate, sulphate and phosphate. The cattle prefer to graze the fertilised area. (Photo. A. Norrgård.)

Skåne in the first and second centuries BC (Jirlow, 1954); according to Klindt-Jensen (1956) they were already in use in the late Stone Age. The plough began to be used in the Viking period (Jirlow, 1958); and the harrow in the Middle Ages, in Sweden, but in the Iron Age in Denmark (Erixon, 1953). Arable cultivation is an intensive practice, in contrast to shifting cultivation: the same area may be in use for hundreds or even thousands of years. Progress was limited by the primitive implements and by the nature of the terrain and of the soil. To try to work wet, clayey or stony soil with wooden implements would hardly have been worth while; archaeological evidence suggests that the soils which were cultivated were light and sandy types (Klindt-Jensen, 1956). The total area of arable land in Sweden increased throughout the nineteenth and early twentieth centuries (see Sjörs, 1954; G. Weimarck, 1952; Figs. 127, 128). This increase was dependent upon the availability of fertilisers. The flattest, least sloping areas with the best soil were chosen, usually areas with a high clay and humus content. Weathering of soil particles was sufficient to maintain at least small yields provided that some stable manure, compost, forest litter or seaweed was supplied. The land was allowed to lie fallow occasionally to increase yields as a result of accumulation of salts from mineral weathering. In the Middle Ages, the soil had to lie fallow every other year (Erixon, 1953). It was only much later that the supply of fertiliser was sufficient to reduce the necessary fallow period to every third or fourth year.

The supply of stable manure was merely a movement of mobile nutrients derived from meadow and forest to arable land. There was no net gain to the soil as a whole. However, this movement was essential for the maintenance of the arable land, since only in this way could losses of nutrients in the harvested crops, and through leaching, be compensated.

Arable cultivation required a certain balance between areas of arable land and of meadow (Tiburtius, 1761, cited by Lindell and Alfons, 1953). If the farmer was tempted to increase the arable area to more than could be maintained by the meadows, the yield went down. This was sometimes interpreted as soil 'ageing', for example by the eighteenth century Swedish chemist, Wallerius. However, an eighteenth century author, Magnus Stridzberg, with the reputation of being a knowledgable farmer, (Olsson, 1955) took a more empirical view and put forward the explanation which is still accepted today. He realised that when the soil was first ploughed it was fertile because it had had a continuous supply of leaves and other litter from the forest. The fertility decreased after cultivation because the material removed was not replaced.

The yields from the arable land were low by presentday standards. Yields of rye and barley were only two to six times the weight of seed sown (G. Weimarck, 1952). In the second half of the nineteenth century the yield was less than six times the weight of seed, sometimes even less than the weight of seed. In the eighteenth century it was only in some parts of Skåne that enough grain was

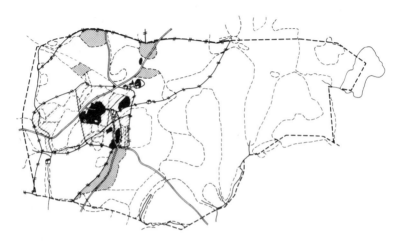

FIG 127. Arable land (black), meadow (oblique lines), burned areas (dotted) and outfields on a farm in north-east Skåne in 1696. (After G. Weimarck, 1952.)

FIG 128. The same farm in 1831. Burning was carried out at intervals of some years.

produced to be liable for taxation by the Crown. Instead, forest products, cattle, hay meadows and grazing land were taxed (Selander, 1955, page 315). However, the yield from shifting cultivation was much better, up to twenty-four times the seed weight, which is more than the average yields of ten to fifteen times the seed weight obtained in presentday agriculture.

Supplying nutrients to the arable land from the meadows and forest was a laborious, complicated and unrewarding task. As a result, the arable soil was continuously deficient in nutrients and there was only a slow increase in the area of arable land. Shifting cultivation following burning remained more popular than arable cultivation over a very long period. In some parts of Finland in the nineteenth century, the total burned area was three to six times greater than the area of arable land and meadows, corresponding to 53 to 61 per cent of the productive forest land (Heikenheimo, 1919). As late as the middle of the eighteenth century, the appearance of the Swedish landscape was scarcely affected by the area of arable land, although it already played an important part in agriculture (Sjöbeck, 1932) (Figs. 127, 128).

Both arable cultivation and shifting cultivation following burning involved exploitation of the mobile nutrient reserves accumulated in the soil and vegetation. It was only at the end of the nineteenth century that it became possible to replace fully the nutrients removed in the harvested crops, when factory production of mineral fertilisers began. The first effect of these fertilisers was to bring about an increase in area of arable land, rather than an increase in yield per unit area. Increase in arable area was also encouraged by the improvement in available implements following the industrial development of the nineteenth century. Steel spades and ploughs, dynamite etc. made it possible to cultivate and drain marshy ground and other land previously used as meadows. The area of arable land in Sweden increased by *c.* 50 per cent between 1870 and 1920 (Fig. 131, page 505). New areas were brought under cultivation with the aid of State loans to farmers for this purpose, and for improvements by drainage or breaking up rocks and boulders. Loans were also offered to people starting new smallholdings. During this period many new areas of Swedish forest were settled, and land was made available for this by the State and by private landowners. The rate of this development reached a maximum about 1920, and then decreased again. Many smallholdings were abandoned, especially those recently set up. This was connected with the decrease in rural population which began about 1880—mainly as a result of emigration to America—and has continued with movement into the towns.

As a result of the use of mineral fertilisers and other practices which improved crop yields, the productive capacity of the arable land was so much higher, despite the decrease in area under cultivation, that it sufficed for hay production and for some grazing. Hay is now mainly produced from the arable land, which is sometimes also used for grazing. Other land which is grazed is prepared by clearing and the application of mineral fertilisers. Forest grazing has become very much less important.

The introduction of mineral fertilisers started a new era in the history of agriculture, soil and vegetation. Before this, exploitation of the original nutrient reserves in southern Scandinavia and other parts of Europe had gone so far that crop plants were generally subject to severe nutrient deficiencies. Despite this, soil impoverishment continued, because of the growing human population, and huge areas of the original deciduous forest had deteriorated to coniferous forest or moorland.

Presentday arable cultivation

1 Nutrients used in the harvested crops. Salts taken up by crop plants from arable soils are carried away in the harvested crops. The quantities depend on the particular element, being significant for K, Ca, Mg, N, P and S, relatively small for Fe, Mn and B, and even smaller for the trace elements. They also depend on the type

FIG 129. Former wooded fens and mowing meadows around the lakes have been cultivated and now are the best arable areas. Västergötland, Sweden. (Photo. A. Norrgård.)

of crop. In the experiments illustrated in Table 102, 40 kg/ha K were contained in an annual crop of autumn-sown wheat and 294 kg/ha K in a crop of carrots. The variation in amounts of nutrients in the different harvested crops is considerable: K 49–274 kg/ha; N 93–268; Ca 14–207; S 9–48; Mg 10–28; P 7–27; Na 0·1–20; Fe 0·22–2·24; B 0·101–1·470; Mn 0·02–1·17; Ca 0·025–0·142.

The amounts of salts harvested also vary with the contents and proportions of the salts in the soils and thus are affected by fertiliser application. Experiments designed to illustrate this (Table 103) show that:

(a) The amount of a harvested salt increases up to a certain limit if the supply of the salt is increased.

(b) Use of one element affects use of another, e.g. use of K increases if the supply of P increases, and vice versa.

(c) The extent to which an element is used up decreases with increasing supply.

(d) Deficiency of K causes poor utilisation of P, whereas K is utilised intensively even if P is in short supply.

As a result of changes in these conditions affecting salt utilisation, the loss of salts from arable soils varies considerably. The figures in Table 102 should merely be regarded as illustrations of the orders of magnitude involved.

If the harvested crops are used in the vicinity of the place where they were grown, the major part of the salts may return to the soil in farmyard manure, but if the products are transported to more densely populated areas, the salts are permanently lost from the arable soil, since they generally pass in the sewers to the sea, where some at least will be sedimented and become unavailable to plants.

Arable land also suffers nutrient losses in other ways. Drainage and cultivation of the soil increases oxygen supply, microorganisms are stimulated and humification processes are more rapid. Thus, the humus content decreases unless there is a compensating supply of litter or farmyard manure. Organic arable soils may be used up so rapidly that the surface sinks several millimetres per year. In ploughed meadow land, the content of organic matter decreased in fifty years from 30 to 40 per cent to about 1 to 2 per cent (Franck, 1951). This process is most rapid on farms with no animals, where the surplus straw is burned. If cattle are kept, humus is generally supplied in amounts sufficient to make a significant contribution to soil maintenance. Franck gives the following example of the normal annual supply of humus:

stubble and roots: 800 kg/ha
farmyard manure and straw: 1400 kg/ha
excess straw: 800 kg/ha

TABLE 102 Nutrients removed in agricultural crops (normal yields) in kg/ha/year. (The figures refer to amounts of the elements themselves.) (After Lundblad, 1951, page 7.)

	YIELD (kg/ha)	K	Na	Ca	Mg	Fe	Mn	Zn	Cu	N	P	S	B
Autumn-sown wheat													
grain	3 000	12	1·1	2	4	0·24	0·17	0·15	0·021	65	12	5	0·125
straw	4 800	37		12	6	0·59	0·35	0·09	0·016	28	5	8	0·234
total		49	1·1	14	10	0·83	0·52	0·24	0·037	93	17	13	0·359
Red clover	6 000	80	4	92	24	1·02	0·60	0·25	0·102	126	13	9	0·220
Potatoes, tubers	20 000	110		3	2	0·21	0·02		0·014	73	4	2	
halm	17 000	119	0·1	42	12				0·011	102	3	7	
total		229	0·1	45	14	0·21	0·02		0·25	175	7	9	
Carrots, roots	50 000	87	18	20	9	0·30	0·03		0·038	115	15	8	0·248
tops	25 000	207	2	187	7	1·94			0·019	125	5		0·094
total		294	20	207	16	2·24	0·03		0·057	140	20	8	0·342
Cabbage	6 000	94	7	33	7	0·30	0·04		0·030	180	16	48	0·157
Tomatoes, fruits	33 600	120	2	7	7	0·11	0·09		0·034	104	12	11	0·011
stem+leaves	4 030	154	1	145	19	0·67	1·08		0·108	76	13	11	0·090
total		274	3	152	26	0·78	1·17		0·142	180	25	22	0·101

TABLE 103 Amounts (kg/ha) of phosphorus and potassium supplied in fertilisers and removed in harvested crop (hay from sown pasture). Mean of four years' results, peat soil, Gisselås, Småland. (After Lundblad, 1952, page 66.)

		P				K			
Series A	Supplied	0	38	74	148	266	266	266	266
	Removed	13	30	36	53	188	303	317	325
	Per cent removed (degree of utilisation)	—	81	49	36	71	114	119	122
Series B	Supplied	74	74	74	74	0	100	199	398
	Removed	21	41	59	50	25	153	230	415
	Per cent removed (degree of utilisation)	28	55	79	57	—	154	115	107

Cultivation also increases leaching (see Chapter 14 and Tables 104, 105). Since a surface covered by vegetation loses more water to the atmosphere than does bare soil, run-off and infiltration of rain increases if the vegetation is removed. Ernest (1946) refers to German experiments that showed that sandy soils, mull soils and clay soils with a cover of vegetation lost an average of 62 kg CaO per year through leaching, and that this figure increased to 213 kg if the vegetation was removed. Arable soil, which is sometimes covered by vegetation, and sometimes not, is leached more than soil with a natural vegetation cover. Cultivation and drainage contribute to this difference, since both favour leaching; so does the application of fertilisers, by increasing the mobile nutrient reserves; and so does the uptake by plants of relatively more basic cations than anions, increasing soil acidity. In a field experiment in Sweden a calcium- and clay-rich mull lost about 67 kg/ha nitrogen in a year by leaching (an average of three years' results), while 62 kg/ha nitrogen was carried away in the harvested crop (Franck, 1946). In the same experiment, there was an annual loss of 100 kg/ha potassium and 650 kg/ha Ca by leaching. Phosphate, however, is not leached.

Detailed analyses of losses by leaching as affected by fertiliser application have been made at Rothamsted, England. Table 104 shows results of a lysimeter experiment in which leaching was intensified after application of lime. The loss is less than in the Swedish experiment, probably because of differences in climate. Other experiments (Table 105), in which water in drainage ditches was analysed, showed that phosphate was leached to some extent, that nitrogen is mostly leached as nitrate, and that some organic matter is also lost.

FIG 130. One of the furthest outposts of cultivation in north Skåne, Sweden, at the end of the nineteenth century. The burned clearing in the forest was used as arable land for some decades. In 1890, the house was abandoned and since then the area has been used as pasture.

2 Replacement of the nutrient and soil reserves. (a) SOILS LEFT FALLOW OR UNDER LEY.

When arable land was exhausted to such an extent that it was no longer worth cultivating, it was left unused. After some years, the nutrient content had increased again as a result of mineral weathering, nitrogen-fixing organisms and humus formation, and soil was again productive. Such experience taught the farmer that soil must 'rest' from time to time, and the practice of allowing soil to lie fallow every few years became established. This is an ancient practice, mentioned in the Old Testament (*Leviticus*, 25, 3–5) in the Law of Moses, with every seventh year fallow. In England it was introduced in the Middle Ages, with fallow years generally every third year (Russell, 1936).

Linnaeus (1732, 1734, 1741, 1749) gives much information about the practice in Sweden, gained from his various journeys. The interval between fallow years depended on the nature of the soil, the amount of available farmyard manure, the type of farm, and the yield required. The fallow period varied between 0 and 85 per cent of the total period under cultivation. In some parts of Skåne, there was no fallow period and barley and rye were grown alternately, but before the barley was sown, a very thin layer of farmyard manure was applied, he writes. In sandy areas in Skåne,

TABLE 104 Leaching from cultivated ground. Mean of eight years' results from a lysimeter experiment in Aberdeen. (From Russell, 1936, page 176.)

	PARTS PER MILLION IN DRAINAGE WATER		
	NO FERTILISER	FERTILISER WITH NO LIME	FERTILISER WITH LIME
CaO	19·3	21·0	29·9
MgO	7·8	7·9	9·4
K_2O	3·2	2·9	3·6
Na_2O	16·6	17·0	22·2
Cl	13·7	14·9	15·3
SO_3	17·0	24·4	28·9
SiO_2	26·0	23·5	21·9
N	3·7	3·6	4·5

kg/ha/year			
	NO FERTILISER	FERTILISER WITH NO LIME	FERTILISER WITH LIME
CaO	85·2	93·0	124·9
MgO	34·4	35·7	39·5
K_2O	14·3	13·8	12·9
Na_2O	74·7	78·7	79·0
Cl	66·6	73·8	74·1
SO_3	75·2	110·8	126·7
SiO_2	119·6	115·5	105·3
N	13·9	13·7	15·8

crops were grown for two years, followed by an eight year fallow period. If this was reduced to six years or less, the soil became quite exhausted and did not even bear weeds (Linnaeus, 1749).

This shows that the farmer used a fallow period of a length determined by experience and depending on the balance between material removed in the harvested crop and new material available from weathering and bacterial activity. This balance could be changed by manuring. Dry farming in some areas of the United

States is analogous, but there the soil lies fallow in order to accumulate water.

A fallow period was particularly important before mineral fertilisers became available. However, even in areas where soil nutrient requirements are now filled by mineral fertilisers, soils

TABLE 105 Leaching from arable soil at Rothamsted, England, lysimeter experiment. (from Russell, 1936, page 176.)

	PART PER MILLION IN DRAINAGE WATER	
	NO FERTILISER	FERTILISER ADDED (MINERAL SALTS)
CaO	98·1	143·9
MgO	5·1	7·9
K_2O	1·7	4·4
Na_2O	6·0	10·7
Fe_2O_3	5·7	2·7
Cl	10·7	20·7
SO_3	24·7	73·3
P_2O_5	0·6	1·54
SiO_2	10·9	24·7
N as NH_3	0·14	0·24
N as NO_3	15·0	32·9
Organic matter	67·7	84·6

may still be allowed to lie fallow since there are other advantages to be gained. Weeds can be eliminated during the fallow period, and the soil can be prepared for early sowing, thus spreading out the seasonal farm work to some extent.

The soil can be kept entirely free of vegetation during the fallow period or it can be kept under ley, being sown with species particularly suited to humus and nitrogen accumulation, such as grasses and legumes. This is an ancient practice: Theophrastus (*c.* 300 BC), Virgil (*c.* 40 BC) and Pliny the Elder (died 79 AD) mention the advantages of allowing arable land to bear legumes before cereals were sown. In the eighteenth century in England, otherwise fallow arable land was sown with selected grasses and clover instead of allowing wild species to colonise it, as happened previously (Russell, 1936). The same was true of Sweden. Linnaeus (1749) wrote that in Skåne rye was sown one year, barley the next, and the third year was a resting period, in which half the area was sown with peas because they 'feed' the soil. In other sections of his journeys he mentions that peas and vetches 'feed' the soil, and that pea roots do this, too.

Special pasture grasses are used nowadays, species of *Phleum*, *Alopecurus*, *Festuca*, *Lolium*, *Poa*, etc., together with legumes such as species of *Trifolium*, *Medicago*, *Lotus*, *Anthyllis* and *Melilotus*. If deep-rooted species like lucerne are used, nutrients are brought up from depths not exploited by most crop plants.

(b) ORGANIC FERTILISERS. *Green manuring.* The greatest gain in humus and salts by the soil is made if the legumes are ploughed in instead of being harvested. This practice was described by Pliny, who recommended lupins for use in this way. Green manuring is common in southern latitudes where the soil is often poor in humus and nitrogen because of the rapid humification. In Sweden it is mostly practised on sandy soils. Permann (1954) gives examples from Swedish experiments showing that green manuring with various legumes added 5000 to 6000 kg/ha of humus-forming dry matter and 100 to 130 kg/ha nitrogen to the soil.

In all these practices, it is mainly the soil processes themselves—mineral weathering and production of nitrogenous compounds—which improve the structure and nutrient status. These alone are insufficient to provide optimum conditions for plant growth and to ensure that the soil does not deteriorate, and an external supply of nutrients is necessary.

Farmyard manure and litter. Farmyard manure has probably been used on arable land from the time the cultivation of such land began. Litter of some kind, straw, leaves, moss, peat etc., is usually mixed with the manure. Plant material may also be applied alone, as seaweed or compost. Nutrient salts and humus are supplied to the soil in this way. Humus, particularly its colloidal constituents, is important to soil structure and for its capacity for storage of water and salts. The application of seaweed, forest litter, or peat, represents a net gain to the arable land. In seaweed, some of the nutrients previously carried away in water from terrestrial plants are returned; peat contains nutrients which have been removed from circulation and become unavailable.

(c) INORGANIC FERTILISERS. *Moraine material, marl.* The simplest way of improving the soil by addition of inorganic materials is by mixing in unweathered, fresh morainic material of a type rich in nutrients. In the nineteenth century, and also later, this method was used to improve cultivated areas of bogs or fens. Such areas become less and less productive with time, even if farmyard manure is supplied. The soil structure deteriorates and the soil becomes much too porous and light, losing its resistance to drying-out. Addition of morainic material improves the structure and the nutrient status, and supplies a mineral fraction subject to weathering. The morainic material already contains a significant quantity of nutrients which are readily available, particularly if it is derived from easily weathered minerals and is rich in silt and clay.

Marl has a similar effect and has long been used to improve all kinds of soil. Pliny, in his *Historia Naturalis*, described several kinds of marl and their use in increasing yields of cereals and pasture grasses. It is well known to presentday farmers that marl, which is often rich in phosphorus and potassium, promotes aggregate formation in the soil, neutralises acids, and increases nutrient uptake. In this respect, marl and morainic material are very different. The morainic material acts as a supply of nutrients as it weathers: the marl makes nutrients available mainly by ion exchange, so that reserves already present in the soil in a bound form are released (see Table 104 and Chapter 15). Marl application therefore results in increased yields only when such reserves are present.

Mineral salts and nitrogen salts. The use of more or less pure salts as fertilisers is relatively recent. It is based on plant physiological experiments carried out at the end of the eighteenth and beginning of the nineteenth centuries, culminating in the discoveries by Ingenhousz (1779), Senebier (1788), de Saussure (1804) and Liebig (1840) (see Chapter 1). It was these scientists who first showed that plants were not built up from the organic matter already present in the soil, but that they themselves synthesised their organic matter chemically from the raw materials, carbon dioxide, water, and certain salts, and that plant development is therefore dependent on the supply of salts.

As early as the sixteenth century, it was well known that some inorganic materials, such as ash and saltpetre, had a favourable effect on arable soil, but the effect was then thought to be on the structure of the soil. The idea that they contained nutrients conflicted with the generally accepted opinion that plants lived on the mull of the soil, and that the materials from which they were made —including the ash (salts)—were formed by the life force (see Chapter 1).

The discovery of the significance of salts carried with it the conclusion that growth could be promoted by the addition of salts to arable soil. Justin von Leibig recommended a mixture of salts for this purpose, and so originated the practice of using mineral fertilisers (see Chapter 1). This practice, more than any other, has brought about the improvement in agricultural production, and enabled agricultural areas to support the large urban population.

At the time when mineral fertilisation began, arable land and meadows had exceptionally large requirements, the soil was usually exhausted and needed to be left fallow for a while, but it was many years before factory-produced fertilisers came on the market in large quantities and even longer before they were widely used. The prices were at first too high for the farmers and cheaper materials such as ash, bonemeal, guano and Chile saltpetre were used. The first factory-produced salt on the market was superphosphate, prepared in a way suggested by Liebig. The raw material was bone or guano at first; nowadays mineral phosphorite is used. Another important salt was ammonium sulphate, made from gas liquor. German potassium salts began to be produced in 1870—the first were made from the minerals karnalite and kainite, from the Stassfurt region; thomas phosphate, a slag product from iron smelting, came somewhat later; and at the beginning of the twentieth century the first artificial nitrogen products were synthesised.

For the first few decades after the introduction of mineral fertilisers, only salts of calcium, potassium, phosphorus and nitrogen were used. After the plant requirements for other elements became known, salts containing boron, manganese and copper were also applied.

Fertilisation with minerals is thus a relatively new practice, first used on a large scale at the end of the nineteenth century, mainly in western Europe. In the twentieth century it has spread to other regions where intensive farming is necessary to support the population. Its use in western Europe has also increased (Tables 106, 107, 108, 109; Fig. 131). New salts and combinations of salts have

TABLE 106   Use of commercial fertiliser in Sweden (in '000 ton). (From *FAO Yearbook*, vols. 9 and 23.)

	N	P_2O_5	K_2O
1938	30	53	55
1948/49–1952/53	61	93	56
1952/3–56/57	80	108	83
1964/65	149	122	103
1965/66	161	122	106
1966/67	644	122	107
1967/68	181	131	119
1968/69	191	150	126

been introduced, the particular requirements of various crops have been tested, and the mobile soil nutrient reserves have been analysed in order to assess the supplementary nutrients required and to determine the economic return resulting from fertiliser application.

It is difficult to make a quantitative assessment of the importance of mineral fertilisers to presentday crop production. The high yields are partly due to other circumstances than fertiliser application, such as intensive soil cultivation, weed control, favourable crop rotation schemes, choice of variety and species, and plant breeding. It is impossible to separate the effects of each of these in increasing yield, since they are all interdependent (see page 27).

Even if it were possible to estimate the contribution of mineral

TABLE 107 Annual use of commercial fertilisers in Scandinavia 1949/50 in kg/ha. (After Franck, 1951, page 228.)

	N	P_2O_5	K_2O
Norway	36·5	36·7	49·8
Denmark	19·0	25·0	26·6
Sweden	16·0	26·5	14·0
Finland	5·4	22·5	10·1

TABLE 108 Use of commercial fertiliser in UK (in '000 ton) (*FAO Yearbook*, vols. 9 and 23).

	N	P_2O_5	K_2O
1938	60	170	75
1948/49–1952/53	209	299	215
1952/53–1956/57	267	329	280
1964/65	596	435	434
1965/66	690	420	436
1966/67	760	403	456
1967/68	909	430	500
1968/69	925	436	485

TABLE 109 Use of commercial fertiliser in USA (in '000 ton) (*FAO Yearbook*, vols. 9 and 23).

	N	P_2O_5	K_2O
1938	346	675	352
1948/49–1952/53	1173	1964	1243
1952/53–1956/57	1787	2154	1680
1964/65	4208	3175	2571
1965/66	4832	3522	2922
1966/67	5468	3905	3304
1967/68	6158	4040	3441
1968/69	6199	4169	3507

fertilisers to the crop yields achieved, this would still not be an estimate of their significance to agricultural production of food for man. Nutrient salts are as essential to plants as light, water and carbon dioxide; they are irreplaceable. Before such salts were produced industrially, crop plant production was dependent on use of the mobile nutrient reserves built up by plants and soil during

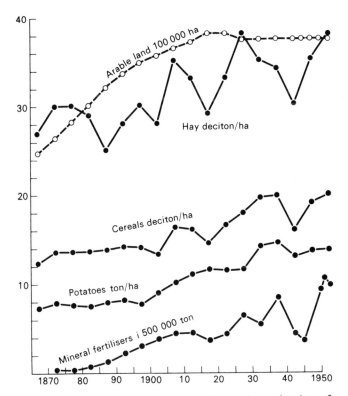

FIG 131. The increase in area of arable land; production of various crops; and imported fertiliser in Sweden, 1870–1950. Note the fall in all these during the two world wars. (From Statistisk Årsbok, 1946 and 1954).

previous centuries. These were used up as a result of cultivation, and the yields of arable land progressively decreased. The practice of shifting cultivation following burning, which opened up the reserves of nutrients in the forests, was the most rewarding. Arable land was developed only in relation to the availability of nutrients from meadows and forests, which became impoverished themselves as a result.

After the development of mineral fertilisers it became possible to prevent or slow down soil impoverishment; nutrient losses could be replaced with fresh reserves derived from mines or from the atmosphere, and the transport of material from meadows and forests was no longer necessary. The productivity of arable land increased so much that it replaced the meadows and forests as the major food producer and shifting cultivation following burning became obsolete with the result that leaching, loss of ash, and burning of humus largely ceased and the humus reserves began to build up again. Consequently, in countries where mineral fertilisers are available and are used in sufficient quantities, it should now be possible to achieve a balance between supply and use of nutrients.

The most important effects of the fertilisers, at least at first, were to double or more than double the yields per hectare (depending on how impoverished the soil was before fertiliser application, see page 492), and to increase the area which could be cultivated (see page 494). This increase in agricultural capacity was very important to human population increase and maintenance; only with it could an increase in urban population be supported, and allow development of industry and commerce.

Without mineral fertilisers, the former agricultural practices would have continued. Population increase would have caused a greater demand for food, food prices would have increased in relation to other products, smaller yields would have become economic, and soil exploitation more severe. Under these conditions, cultivated areas would unavoidably have been degraded to heaths. Production could not have been sufficient to maintain presentday populations and towns could not have grown so large. Hence, the development of industry and commerce would have been considerably restricted.

3 Irrigation, tools, improvement of varieties of crop plants. Control of weeds, parasites and diseases. Agricultural development in the last hundred years or so, particularly in western Europe, has been favoured by many practices other than the application of mineral fertilisers. These include irrigation, development of improved tools and implements, and breeding for improved plant characteristics.

Arable production in semiarid and arid regions is limited just as much by deficiency of water as of mineral nutrients: to supply water is therefore at least as economic as to supply fertilisers. Irrigation systems have been used since ancient times in places with suitable topography and river systems.

Effective, labour-saving tools and implements were the result of technical development: manpower was replaced by machines; soil cultivation became more intensive; drainage and transportation became easier.

Varietal yields have increased through genetic selection by plant breeders; and so has resistance to extremes of climate and to pests and diseases. Disease and weed control have intensified as more knowledge has been gained of the nature and ecological requirements of the various species, and as effective chemical means of control have been discovered. (See page 538 for discussion of their importance.)

Forestry

Forestry in ancient times

The peat record shows that the Scandinavian countries were covered to a large extent by deciduous forests at the time when man moved into the area, in the early Stone Age: these were probably forests with numerous old and full-grown trees such as oaks. Trees are now rarely left undisturbed long enough to attain a very great age. The approximate maximum ages for European deciduous tree species are given by Möller, 1957, e.g. *Betula* 100–150 years, *Carpinus betulus* 150–250, *Ulmus* 500–600, *Acer pseudoplatanus* 500–600, *Fagus sylvatica* 600–900, *Tilia* 800–1000, and *Quercus robur* 1500–2000 years.

There were also virgin forests with large trees in other parts of Europe in the early Stone Age. Tacitus (98 BC) wrote that Germany was filled with wild forests and terrifying swamps. According to Theophrastus (third century BC), in the plains of Latium (western Italy), there were beeches so large that a single trunk was enough for the keel of a great Tyrrhenian ship (cited by Hyams, 1952). Caesar (*Gallic Wars*, 6) wrote that the breadth of the forests which extended eastwards from the area of the presentday Black Forest was equivalent to nine days march for lightly-loaded men; no man in that part of Germany could say that he had come to the end of the forest, even after sixty days march, nor that he had heard where the forest began.

Forest exploitation began much earlier in southern Europe than in Scandinavia. In the deciduous forests in south Sweden, it was still possible to find oaks, beeches and limes with trunks one to two metres in diameter in the nineteenth century, in such large numbers that they made up the majority of the trees in small stands. The few remaining examples, now mostly protected by law, are all that remains of the original forest. For the first human occupants, living by hunting and fishing, these rich deciduous forests must have offered a good chance for survival. When animal husbandry began, the forests were a hindrance, and were cleared to make space where the animals could be kept. The natural meadows along the seashores and river banks provided extensive but not particularly good grazing: grazing animals tend to prefer

FIG 132. An old oak tree. Circumference at breast height *c*. 9·5 m. Uppland, Sweden.

dryer pastures, which are more nutritious and give more variation, and forest clearance probably started very early to provide it. As the population became settled, land was also cleared for building.

As settlement went on, and handcrafts developed, men learned to use the products of the forests in various ways: from being hunting grounds, to being merely a hindrance, the forests became an important source of food and raw materials. Because of the large number of tree species, there were many kinds of products, useful both directly and indirectly.

Grazing of beech nuts. Beech nuts were an important food for pigs; they are large and palatable and have a high oil content, and they are produced in large quantities by some trees, quickly found by the wandering pigs. The bare soil that the pigs exposed provided better germination conditions for the beech nuts than the undisturbed thick layers of beech litter, so there was some advantage to the plant community as well as to the grazing pigs. This type of grazing was of particular importance because it could continue through the winter and no indoor winter fodder was necessary. Pig-keeping up to the nineteenth century in Swedish beechwood areas was mainly dependent on this beechnut grazing, which is often mentioned in old records; Linnaeus (1749) described how the pigs fed on beechnuts became fat and lazy, especially in cold weather. South Swedish beechwoods were used in this way until the middle of the nineteenth century, and in a few places the system is still operated today. As long as there was only a small area of low-yielding arable land, it had to be used for cereals needed to make bread. Animal production had to rely largely on beechnuts, which were thus equivalent to presentday crops of oats. In some places, beechnuts were collected and sold for pig food (Linnaeus, 1741). The beechnut grazing was taxed, and thus provided a source of revenue to the State (Wibeck, 1909; Malmström, 1939). The same thing happened in the countries to the south: Wibeck (1919) records that there were up to 30 000 pigs in some Danish forests, generally two pigs to the acre.

Edible products were collected from other trees and bushes, such as hazel, oak and crab apple, and whitebeam. Whitebeam berries were used in brewing or eaten after they had been frozen or cooked (Linnaeus, 1741). Flour was made from acorns, eaten by man up to the Middle Ages, but later used only as animal fodder (M. Fries, 1958). Trees with edible products were protected by law (Wibeck, 1909). In the late Stone Age site at Dagsmosse in Sweden, large quantities of the fruits of *Thelycrania sanguinea* were found, which had possibly been collected for oil extraction (oil content is 35 to 45 per cent) (Berggren, 1956).

Timber. Before industrial development, furniture and various household equipment was always made by village craftsmen or by the user himself, and wood was the major raw material. If the forests were rich in species, it was easy to select suitable raw material for particular purposes. Beech was perhaps the most widely used and the beech forests were severely exploited for timber. The trunks of beeches can easily be split into planks with an axe; this was particularly important before the development of saws—the first records of water-driven saws in Sweden date from the middle of the fifteenth century (Näslund, 1937).

Some of the most important uses of the various kinds of timber available, as recorded by Linnaeus (1750 and earlier) and Sjöbeck (1927), were as follows.

Beech. Barn floors, wheels, furniture, wooden plates and dishes, weaving and spinning equipment, butter churns, barrels, etc.; keels of ships in England and southern Europe.

Hornbeam. Cogs, axles and other wooden components subjected to severe mechanical stress, since the wood is hard, tough and resistant to bending.

Oak. Shipbuilding, houses, furniture, wheels and other parts of carts, bark for tanning.

Elm. Furniture, keels of ships (resistant to rot), mill wheels, axes, presses and mangles, since the wood is hard and tough.

Ash. Axles, handles of axes, etc., planks, since the wood will not break easily.

Lime. Scythe handles, wood for carving, ropes.

Birch. Wood for carving, roof tiles, birch brooms, whisks.

Alder. Wooden shoes, mustard pots and other articles which are hollowed out, since the wood is easily cut in all planes.

Aspen. Milk containers, bottles, troughs, etc.; the wood is white in colour; meal for fodder from the bark.

Poplar (*Populus nigra*). Rifle butts, since the wood is tough, hard to split, and light in weight; fishing floats from the bark.

Crab apple. Cogs for mill wheels, axles (hard, tough wood).

Hazel. Pins for roof thatching with straw, baskets, sieves, barrel hoops, hay rake pins.

Spindle (*Euonymus europaeus*). Hay rake pins, pegs.

Hawthorn. Cogs in mill wheels, wooden nails, blocks.

Honeysuckle (*Lonicera*). Teeth in combs for weaving.

Alder buckthorn (*Frangula alnus*). Charcoal for making gunpowder.

In comparison, coniferous species had more limited use as timber. Timber of pine and spruce was used for fencing, as it can relatively easily be split with an axe into long planks. Tar was made from pine, the heart wood of which is rich in resins especially in old, slow-growing trees, such as those in bogs. Wood tar was the most important wood preservative. Juniper was used for gate-latches, baskets and harness connections (the wood is tough and can be twisted), and for jugs and mugs, which retained the aromatic scent of the wood for many years. Linnaeus, 1750, wrote that these were pleasanter to drink out of than silver vessels.

FUEL. Wood has always been used for fuel and the requirement increased with the development of smelting of ores, saltpetre production (see page 149) etc. For example, 20 m^3 of wood was burned to produce a barrel of salt by boiling sea water. In the seventeenth century, the use of wood in Sweden increased enormously, mainly because of iron ore smelting. Wood alone, preferable beech or birch, was used at first. Charcoal was made from pine and spruce; charcoal for gunpowder from deciduous trees such as lime, alder and birch; birch and beech provided tinder.

POTASH. Potash (potassium carbonate) was necessary for making gunpowder and many other products. It was an important raw material in the newly developing chemical industry. Potash was produced in Sweden as early as the thirteenth century, and up to the end of the nineteenth century (Wibeck, 1909; Bodman, 1950, cited by Malmström, 1951). It was mostly made from beech, because of the plentiful supply and the high energy content of the wood, which resulted in ash hot enough for complete combustion, leaving a solid fused mass after cooling. The potassium content of the wood is no higher than that of other trees: the K_2O content of ash of the wood of beech is 13, ash 15, oak 12 and aspen 10 per cent (Henry, 1908, cited by Lutz and Chandler, 1946).

BARK AS FOOD. In times of need, bread was made from substitute flour made from various materials, including bark. The cambial zone, with the adjacent phloem, was scraped off, dried and ground, preferably in the spring when the sap was rising; bark of pine, and also of aspen, was used in this way (Linnaeus, 1737). After removal of the bark, which was dried, broken up and fed to pigs and cattle, a gelatinous layer of the young, newly formed xylem, together with some remaining cambial cells, was left on the tree. This was scraped off with a fine brass thread and used to make flour. This was nutritious and tasty according to Linnaeus (1750), but the tree was often killed. The bark was stripped from the lower parts of the growing trees; it was the most valuable product of the coniferous forest.

FOREST PRODUCTS USED FOR TRADE AND EXPORTS. The products of the forests nearest the sea were shipped to other parts of the country or to foreign lands. Timber export from south Sweden (Halland and Bohuslän) began at the end of the Middle Ages (Sjöbeck, 1933) and was so extensive that the land was rapidly denuded of forest. There were other exports, too: tar (from pine), potash, fibres, and nuts. The demand for lime tree fibres was so great that the fibres were taken all the year round, in south Sweden (Sjöbeck, 1927, 1936, 1950). Wood tar was the most important lubricant right up to the early part of the nineteenth century; until the end of the eighteenth century tar and pitch were the third most important Swedish exports (Selander, 1955).

THE DISAPPEARANCE OF THE FORESTS. Because of their many useful products, the forests were severely exploited from the times of early habitation. The deciduous forests, particularly of oak and beech, gradually disappeared as a result of this exploitation or because they were burned to allow shifting cultivation. The possibility of regeneration diminished at the same time, because of the grazing animals, and because the soil had become so impoverished that it could no longer support the more demanding species. There now remain only fragments in places which could not be cultivated and which were unsuitable for grazing or burning—on steep slopes

FIG 133. Grazed coastal heath with juniper bushes and fescue. Skåne, Sweden.

and screes and in ravines (Selander, 1955; see Chapter 11, page 236). There are some remnants very close to farms where there was protection from the grazing animals and sometimes a supply of nutrients derived from land further away. When the in-fields were cleared, some of the best deciduous trees were left. Because of the division into in-fields and out-fields, the farms and villages were often surrounded by deciduous trees, and fragments of the original open deciduous woodland now remain in their vicinity.

Despite the farmers' attempt to husband the resources of the forest, and protective legislation which dates back to the fifteenth century in Sweden (Wibeck, 1909), the deciduous forests continued to diminish. In two districts in Småland, studied by Wibeck (1909), there were large beech forests at the end of the seventeenth century, but only 5 per cent remained in 1900. This decimation was generally most severe in the mid nineteenth century, at about the time when mineral fertilisers were developed. By that time, most of south Sweden was treeless. The heathlands there, and on the islands of Gotland and Öland, reached their maximum extent. In some areas there were so few trees that heather (Malmström, 1939, 1951; Sjöbeck, 1950), or even stable manure (Linnaeus, 1741, 1749; Walfridsson, 1956), had to be used as fuel.

TABLE 110 *Calluna* heath in Halland during the past 100 years. (After Malmström, 1952, page 330.)

YEAR	FOREST (INCLUDING ENCLOSED PASTURE) (ha)	*Calluna* HEATH
1850	max. 99 500	150 400
1928	155 350	57 750
1946	196 400	13 500

Elias Fries (1852) described Västergötland and Halland as having stony deserts on the plains which once were fertile, bearing verdant beech forests.

Presentday forestry

Forests are still the source of timber, fuel, and raw materials for chemical and technical industrial processes. Wood as a raw material has to some extent been replaced by metal, but on the other hand its use has increased in the production of pulp and some chemicals. Man is just as dependent on forests as he has always been.

The area of forests in Sweden has increased again in recent decades, either by natural regeneration or by planting (Fig. 134). However, the increase is of coniferous forest, not deciduous: presentday deciduous forests are usually of less-demanding species such as birch and aspen, since forest soils are now too poor to support more demanding trees, which can no longer compete with spruce, and which were anyway prevented from establishing themselves by the grazing which still went on in the forests in the early twentieth century. The fences separating in-fields and out-fields in many places still correspond with the boundary between deciduous and coniferous woodland (Fig. 135).

Like agriculture, the starting-point for presentday forest practice was determined by the history of the vegetation and the soil and by man's interference. Soil impoverishment by burning, removal of timber and litter, increased leaching resulting from clear

FIG 134. Spruce planted on *Calluna* heath. Halland, Sweden.

felling and the increasing oligotrophy of the flora have together brought about the deficiency of mineral nutrients, which now generally limits forest productivity. Increase of nutrient supply and improvement of the condition of the soil is an important aim of modern forestry.

Forest trees use relatively small amounts of mineral nutrients compared with agricultural crops (Table 111). A large proportion of the nutrients taken up returns to the soil in the leaf litter. However, a significant amount of nutrients has accumulated over the

TABLE 111 Use of mineral nutrients by forest trees (kg/ha/year). (According to Ebermayer, 1876, page 116; Aaltonen, 1948, page 269.)

	K_2O	CaO	P_2O	MgO
Tree species				
Spruce (wood and leaves)	8·9	70·1	7·9	9·0
Pine (wood and leaves)	7·4	28·9	4·8	6·5
Beech (wood and leaves)	14·5	96·3	13·3	16·1
Birch (wood)	2·5	4·0	1·4	1·8
Oak (wood)	9·4	31·9	6·3	5·9

FIG 135. A stone wall boundary between ungrazed, overgrown meadow and outfields which were burned during the nineteenth century, next used for grazing and now reverted to spruce forest. North Skåne, Sweden.

years in the trunks removed from the area after felling, so there is some permanent depletion of the mobile nutrient reserves. This effect of felling varies according to the distribution of the nutrients. If the soil is deep, with a large content of humus or of fine, easily weathered material, felling may have little effect on productivity, but if the greater part of the mobile nutrients is in the vegetation, then the soil may be left with insufficient resources to support growth of a new forest (see page 443). Malmström's (1947) experimental results suggest that the nutrient distribution in some pine heaths in north Sweden may be of this latter type. His results showed that seedling growth was primarily dependent on nutrient supply, which, in turn, was mostly dependent on litter fall. Development was therefore limited by the rate at which nutrient reserves in the vegetation became available to the seedlings. If thinning was too early, gaps arose which were not readily filled with new trees. However, thinning of a mature stand resulted in increased litter, of branches, leaves, roots, mycorrhiza, etc., and increased growth of the remaining vegetation. The reserves of mobile nutrients were so small that they sufficed only for a single generation of trees; regeneration of the stand could take place

FIG 136. After cultivation of lupins in a stand of pine, the width of the annual rings increased considerably. (After Wittich, 1957.)

only if the nutrient reserves in the older trees became available to the developing seedlings.

This experiment was concerned with the minimum requirements of the trees, and there is very little information about the optimum nutrient requirements of forest trees, either quantitatively or qualitatively, but this is being obtained by current research (Ingestad, 1970, 1971). Malmström (1935, 1943) describes experiments showing that supplying wood ash to areas of drained peat in north Sweden had a large effect on growth of tree seedlings and other plants. Mosses died out, colonisation by grasses and herbs was favoured and bushes and trees grew faster (Fig. 48, page 216). Romell (1950) set up experiments to analyse this effect of wood ash and showed that complete fertiliser application of Ca, Mg, K, S, P, Cu, B, Mn and Zn had a similar effect to wood ash application; that a mixture containing Ca, Mg, K, S and P was less effective; and that this mixture without P was not effective. There was no reaction to N alone by the plants on the peat, unlike forest trees on mineral soil (Malmström, 1949, 1952).

Such experiments demonstrate that forests in some areas require the application of mineral fertilisers: forest growth is often limited by lack of soil nutrients. However, no economic method of fertilising forests has yet been worked out, nor is it always known which nutrients are lacking, and, as a result, forest land is still exploited without replacement of the mineral nutrients utilised from the soil. Compared with those of central Europe and the United States, soils in northern Europe are relatively young (C. O. Tamm, 1954), so that nutrient deficiencies are less acute.

The nutrients used by the herbs and shrubs in the forest affect the nutrient economy of the forest trees. Mosses, which are a major constituent of the field layer, particularly in coniferous forests, use mineral nutrients to almost the same extent as the trees (Table 112).

TABLE 112 Use of mineral nutrients by mosses (*Hylocomium* sp.) kg/ha/year. (According to C. O. Tamm, 1953, page 129.)

	K	K$_2$O	Ca	P	P$_2$O$_5$	N
Vestlandet (Norway)	4	4·8	of same order	1·1	2·5	10
Roslagen (Sweden)	6	7·2	as K	1·5	3·4	*c.* 10

This has important consequences for peat accumulation in coniferous forests and for breakdown of the peat after clear felling, or removal by burning, because a significant fraction of the total mobile nutrient reserves is involved. The more raw humus and peat, the more nutrients which have become unavailable (C. O. Tamm, 1947).

There have in recent times been attempts to encourage forest regeneration in places where there is a thick raw humus layer by burning. The top of the raw humus is burnt away, leaving a layer of ash to supply nutrients to the seedlings, and facilitating root penetration to mineral soil. There are many places in north Sweden where otherwise unavailable reserves of nutrients can be mobilised in this way.

The amounts of mineral nutrients released by burning can be estimated to some extent from results of Mork's (1946) experiments (see page 470). The annual nutrient consumption by trees is small compared to the amounts of nutrients in the ash, so the nutrients should have an effect over a long period (Table 111). However, some of the nutrients are lost in the wind and by leaching, so the remaining nutrients, together with those released by weathering, may or may not be sufficient to maintain satisfactory growth of the trees to maturity. The risk of developing nutrient deficiency is, of course, greatest in places which were deficient before felling and burning (C. O. Tamm, 1947).

Utilisation of natural reserves

Before the effect of interference by man on ecological equilibria became significant, the natural reserves were building up continuously; nitrogen, humus and soil, and the vegetation cover, all increased, and there was a general shift in the direction of eutrophy. Productivity of plants increased until there was an equilibrium between building-up and breaking-down processes. Recent exploitation by man has used up not only the annual increase in the reserves, but is leading to the depletion of the total.

New ground is used for building, for power lines, roads, etc., every year. In 1956, the urban areas of Sweden covered 6 to 7 per cent of the land surface, and this figure is higher in more densely populated countries. In addition, significant amounts of land are used in quarries, brickmaking, etc.

The effect of man is to accelerate the processes of breakdown which occur naturally but which are balanced by the building-up processes in situations where there is no interference.

Deciduous forests, with a nutrient economy leading to increased reserves of mineral nutrients and mull, have to a large extent been replaced by coniferous forests or heaths with large quantities of dwarf shrubs or mosses, whose acid products accelerate podsolisation. In tropical and subtropical regions, the change has been towards steppe and semidesert.

Humus—important for retention of water and nutrients—has been destroyed to a large extent by fire and by cultivation: working and mixing the soil brings about more rapid decomposition of the humus.

Wood, peat and litter have been used for domestic fuel since prehistoric times, and more recently for industrial purposes. The first methods of production of iron and saltpetre required large quantities of wood and charcoal; potash was also obtained from the forest by burning. Living vegetation has also been burned on occasion.

Nitrogen, fixed from the atmosphere by plants themselves, and often a factor limiting the growth of the vegetation, has been used up by man in many different ways, perhaps most by fire. The huge nitrogen reserves in the virgin forests, accumulated over thousands of years, were mostly used up as the forests were cleared and the land became covered by heaths or other oligotrophic communities.

Mineral nutrients were carried away from arable land in the harvested crops, often to towns and other densely populated areas. There have been other losses in ash blown from burned areas by the wind, and by leaching.

Interference by man has often led to erosion by wind and

water: removal of the vegetation, which acts as a buffer against erosion, is one of the causes of this movement of soil. Erosion may take place in a natural, closed plant community, but it is prevented from spreading by the network of roots and rhizomes permeating the soil. However, cultivation of crops involves removing the natural vegetation and replacing it by a different plant cover. This destroys the soil structure, making it necessary to plough and harrow the soil to maintain the soil porosity. Fine soil is susceptible to the action of wind and water, the particles are loose and may be saturated; there is no root network to retain them. The loose particles may form an impermeable surface layer, increasing run-off and erosion. The erosion channels are deepest in large areas of arable soil, particularly on sloping ground. In the United States, special precautions must be taken if land with a slope of more than 5 per cent (1 in 20) is to be cultivated (Vogt, 1948). Wind erosion is increased by cultivation, particularly harrowing. Cultivation following burning brings about loss of ash in the wind.

Overgrazing by cattle may destroy the vegetation, so that bare soil is exposed, while plants which are grazed off too severely cannot maintain their root systems, so the soil becomes loose, and trampling by the animals also damages the vegetation.

Forest clearance has often been the primary cause of erosion, since the forest retains the precipitation and run-off is minimal. In a closed spruce forest in south Sweden, up to half the annual precipitation of *c*. 800 mm was retained by the tree canopy; about 10 per cent of the rest remained in the surface vegetation and the remainder penetrated slowly into the soil (see Chapter 3). Felling the trees would result in all the precipitation reaching the ground, and run-off would increase proportionately, while the loss of the windbreak effect of the trees might also lead to wind erosion.

The demands made on the plant world

Human population and food requirements

The search for food by man has always met with variable success. The luckier section of the population have had sufficient: others, usually the majority, have suffered occasional food shortage, either because food was difficult to obtain, or because the population increased more rapidly than the available resources. Success in hunting and fishing varies, as does that in agriculture or rearing cattle. Long periods of drought or calamities such as war, which prevents sufficient attention being given to farming, have often led to lack of food. As long as means of communication and transport were slow, lack of food in one region could not be compensated by excess in another.

Sweden often suffered from bad years for crops and consequent famine. There were fourteen famine years in Dalsland in the seventeenth century, ten in the eighteenth century and seven in the first three decades of the nineteenth century (Anders Lignell, 1851–1852, cited by Sjöbeck, 1953). In these years, the population had to rely on bark bread and other substitutes: in the poorer villages bark, chaff, etc. were a normal constituent of the bread. Despite this, there were many months when no bread was available and many people died of starvation (Linnaeus, 1734, 1741). The increase in population was reversed by starvation and disease about every fifth to tenth year in Sweden between 1750 and 1850 (Thompson, 1950). Undernourishment and consequent disease was still common in the last half of the nineteenth century, and the same was true of other European countries. In England, from 1200 to 1600, there were famine years about every fifteenth year (Vogt, 1948).

Food was in no less short supply outside Europe. From AD 108 to 1911, China had an average of about one period of famine each year. In Bengal, one-third of the population starved to death in a single famine (Vogt, 1948). There is still insufficient food in these countries: good and bad harvests alternate, reflected by the increased birth and death rates, respectively (Thompson, 1950). In India, in the 1930s, 43 per cent of children died before they reached the age of ten and 33 per cent died in their first year. The intake of calories in India is less than in most other countries. In 1951 to 1968, the average daily intake was about 1700 to 2000 calories per person (*FAO Yearbook*, 1969).

In 1968 the daily calorie consumption per person was 2800 or more in Europe, Canada, USA, Australia, New Zealand, Israel, Turkey, UAR, Argentina and Uruguay; 2800 to 2200 in the rest of South America, Japan and Africa; and less than 2200 in China, India, South-East Asia, Indonesia and the Philippines (*FAO Yearbook*, 1969). A large part of the world population comes into the third category (Tables 113a and b).

World population in the past has been regulated mainly by rate of production of food. Its increase has been a function of the equilibrium between birth rate and food production. Other circumstances, such as war, plagues, changes in standard of living, changes in mean length of life, have generally had effects which have been only temporary and local. The two World Wars in the twentieth century, for example, despite the tremendous loss of life involved, hardly made a kink in the rising curve of world population (Fig. 137). Birth rate is less subject to change than rate of food production, so it is primarily food production which has a determining effect on population growth. The comparatively small proportion of presentday world population enjoying a high standard of living, and practising birth control, is, of course, an exception.

TABLE 113a Calories consumed by the world population (excluding USSR, China and some other Asian countries). (From *FAO Yearbook*, Vol. 9.)

CALORIES PER DAY	NO. OF PERSONS (millions)		PER CENT	
	BEFORE 1939	1953–54	BEFORE 1939	1953–54
> 2800	367	636	23	43
2800 to 2200	631	318	38	21
< 2200	617	545	39	36
Total	1615	1499	100	100

TABLE 113b Calories consumed per head. (*FAO Production Yearbook*, Vol. 23.)

	1948–50	1964–66	1968–69
Sweden	3110		2880
UK	3130		3180
Canada	3110		3180
USA	3200		3240
Chile	2420		2720
Peru			2300
Turkey	2510		2860
UAR	2360		2960
China (mainland)			2050
Taiwan	1980		2510
India	1700		1900
Japan			2460
Pakistan	2020		2230
Tanzania		2140	
Zambia		2290	
Australia	3240		3110
New Zealand	3360		3290

When food production failed to keep pace with the demands made by a high birth rate, some areas became overpopulated, and this generally led to settlement in new areas. When previously un-exploited areas had been settled and still more space was required, this was generally obtained by further migration, sometimes by desire, sometimes forced by hunger. Caesar (*Gallic Wars*, 6) describes how the Gauls, because of overpopulation and shortage of arable land, overcame the German tribes and took over the most fertile tracts in Germany, to the east of the presentday Black Forest.

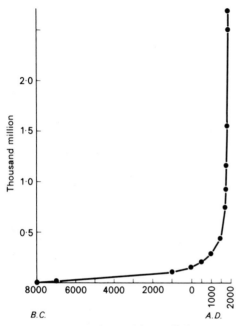

FIG 137. Growth in world population.
(After Bethke, 1957.)

Under the primitive conditions which first prevailed, a farmer could maintain only his own family. Subsequently, agricultural capacity increased gradually so that people not involved in farming could be supported. It has been calculated that in northern Europe it was not until the beginning of the eighteenth century that four families could produce enough food for five (Thompson, 1950). Production was extensive rather than intensive, and improvements in methods were rare. World population therefore increased only slowly, and the increase was a result of exploitation of new areas rather than of improvement of methods of production. There was a radical change in rate of population growth in the seventeenth century (Thompson, 1950). At the beginning of the seventeenth century, the world population was *c.* 400 million; in 1830 it was 800 million; in 1900, 1600 million; in 1937, 1950 million (excluding USSR). The figure has increased five times in the last three centuries, and doubled in the last one hundred years. At present the increase is *c.* 2·0 per cent per year, or *c.* 69 million people (*FAO Yearbook*, 1969).

1 Increase in rate of food production in recent centuries. An increase in rate of population growth such as that at the beginning of the seventeenth century is dependent on radical improvements in food production made possible by a number of circumstances which enabled man to exploit natural resources in new ways during the succeeding centuries. There were two major sources of such improvements: the discovery of America and the growth of scientific knowledge. The latter led to technical developments and that, in turn, to the development of industry. Previously unexploited natural resources were utilised and food production increased many times, so that the population increased as a result. With the discovery of America, the white races, especially the Europeans, gained access to apparently inexhaustible riches. There was undisturbed cultivable soil, good grazing, forests, game, etc.

Exploitation was rapid because of the extensive methods used by the colonists. The New World could produce and export food in increasing quantities and could support immigrants from other countries with excess population.

The development of industry and trade resulted in better tools on the market. Ploughs, harrows and rollers were among the agricultural implements made of iron and steel, together with machines for sowing, harvesting and threshing, tractors, tractor-driven ploughs, etc. As a result, there was an increase in productivity per man, and in the total area under cultivation.

In the last few decades, horses have been replaced more and more by tractors. Arable land previously needed to maintain the horses became available for production of cereals and other crops. In the United States, the number of tractors increased from 2·2 to 4·8 million between 1944 and 1968: the number of horses decreased from 10·6 million in 1939 to 4·9 million in 1956 and has since increased (*FAO Yearbook*, 1969).

With the new tools, soil could be worked more thoroughly and to a greater depth; more land could be drained, and the soil nutrient reserves could be utilised more thoroughly. Timber was transported from the forests by railways and then by motor transport as well. The main advantage of the new tools was that they enabled increased exploitation of natural resources.

Transport of food was revolutionised by the development of steam and petrol engines. Goods could rapidly be delivered to areas of dense population, and food could be moved from areas with surplus production to areas of short supply.

Crop yields could be increased several times (see page 502) as a result of application of mineral fertilisers, and exhausted soil which would otherwise have been useless could be made productive again. The proportion of arable land lying fallow could be reduced steadily, and more land could be cultivated for arable use since lack of plant nutrients was no longer limiting. Furthermore, new varieties of crop plants were introduced, and plant breeders developed improved yield, quality and resistance.

At first, animal food was most readily available. As agriculture developed, and cereals and root crops were produced in large quantities, this situation changed. Vegetable food became cheaper than animal food, and the populations in poor regions and over-populated countries changed more and more to a vegetable diet. Direct production of edible crops from the land is much more economical of soil and vegetation than conversion of the crops to animal material before they are eaten. Production of a certain weight of meat requires fifteen to thirty times more land than does production of the same weight of bread (Russell, 1954).

Industry also released manpower in the country districts by taking over such work as manufacture of clothing, household equipment, furniture and other articles once made by the country dwellers themselves.

2 Population growth and consumption. The new methods of production enabled exploitation of plant and animal life to be more intensive than ever before. More food came on the market, allowing the non-farming fraction of the population to grow. The mutual interaction of farming and industry led to the quick development of both, and the human population increased rapidly, especially in industrial communities and industrial countries; this growth has been most rapid in the nineteenth and twentieth centuries. World population doubled between 1800 and 1940, but in about the same period (1800 to 1950) the population of England quadrupled and that of Germany and Japan trebled (Vogt, 1948). Population growth was less in primarily agricultural countries. The manpower needed by industry was to a large extent supplied by population movement from the country to the towns, so that in many countries, including Sweden, the rural population decreased (Figs. 138, 139).

Despite the small increase, or even decrease, in the number of farmers, food production was greater than before. More land was cultivated, and that which was under cultivation was used more intensively, so that previously unavailable mobile nutrient reserves in the soil were brought into circulation. However, if the nutrients were not replaced, the increased production represented only a greater depletion of natural reserves. Even if mineral fertilisers and other manure were applied, an overall depletion of reserves might still occur if soil erosion took place.

Despite this rapid increase in production, there is still an overall world food shortage (Table 113, page 521). Only in the countries

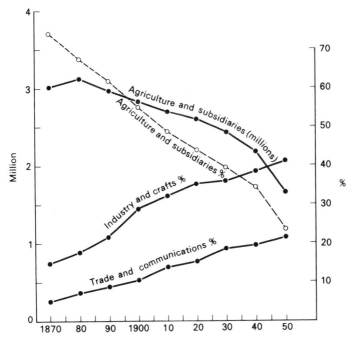

FIG 138. Population growth in Sweden, 1870–1950. Comparison with change in the population involved in agriculture, industry and communications. (From Statistisk Årsbok, 1946, 1951 and 1954.)

and regions which have been most affected by scientific and technical advances—mainly the industrial countries—is there sufficient food. In the other countries, with the larger part of the world population, there is still a shortage. To remove the shortage by increasing the utilisation of natural resources is impossible if this increased exploitation is always followed by an increase in population. The birth rate must be kept down to a level which is in balance with world food production.

More intense exploitation of natural reserves may decrease the chances of future maintenance of the population in the region, and man has sometimes had to withdraw from populated areas. Indeed, significant areas of cultivated land have degenerated into waste land and desert, and many other areas are still being cultivated but

are rapidly degenerating. Even at the present time, the human
population is maintained only at the expense of the natural reserves.

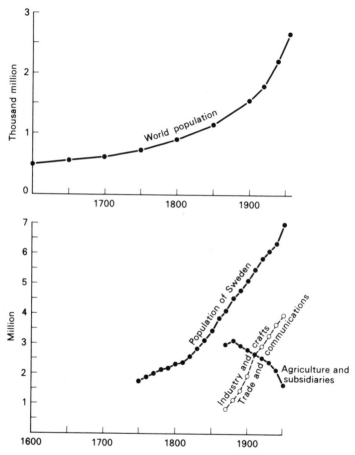

FIG 139. **Population growth in Sweden and the world as a whole
during recent centuries (From Statistisk Årsbok, 1946, 1954 and
FAO, 1953.)**

Depletion of natural resources of soil and vegetation

Throughout the world, the first stages of exploitation have been
similar to those in northern Europe. Shifting cultivation following
burning, arable cultivation, and grazing have succeeded each other,
so that the original vegetation, the game, and some of the mobile
nutrient reserves have been used up; sometimes soil erosion has

taken place. Nevertheless, there has been a change from extensive to intensive forms of exploitation, in which a balance between depletion and replacement of reserves has been sought.

Soil erosion is perhaps the worst of the destructive processes, since it may render the land completely unproductive: if a layer of soil is present, nutrient reserves can be built up again from external sources, and vegetation and animal life can return, but if all the soil is lost no plants can grow. The destruction of plant and animal life, nutrient reserves and soil has proceeded to various degrees in different parts of the world, depending on climate, edaphic conditions, population density, duration of habitation, etc. Some examples follow, arranged more or less in order of increasing severity of the damage.

Sweden. Soil erosion in northern Europe has not been sufficient to affect crop production as a whole. In central Sweden, it has been calculated that erosion by wind and water is so slight that it would take many thousands of years before a layer of soil one metre thick would be lost (Hjulström, 1940). However, in some areas, where there are special conditions of climate, soil and topography, much of the fine soil has been eroded away and only sand remains. In these places, a single sand storm may result in the loss of half of the remaining fine particles (Troedsson, 1949/50). These storms come every second year, on average, so that the arable land gradually degenerates to sand. The damage has increased in the last few decades, because of thorough cultivation, weed-suppressing crops and disappearance of pasture (Troedsson, 1949): drainage accentuates the erosion. The area damaged by erosion in south Sweden was estimated as *c.* 35 000 hectares in 1957. In Skåne, soil erosion originated some centuries ago. The three oldest areas of bare sand are close to mediaeval towns, probably because of removal of the forest and other vegetation at that time.

In cultivated hilly regions, particularly areas of moraine, topsoil has been lost from the higher ground, so that boulders and stones are exposed.

Germany. In more southern parts of Europe, there is a greater tendency to erosion, because of the drier soil and the greater population density. However, there is still relatively little erosion in Germany. At the present rate, it would take 10 000 years for erosion to remove the equivalent of a layer of soil 1 m thick from the whole country (Boresch, 1931).

South and Central America. In the relatively short period of exploitation of the American continent by white people, soil erosion has destroyed large areas of cultivated soil (Vogt, 1948; Borgström, 1952). In eleven South American countries, about one-tenth of the total land area of 175 million hectares has been used for cultivation or grazing; about one-third of this has now been seriously damaged or destroyed by erosion. There is arable cultivation on the plains, particularly the Argentine pampas, as well as near the mountains and on mountain slopes. In the mountain areas, practices are primitive, and include shifting cultivation following burning—an ancient practice in this area as well as in parts of Europe. After a few harvests, the ground is left for some years, then burned and cultivated again. A considerable increase in population has forced the farmer to reduce the fallow period and be satisfied with poorer yields as a result. The area under cultivation has increased more and more, and this, together with cultivation of sloping surfaces, has encouraged erosion by water. The present extent of erosion is probably mainly the result of forest clearance. The forests on the mountain slopes have been clear-felled or destroyed by fire, allowing increased run-off and transport of soil down into the valleys, to accumulate in lakes, estuaries, harbours etc., and increase the frequency of flooding of the surrounding areas. Erosion has dissected the slopes into ravines and made parts of them unusable. A further consequence is the lower water table and dryer soil (Vogt, 1948; Wallén, 1955).

Even in the valleys and on the plains, cultivation is extensive rather than intensive; mineral fertilisers are little used and new areas are cultivated when yields get too low. In many coastal areas of Central America and Mexico, the sequence from virgin forest to banana cultivation to abandoned land is only seven to eight years. Herds of cattle are large in relation to the available grazing, the grass cover is broken up and the soil exposed to the eroding effect of the wind.

The low tolerance of the soil is probably partly due to previous exploitation by ancient civilisations. The rich societies which flourished in Mexico and Central America in ancient times, and up to the Middle Ages, would have required considerable supplies of food from the surrounding countryside, and could not have developed without accompanying soil impoverishment. The impoverishment was the more rapid as agricultural practices were extensive rather than intensive, with no form of replacement of the reserves which were being used up. The Central American Maya civilisation was dependent on shifting cultivation following burning, unlike the ancient Old World civilisations, which depended on arable cultivation with soil fertility maintained by silt carried down in the rivers. The Maya farmer had no cattle, and no real agricultural implements. He used the simplest tools to cultivate maize on land which had previously been burned. The growth of the towns meant that the productivity of the nearest land deteriorated, and the farmers had to move further away. After a few centuries the nearest productive land was so far away that the food supply to the towns could no longer be maintained. It was probably for this reason that the townspeople abandoned their towns, with their treasures, temples, palaces and other buildings, moved to areas of virgin soil and built new towns (Ceram, 1955). The civilisations based on this type of agriculture seem to have developed to as great an extent as the Old World civilisations, but their foundations were less firm, since the soil and its reserves were destroyed, and they were forced to be nomadic to some extent.

The United States. Three hundred years ago the soil on the North American continent was stabilised by forests, which covered about 40 per cent of the total land area, and by the grasses and herbs of the prairies (Osborn, 1949). The forests and the plains had abundant game such as buffalo, antelopes and turkeys (Vogt, 1948). The colonists exploited the land, cut down or burned the forests, cultivated the soil and rapidly converted the soil reserves into plant crops and meat: as the soil in one place became impoverished they moved to a new area. The soil reserves were used up so rapidly that a maize yield of 1600 kg/ha in the first year of cultivation might have decreased to only about 50 kg/ha in the third year (Vogt, 1948). Only the best soil would give satisfactory yields for a period as long as thirteen years.

The building of the railways across the continent in the second half of the nineteenth century led to a rapid increase in exploitation. The area of arable land increased at the expense of forest and prairie—the forest by then covered only 20 per cent of the land area (Osborn, 1949, 1963). The dry farming method was used in the prairie soil areas, in which the soil was allowed to lie fallow every second year or so, to accumulate water and nutrients. The game was destroyed and replaced by herds of cattle. The extensive methods made food production cheap, and exports to Europe increased more and more (Hjulström, 1940).

The large extent of the plains, the low rainfall, long dry periods and the torrential nature of the rain, all combined to bring about erosion by wind and by water, erosion which was made possible by the process of cultivation, which took place at an increasingly rapid rate. Between 1914 and 1928, the area of arable land increased more than 50 per cent, from 90 to 139 million hectares (Hjulström, 1940), and to 193 million hectares by 1950, but is now declining (*FAO Yearbook*, 1955, 1969).

In the 1930s, when there were long periods of drought, wind erosion reached a hitherto unknown intensity: the dust storms were worst on the large open spaces of the Great Plains, particularly in Oklahoma, which had large new arable areas. Soil from Oklahoma and the neighbouring states of Kansas, Nebraska, Wyoming and

Colorado blew into the eastern states and out to sea. Some of the soil was deposited on grassy plains, forming a layer up to 5 cm thick, with drifts up to 1 m deep. When this loose soil was shifted again by fresh storms, the underlying turf, which had been killed, tended to break up and blow away with it. In this way, wind erosion gets worse and worse, once it has started. A contributory cause of the lack of resistance of the vegetation was often the damage done by severe overgrazing (Osborn, 1949, 1963).

FIG 140. Erosion damage in Tennessee, USA. (USIS.)

Weaver and Albertson (1940) studied the changes in the flora of steppes subject to erosion and showed that there was a change in a xerophytic direction. Species which lacked xerophytic features were decimated, or disappeared completely, whereas *Cactus* species increased in frequency, together with other succulents such as *Portulacea*, *Amaranthus* and *Opuntia*, and the vegetation became more open. In eighty-eight affected areas investigated, the percentage cover of 16 per cent of the total area was 21 per cent; of a further 16 per cent it was 11 to 20 per cent; of 18 per cent it was 6 to 10 per cent; of 16 per cent it was 2 to 5 per cent; and of 24 per cent it was less than 1 per cent (desert). So about a quarter of the eroded areas had already become desert, and a large proportion of the remainder was semidesert.

Water erosion has also been disastrously extensive in the United States. In the 1930s it was estimated to have carried away 3000 million tons of solid material—an amount which would fill a goods train nineteen times as long as the Equator (Hjulström, 1940): this silt is redeposited in reservoirs, harbours and estuaries. Lake Austin in Texas was 90 per cent filled in twenty years, and the Billesbay reservoir in Virginia 60 per cent filled in twenty-three years (Hjulström, 1940). In less than a century (1864–1938), 60 million m³ of sediment was deposited in Chesapeake Bay, so that the average depth over 80 km² decreased by 75 cm; at a place in Maryland, formerly with two harbours open to ocean-going traffic, the quays are now 4 km from navigable water (Vogt, 1948).

Destruction of the soil has been rapid in the United States, and has already rendered large areas of land useless. Of a total of about 770 million hectares of land, there are 400 million hectares of arable land and sown and natural pasture. By the end of the 1940s, 100 million hectares of this had been damaged or completely destroyed. According to some estimates, the amount of plant nutrients lost in erosion is twenty-one times that in a normal annual harvest (Osborn, 1949 and 1963).

Italy and North Africa. In the countries around the Mediterranean, where cultivation and destruction of soil has gone on for thousands of years, plant communities have become impoverished over large areas, and may even have reached final stages of semidesert or desert. Before Rome attained greatness, the Italian mountains were covered by forests, and only the valleys and plains were cultivated, but after iron implements became available (700 to 800 BC), the forests were felled and burned and used for grazing or cultivation. This interference triggered off the processes of erosion. Changes in the soil and the vegetation were of the same type as those which have taken place more recently in the United States.

In the Italian mountains, and in steeply sloping terrain generally, there is a movement of soil down the slopes, brought about by use of implements and by water. The highest places have long since lost their fine soil, and stones and gravel from the weathered bedrock are exposed. When cultivation was no longer worth while, these areas were used for grazing; they are now semideserts, with a xerophytic flora of shrubs and grasses. Grazing still contributes to the increasing oligotrophy of the land. Trampling by sheep and goats breaks up the surface and the surviving vegetation, and erosion by wind and water begins in the bare patches.

The soil carried downwards from the slopes collects in the valley bottoms and is transported further by the streams or rivers running there. The course of the rivers changes from one side of the valley to the other, scouring away the soil. Construction of dams and side walls is being undertaken to help prevent this damage. Where the forests were allowed to remain, e.g. around certain springs which had some religious significance or were used for medicinal purposes, there may still be a layer of fine soil several metres thick.

FIG 141. Terraced areas. The material is now gravel and stones. Erosion by water and by agricultural implements has carried the fine material down into the valley, into a stream which has transported it further. Fiuggi, central Italy. 700 m.

On its way to the sea, some of the fine soil has been deposited and has silted up water courses and estuaries and caused flooding. The Pontine plain, south of Rome, an area of about 1200 km², which in prehistoric times was a fertile area with sixteen towns, deteriorated into marsh as a result of floods which caused fevers and made it uninhabitable (Osborn, 1949). For two thousand years, attempts have been made in various ways to reclaim the area, but it was only in 1930 that development of modern techniques gave a measure of success.

The town of Paestrum and the surrounding area in Campania met the same fate. At the time of the birth of Christ, the situation had deteriorated so badly that the town was notorious for its 'bad air', and fevers were rampant. The number of inhabitants, formerly about one hundred thousand, was decimated, and by *c*. AD 1000 the town was deserted and the surrounding area was marshy. The mountains in the region had lost so much soil that they were mostly semidesert, as in many other places in Italy and other Mediterranean countries (C. Fries, 1954a).

The same tendencies occurred throughout the Roman Empire and neighbouring countries. Carthage and other Phoenician colonies along the north coast of Africa were supported by rich land with forests and fertile soil. Before the area was conquered by the Romans, the inhabitants of Carthage had cleared the forest, cultivated the soil, and had even exploited the areas bordering the Sahara (Hyams, 1952). The destruction of the natural vegetation surrounding the Sahara allowed the desert to spread further. Later, sand storms devastated inhabited areas to the north and buried several towns: Thimgad, for example, was destroyed in AD 100: it had had piped water and a sewage system, and several public buildings, including a library and a theatre. Cuicul was also a wealthy town with many temples, two squares, and a store for cereals and oil. Thydrus, on the Tunisian coastal plain, had a colosseum for 60 000 spectators, the largest outside Rome (Vogt, 1948). The food requirements of these and other towns brought about increased exploitation of the soil and accelerated erosion. Terracing and irrigation were widespread, but only served to delay the destruction of the soil. After the conquest of the area by the Arabs in the seventh century, agriculture ceased and nomads with their herds of animals took over. Formerly cultivated areas became pasture and therefore went a step further on the path to ruin (C. Fries, 1954a).

Syria. Syria was originally a rich area with extensive forests of oak, stonepine, cypress and cedar, with a layer of soil one to two metres thick covering the bedrock. During the first century BC and the first century AD, when the Roman Empire was at its greatest, the forests were felled to provide timber for export to Egypt and were replaced by vineyards and olive groves. Evidence of the magnitude of the exports from Syria and other Mediterranean areas, and of the dependence of the Imperial capital on products from the Empire is found in one of the Roman hills—the Monte Testacio on the Tiber. This is 35 m in height and is formed mainly of pieces of the clay pots in which wine and oil was carried to Rome from Syria, Africa and Spain (C. Fries, 1954a). While this trade continued, Syria was a wealthy land with a hundred or so towns and flourishing agriculture. The soil was protected by terracing and irrigation. After the Arab invasion in AD 630, the settled population was ousted by nomads and the vineyards and olive groves were neglected and replaced by arable land and pasture. This caused erosion. However, the towns were not buried by sand as they were in North Africa; instead, the soil blew away, exposing the bedrock. The foundations of ruined buildings are now several feet above ground level. Pockets of soil remain among the ruins even today, and vines, olives and other cultivated plants can be found in such places. Their disappearance is therefore not a result of change in climate, but of the destruction of the soil (Hjulström, 1940; C. Fries, 1954a).

Greece. About 6 per cent of the land was probably originally forest-covered, and hunting was a common occupation. There is now very little forest left—covering only a few per cent of the land area. Birds and other game have almost completely disappeared (Osborn, 1949, 1963). As early as 500 BC, much of the forest had been felled and the land used as pasture. This pasture was parklike, with trees (C. Fries, 1954a), a transition stage between the forest and the presentday desert. The clearance of the forest and the trampling of grazing animals destroyed the vegetation and made soil erosion probable. In Greece, erosion by wind predominates in the summer and by water in autumn and winter, when there is heavy rain. The soil has long since been washed down into the valleys and the sea and silted up the water courses, causing flooding. Attempts were made to lead the water away in stone-walled canals; but some of these now lie up to 3 km from the course of the river (Osborn, 1949).

The mountain slopes lie bare, or covered here and there by sparse xerophytic scrub, run-off is at a maximum, since both water-storing and water-using media, soil and trees, have gone. This accentuates flooding, which is now common in these areas.

Not even the soil on the arable plains has escaped erosion; the fine material has blown away and exposed the subsoil. Humus production is small because of the rapid decomposition. Stable manure was not supplied in large quantities, partly because it was used as fuel since wood was scarce. According to Osborn (1949, 1963), only 2 per cent of the land has its original topsoil left, and this is in isolated places where there is still forest.

Since time immemorial, in Greece and elsewhere, terracing has been used in an attempt to retain the soil in situations where erosion is prevalent. However, this method only has a delaying effect, and eventually the soil is carried away by erosion: mostly only the stone walls of the terraces now remain, and some of these still retain remnants of soil along the walls, where a little barley or wheat may grow (C. Fries, 1954a).

The use of terraces, and of the canals mentioned earlier, make it clear that men were very early aware of soil deterioration and did what they could to prevent it, indeed, the following description in one of the writings (*Kritias*) by Plato (died 347 BC) shows that the causes of erosion, and the way in which it started, were known then. He described how soil flowed down from the heights and disappeared into the depths. He says that in comparison with their previous state, the small islands were like the bones of a body racked by disease; the rich, soft soil had trickled away and only the skeleton of the land remained. Before the land was exploited, the mountains were high heaps of soil, and the plains were full of rich mull. On the mountains there were forests in plenty, and signs of this still remained. Although some of the mountains then provided only bees with food, it was not long since fine roof timbers, sufficient for large buildings, had been cut from the trees in the forests. There were also many tall fruit trees and plenty of grazing for the animals. Every year the country rejoiced in the rain from Zeus' sky: it did not run away useless into the sea from the bare rocks of the land, as it does now: it was absorbed by the land, and retained in the protective layers of clay. The water flowed down into hollows, and was the source of springs and rivers.

This description by Plato, made about 2500 years ago, clearly shows that Greece was a severely exploited land even then. It also shows that the people at the time understood the relation between the topsoil, the vegetation and water, together with the consequences of man's interference.

The Babylonian Empire. The Babylonian Empire, and the preceding Sumerian Empire, which lay between the Tigris and Euphrates, the Syrian desert and the Persian gulf, with a history going back more than 6000 years, now exemplifies the final stages of the ruin of the land. The same is true of Assyria, or Mesopotamia, which lay north-west of the Babylonian Empire, with Nineveh as its capital. The Tigris flowed through Assyria, and there was a plain in the south. When these Empires were at their greatest, agriculture played an important part: productivity was high, and irrigation systems were in operation. The climate was warm and semiarid. Water from the rivers was diverted over the plain in canals. Because of erosion at the sources and upper reaches of the rivers, the water was laden with silt. Terraces were constructed on sloping ground: the hanging gardens described by Herodotus (*c.* 485–425 BC), and considered to be one of the seven wonders of the world, were probably just such terraced slopes (Osborn, 1949).

Because of the never-failing supply of water, bearing silt rich in nutrients, agriculture was able to support a settled population for thousands of years, so that cultural development could proceed.

Both empires fell, some centuries before the birth of Christ. Babylon, which was the capital of the Babylonian Empire as early as 1900 BC, still stood in about 300 BC. Now only ruins of Babylon and of the other towns remain, and the plains which were once fertile arable land have become river delta, subject to flooding. Remains of temples 21 m below the present surface were found in excavations at the site of Babylon.

There are several possible reasons for this catastrophe: war may have caused misuse of the irrigation system, directly or indirectly, with consequent flooding; or irrigation in conjunction with the arid climate may have led to such a high concentration of salts in the topsoil that it eventually became unproductive (Jacobsen and Adams, 1958); or erosion and flooding may have been caused by clearance of forests around the sources of the rivers: kilns for brickmaking required large amounts of timber for fuel, and more land was needed for cultivation and grazing. Another possibility is that increased flooding was the result of the climatic deterioration around 500 BC.

There were probably floods even before the time of the Babylonian Empire. One of the towns destroyed was Ur, on the lower reaches of the Euphrates. The river changed its course so that it lay 15 km away from the town, the irrigation system was ruined and the area deteriorated to desert. Excavation revealed a layer of debris *c.* 12 m thick under the town, and under this was a 2 m layer of clay. Beneath this clay were the ruins of an even earlier town, probably also destroyed by floods. The Euphrates is still subject to changes of course, and there are floods to a depth of up to 4 m. The story of the Flood in the Old Testament is based on these early calamitous floods. There are similar legends among other peoples, especially in America, Polynesia and India.

Like the Sumerian-Babylonian-Assyrian cultures, ancient Egyptian civilisation was based around the mouth of the Nile, and the ancient Indian civilisations were around the mouth of the Indus, developing on land which was kept fertile by riverborne silt, i.e. by material eroded from other areas. This was the only possible way of obtaining harvests every year for thousands of years from the same land, thus allowing people to build towns and settle for long periods in the same place so that cultural ideas and traditions had time to develop (Hyams, 1952).

Seen in a wide context, erosion is both a constructive and a destructive force. It has brought about the disappearance of civilisations which were only enabled to develop because of its operation. Its action as a constructive force is of the same kind as that of the early arable farmer in bringing about movement of nutrients from forests and meadows to the arable soil, but took place over a longer period. The fertility of water meadows and fens along the courses of rivers depends on water erosion, like that of the delta soils. Such areas have always been one of the most valuable resources of the farmer. Loess soils built up by wind erosion may also attain considerable thickness.

Remaining resources

Despite all the exploitation of the world's natural resources which has taken place, some resources are still available, for food production, for example. It is very difficult to make an evaluation, but it is possible to indicate some of the possible ways of extending agriculture.

1 New areas for agriculture. In the 1960s, the total land area was as follows (*FAO Yearbook*, 1969):

Total area	13 392 million hectares
Cultivated land	1 406
Meadows, pasture	3 001
Forest	4 068

About 10 per cent of the world land surface is cultivated, and about 17 per cent is used for grazing. These figures represent the land most easily available and most productive. There are other areas which could be cultivated, but have not been, for economic or other reasons. There are large areas of this type in South America, especially in the vicinity of the Amazon, for example. Estimates of the total area of such land vary between about 400 and 2800, or even 6000, million hectares, depending on the criteria used (Russell, 1954). Food prices are important in determining whether such areas should be brought under cultivation. If these are high in relation to prices of other commodities, fairly poor areas can be cultivated and give satisfactory returns. These potential agricultural areas are generally situated so that they are difficult to exploit for reasons of climate, health, drainage, lack of water, etc. The soil may be poor in nutrients, in tropical areas because humus is so rapidly lost, and so the cost of production may be high. Advances in agricultural techniques, medicine, plant nutrient physiology and other sciences should make it possible to overcome some of these difficulties.

2 Increase in area within presentday agricultural areas. Within presentday agricultural areas, the most easily available and most productive land is already under cultivation: there is still more land which could be cultivated, although only at relatively high cost. From 1870 to 1920, the total area under cultivation increased by about 18 per cent (*c.* 182 million hectares), an increase which continued up to about the 1930s; it was about 45 million hectares from 1910 to 1920. After the 1930s, there was no further increase, or even a slight decrease, as in Sweden, for example, probably because there was overproduction of food in several countries in the 1930s, and prices fell (relative rises in food prices will cause increases in the area of cultivated land).

3 Restoration of land. Technical advances in development of machines for digging and soil movement generally have made it possible to restore land ruined by erosion. Whether this would be practicable depends on food prices.

4 Irrigation. The ancient method of making arid and semiarid land productive by leading river water or other fresh water over the area is still used to enable exploitation of new areas and extension of old ones, e.g. in Spain, India, China, Australia, Egypt and America: in 1967 Spain had $c.$ 2·24 million hectares of cultivated, artificially irrigated land; USA 12·8; USSR 10·1; China 74·0; India 27·5; Japan 3·2; and Pakistan 12·0 million hectares (*FAO Yearbook*, 1969). The supply of water can, of course, be regulated, in contrast to natural rainfall. Water supply very commonly limits crop production, and yields vary according to chance variations in climate. Water supply by tank culture has been tried in America: plants are grown on a substrate which is periodically lowered into tanks containing nutrient solution. This technique does not give higher yields than other methods of cultivation of plants under optimum conditions (Hoagland and Arnon, 1948), but it has the advantage of allowing salt and water supply to be controlled (Baumeister, 1954). It would be expected to give good results, especially in arid tropical and subtropical regions (Douglas and Sholto, 1955).

There are other ways in which arid areas could be made more productive. White *et al.* (1956) recommended several, e.g. selection of more drought-resistant varieties of cultivated plants, increased planting of food-producing trees, such as olives, almonds, figs, dates, together with detailed research into their water economy; desalination of salt water; control of the balance between growth of vegetation and its utilisation for grazing or other purposes, etc.

5 Increased supply of mineral fertilisers. Although they have been available for over a century, mineral fertilisers have so far been used extensively in only a few of the world's agricultural countries (Table 114), mostly in western Europe, but even there only in small areas are they used in optimum amounts. In other parts of the world they are not used at all, or only to an insignificant extent. Yields on some of the largest arable areas are still largely limited by the rate of mineral weathering. In the oldest cultivated areas, some sort of equilibrium between weathering and crop yields has developed in the course of time. The soil must often be allowed to lie fallow, sometimes every second year, in order that sufficient mineral salts can accumulate to support a sparse harvest. It is not only the elements commonly applied in fertilisers which are deficient, nitrogen, phosphorus and potassium, but also sulphur and trace elements. One of the most important requirements of agri-

TABLE 114 World use of mineral fertilisers 1968–69. (*FAO Fertilizers*, 1969.) The calculations are based on the assumption that all fertilisers were used on cultivated areas.

	CULTIVATED SOIL (1000 ha)	N 1000 Mg	N kg/ha	P_2O_5 1000 Mg	P_2O_5 kg/ha	K_2O 1000 Mg	K_2O kg/ha
Europe	149 000	8 754	58·7	6 818	45·7	6 544	45·7
North and Central America	253 000	7 291	28·8	4 852	19·2	4 019	15·9
South America	89 000	387	4·4	480	5·4	299	3·4
Asia	335 000	3 785	11·3	1 651	4·9	1 227	3·7
Africa	204 000	675	3·3	512	2·5	240	1·2
Oceania	43 000	172	4·0	1 214	28·2	174	4·0
USSR	224 300	3 454	14·5	1 748	7·8	2 210	9·9
Total	1 406 000	24 520	17·2	17 280	12·3	14 710	14·8

cultural research is to establish a good nutrient economy in these areas.

World agriculture in the 1950s was able to maintain the same rate of food production per head as before the Second World War, despite the large increase in population, because of increased use of mineral fertilisers (Lamer, 1957). World consumption of mineral fertilisers was 8·9 million tons in 1939, and 20·9 million tons in 1957. From 1956 to 1961 there was a 30 per cent increase in the amount used (*FAO Ann. Rev.*, 1961).

The amounts of possible increases in yield through increased fertiliser application are difficult to predict, but this is probably the most hopeful way of increasing yield at present available. Of the fundamental requirements for plant development—radiation, heat,

TABLE 115 Wheat production and use of mineral fertilisers (N, P_2O_5, K_2O). Mean for all kinds of cultivated soil (*FAO Production Yearbook*, Vol. 23.)

	WHEAT PRODUCTION (Mg/ha)			MINERAL FERTILISER (kg/ha)	
	1948–52	1964	1968	1963	1968–69
Denmark	3·63	4·23	4·81	164	204
Holland	3·66	4·71	4·42	609	622
Sweden	2·10	3·75	4·33	109	150
France	1·83	3·15	3·66	147	204
UK	2·72	4·25	3·55	202	255
Italy	1·52	1·95	2·24	57	76
Spain	0·87	0·96	1·34	36	52
Europe (excl. USSR)	1·47	2·10	2·53	106	148
USSR	0·84	1·10	1·39	14	33
USA	1·12	1·73	1·92	56	79
Canada	1·28	1·36	1·49	12	23
Brazil	0·74	0·88	0·88	8	18
N and C America	1·16	1·60	1·78	43	64
S America	1·07	1·64	1·03	8	13
Africa	0·70	0·83	1·02	4	7
Australia	1·12	1·38	1·36	18	28
Japan	1·85	2·45	3·14	320	400
Pakistan	0·87	0·83	1·07	3	14
India	0·66	0·73	1·10	4	10
Asia (excl. China)	0·82	0·85	1·07	18	33
World total	0·99	1·28	1·46	26	40

water and mineral salts—the first two do not limit production in subtropical and tropical regions, and the last two often do. Only by increasing the supply of water and nutrients can the good light conditions, high temperatures and often long growing period be utilised fully. Two or three harvests a year can be obtained in areas with a long growing period, if there is a good nutrient supply and if soil condition is maintained satisfactorily. However, present practices allow arable soil to lie unproductive through a large part of the year.

Virgin tropical forest is the most eutrophic land community in the plant world. It should be possible for tropical agriculture to be at least as productive, provided that the water economy is satisfactory.

6 New types of crop. Among the vast number of wild plant species, there are probably many which could be used for food in some way. Now and then, the products of new crop species come on the market, mostly fruit and vegetables, and further research will surely yield more, but this possibility is continually reduced as more and more species become extinct.

Cultivation of food-producing trees, such as oil palms and cacao, is less destructive of the soil than cultivation of annual crops, and would be advantageous in areas subject to erosion. Trials on these lines are going on in West Africa, for example (Russell, 1954). Food-producing trees were previously much more important in Scandinavia than they are now. Some return to this type of economy is now in progress, as fruit growing has expanded very considerably in recent decades.

7 Genetic improvement of varieties of crop plants. Yield, frost and drought hardiness, and disease resistance, can be improved by crossing and selection. The results of such improvement were at first of benefit to agriculture in only limited areas of the world. In areas where agriculture is of an extensive rather than an intensive type, the advantage of more productive varieties is small since production is limited by lack of nutrients and of water.

In 1970, the Nobel Peace Prize was awarded to Dr Borlaug for his work in developing the high-yielding varieties of Mexican dwarf wheats, which have recently dramatically increased crop yields in many countries, including the Indian–Pakistan sub-continent. India's wheat crop, for example, increased from 12 million tons in 1965 to 21 million tons in 1970, so its overall cereal crop is now expanding more rapidly than population. High-yielding dwarf rice varieties were developed at the International Rice Research Institute in the Philippines in the mid 1960s, with the dwarf wheats as a prototype. These new wheats and rices are more responsive to irrigation and fertilisers than the varieties formerly available. They have greatly improved the well-being of large numbers of people, although they will be of little long-term benefit unless world population growth can be stabilised (Brown, 1970).

8 Technical development. Some of the increased production in the last century has been the result of technical development, particularly the contribution of machines. However, this increase represents increased exploitation of natural resources, as well as better utilisation of the products. Continuation along these lines may lead to more and more rapid depletion of natural resources, particularly of the soil. Some natural resources, such as the game and other wild life of the forests, which were originally an important source of food for man, are already so depleted as to be no longer significant.

In other respects, technical development may allow better husbandry of natural resources and restoration of damaged areas. The transport requirements of agriculture can be met; irrigation systems can be set up; mineral fertilisers can be manufactured; soils can be mixed to improve their structure; and difficult terrain can be cultivated, e.g. by draining marshes and building roads.

9 The food industry. The products of the land and the sea have often been treated wastefully. Hay and cereals have been harvested by primitive methods, the results of which are dependent on the weather, and the highest yields often suffer the most wasteful treatment. Wastage is particularly high in wet climates. Large catches of fish or large crops of fruit used to be difficult to handle, but relatively recent advances in industrial techniques have helped considerably in this respect. Canning, freezing, silage-making and artificial drying of cereals and fodder have already cut down wastage considerably, but continued development of such methods is of great importance.

Factory treatment of foods has previously been concerned with preparation and packing of natural products; food production as such has been of very rare occurrence. However, it is theoretically possible to manufacture food from organic material which cannot be eaten in its natural state, from inorganic substances such as carbon, nitrogen and water, and there has already been some experimental work with this aim in view.

The hydrolysis of cellulose yields sugars which can replace pure sugar as raw material for fermentation in some circumstances. This hydrolysis also yields other carbohydrates, more or less digestible and useful as food. Such 'fodder cellulose' was used to a considerable extent for animals in Sweden in the 1940s. Large quantities of raw material are available, e.g. waste products from forests and saw mills, and wood now used as fuel but being replaced more and more by other fuels.

The production of fats and proteins by growing microorganisms on sugars made from wood, with the addition of the necessary salts, was tested in Germany during the First World War, and edible products were obtained (see Chapter 7, page 134). However development along these lines can only be economic if there are cheap methods of production, or high food prices (see page 134). Production of vitamins by microphytes has become more and more important (Mothes, 1954; Table 116).

Research is now being done in many countries on protein production by microbial growth on substrates such as petroleum, for example by BP in France (Champagnat, 1965). Coconuts and leaves from many species of plants are also potential protein sources, but contain so much fibre that the protein must be separated. Research into methods of separation is going on at Rothamsted and elsewhere. Methods of production of plant protein by cultivating algae in nutrient solutions in large tanks are also being tried, in Japan and Czechoslovakia, for example.

10 The sea. Steeman Nielsen (1957) calculated that an average of only about 0·01 per cent of the carbon dioxide assimilated by marine plankton, and then used as animal food, or sedimented, is ever made available to man in the form of fish as food. In areas of intensive fishing, such as the North Sea, the value may reach 0·2 to 0·3 per cent. It is certainly theoretically and practically possible to use marine plants for food in other ways, for example, by cultivation of marine algae. The problem is to make such production economic and on a sufficiently large scale to make a significant contribution to the food supply. Because water is not limiting, as it often is for land plant cultivation, and because mineral salts can be supplied in optimum amounts, yields per hectare can be many times greater than for land plants. Ten tons of protein-rich dry matter per hectare can be obtained in eight days in cultures of *Chlorella*, *Scenedesmus* and other algae (Mothes, 1954).

The high productivity of microorganisms is largely the result of their primitive organisation. All parts of the organism are engaged in dry matter production: there is no division into different organs as there is in higher plants, which have organs for carbon dioxide assimilation, others for the uptake of minerals and water, and yet others for storage. A major part of production consists of transport of raw materials and of the final product. Such organisms use more respiratory energy and require more respiratory substrate for their maintenance than do primitive organisms consisting only of assimilatory cells. Higher plants require special materials such as cellulose, lignin, suberin, etc. The microphytes avoid all this expenditure (Table 116). Another reason for the high productivity of microphytes is the close contact between the organisms and the raw materials they use. Yeast, for example, takes up sugar through all parts of the cell wall, and the sugar is used immediately; little transport of raw materials is necessary. In addition, the area of contact between the organism and the external medium is very large in relation to the weight of the organism. Fungi and other heterotrophic organisms are independent of light and so synthesis can proceed continuously. As a consequence, the productivity per unit

TABLE 116 The composition of *Scenedesmus obliquus*. (After Meffert and Stratman, 1954.)

	PER CENT (dry wt)		PARTS PER MILLION (dry wt)
Total N	7·16	Vitamin B_2	38·3
Raw protein	44·8	,, B_6	1·81
Raw fat	4·7	,, B_{12}	0·024
Ash	24·8	Pantothenic acid	12·00
P_2O_5	9·73	Nicotinic acid	72·70
Ca	4·96	Folic acid, free	0·80
Mg	1·33	Folic acid conjugate	6·00
Fe	0·30	Leucovorin free	3·00
K	1·22	Leucovorin conjugate	19·00
Na	0·16	Biotin	0·20
S	0·66		
Cu	$5·9 \times 10^{-3}$		
Pb	$8·5 \times 10^{-3}$		
As	$0·6 \times 10^{-3}$		

weight of microorganisms is considerably greater than that of other organisms. A bullock weighing a half ton produces *c.* 0·4 kg protein per day. Soya beans growing in optimum conditions produce about 36 kg per half ton per day, whereas the corresponding figure for yeast (*Torulopsis*) of the type usually used for protein production is up to 5000 kg (Thaysen, 1953, cited by Gilbert and Robinson, 1957).

An important aim in the husbandry of natural resources of the sea is to encourage the production already going on in the marine world of plants and animals, to protect it from fish predators and to ensure that the more valuable species do not disappear. The oceans and other bodies of water are the largest productive areas of the world: even small changes in conditions—for better or worse—will have a very large effect.

11 Increase of vegetable food. There are potential reserves of food which can be tapped by change from animal to vegetable food. A high living standard, or availability of large numbers of cattle, is usually followed by consumption of large amounts of animal food and consequent high demands on areas of productive land (Fig. 142). Because of the large proportion of animal food, food supply

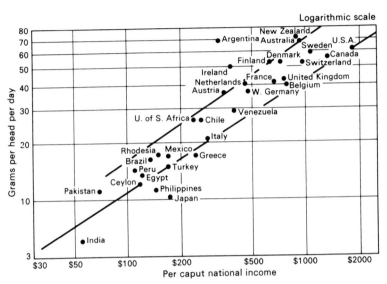

FIG 142. The relation between the consumption of animal protein (grams per person and day) and income (dollars per person and year). (FAO, 'The State of Food and Agriculture 1954', Rome, 1954.)

in Great Britain requires 0·53 to 0·57 hectare per person, whereas in India, where the diet is mainly vegetable, only 0·24 hectare per person is needed. More than half of the world population is supported by an average of 0·24 hectare per person, and this low figure is possible only because of the very low consumption of meat and milk (Russell, 1954).

It is therefore still possible to increase food production. There are still reserves of soil and raw materials as yet unused. In addition, scientific and technical developments may open up new possibilities.

The rate and extent of exploitation of these reserves depends mostly on changes in prices, or, more precisely, on the capacity of food production to compete with other branches of the economy, and on the trade between the countries with excess food and those with not enough. This, in turn, depends on population changes.

While total annual production of food continues to increase and to benefit the countries without enough food, world population can still go up. If food production fails to increase, population growth must cease, provided that it continues to be regulated mainly by food supply and birth rate.

Whether food production and population have reached their limit, or whether growth will continue, is a question which is often discussed. To predict the future by extrapolation of presentday trends is of doubtful value. Previous predictions made at the end of the eighteenth and nineteenth centuries have proved wrong. Russell (1954) mentions a calculation made in 1898 by the president of the British Association, Sir William Cookes. At that time, according to Sir William, there were 516 million wheat eaters in the world, and this number would increase to 746 million in thirty years time. There would have to be an increase in wheat production from 2·3 to 3·3 thousand million bushels to feed this increased population, and since this was thought to be impossible, there would be severe risk of famine. Russell points out how difficult it is for the present generation to appreciate what a shock this was. During the nineteenth century there had been a considerable increase in population and a general optimism about the future; to learn that it would all end in famine was a severe blow.

However, the prediction was not fulfilled. When the 1930s came there was such a large surplus of cereals that wheat, for example, was used as fuel in Argentina. The market was flooded with food, prices were depressed, and agriculture was beset with great economic difficulties. This rapidly accumulated surplus was particularly surprising in view of the shortages and unprecedented destruction caused by the First World War.

During that war, there was a shortage of food in Europe and in other parts of the world. Home-produced and imported foods were rationed, and were still sometimes insufficient; the shortage of fat was particularly acute. There was very little or no coffee, tea and other imported goods. There was a shortage of imported raw materials in industry and agriculture, e.g. mineral oils and mineral fertilisers. However, production recovered rapidly after the war had ended and at the beginning of the 1930s the market was already flooded with most essential goods. After some years, production was so high that it exceeded consumption even in countries with the highest consumer spending, warehouses and stores were full and prices sank. Then production was cut back and there was widespread unemployment. Terms such as *wheat surplus, butter surplus* and, in subtropical countries, *coffee surplus*, began to be used in discussions of the situation.

When the Second World War broke out in 1939 the same series

of changes began: first, shortages and rationing during the war; then rapid recovery; then, after ten to fifteen years, a surplus, bringing about lowered prices and stagnation.

The food surplus partly resulted from the replacement of horses, one of the main groups of cereal consumers, by tractors. In Sweden, oats were replaced by oil plants such as rape, yielding the raw material for margarine production.

Although food production increased more than was expected, and the increase during the previous century had been enormous, a large proportion of the world's population suffered starvation. The gain in production was used up as a result of the increased birth rate, or else could not be bought by those who needed it. The relative numbers of people starving have remained unchanged. While size of population is regulated by food shortage in large parts of the world, starvation cannot be prevented by intensified food production or by new ways of producing food. If, for example, all the open water surfaces in densely populated parts of the world were used for algal cultivation, so that more food was available, the reservoirs and recreational opportunities would be destroyed with no overall improvement in the food situation, because the population would increase (Cole, 1955). Continued intensification of the type of food production which uses up natural resources will therefore only lead to further deterioration or destruction of the land on which man is living.

Conservation

In most countries in which land has been damaged by erosion, various methods of preventing or limiting the damage are employed. Some of the most important are the following:

Terracing, used now, as it has been for centuries, to protect soil on sloping ground.

Strip cultivation. Narrow strips of open soil are cultivated between grass or other vegetation-covered surfaces which can absorb the trickles of water.

Contour ploughing, involves positioning of the cultivated strips at right-angles to the slope, sometimes in curves or angles, so that long lines of erosion are prevented.

Irrigation prevents wind erosion. Large areas of cultivated land in arid and semiarid regions are maintained by artificial irrigation (see page 538).

Wind breaks decrease the effect of the wind; planting sand-binding grasses and other plants lessens the movement of sand dunes.

Improvement of soil structure decreases the tendency for movement of soil particles.

Mulching is a technique, used particularly in America, whereby the soil is covered by straw, leaves or other litter, to protect it from both wind and rain. The litter has a beneficial effect as a fertiliser, on animal life in the soil and, consequently, on soil structure.

FIG 143. Contour ploughing on sloping land. (Iowa, USA. (USIS).)

Temperature fluctuations are damped down, and the soil freezes less readily. The effect on soil moisture is more complicated. Small amounts of precipitation are retained in the litter and do not reach the soil; large amounts are conserved in the soil, because there is less evaporation (see Millar, 1955).

Research on the nature of erosion, on the effect of various methods of working the soil, on the action of the eroding forces, and on the way in which the soil and plants react in response to the wind and water, is now going on in several countries. Institutes for this purpose were set up in Russia in 1921 (Borgström, 1952) and in the USA in 1929. The US Soil Conservation Service was inaugurated in 1933. There are laws aimed at preventing erosion in the USA, and in other countries. In the USA in 1952 there were 2400 conservation districts, comprising 76 per cent of the total arable land area (Russell, 1954).

Soil conservation is just one branch of nature conservation as a whole. The modern conservation movement, which started at the end of the nineteenth century, aims at saving the soil, plant and animal species for the future, together with some of the old original plant communities, natural geological formations and certain characteristic or unique types of landscape or details of landscape. It also aims at preserving parts of former types of landscape which may disappear as a result of changes in agricultural practices. This is an urgent task, since exploitation of natural resources is rapidly spreading to areas hitherto unused or little used and methods now being employed often cause changes in topography as well as in geological and biological characteristics. The vegetation and soil may even be removed completely, from peat bogs, for example.

Even moderate exploitation of a plant community has far-reaching biological effects. Plants and animals find themselves in an environment strange to them and they may not survive. There is already a long list of plant and animal species which have become extinct, and the list grows longer every year. As far as possible, threatened species are protected, in order to save them, but, unfortunately, this often only delays their disappearance. Species threatened in this way can generally only be saved by preserving their natural environment as a whole, i.e. the plant community and landscape, and the balance between all the constituent organisms and the soil. Man is the poorer for every such loss. A plant or animal species which disappears, although it may not seem to have any economic value at present, may well have possibilities which might have been discovered and developed in the future.

Bibliography

AALTONEN, V. T. (1948). Boden und Wald. Berlin.

AARNIO, B. (1935 no. 39) 'On the factors acting upon the qualities of the humus containing layer of natural soils'. Soil division of the Central agricult.-exp. station of Finland Agrogeol. julkaisuja.

ACHARYA, C. N. (1935) *Biochem. J.*, **29**, 528, 953, 1116, 1459.

ADDISON, E. LEE and WHALEY, W. G. (1953) 'Effects of thiamin, niacin and pyridoxine on interval growth of excised tomato roots in culture'. *Bot. Gaz.*, **114**, 343.

ADEL, A. (1951) 'Atmospheric nitrous oxide and the nitrogen cycle'. *Science*, **113**, 624.

AGERBERG, L. S. (1958) 'Bor- och kopparhalt i några norrländska åkerjordar'. *K. Skogs och Lantbr. akad. tidskr.*, **5**, 281.

ÅKERMAN, A. (1951) *Stråsädesodling.* Ernest Jordbrukslära. Stockholm.

ALSTERBERG, G. (1922) 'Die respiratorischen Mechanismen der Tubificiden'. *Lunds Univ. årsskrift N. F. Avd.*, **2**, 18.

ALWAY, F. J., MALLI, T. E. and METHLEY, W. J. (1934) 'Composition of the leaves of some forest trees'. *Amer. soil survey assn. Bull.*, **15**, 81.

ANDERSSON, G. and BIRGER, S. (1912) 'Den norrländska florans geografiska födelning och invandringshistoria med särskild hänsyn till dess sydskandinaviska arter'. *Norrländskt handbiblioteck* 5, Uppsala.

ANDERSSON, S. (1954) 'Markens struktur och om en metod att analysera makroaggregering'. *Grundförbättring*, **2**, 114.

—— (1955) 'Markfysikaliska undersökningar i odlad jord VIII. En experimentell metod'. Ibid., **8**.

ANDERSSON, S. O. and ENANDER, J. (1945) 'Om produktionen av lövförna och dess sammansättning i ett mellansvenskt aspbestånd'. *Sv. Skogsvårdsfören. tidskr.*, **46**, 265.

ANDRÉ, P. (1947) 'Bärrisens och mossornas förnaproduktion i ett mellansvenskt barrskogsbestånd'. Ibid., **46**, 122.

ÅNGSTRÖM, A. (1925) 'The albedo of various surfaces of ground'. *Geogr. Ann.*, 323.

ÅNGSTRÖM, A. and HÖGBERG, L. (1952) 'On the content of nitrogen (NH_4-N and NO_3-N), in atmospheric precipitation'. *Tellus*, **4**, 31.

ARISZ, W. H. (1953) 'Active uptake, vacuole, secretion and plasmatic transport of chloride-ions in leaves of *Valisneria spiralis*'. *Acta bot. neerl.*, 1:506.

ARISZ, W. H., CAMPHEUS, I. J., HEIKENS, H. and VAN TOOREN, A. J. (1955) 'The secretion of the salt glands of *Limonium latifolium*'. *Acta bot. neerl.*, **4**, 322.

ARNOLD, A. (1936) 'Beiträge zur ökologischen und chemischen Analysen des Halophytenproblems'. *Jb. wiss. Bot.*, **83**, 105.

ARNON, D. I. (1950a) *Functional aspects of copper in plant copper metabolism.* John Hopkins, Baltimore.

—— (1950b) 'Criteria of essentiality of inorganic micronutrients for plant trace elements in plant physiology'. *Chronica botanica.*

ARNON, D. I., FRATZKE, W. E. and JOHNSON, C. M. (1942) 'Hydrogen ion concentration in relations to absorption of inorganic nutrients by higher plants'. *Plant Physiol.*, **17**, 515.

ARNON, D. I. and JOHNSON, M. (1942) 'Influence of hydrogen ion concentration on the growth of higher plants under controlled conditions'. *Plant Physiol.*, **17**, 525.

ARNON, D. I. and WESSEL, G. (1953) 'Vanadium as an essential element for green plants'. *Nature, Lond.*, **172**, 1039.

ARRHENIUS, G. (1950) 'The Swedish deep sea expedition. The geological material: its treatment with special regard to the eastern Pacific'. *Geol. fören. förh.*, **72**, 185.

ARRHENIUS, O. (1931) 'Markanalysen i arkeologiens tjänst'. *Geol. fören. förh.*, **53**, 47.

ARVIDSSON, I. (1951) 'Austrocknungs- und Dürresistenzverhaltnisse eineger Wasserabsorption durch oberirdische Organe'. *Oikos, Supplement I.*

ASHBY, E. (1936) 'Statistical ecology'. *Bot. Rev.*, **2**, 232.

ÅSLANDER, A. (1929) 'Concentration of the nutrient medium versus its hydrogen ionconcentration as manifested by plant growth'. *Sv. bot. tidskr.*, **23**, 96.

—— (1932a) 'Die Abhängigkeit unserer Kulturpflanzen von der Reaktion und dem Nährstoffgehalt des Bodens'. *Z. f. Pflanzenern., Düngung und Bodenk.*, **23**, 362.

—— (1932b) 'Markreaktion, jordens näringsinnehåll och skördeutbytet'. *K. Lantbruksakad. vet. avd.*

—— (1952) 'Standard fertilisation and liming as factors in maintaining soil productivity'. *Soil Sci.*, **74**, 181.

ATKINSON, M. R. *et al.* (1967) 'Salt regulation in the mangroves *Rhizophora mucronata* Lam. and *Aegialitis annulata*' R.Br. *Aust. J. Biol. Sci.*, **20**, 589–599.

BAKER, F. and MARTIN, R. 'Studies in the microbiology of the caecum of the horse'. *Cbl. f. Bakt. Abt. II.* **99**, 400.

BARTHEL, CHR. (1949) *Mikroorganismerna i lantbrukets och industriens tjänst.* Stockholm.

BARTELS, J., SCHUBERT, J. and GEIGER, R. (1935) 'Witerung und Bodenfeuchtigkeit'. *Forst. u. Jagdwesen*, **65**, 220.

BAUMEISTER, W. (1954) *Mineralstoffe und Pflanzenwachstum.* Stuttgart.

BAVENDAMM, W. (1928) 'Über das Vorkommen und der Nachweis von Oxidasen bei holzzerstörenden Pilzen'. *Z. f. Pflanzenkrankh. u. Pflanzenschutz*, **38**, 257.

BAVER, L. D. (1956) *Soil Physics.* New York and London, 1956. 3rd Ed. 1970.

BERGER-LANDEFELDT, M. (1957) 'Beiträge zur Ökologie der Pflanzen Nord-afrikanischer Salzpflanzen'. *Vegetatio Acta geobot.*, **7**, 169.

BERGGREN, G. (1956) 'Växtmaterial från träskboplatser i Dagsmosse'. *Sv. bot. tidskr.*, **50**, 97.

BERNSTEIN, L. (1962) 'Salt-affected soils and plants'. *UNESCO Arid Zone Research*, **18**, 139–174.

BERNSTEIN, L. and HAYWARD, H. E. (1958) 'Physiology of salt tolerance'. *Ann. Rev. Plant Physiol.*, **9**, 25.

BETHKE, S. (1957) 'Gegenwärtige und zukünftige Probleme der Welternährungswirtschaft'. *Ber. üb. Landwirtsch.*, **35**, 150.

BIEBL, R. (1937) 'Ökologische und Zellphysiologische Studien an Rotalgen der englischen Südküste'. *Beih. bot. Cbl.*, **57**, 381.

—— (1947) 'Die Resistenz gegen Zink, Bor und Mangan als Mittel zur Kennzeichnung verschiedener pflanzlichen Plasmasorten'. *Sitzungsber. Akad. Wiss. Wien. Mat. nat. Kl.*, *Abt.* **1. B.** 155, 149.

—— (1949) 'Über die Reizung pflanzlicher Plasmen gegen Vanadium'. *Protoplasma*, **39**, 251.

—— (1962) 'Protoplasmatische Ökologie der Pflanzen. Wasser und Temperatur'. *Protoplasmologia* (Vienna), **12**, 1.

BILLINGS, W. D. (1965) *Plants and the Ecosystem.* Macmillan, London.

BILLINGS, W. D. and MOONEY, H. A. (1968) 'The ecology of arctic and alpine plants'. *Biol. Rev.*, **43**, 481–529.

BIRSE, E. L. and DRY, F. T. (1970) Assessment of climatic conditions in Scotland 1. Based on accumulated temperature and potential water deficit. (Map and explanatory pamphlet.) *Soil survey of Scotland.* Macaulay Institute for Soil Research, Aberdeen.

BJÖRKMAN, E. (1941) 'Mykorrhizans utbildning och frekvens hos skogsträd på askgödslade och ogödslade delar av dikad myr'. *Medd. Stat. skogsförs. anst.*, page 255.

—— (1942) 'Über die Bedingungen der Mykorrhizabildung bei Kiefer und Fichte'. *Symb. bot. upsal.*, 6.

—— (1944–1946) 'Skogsplantering ur näringsbiologisk synpunkt'. *Skogsodling. Sv. skogsvårdsfören.*

—— (1948) 'Studier över snöskyttesvampens (*Phacidium infestans* Karst.) biologi samt metoder för snöskyttets bekämpande'. *Medd. Stat. skogsforskn. inst.*, **37**, 1.

—— (1956) 'Über die Natur der Mykorrhizabildung unter besonderer Berücksichtigung der Waldbäume und die Anwendung in der forstlichen Praxis'. *Forstwiss. Cbl.*, **75**, 265.

BLACKMAN, F. F. (1905) 'Optima and limiting factors'. *Ann. Bot.*, **19**.

BLISS, L. C. (1962) 'Net primary production of tundra ecosystems'. In: *Die Stoffproduktion der Pflanzendecke* ed. H. Lieth. Fischer Verlag, Stuttgart.

BLOMFIELD, C. (1954) 'A study of podsolisation'. *Soil. sci.*, **5**.

BOJARSKI, T. (1948) 'On cuticular excretion of mineral salts by green parts of plants'. *Acta soc. bot. Pol.*, **19**, 213.

BOLIN, B. (1970) 'The carbon cycle'. *Scient. Amer.*, **223** (3) 125–32.

BOLLARD, E. G. and BUTLER, G. W. (1966) 'Mineral nutrition of plants'. *A. Rev. Pl. Physiol.*, **17**, 77–112.

BOND, G. (1957) 'Isotopic studies of nitrogen fixation in non legume root nodules'. *Ann. Bot.*, **21**, 513.

BORESCH, K. (1931) *Kreislauf der Stoffe in der Natur*. Handbuch d. Pflanzenernährung und Düngerlehre, vol. **1**, 285.

BORGSTRÖM, G. (1952) 'Jordförstöringen och den odlade jordens framtid'. *Ymer*, **72**, 110.

BORMANN, F. H. and LIKENS, G. E. (1970) 'The nutrient cycles of an ecosystem'. *Scient. Amer.*, **223** (4), 92–101.

BOUSSINGAULT, J. B. (1841) *Ann. Chim. Phys.* (III) **1**, 208.

BOYCE, S. G. (1951) 'Source of atmospheric salts'. *Science*, **113**, 620.

BOYER, J. S. (1965) 'Effects of osmotic water stress on metabolic rates of cotton plants with open stomata'. *Plant Physiol.*, **40**, 229–34.

BOYKO, H. (1967) 'Salt-water agriculture'. *Scient. Amer.*, **216** (3), 89–96.

BOYLE, R. (1661) *The Sceptical Chemist*, Pt. II.

BOYSEN-JENSEN, P. (1932) *Die Stoffproduktion der Pflanzen*. Jena.

BRAARUD, T. (1951) 'Salinity as an ecological factor in marine phytoplankton'. *Physiol. plantarum*, **4**, 28.

BRAY, J. R. and GORHAM, E. (1964) 'Litter production in forests of the world'. In: *Adv. Ecol. Res. 2*, ed. J. B. Cragg. London.

BREITENFELD, E. and MOTHES, K. (1940) 'Bestandesgeschichtliche Untersuchungen an Masurischen Wäldern'. *Schriften d. Physik. Ökonom. Ges.*, **71**, 239.

BRENCHELEY, W. E. (1947) 'The essential nature of certain minor elements for plant nutrition'. *Bot. Rev.*, **13**, 169.

BRENNER, W. (1930) 'Beiträge zur edaphischen Ökologie der Vegetation Finlands'. *Acta bot. fenn.*

BROADBENT, F. E. and NORMAN, A. G. (1947) *Proc. Soil Sci. Soc. Amer.*, **11**, 264.

BROMFIELD, S. M. (1954) 'Reduction of ferric compounds by soil bacteria'. *J. gen. microbiol.*, **11**, 1.

BROWN, L. R. (1970) 'Dwarf high yield wheats and maize'. *Science*, **170**, 518–19.

BURD, J. and MARTIN, J. C. (1924) 'Secular and seasonal changes in the soil solution'. *Soil Sci.*, **18**, 151.

BURGER, H. (1954) 'Der Einfluss des Waldes auf den Wasserhaushalt'. *Mitt. d. Arbeitskreises Wald u. Wasser*, **1**, 8.

BURKHOLDER, P. R. and EVANS, A. W. (1945) 'Further studies on the activity of lichens'. *Bull Torrey Bot.* Club **72**.

BURSTRÖM, H. (1934) 'Über antagonistische Erscheinungen bei der Kationenaufnahme des Hafers'. *Sv. bot. tidskr.*, **28**, 157.

—— (1939) 'Tungmetallernas inverkan på växtens nitratassimilation'. *K. Lantbruksakad. tidskr.*, **78**, 35.

—— (1939) 'Mineralstoffwechsel'. *Fortschr. Bot.*, **12**, 216.

CALLENDAR, G. S. (1958) 'On the amount of carbon dioxide in the atmosphere'. *Tellus*, **10**, 243.

CARBONNIER, CH. (1945–46) 'De ädla lövträdens fördelning pä boniteter i Halland, Skåne och Blekinge enligt rikstaxeringen'. 1945–1946. *Medd. Stat. skogsforskn. inst.*, **37**, 19.

CARLISLE, A., BROWN, A. H. F. and WHITE, E. J. (1966) 'The organic matter and nutrient elements in precipitation beneath a sessile oak (*Quercus petraea*) canopy'. *J. Ecol.*, **54**, 87–98.

CEDERHOLM, Y. (1946) *Biskötsel i övre Norrland*. Norrbottens läns biodlareförb. Boden.

CHAMPAGNAT, A. (1965) 'Protein from petroleum'. *Scient. Amer.*, **213**, 13–17.

CHAPMAN, V. J. (1964) *Coastal Vegetation*. Pergamon Press. Oxford.

CHODAT, R. (1920) *Principes de Botanique*. Geneva.

CLARKE, C. (1958) 'World population'. *Nature, Lond.*, **181**, 1235.

CLARKE, F. W. (1920) *The data of geochemistry*. Washington, DC.

CLARKE, G. L. (1954) *Elements of ecology*. New York.

COLE, L. C. 'Ecology and the population problem'. *Science*, **122**, 831.

COLLANDER, R. (1941) 'The distribution of different cations between root and shoot'. *Acta bot. fenn.*, **29**.

COPPER, R. and RUDOLPH, E. D. (1953) 'The role of lichens in soil formation and plant succession'. *Ecology*, **34**, 805.

CROCKER, R. L. and MAJOR, J. (1955) 'Soil development in relation to vegetation and surface age at Glacier Bay, Alaska'. *J. Ecol.*, **43**, 422.

CURTIS, J. T. (1955) 'A prairie continuum in Wisconsin'. *Ecology*, **36**, 558.

CUTLER, D. W. and CRUMP, L. M. (1935) *Problems of soil microbiology*. London.

DARKEN, M. A. (1953) 'Production of vitamin B_{12} by microorganisms and its occurrence in plant tissues'. *Bot. Rev.*, **19**, 99.

DARWIN, CH. (1881) *Earthworms and vegetable mould*. London.

DAUBENMIRE, R. F. (1947) *Plants and environment*. New York. 2nd ed. 1959.

DELWICHE, C. C. (1970) 'The nitrogen cycle'. *Scient. Amer.*, **223**, 136–46.

DOIARENKO, A. G. (1926) Ausgewählte Arbeiten und Aufsätze, Moscow. Ref. Krauze, M. von (1931) 'Russische Forschungen auf dem Gebiete der Bodenstruktur'. *Landwirtsch. Jb.*, **73**, 603–90.

DOUGLAS, J. W. and SHOLTO, E. H. (1955) 'Soilless cultivation of crops in India'. *Nature, Lond.*, **175**, 884.

DRISCHEL, H. (1940) 'Chlorid-Sulfat- und Nitratgehalt der atmosphärischen Neiderschläge in Bad Reinerz und Oberschreiberhau im Vergleich zu bisher bekannten Werten anderer Orte'. *Der Balneologe*, **7**, 321.

DIMBLEBY, G. W. (1952) 'Soil regeneration on the north-east Yorkshire moors'. *J. Ecol.*, **40**, 331.

DU RIETZ, G. E. (1917) 'Ekskogen vid Borgholm'. *Sveriges Natur*.

—— (1925) 'Gotländska vegetationsstudier'. *Sv. växtsoc. sällsk. handl.*, **2**.

—— (1925) 'Die regionale Gliederung der skandinavischen Vegetation'. Ibid., **8**.

—— (1949) 'Huvudenheter och huvudgränser i svensk myrvegetation'. *Sv. bot. tidskr.*, **43**, 274.

—— (1952) 'Vegetations- och odlingsregionerna som uttryck för klimat och jordmån'. *Trädgårdskonst*. Stockholm.

—— (1953) *Växtvärlden i Älfsborgs län*. Sveriges bebyggelse, landsbygden, Älfsborgs län., Uddevalla.

—— (1954) 'Sydväxtberg'. *Sv. bot. tidskr.*, **48**, 174.

DU RIETZ, T. (1929) 'Studies on the rocks of Rådmansö'. *Sv. Geol. fören. förh.*, **51**, 516.

DVŎRAK, J. (1912) 'Studien über die Stickstoffanhäufung im Boden durch Mikro-Organismen'. *Z. Landw. Vers. stat.*, Austria, **15**, 1077.

EATON, F. M. (1942) 'Toxicity and accumulation of chloride and sulfate salts in plants'. *J. Agric. res.*, **64**, 357.

EBERMAYER, E. (1876) *Die gesamte Lehre der Waldstreu mit Rücksicht auf die chemische Statik des Waldbaues*. Berlin.

EGLE, K. and MUNDING, H. (1954) 'Ausbildung und Funktion von Hämoglobin in den Wurzelknollchen von Leguminosen'. *Biol. Zbl.*, **73**, 577.

EGNÉR, H. (1953) 'Atmosfärens innehåll av växtnäring'. *Sv. jordbruksforskn., årsb.*, 30.

—— (1954) 'Växtnäringsämnen från luften'. *Växtnäringsnytt*, **10**.

EGNÉR, H. and ERIKSSON, E. (1958) 'Current data on the chemical composition of air and precipitation'. *Tellus*, **10**, 281.

EKLUND, B. (1943) 'Sambandet mellan berggrundens beskaffenhet och frekvensen av olika trädslag och skogssamhällen inom Jämtlands län'. *Sv. Skogsvårdsfören. tidskr.*, **41**, 20.

EKSTRÖM, G. (1946; 1951) *Marklära.* (In ERNEST *et al. Jordbrukslära*, Stockholm.)

EMANUELSSON, A., ERIKSSON, E. and EGNÉR, H. (1954) 'Composition of atmospheric precipitation in Sweden'. *Tellus*, **6**, 261.

EMERSON, R. and GREEN, L. (1938) 'Effect of hydrogen-ion concentration on *Chlorella* photosynthesis'. *Pl. Physiol.*, **13**, 157.

ERIKSSON, E. (1952–53) 'Composition of atmospheric precipitation.—I. Nitrogen compounds'. *Tellus*, **4**, 215.

ERIXON, S. (1953) *Byar och bykultur.* A book on the Mälar countryside, published by *Mälarprovinsernas hypoteksförening.* Stockholm.

—— (1956) 'Lantbruket under historisk tid med särskild hänsyn till bondetraditionen'. *Nordisk kultur*, **13**. Published by Sigurd Erixon. Stockholm.

ERNEST, E. *et al. Jordbrukslära.* (1946) Stockholm.

FABER, F. C. V. (1913) 'Uber die Transpiration und den osmotischen Druck bei Mangroven'. *Ber. d. d. bot. Ges.*, **31**, 277.

—— (1923) 'Zur Physiologie der Mangroven'. Ibid., **41**, 227.

—— (1925) 'Untersuchungen über die Physiologie der jawanischen Solfataren-Pflanzen'. *Flora*, **118–19**, 89.

FAGERSTRÖM, L. (1954) 'Växtgeografiska studier i Strömfors—Pyttis skärgård i östra Nyland'. *Acta bot. fenn.*, **54**.

FALCK, R. (1926) 'Über korrosive und destruktive Holzzersetzung und ihre biologische Bedeutung'. *Ber. d. d. bot. Ges.*, **44**, 652.

—— (1930) 'Nachweise der Humusbildung und Humuszehrung durch bestimmte Arten höherer Fadenpilze im Waldboden'. *Forstarchiv.*, **6**, 366.

FAO (1953, 54, 55) *Yearbook of food and agricultural statistics*, Vol. 7, parts 1 and 2. Vol. 8, part 1. Vol. 9, part 1.

FEHÉR, D. (1933) 'Untersuchungen über die Pilzflora der Waldböden'. *Erdészeits Kisérletek*, **35**, 55.

—— (1948) *Researches on the geographical distribution of soil microflora.*

—— Comm. Bot. Inst. Hungarian Univ. of Techn. and Economic Sci. Sopron 21.

FEHÉR, D. and VARGA, L. (1929) 'Untersuchungen über die Protozoen-Fauna des Waldbodens'. *Cbl. f. Bakt. Parasiten und Infekt. Krankh.*, **77**, 524.

FEUSTAL, I. C. and BYERS, H. G. (1936) 'The comparative moisture-absorbing and moisture-retaining capacities of peat and soil mixtures'. *US Dept. agr. Techn. bull.*, **522**.

FIRBAS, F. (1931) 'Untersuchungen über den Wasserhaushalt der Hochmoorpflanzen'. *Jb. wiss. Bot.*, **74**, 459.

—— (1952) 'Einige Berechnungen über die Ernährung der Hochmoore'. *Veröff. d. Geobot. Inst. Rübel*, Zürich, **25**.

FISCHER, G. (1953) 'Untersuchungen über den biologischen Abbau des Lignins durch Mikroorganismen'. *Arch. für Mikrobiol.*, **18**, 397.

FLETCHER, J. E. and MARTIN, W. P. (1948) 'Some effects of algae and molds in the raincrust of desert soils'. *Ecology*, **29**, 95.

FORSSLUND, K.-H. (1948) 'Markfaunan och dess betydelse för humusens nedbrytning'. *Sv. Skogsvårdsfören, tidksr.*, **201**.

FOX, SIR CYRIL (1950) 'Land erosion'. *Endeavour*, **9**, 182.

FRANCK, O. (1951) *Handelsgödseln och dess användning.* Jordbrukslära published by V. Lengdén. Stockholm.

FRENZEL, B. (1957) 'Zur Abgabe von Aminosäuren und Amiden an das Nährmedium durch die Wurzeln von Helianthus annuus'. *Planta*, **49**, 210.

FREY-WYSSLING, A. (1935a) 'Die Stoffausscheidung der höheren Pflanzen. Berlin. 1935.

—— (1935b) 'Die unentbehrliche Elemente der Pflanzennahrung'. *Naturw.*, **23**, 767.

—— (1943) 'Die Blattfermentation als Autolyse'. *Vierteljahrsschr. Nat. Ges.* Zurich, **88**, 176.

—— (1945) *Ernährung und Stoffwechsel der Pflanzen.* Zürich, 1945.

—— (1953) *Submicroscopic morphology of protoplasm.* Amsterdam.

FRIES, C. (1954a) *Vägen till Rom.* Stockholm.

—— (1954b) 'Det svenska landskapet'. *Sveriges Natur.*, **9**.

FRIES, E. (1852) *Botaniska utflygter.* Vol. 2. Stockholm.

FRIES, M. (1951) 'Pollenanalytiska vittnesbörd om senkvartär vegetationsutveckling, särskilt skogshistoria i nordvästra Götaland'. *Acta phytogeogr. Suec.*, **29**.

—— (1958) 'Vegetationens utvecklingshistoria i Varnhemstrakten'. Ibid., **39**.

FRIES, N. (1938) 'Über die Bedeutung von Wuchs-Stoffen für das Wachstum verschiedener Pilze'. *Symb. bot. Upsal.*, **3**.

—— (1943) 'Die Einwirkung von Adermin, Aneurin und Biotin auf das Wachstum einiger Ascomyceten'. Ibid., **7**.

—— (1948) 'The nutrition of fungi from the aspect of growth factor requirements'. *Trans. Brit. mycol. soc.*, **30**.

—— (1950) 'Growth factor requirements of some higher fungi'. *Sv. bot. tidskr.*, **44**, 379.

—— (1952) 'Variations in the content of phosphorus, nuclei acids and adenine in the leaves of some deciduous trees during the autumn'. *Pl. Soil*, **4**, 29.

—— (1953) 'Further studies on mutant strains of *Ophiostoma* which require guanine'. *J. biol. Chem.*, **200**, 325.

FRIES, N. and FORSMAN, B. (1951) 'Quantitative determination of certain nucleic acid derivatives in pea root exudate'. *Physiologia Pl.*, **4**, 410, 410.

FRÖDIN, J. (1919) 'Über nitrophile Pflanzenformationen auf den Almen Jämtlands'. *Bot. Not.*, 271.

FÅHRAEUS, G. (1944) 'The course of cellulose decomposition by *Cytophaga*'. *Ann. Agric. coll.*, *Sweden*, **12**, 1.

—— (1947) 'Studies in the cellulose decomposition by *Cytophaga*'. *Symb. bot. upsal.*, **9**, 2.

GALE, E. F. (1940) 'Enzymes concerned in the primary utilization of amino acids by bacteria'. *Bact. Rev.*, **4**, 134.

GARRETT, S. D. (1951) 'Ecological groups of soil fungi: A survey of substrate relationships'. *New Phytologist*, **50**, 149.

GEIGER, R. (1950, 1965) *Das Klima der bodennahen Luftschicht*. Braunschweig.

GESSNER, F. (1933). 'Die Produktionsbiologie der Ostsee'. *Die Naturwiss.*, 21.

—— (1940) 'Untersuchungen über die Osmoregulation der Wasserpflanzen'. *Protoplasma*, **34**, 593.

—— (1943) 'Die assimilatorische Leistung des Phytoplanktons, bezogen auf seinen Chlorophyllgehalt'. *Z. Bot.*, **38**, 414.

GESSNER, F. and SCHUMAN-PETERSON (1948) 'Untersuchungen über den Wasserhaushalt der Pflanzen bei Stickstoffmangel'. *Z. f. Naturforsch.*, **36**, 36.

GILBERT, F. A. (1951) 'The place of sulfur in plant nutrition'. *Bot. Rev.*, **17**, 671.

GILBERT, F. A. and ROBINSON, R. F. (1957) 'Food from fungi'. *Economic Bot.*, **11**, 126.

GINZBERGER, A. (1925) 'Der Einfluss des Meereswassers auf die Gliederung der süddalmatischen Küstenvegetation'. *Öst. bot. Z.*, **74**, 114.

GLIMBERG, C.-F. (1955) 'Sandflykten i Skåne—ett gammalt naturskyddsproblem'. *Sveriges natur.*, 124.

GOOR, C. P. V. (1951). 'The influence of nitrogen on the growth of japanese larch (*Larix leptolepis*)'. *Pl. Soil*, **5**, 29.

GORHAM, E. (1953) 'The development of the humus layer in some woodlands of the English Lake District'. *J. Ecol.*, **41**, 123.

GRADMANN, H. (1928) 'Untersuchungen über die Wasserverhältnisse des Bodens als Grundlage des Pflanzenwachstums'. *Jb. wiss. Bot.*, **69**.

GREB, H. (1957) 'Der Einfluss tiefer Temperatur auf die Wasser- und Stickstoffaufnahme der Pflanzen und ihre Bedeutung für das Xeromorphieproblem'. *Planta*, **48**, 523.

GREIG-SMITH, P. (1964) *Quantitative Plant Ecology*. 2nd edition. Butterworths. London.

GRÜMMER, G. (1955) *Die gegenseitige Beeinflussung höherer Pflanzen. Allelopathie*. Jena.

GUNDERSEN, K. (1956) 'Biokjemisk paleontologi eller 100 000 år gammelt klorofyll'. *Naturen*, **80**, 382.

GUSTAFSON, F. G. (1953) 'Influence of photoperiod on thiamine, riboflavin and niacin content of green plants'. *Amer. J. Bot.*, **40**, 256.

GUSTAFSSON, K. A. (1954) 'Skötseln av ängen och hagen'. *Sveriges natur*, **26–34**.

GYLLENBERG, H. (1956) 'The "rhizosphere effect" of graminaceous plants in virgin soils. II. Nutritional characteristics of nonsporogenous bacteria associated with the roots'. *Physiologia Pl.*, **9**, 119.

—— (1956) 'III. Comparison with the effect of other plants'. Ibid., **9**, 441–5.

GÜTTLER-TETSCHEN-LIEBWERD, R. (1941) 'Über den Gehalt wildwachsender Pflanzen an Kalium, Phosphor, Eisen und Mangan'. *Der Forschungsdienst*, **11**, 490.

HAECKEL, E. H. (1866) *Generelle Morphologie der Organismen*. Berlin.

HALDEN, B. (1923) *Svenska jordarter*. Stockholm.

—— (1926) 'Studier över skogsbeståndens inverkan på markfuktighetens fördelning hos skilda jordarter'. *Sv. Skogsvårdsfören, tidksr.*, 125.

—— (1941) *Skogsmarken* (in *Kort handledning i skogshushållning* published by *Norrlands skogsvårdsförbund*). Stockholm.

—— (1950) 'Fragment av en studie i kalkväxtfrågan, speciellt floran i "kalkbranterna"'. *Svensk bot. Tidskr.*, **44**, 534.

HALE, J. B. and HEINTZE, S. G. (1946) 'Manganese toxicity affecting crops on acid soils'. *Nature, Lond.*, **157**, 554.

HALES, S. (1727) *Vegetable Statiks*. London.

HANNIG, E. (1912) 'Untersuchungen über die Verteilung des osmotischen Drucks in der Pflanze in Hinsicht auf die Wasserleitung'. *Ber. dtsch. bot. Ges.*, **30**, 194.

HANSEN, C. (1926). 'The water-retaining power of the soil'. *J. Ecol.*, **14**, 111–119.

HANSON, H. C. (1958) 'Principles concerned in the formation and classification of communities'. *Bot. Rev.*, **24**, 65.

HARDER, R. (1921) 'Kritische Versuche zu Blackmans Theorie der "begrenzenden" Faktoren bei der Kohlensäureassimilation'. *Jb. wiss. Bot.*, **60**.

HAYWARD, H. E. and BERNSTEIN, L. (1958) 'Plant-growth relationships on salt-affected soils'. *Bot. Rev.*, **24**, 588.

HEIKENHEIMO, O. (1919) 'Der Einfluss der Brandwirtschaft auf die Wälder Finlands'. (Ref.) *Acta forest. fenn.*, **4**.

HEIKURAINEN, L. (1957) 'Über Veränderungen in der Wurzelverhältnissen der Kiefernbestände auf Moorböden im Laufe des Jahres'. Ibid., **65**, 1.

HELLRIEGEL, H. and WILFARTH, H. (1883) *Ztschr. Pübenzucker-Ind.* (supplement).

HELLQUIST, E. (1948) *Svensk etymologisk ordbok*. Lund.

HENRICI, A. T. (1947) *Moulds, yeasts and actinomycetes.* New York.

HENRICI, M. (1946) 'Transpiration of S. African plant-associations.— Indigenous and exotic trees under semiarid conditions'. *Dep. Agr. Sc. Bull.*, 248.

HESSELMAN, H. (1904) 'Zur Kenntnis des Pflanzenlebens schwedischer Laubwiesen'. *Beih. bot. Cbl.*, **17**.

—— (1905) 'Svenska löfängar'. *Sv. Skogsvärdsfören, tidskr.*, **3**, 1.

—— (1908) 'Skogsväxten och vegetationen på Gotlands hällmarker'. *Svensk. bot. Tidskr.*, **2**.

—— (1915) 'Om förekomsten av rutmark på Gotland'. *Geolog. fören. förh.*, **37**, 481.

—— (1917a) 'Studier över salpeterbildningen i naturliga jordmåner och dess betydelse i växtekologiskt avseende'. *Medd. Stat. skogsförsknings-anst.*, **12**, 297.

—— (1917b) 'Om vissa skogsföryngringsåtgärders inverkan på salpeterbildningen i marken och dess betydelse för barrskogens föryngring'. Ibid., **13–14**, 923.

—— (1926) 'Studier över barrskogens humustäcke, dess egenskaper och beroende av skogsvården'. Ibid., 169.

—— (1932) 'Om klimatets humiditet i vårt land och dess inverkan på mark, vegetation och skog'. Ibid., 515.

—— (1937) 'Om humustäckets beroende av beståndets ålder och sammansättning etc. Ibid., 533.

HEVITT, E. J. (1948) 'Relation of manganese and some other metals to the iron status of plants'. *Nature, Lond.*, **161**, 489

—— (1963) 'The essential nutrient elements: requirements and interactions in plants'. In *Plant Physiology* Vol. 3, Ch. 2, 137–60, ed. Stewart, F. C. Academic Press. New York.

HJELMQVIST, H. (1952) 'Några sädeskornsavtryck från Sydsveriges stenålder'. *Bot. Not.*, 331.

HJULSTRÖM, F. (1940) 'Jordförstöringen i Förenta Staterna och andra länder—ett aktuellt problem'. *Ymer*, **60**, 2.

HOAGLAND, D. R. and ARNON, D. L. (1948) 'Some problems of plant nutrition'. *Scient. Mon.*, **67**, 201.

HOFMAN-BANG, O. (1905) 'Studien über schwedische Fluss- und Quellwässer'. *Bull. geol. inst. univ. Upsala*, **6**, 101.

HOLLEMAN, A. F. (1952) (Wiberg, E.) *Lehrbuch der anorganischen Chemie.* Berlin.

HOME, F. (1757) *The Principles of Agriculture and Vegetation.* Edinburgh.

HOMÉN, TH. (1893) *Om nattfrosten.* Helsingfors.

HONERT, T. H. VAN DEN (1930) 'Carbon dioxide assimilation and limiting factors'. *Rec. d. trav. bot. neerl.*, **27**, 149.

HYAMS, E. (1952) *Soil and civilisation.* London.

HÅRD AF SEGERSTAD (1924) *Sydsvenska florans växtgeografiska huvudgrupper.* Malmö.

HOFLER, K. (1951) 'Plasmolyse mit Natriumkarbonat'. *Protoplasma*, **40**, 426.

—— (1952) 'Zur Kenntnis der Plasmahautschichten'. *Ber. d. d. bot. Ges.*, **65**, 391.

HÖFLER, K. and WEIXL-HOFMANN, H. (1939) 'Salzpermeabilität und Salzresistenz der Zellen von *Suaeda maritima*'. *Protoplasma*, **32**.

HÖGBOM, A. G. (1899) 'Ragundadalens geologi'. *Sveriges geol. unders. förh.*, C **182**.

HORT, A. (1949) *Theophrastus: Enquiry into Plants* (with an English translation). London.

HUGHES, A. P. and COCKSHULL, K. E. (1969) 'Effect of carbon dioxide concentration on the growth of *Callistephus chinensis* cultivar Johannistag *Ann. Bot.*, **33**, 351–65.

HURD, R. G. (1968) 'Effects of CO_2-enrichment on the growth of young tomato plants in low light'. *Ann. Bot.*, **32**, 531–42.

ILJIN, W. S. (1923) 'Über verschiedene Salzbeständigkeit der Pflanzen'. *Sitzungsber. d. K. böhm. Ges. d. Wiss. Kl.*, **II**.

—— (1933) 'Zusammensetzung der Salze in den Pflanzen auf verschiedenen Standorten. Kalkpflanzen'. *Beich. bot. Cbl.*, **50**, 95.

—— (1938 and 1940) *Bull. Assoc. Russ.*, Prague. 7:43. Ref. in *Fortschr. Bot.*, **8**, 245; **10**, 230.

INGENHOUSZ, J. (1779) *Experiments upon vegetables, discovering their great power of purifying common air in the sunshine and of injuring in the shade and at night*. London.

INGESTAD, T. (1970) 'A definition of optimum nutrient requirements in birch seedlings I'. *Physiol. Plant.*, **23**, 1127–38.

—— (1970) 'A definition of optimum nutrient requirements in birch seedlings II'. *Physiol. Pl.*, **24**, 118–25.

IVERSEN, J. (1929) 'Studien über die pH-Verhältnisse dänischer Gewässer und ihren Einfluss auf die Hydrophyten-Vegetation'. *Bot. Tidsskr.*, **40**, 277.

—— (1941) 'Landnam. Danmarks Stenalder. En pollenanalytisk Undersögelse over det förste Landbrugs Indvirking paa Vegetationsudviklingen'. *Danmarks geol. undersögelse*, II, 66. Köpenhamn.

JACOBSEN, T. and ADAMS, R. M. (1958) 'Salt and silt in ancient Mesopotamian agriculture'. *Science*, **128**, 1251.

JANKE, A. (1949) 'Der Abbau der Zellulose durch Mikroorganismen'. *Österr. bot. Z.*, **96**, 399.

JENNY, H. and GIESEKING, J. E. (1936) 'Behaviour of polyvalent cations in the base exchange'. *Soil Sci.*, **42**, 273.

JENSEN, H. L. and BETTY, R. C. (1943) 'Nitrogen fixation in leguminous plants III. The importance of molybdenum in symbiotic nitrogen fixation'. *Proc. Linn. Soc. NSW*, **68**, 1.

JESSEN, K. F. W. (1948) *Botanik der Gegewart und Vorzeit*. Waltham. Mass, USA.

JIRLOW, R. (1954) 'Årder och plog, Skåne'. *Skånes hembygdsförbunds årsbok.*

—— (1958) 'Den svenska plogens historia'. *K. Skogs. och lantbr. akad. tidskr.*, **97**, 121.

JONASSEN, H. (1950) 'Recent pollen sedimentation and Jutland heath diagrams'. *Dansk bot. Ark.*, **13**.

JOHANSSON, K. (1897) 'Hufvuddragen af Gotlands växttopografi och växtgeografi grundade på en kritisk behandling af dess kalkväxtflora'. *K. Sv. Vet.akad. handb.*, **29**, 25.

JOHNSSON, P. (1921) *Loushultsbranden.* Osby. 1921.

JOHNSON, J. M. and BUTLER, G. W. (1957) 'Iodine content of pasture plants'. *Physiologica Pl.*, **10**, 100.

JULIN, E. (1948) 'Vessers udde'. *Acta phytogeogr. Suec.*, **23**.

KAINDL, K. (1953) 'Untersuchung über die Aufnahme von P^{32} markiertem primärem Kaliumphosphat durch Blattoberfläche' *Die Bodenkultur*, **7**, 324.

KALELA, E. K. (1949) 'On the horizontal roots in pine and spruce stand I'. *Acta forest. fenn.*, **57**, 2.

—— (1957) 'Über Veränderungen in den Wurzelverhältnissen der Kiefernbestände im Laufe der Vegetationsperiode'. Ibid., **65**, 1.

KARLSSON, N. (1952) 'Kalium i marken'. *K. Lantbruksakad. tisdkr.*, **91**, 297.

KASSAS, M. (1952) 'Habitat and plant communities in the Egyptian desert'. *J. Ecol.*, **40**, 342.

KAUSCH, W. (1954) 'Saugkraft und Wassernachleitung im Boden als physiologische Faktoren'. *Planta*, **45**, 217.

KELLER, B. (1925a) 'Halophyten und Xerophytenstudien. *Planta*, **13**, 224.

—— (1925b) 'Die Vegetation auf den Salzböden der russischen Halbwüsten und Wüsten'. *Z. Bot.*, **18**, 113.

KHAN, M. I. (1955) 'Water relations of plants in arid regions'. *Empire Forest Rev.*, **33**, 124.

KIENDL, J. (1954) 'Zur Transpirationsmessung an Sumpf- und Wasserpflanzen. *Ber. d. d. bot. Ges.*, **67**, 243.

KILLIAN, C. and FAUREL, L. (1933) 'Observations sur la pression osmotique des végétaux désertique et subdésertiques de l'Algérie. *Bull. Soc. Bot.*, **80**, 775.

KIVENHEIMO, V. J. (1947) 'Untersuchungen über die Wurzelsysteme der Samenpflanzen in der Bodenvegetation der Wälder Finnlands'. *Annal. bot. Soc. zool. bot. fenn.* Vanamo, 22.

KLAPP, E. L. (1930) 'Studien über Zusammenhänge von Bodenreaktion, Verbreitung der Wiesen-Pflanzen und Wiesenerträgen'. *Landw. Jb.*, **71**, 807.

KLINDT-JENSEN, O. (1956) 'Landbrug i Norden i fornhistorisk tid.

Nordisk Kultur 13. Lantbruk och bebyggelse, published by Sigurd Erixon. Stockholm.

KNUDSEN, R. and MAURITZ-HANSSON, H. (1939) 'Om, produktionen av lövförna och dennas sammansättning i ett mellansvenskt björkbestånd'. *Sv. skogsvårdsfören. Tidskr.*, **37**, 339–49.

KOLKWITZ, R. (1917; 1918; 1919) 'Über die Standorte der Salz-Pflanzen'. *Ber. d. d. bot. Ges.*, **35, 36, 37**.

KOLLMANNSPERGER, F. (1951–52) 'Über die Bedeutung der Regenwürmer für die Fruchtbarkeit des Bodens. Decheniana'. *Verh. d. Nat. hist. Ver. d. Rheinlande und Westfalens'*, **105–6**.

KÖLREUTER, J. G. (1761) *Vorlaüfige Nachricht von einigen das Geschlecht der Pflanzen betreffenden Versuchen und Beobachtungen.* Leipzig. Fortsetzungen 1, 2, und 3. Leipzig 1763, 1764, 1766.

KOSMAT, H. (1956) 'Beurteilung der Bodenstruktur mit Hilfe des Aufgusstestes'. *Die Bodenkultur*, **9**, 4.

KOSTYTSCHEV, S. (1926) 'Lehrbuch der Pflazenphysiologie'. Berlin.

KOTILAINEN, M. J. and SALMI, V. (1950) 'Two serpentinicolous forms of *Cerastium vulgatum* L. in Finland'. *Acta. Soc. Vanamo*, **5**, 1.

KRAMER, P. J. (1944) 'Soil moisture in relation to plant growth'. *Bot. Rev.*, **10**, 525.

—— (1945) 'Absorption of water by plants'. Ibid., **11**, 210.

—— (1969) *Plant and Soil Water Relationships: a modern synthesis.* McGraw-Hill, New York.

KRAUSE, M. (1931) 'Russische Forschungen auf dem Gebiete der Bodenstruktur'. *Landw. Jb.*, **73**, 603.

KROGH, A. (1904) *On the tension of carbonic acid in natural waters and especially in the sea. Meddelelser om Grönland.* Copenhagen.

KROOK, J. (1765) *Tankar om Swedjande och huruwida det tål inskränkning uti norra delen av Tavastland, Savolax och Carelen.* Stockholm.

KRUCKEBERG, A. R. (1951) 'Intraspecific variability in the responds of certain native plant species to serpentine soil'. *Am. J. Bot.*, **38**, 408.

KUBIËNA, W. L. (1953) *Bestimmungsbuch und Systematik der Böden Europas.* Stuttgart.

KÖGL, F. and FRIES, N. (1937) 'Über den Einfluss von Biotin, Aneurin und Meso-Inosit auf das Wachstum verschiedener Pilzarten'. *Z. physiol. Chemie.*, **249**, 93.

KÖIE, M. (1951) 'Relation of vegetation, soil and subsoil in Denmark'. *Dansk bot. ark.*, **14**, 7.

LAATSCH, W. (1948) 'Untersuchungen über die Bildung und Anreicherung von Humusstoffen'. *Beitr. z. Agrarwiss.*

—— (1954) *Dynamik der mitteleuropäischen Mineralböden.* Dresden and Leipzig.

LÅG, J. and BERGSETH, H. (1957) 'Influence of salt concentration and of some of the components of the sea-salt on the flocculation of Norwegian clay material'. *Medd. fra Norges Landbrukshögskole*, **36**, 5.

LÅG, J. and EINEVOLL, O. (1954) 'Preliminary studies on the water permeability of raw humus in podzol profiles in the western part of Norway'. Ibid., **34**, 525.

LAITAKARI, E. (1927) 'The root system of pine (*Pinus silvestris*)'. *Acta for. fenn.*, **33**, 1.

—— (1935) 'The root system of birch (*Betula verrucosa* and *odorata*). Ibid., **41**, 1–168.

LAMER, M. (1957) *The world fertiliser economy*. Stanford, Calif.

LANGE, O. L. (1965) 'Der CO_2-Gaswechsel von Flechten bei tiefen Temperaturen'. *Planta*, **64**, 1–19.

LANGLET, O. (1936) 'Studier över tallens fysiologiska variabilitet och dess samband med klimatet'. *Medd. Stat. skogsforskn.inst.*, **29**, 219.

LAVOISIER, A. L. (1789) *Traité élémentaire de la Chimie*. Paris.

LAW, F. (1958) 'Measurement of rainfall, interception and evaporation losses in a plantation of Sitka spruce trees'. *Intern. Assoc. Sci. Hydrol. Gen. Assembly* (Toronto), **2**, 397.

LEBEDEFF, A. F. (1927) 'The movement of ground and soil waters'. *Proc. 1st intern. congr. soil sci.*, **1**, 459.

LEPESCHKIN, W. (1948) *Am. J. Bot.*, **35**, 258.

LEVRING, T. (1945) 'Some culture experiments with marine plankton diatoms'. *Göteborgs K. vet. och. vitt. samh. handl. 6 ser. B.*, **3**.

LIEBIG, J. V. (1862) 'Die Chemie in der Anwendung auf Agrikultur 1840'. 7. Aufl. Braunschweig.

LINDBERG, S. and NORMING, H. (1943) 'Om produktionen av barrförna och dennas sammansättning i ett granbestånd invid Stockholm'. *Sv. Skogsvårdsfören. tidskr.*, 353.

LINDEBERG, G. (1944) 'Über die Physiologie Ligninabbauenden Bodenhymenomyceten'. *Symb. bot. upsal.*, **VIII**.

—— (1946) 'Thiamin and growth of litter-decomposing hymenomycetes'. *Bot. Not.*

—— (1948a) 'On the occurrence of polyphenol oxidases in soil inhabiting basidiomycetes'. *Pl. Physiol.*, **1**, 196.

—— (1948b) 'Some properties of the catecholases of litter-decomposing and parasite hymenomycetes'. Ibid., **1**, 401.

LINDEBERG, G. and KORJUS, M. (1949) 'Gallic acid and growth of *Marasmius foetidus*'. *Pl. Physiol.*, **1**, 103.

—— (1955) 'Ligninabbau und Phenoloxydasebildung der Bodenhymenomyceten'. *Zeitschr. f. Pflanzenernähr., Düngung, Bodenkunde*, **69**, 142.

LINDELL, TH. and ALFONS H. (1953) *I Östergötland*. Uddevalla.

LINDNER, E. (1949/50) 'Zellphysiologische Resistenzuntersuchungen an Ruderal-, Wiesen- und Kulturpflanzen'. *Protoplasma*, **39**, 507.

LINDQUIST, B. (1938) 'Dalby söderskog'. *Acta phytogeogr. Suec.*, **10**.

LINNAEUS, C. (1732) *Lapplandsresan 1732*. Edit. Stockholm, 1957.

—— (1734) *Dalaresan 1734*. Edit. Uppsala, 1953.

—— (1741a) *Öländska resan 1741*. Edit. Stockholm, 1957.

—— (1741b) *Gotländska resan 1741*. Edit. Stockholm, 1957.

—— (1746) *Västgötaresan 1746*. Edit. Stockholm, 1947.

—— (1749a) *Skånska resan 1749*. Edit. Lund, 1874.

—— (1749b) *Exercitatio Botanico-Physica 1749*. Edit. Uppsala, 1828.

—— (1749c) *Oeconomia Naturae*. Stockholm, 1749. Svensk översättning, 1750.

—— (1751) *Philosophia Botanica*. Stockholm.

LOHAMMAR, G. (1938) 'Wasserchemie und höhere Vegetation schwedischer Seen'. *Symb. bot. upsal.*, **III**, 1.

LOVE, L. D. (1955) 'The effect on stream flow of the killing of spruce and pine by the Engelmann spruce beetle'. *Trans. Amer. geophys. Un.*, **36**, 113.

LUKKALA, O. J. (1942) 'Niederschlagsmessungen in verschiedenartigen Beständen'. *Acta forest. fenn.*, **50**, 23.

LUNDH, A. (1951) 'Studies on the vegetation and hydrochemistry of scanian lakes'. *Bot. Not.*, Supplement 3:1.

LUNDBLAD, K. (1924) 'Ett bidrag till kännedomen om brunjords- eller mulljordstypens egenskaper och degeneration i Södra Sverige'. *Medd. Stat. skogsförs.anst.*

—— (1934) 'Studies on podsols and brown forest soils'. *Soil Sci.*, **37**, 137.

—— (1936) 'Studies on podsols and brown forest soils II, III. Ibid., **41**, 295 and 383.

—— (1946) 'Mikroelement och bristsjukdomar hos odlade växter'. *K. Lantbruksakad Tidssker.*, **435**.

—— (1951) 'Olika lantbruks- och köksväxters förbrukning av växtnäringsämnen'. *Växtnäringsnytt*, 3. (special number).

—— (1952) 'Gödslingens inverkan på vegetation och mark'. *Statens Jordbruksförsök. Medd.*, **42**.

LUNDEGÅRDH, H. (1921) 'Ecological studies in the assimilation of certain forest-plants'. *Sv. bot. tidskr.*, **15**, 46.

—— (1924) *Der Kreislauf der Kohlensäure in der Natur*. Jena.

—— (1949) 'Klima und Boden'. Jena.

LUNDEGÅRDH, H. and BURSTRÖM, H. (1933) 'Atmung und Ionenaufnahme'. *Planta*, **18**, 683.

LUNDEGÅRDH, H. and STENLID, G. (1944) 'On the exudation of nucleotides and flavanone from living roots'. *Ark. f. bot.*, **31**.

LUNDQVIST, J. (1965) 'South-facing hills and mountains'. *Acta phytogeogr. suecica*, **50**, 216.

LUNDQUIST, L. O. (1956) 'Xeromorphose in Beziehung zur Wasser- und Stickstoffaufnahme'. *Svensk bot. Tidskr.*, **50**, 361.

LUTZ, H. J. and CHANDLER, R. F. (1946) *Forest Soils*. New York.

MCELROY, W. D. and NASON, A. (1954) 'Mechanism of action of micronutrient elements in enzyme systems'. *A. Rev. Pl. Physiol.*, **5**, 1.

MAGISTAD, O. C. (1945) 'Plant growth relations on saline and alkali soils'. *Bot. Rev.*, **XI**, 181.

MAGNUSSON, N., LUNDQUIST, G. and GRANLUND, E. (1957) *Sveriges geologi*. Stockholm.

MALMER, N. and SJÖRS, H. (1955) 'Some determinations of elementary constituents in mire plants and peat.' *Medd. fr. Lunds bot. museum*, **109**.

MALMSTRÖM, C. (1923) 'Degerö stormyr'. *Medd. Stat. skogsförs.anst.*, **20**, 69.

—— (1928) 'Våra torvmarker ur skogsdikningssynpunkt'. Ibid., 251.

—— (1935) 'Om näringsförhållandenas betydelse för torvmarkers skogsproduktiva förmåga'. Ibid., **28**, 571.

—— (1935) 'Hallands skogar under de senaste 300 åren'. Ibid., 174.

—— (1940) 'Korta anvisningar för bedömning av torvmarkers lämplighet för skogsdikning'. *Skogliga rön nr 6. Stat. skogsförs. anst.*

—— (1943) 'Skogliga gödslingsförsök på dikade torvmarker'. *Norrl. skogsv. fören. tidskr.*

—— (1947) 'Några markförbättringsförsök på nordsvenska tallhedar. *Medd. Stat. skogsforskn.inst.*, **36**, 5.

—— (1949) 'Studier över skogstyper och trädslagsfördelning inom Västerbottens län'. Ibid., **37**, 1.

—— (1957) 'Om den svenska markens utnyttjande för bete, åker, äng och skog genom tiderna, och orsakerna till rörligheten i utnyttjandet'. *K. Lantbruksakad. Tidskr.*, **90**, 292.

—— (1952) 'Svenska gödslingsförsök för belysande av de näringsekologiska villkoren för skogsväxt på torvmark'. *Communicationes Inst. Forest. Fenn.*, **40**.

MÅRTENSSON, O. and BERGGREN, A. (1954) 'Some notes on the ecology of the "Copper mosses"'. *Oikos*, **5**, 99.

MARTHALER, H. (1937) 'Die Stickstoffernährung der Ruderalpflanzen'. *Jb. wiss. Bot.*, **85**, 76.

MATTSON, S. (1931). The laws of soil colloidal behaviour: VIII. Forms and functions of water'. *Soil Sci.*, **33**, 301.

—— (1938) 'The constitution of the pedosphere'. *Ann. Agric. coll. Sweden*, **5**, 261.

—— (1942) 'Laws of ionic exchange'. Ibid., **10**, 56.

—— (1946) *Kalkbehov och kalkning*. In Ernest *et al. Jordbrukslära*. Stockholm.

MATTSON, S., ERIKSSON, E. and VAHTRAS, K. (1948) 'Effects of excessive liming on leached acid soils'. *Ann. Agric. coll. Sweden*, **15**, 291.

MATTSON, S. and GUSTAFSSON, Y. (1934) 'The chemical characteristics of soil profiles I. The podzol'. Ibid., **1**, 33.

MATTSON, S. and KARLSSON, N. (1944) 'The pedography of hydrologic soil series: VI. The composition and base status of the vegetation in relation to the soil'. Ibid., **12**, 186.

MATTSON, S. and KOUTLER-ANDERSSON, E. (1941). 'The acid-base condition in vegetation, litter and humus. I. Acids, acidoids and bases in relation to decomposition'. Ibid., **9**.

—— (1943) 'The acid-base condition in vegetation, litter and humus. VI. Ammonia fixation and humus nitrogen. Ibid., **11**, 107.

—— (1944) 'The acid-base condition in vegetation, litter and humus. VIII. Forms of acidity'. Ibid., **12**, 70.

—— (1945) 'The acid-base condition in vegetation, litter and humus. IX. Forms of bases'. Ibid., **13**, 153.

—— (1954) 'Geochemistry of a raised bog'. *K. Lantbruksakad. ann.*, **21**, 321.

—— (1955) 'Some nitrogen relationships'. Ibid., **22**, 219 and 224.

MATTSON, S., SANDBERG, G. and TERNING, P. E. (1944) 'Atmospheric salts in relation to soil and peat formation and plant composition'. *Ann. Agr. Coll. Sweden*, **12**, 101.

MATTSON, S., WILLIAMS, E. G., KOUTLER-ANDERSSON, E. and BARKHOFF, E. (1950) 'Forms of P in the Lanna soil'. Ibid., **17**, 130.

MEAGHER, AV. R., JOHNSSON, C. and STOUT, P. R. (1952) 'Molybdenum requirement of leguminous plants supplied with fixed nitrogen'. *Plant Physiol.*, **27**, 223.

MEETHAM, A. R. (1952) *Atmospheric Pollution*. London.

MEFFERT, M. E. and STRATMANN H. (1954) 'Über die unsterile Glasskulturen von *Scenedesmus obliquus*'. *Cbl. f. Bakt. Parasitenk. Inf. Krankh. und Hygiene*, **108**, 154.

MEINECKE, TH. (1927) *Die Kohlenstoffernährung des Waldes*. Berlin.

MELIN, E. (1924) 'Die Phosphatiden als ökologischer Faktor im Boden'. *Sv. bot. Tidskr.*, **18**, 459.

—— (1930) 'Biological decomposition of some types of litters from north American forests'. *Ecology*, **11**, 72.

—— (1941) 'Växthormoner och deras betydelse'. *K. Lantbruksakad. Tidskr.*, **80**, 413.

—— (1946) 'Der Einfluss von Waldstreuextrakten auf das Wachstum von Bodenpilzen, mit besonderer Berücksichtigung der Wurzelpilze von Bäumen'. *Symb. bot. upsal.*, **VIII**.

—— (1952) 'Transport of labelled nitrogen from an ammonium source to pine seedlings through mycorrhizal mycelium'. *Sv. bot. Tidskr.*, **46**, 281.

MELIN, E. and LINDEBERG, G. (1939) 'Über den Einfluss von Aneurin und Biotin auf das Wachstum einiger Mykorrhizenpilze'. *Bot. Not.*

MELIN, E. and NILSSON, H. (1950) 'Transfer of radioactive phosphorous to pine seedlings by means of mycorrhizal hyphae'. *Physiologica Pl.* **3**, 88–92.

—— (1952) 'Transport of labelled nitrogen from an ammonium source to pine seedlings through mycorrhizal medium'. *Sv. bot. Tidskr.*, **46**, 281.

—— (1953) Transfer of labelled nitrogen from glutamic acid to pine seedlings through the mycelium of *Boletus variegatus* (Sw.) Fr.' *Nature*, **171**.

—— (1954) Transport of labelled phosphorous to pine seedlings through the mycelium of *Cortinarius glaucops*'. *Sv. bot. Tidskr.*, **48**, 555.

—— (1957) 'Transport of C^{14}-labelled photosynthate to the fungal associate of pine mycorrhiza'. Ibid., **51**, 166.

MELIN, E. and NORKRANS, B. (1942) 'Über den Einfluss der Pyrimidin- und Thiazolkomponente des Aneurins auf das Wachstum von Wurzelpilzen'. Ibid., **36**, 271.

MELIN, E. and NYMAN, B. (1940) 'Weitere Untersuchungen über die Wirkung von Aneurin und Biotin auf das Wachstum von Wurzelpflanzen'. *Arch. Mikrobiol.*, **11**, 1.

MELIN, E. and RAMA, D. V. S. (1954) 'Influence of root-metabolites on the growth of tree mycorrhizal fungi'. *Pl. Physiol.*, F. 851.

MELIN, R. (1957) *Atlas över Sverige 37–38*. Stockholm.

METSÄVAINIO, K. (1931) 'Untersuchungen über das Wurzelsystem der Moorpflanzen'. *Ann. bot. Soc. zool. bot. fenn.*, Vanamo, **1**, 1–417.

MEVIUS, W. (1921) 'Beiträge zur Physiologie "kalkfeindlicher" Gewächse'. *Jb. wiss. Bot.*, **60**, 148.

—— (1931) 'Die Bestimmung des Fruchtbarkeitszustandes des Bodens auf Grund des natürlichen Pflanzenbestandes'. *Handb. d. Bodenlehre*. (E. Blanck), **8**, 61.

MEVIUS, W. and ENGEL, H. (1929) 'Die Wirkung der Ammoniumsalze in ihrer Abhängigkeit von der Wasserstoffionenkonzentration'. *Planta*, **9**, 1.

MEYER, L. (1942) 'Experimenteller Beitrag zu makrobiologischen Wirkungen auf Humus und Bodenbildung'. *Bodenk Pfl. ernähr*, **29**, 119.

MEYER, B. S. and ANDERSON, D. B. (1949) *Plant Physiology*. New York.

MILLAR, C. E. (1955) *Soil Fertility*. New York.

MILLER, E. C. (1938) *Plant Physiology*. New York and London.

MITSCHERLICH, E. A. (1909) 'Das Gesetz des Minimums und das Gesetz des abnehmenden Bodenertrages'. *Landw. Jb.*, **38**, 537.

MOHR, E. C. J. and VAN BAREN, F. A. (1954) *Tropical Soils*. Haag.

MOLCHANOV, A. A. (1963) *The Hydrological Role of Forests*. Translated from Russian. Published by the Israel program for scientific translations. Jerusalem.

MÖLLER, C. M. (1945) 'Untersuchungen über Laubmenge, Stoffverlust und Stoffproduktion des Waldes'. *Det forstl. forsögsv. i Danmark*, **17**.

—— (1945) 'Untersuchungen über Laubmenge, Stoffverlust und Stoffproduktion des Waldes'. *Forstl. Forsøksv. Danm.*, **17**.

—— (1957) 'Über die Bedingungen für das Erreichen hohen Alters bei Waldbäumen'. *Forstw. Cbl.*, **76**, 321.

MÖLLER, C. M. MÜLLER, D. and NIELSEN, J. (1954) 'Respiration in stem and branches of beech'. *Forstl. Forsøksv. Danm.*, **21**, 273–301.

—— (1954) 'Ein Diagramm der Stoffproduktion im Buchenwald'. *Ber. schweiz. bot. Ges.*, **64**, 487.

—— (1957) Stormfaldets betydning for dansk skovbrug'. *Dansk skovforen. tidskr.*, **42**, 526.

MONTFORT, C. (1926) 'Einfluss ausgeglichener Salzlösungen auf Mesophyll- und Schliesszellen; Kritik der Iljin'schen Hypothese der Salzbeständigkeit'. *Jb. wiss. Bot.*, **65**, 502.

—— (1927) 'Über Halobiose und ihre Abstufungen. Versuch einer synthetischen Verknüpfung isolierter analytischer Probleme'. *Flora* NF, **121**, 434.

MONTFORT, C. and BRANDRUP, W. (1926/27) 'Ökologische Studien über Keimung und erste Entwicklung der Halophyten'. *Jb. wiss. Bot.*, **66**.

MORK, E. (1942) 'Om ströfallet i våre skoger'. *Meddelelser fra det norske skogforsöksvesen*, **29**.

—— (1946) 'Om skogsbunnens lyngvegetasjon'. Ibid., **33**.

MOTHES, K. (1932) 'Ernährung, Struktur und Transpiration'. *Biol. Zbl.*, **52**, 193.

—— (1939) *Die Bedeutung der Spurenstoffe für die Entwicklung und Vergesellschaftung der Pflanzen. Organismen der Umwelt.* Dresden and Leipzig.

—— (1954) Bakterien Pilze und Algen: Ein neuer Typ von Kulturpflanzen'. *Die Kulturpflanzen*, **2**, 237. Berlin.

MUIR, A. (1934) 'The soils of the Teindland State forest'. *Forestry*, **8**, 25.

MULDER, E. G. (1954) 'Molybdenum in relation to growth of higher plants and microorganisms'. *Plant and Soil*, **I**, 368.

MÜLLER, D. (1948) *Plantefysiologi.* Copenhagen.

MÜLLER, D. and LARSEN, P. (1935) 'Analyse der Stoffproduktion bei Stickstoff- und Kaliummangel'. *Planta*, **23**, 501.

MÜLLER, K. (1934). *Aschenanalysen über den standörtlich verschiedenen Mineralstoffgehalt der Fichtennadeln bei vergleichbarer Probenahme.* Diss. Dresden.

MÜLLER, P. E. (1879) 'Studier over Skovjord. I–II'. *Tidskr. for Skovbrug*, **3**, 1–124.

MÜLLER-STOLL, W. R. (1947) 'Der Einfluss der Ernährung auf die Xeromorphie der Hochmoorpflanzen'. *Planta*, **35**, 225.

MURÉN, A. (1934) 'Untersuchungen über die Wurzeln der Wasserpflanzen'. *Ann. bot. Soc. zool. bot. fenn.*, Vanamo, **5**, 1–56.

NÄSLUND, O. J. (1937). *Sågar. Bidrag till kännedomen om sågarnas uppkomst och utveckling.* Stockholm.

NASON, A. and MCELROY, W. D. (1963) 'Modes of action of the essential mineral elements'. In *Pl. Physiol.*, Vol. 3, ed. Steward, F. C., Academic Press, New York.

NEEMAN, M. (1955) 'Adaptability of wheat varieties to acid soils'. *Nature, Lond.*, **175**, 1090.

NICHOLAS, D. J. D. and THOMAS, W. D. E. (1953) 'The effect of cobalt on the fertiliser and soil phosphate uptake and the iron and cobalt status of tomato'. *Plant and Soil*, **5**, 67.

NILSSON, P. E. (1950). 'Något om relationerna mellan växten och markens mikroflora'. *K. Lantbruksakad. tidskr.*, **89**, 262.

NILLSSON, R., BJÄLVE, G. and BURSTRÖM, D. (1938 and 1939) 'Über Zuwachsfaktoren bei Bact. radicicola'. *Ann. Landwirsch. Hochschule Schwedens*, **6**, 299, and **7**, 301.

NÖMMIK, H. (1954) 'Faktorer inverkande på kvävehushållningen i marken'. *Växtnäringsnytt*, **10**.

NORD, F. F. and VITUCCI, J. C. (1948) 'Certain aspects of the microbiological degradation'. *Adv. enzymol.*, **8**, 253.

NORDGAARD, O. (1903) 'Studier over naturforholdene i vestlandske fjorde. I Hydrografi'. *Bergens Museums Aarbog*, **8**.

NORKRANS, B. (1950) 'Studies in growth and cellulolytic enzymes of *Tricholoma*'. *Symb. bot. upsal.*, **11**.

NYKVIST, N. (1963) 'Leaching and decomposition of water-soluble organic substances from different types of leaf and needle litter'. *Studia forest. Suec.*, **3**, 31.

OLSEN, C. (1921) 'The ecology of *Urtica dioica*'. *J. Ecol.*, **9**, 1.

—— (1923) 'Studies on the hydrogen ion concentration of the soil and its significance to the vegetation, especially to the natural distribution of plants'. *Compt. Rend. lab. Carlsberg*, **15**, 1–166.

—— (1936) 'Absorption of manganese by plants'. *Compt. Rend. lab. Carlsberg. Ser. chem.*, **21**, 129 (448).

—— (1938) 'Growth of *Deschampsia flexuosa* in culture solutions (water culture experiments) and in soils with different pH'. *Ibid.*, **22**, 405.

—— (1950) 'The significance of concentration for the rate of ion absorption by higher plants in water culture'. *Physiol. plantarum*, **3**, 152.

—— (1953a) 'II. Experiments with water plants'. *Ibid.*, **6**, 837.

—— (1953b) 'III. The importance of stirring'. *Ibid.*, **6**, 844.

—— (1953c) 'IV. The influence of hydrogen ion concentration'. *Ibid.*, **6**, 848.

OLSSON, B. (1954–55) 'Mundus senescens. En skiss om tron på en åldrande värld i svenskt folkliv och litterature'. *Lychnos*, 66.

OLTMANNS, F. (1892) 'Kultur und Lebensbedingungen der Meeresalgen'. *Jb. wiss. Bot.*, **23**.

OSBORN, F. (1949) *Our plundered planet*. Little, Brown, Boston.

—— (1963) *Our crowded planet*. Allen and Unwin Ltd., London.

OSTERHOUT, W. J. V. (1922) 'Some aspects of selective absorption'. *J. gen. Physiol.*, **5**, 225.

OSWALD, H. (1935) 'Betans hjärtröta och dess bekämpande' *Sv. Socker-fabr. ab. Odlaremedd*, **10**.

OVINGTON, J. D. (1957) 'Dry matter production by *Pinus sylvestris* L.'. *Ann. Bot. Lond. N.S.*, **21**, 287–314.

OVINGTON, J. D. (1965) *Woodlands*. English Universities Press, London.

OVINGTON, J. D. and PEARSALL, W. H. (1956) 'Production ecology II; Estimates of average production by trees', *Oikos*, **7**, 202–205.

PAECH, K. (1940) 'Stoffwechsel organischer Verbindungen' *Fortschr. d. Bot.*, **10**, 209.

PEARSALL, W. H. and GORHAM, E. (1956) 'Production ecology I'. *Oikos*, **7**, 193–201.

PEDERSEN, R. (1883) *Planternes næringsstoffer*. Copenhagen.

PENMAN, H. L. (1963). 'Vegetation and Hydrology'. *Tech. Comm. No. 53*, Commonwealth Bureau of Soils.

PERMANN, O. (1954) *Stallgödseln och dess användning*. (In *Jordbrukslära för ungdomsskolor, jordbrukskurser och självstudier*. Published by *V. Lengdén*. Stockholm.

PETTERSSON, H. (1950) 'Resultat från den svenska djuphavsexpeditionen'. *Ymer*, **70**, 207.

PHILIPSON, TH. (1953) 'Boron in plant and soil'. *Acta agric. scand.*, **3**, 121.

PIRSCHLE, K. (1939, 1940, 1941) 'Mineralstoffwechsel'. *Fortschr. d. Bot.*, **9**, 160; **10**, 189; **11**, 167.

PIRSON, A. (1939–1946) *Stoffwechsel der Pflanzen. Naturforschung und Medizin in Deutschland*.

—— (1955) 'Functional aspects in mineral nutrition of green plants'. *A. Rev. Pl. Physiol.*, **6**, 71.

PIRSON, A. and SEIDEL, F. (1950) 'Zell- und stoffwechselphysiologische Untersuchungen an der Wurzel von *Lemna minor* L. unter besonderer Berücksichtigung von Kalium- und Calcium-Mangel'. *Planta*, **38**, 431.

PISEK, A. and BERGER, E. (1938) 'Kutikuläre Transpiration und Trocken-resistenz isolierter Blätter und Sprosse'. *Planta*, **28**, 124.

PISEK, A. and CARTELLIERI, E. (1939) 'Zur Kenntnis des Wasserhaus-haltes der Pflanzen'. *Jb. wiss. Bot.*, **88**, 22.

—— (1941) 'Der Wasserverbrauch einiger Pflanzenvereine'. *Jb. wiss. Bot.*, **90**, 225.

PISEK, A. and TRANQUILLINII, W. (1951) 'Transpiration und Wasser-haushalt der Fichte (*Picea excelsa*) bei zunehmender Lufttrocken-heit'. *Physiol. plantarum*, **4**, 1–27.

POEL, L. W. (1949) 'Germination and development of heather and the hydrogen ion concentration of the medium'. *Nature, Lond.*, **163**, 647.

POLSTER, H. (1950) *Die physiologischen Grundlagen der Stofferzeugung im Walde*. Munich.

POMPE, E. (1940) 'Beiträge zur Ökologie der Hiddenseer Halophyten'. *Bot. Zbl. Beihefte Bd.*, **60**, 223.

POST, L. V. and GRANLUND, E. (1926) 'Södra Sveriges torvtillgångar'. *Sveriges geol. unders.* Ser. C. 335.

POWERS, W. L. and BOLLEN, W. B. (1935) 'The chemical and biological nature of certain forest soils'. *Soil Sci.*, **40**, 321.

PRIESTLEY, J. (1776) *Experiments and observations on different kinds of air.* London.

PRINGSHEIM, E. G. (1949) 'Iron bacteria'. *Biol. Rev.*, **24**, 200.

PURI, G. S. (1949) 'Ecological problems of the humus layer in English forests'. *Proc. Indian. Sci. Congr.*, **148**.

RABINOVITCH, E. (1945) *Photosynthesis.* New York.

REPP, G. (1951) 'Kulturpflanzen in der Salzsteppe'. *Bodenkultur*, (Vienna), **5**, 249.

—— (1958) 'Die Salztoleranz der Pflanzen. I Salzhaushalt und Salzresistenz von Marsch-pflanzen der Nordseeküste Dänemarks in Beziehung zum Standort'. *Österr. bot.*, Z. **104**, 454.

REVELLE, R. and SUESS, H. E. (1957) 'Carbon dioxide exchange between atmosphere and ocean and the question of an increase of atmospheric CO_2 during the past decades'. *Tellus*, **9**, 18.

RICHARDS, P. W. (1952) *The tropical rain forest.* Cambridge.

RINDELL, A. (1919) *Lärobok i agrikulturkemi och agrikulturfysik.* Helsingfors.

ROBINSON, G. W. (1949) *Soils, their origin, constitution and classification.* London.

RODIE, W. (1948) 'Environmental requirements of fresh-water plankton algae'. *Symb. bot. upsal.*, **10**, 1.

ROKITZKA, A. (1949) *Allgemeine Mikrobiologie.* Munich.

ROMELL, L. G. (1922) 'Luftväxlingen i marken som ekologisk faktor'. *Medd. Stat. skogsförs.anst.*, **19**.

—— (1926) 'Über das Zusammenwirken der Produktionsfaktoren'. *Jb. wiss. Bot.*, **65**, 739.

—— (1928) 'Studier över kolsyrehushållningen i mossrik tallskog'. *Medd. Stat. skogsforskningsanst.*, 1.

—— (1932) 'Mull and duff as biotic equilibria'. *Soil Sci.*, **34**, 161.

—— (1935) 'Ecological problems of the humus layer in the forest'. *Cornell Univ. Agric. exp. sta. Memoir*, 170.

—— (1939) 'Den nordiska blåbärsskogens produktion av ris, mossa och förna'. *Sv. bot. Tidskr.*, **33**, 366.

—— (1945) 'Lien och landskapet *Ymer*, 284.

—— (1946) 'Organic dust in the air and the ammonia found in atmospheric water'. *Sv. bot. Tidskr.*, **40**, 1.

—— (1950) *Excursion to upper Norrland.* 7th Int. Bot. Congress. Stockholm 1950. Excursion guide's Sect. EXE. Uppsala.

—— (1952) *Heden. Natur i Halland.* Stockholm.

—— (1964) 'Skog och odling i svensk "natur"'. *Sveriges Natur. Årsbok.*

ROMELL, L. G. and HEIBERG, S. O. (1931) 'Types of humus layer in the forests of North-eastern United States'. *Ecology*, **12**, 567.

ROMELL, L. G. and MALMSTRÖM, C. (1945) 'Henrik Hesselmans tallhedsförsök åren 1922–42'. *Medd. Stat. skogsförskningsanst.*, **34**, 543.

ROMOSE, V. (1940) 'Ökologische Untersuchungen über *Hömalothecium sericeum*, seine Wachstumsperioden und seine Stoffproduktion'. *Dansk bot. ark.*, **10**.

ROOTS, B. (1956) 'The water relations of earthworms, II Resistance to desiccation and immersion and behaviour when submerged and when allowed a choice of environment'. *J. exp. Biol.*, **33**, 29.

ROSSBY, C.-G. and EGNÉR, H. (1955) 'On the chemical climate and its variation with the atmospheric circulation pattern'. *Tellus*, **7**, 118.

RUHLAND, W. (1915) 'Untersuchungen über die Hautdrüsen der Plumbaginaceen. Ein Beitrag zur Biologie der Halophyten'. *Jb. wiss. Bot.*, **55**, 409.

RUHLAND, W. and WETZEL, K. (1926, 1927) Zur Physiologie der organischen Säuren in grünen Pflanzen'. *Planta*, **1**, 558; **3**, 765.

RUNE, O. (1953) 'Plant life on serpentines and related rocks in the north of Sweden'. *Acta phytogeogr. Suec.*, **31**.

RUSSELL, E. J. (1936) *Boden und Pflanze*. Dresden and Leipzig.

RUSSELL, SIR E. J. (1952) *Soil conditions and plant growth*. 8th Edn rewritten by E. W. Russell. London.

—— (1954) *World population and world food supplies*. London.

RUSSELL, SIR E. J. and APPLEYARD, H. (1915) 'The atmosphere of the soil, its composition and the causes of variation'. *J. agric. Sci.*, **7**, 1–48.

RUTTER, A. J. (1963) 'Studies in the water relations of *Pinus sylvestris* in plantation conditions. 1. Measurements of rainfall and interception'. *J. Ecol.*, **51**, 191.

—— (1967) 'Studies in the water relations of *Pinus sylvestris* in plantation conditions. 5. Responses to variation in soil moisture'. *J. appl. Ecol.*, **4**, 73.

RUTTER, A. J. (1968). 'Water consumption by forests'. In *Water Deficits and Plant Growth*. II. ed. T. T. Kozlowski. Academic Press, New York. Pp. 23–84.

RYBERG, M. (1956) 'Kolsö, ett sörmländskt lundområde och dess utveckling'. *Sv. bot. Tidskr.*, **50**, 163.

SACHS, J. (1860) 'Physiologische Untersuchungen über die Abhängigkeit der Keimung von der Temperatur'. *Jb. wiss. Bot.*, **2**, 338.

—— (1875) *Geschichte der Botanik*. Munich.

SALLE, A. J. (1943) *Fundamental principles of bacteriology*. New York.

DE SAUSSURE, T. (1804) *Reserches chimiques sur la végétation*. Paris, V. Nyon.

SCHARRER, K. (1955) *Biochemie der Spurenelemente*. Berlin.

SCHEFFER, F. and SCHACHTSCHABEL, P. (1956) *Bodenkunde*. Stuttgart.

SCHIMPER, A. F. W. (1898) *Pflanzengeographie auf physiologische Grundlage*.

SCHLICHTING, E. (1958) 'Kupferbindung und -fixierung durch Humusstoffe'. *Acta Agric. scand.*, **5**, 313.

SCHLOESING, T. and MÜNTZ, A. (1837) 'Sur la nitrification par les ferments organisés'. *Compt. rend.*, **84**, 301.

SCHMIDT, W. (1949) 'Waldverwüstung und Bodenzerstörung'. *Umschau*, **49**, 171.

SCHMIED, E. (1953) 'Spurenelementdüngung und Wasserhaushalt einiger Kulturpflanzen'. *Österr. bot. Z.*, **100**, 552.

SCHOPFER, W. H. (1934) 'Les vitamines cristallisées B comme hormones de croissance chez un microorganisme'. *Ach. f. Mikrobiol.*, **5**, 513.

—— (1943) *Plants and vitamins*. Waltham, Mass.

SCHRATZ, E. (1936) 'Beiträge zur Biologie der Halophyten'. *Jb. wiss. Bot.*, **83**, 133.

—— (1937) 'Beiträge zur Biologie der Halophyten. IV. Die Transpiration der Strand- und Dünenpflanzen'. *Ibid.*, **84**, 593.

SCHROEDER, H. (1919) 'Die jährliche Gesamtproduktion der grünen Pflanzendecke der Erde'. *Die Naturwiss.*, **7**, 8.

SCHUBERT, A. (1939) 'Untersuchungen über den Transpirationsstrom der Nadelhölzer und der Wasserbedarf von Fichte und Lärche'. *Tharandt forstl. Jahrb.*, 821.

SCHWARZ, A. R. V. (1879) 'Vergleichende Versuche über die physikalischen Eigenschaften verschiedener Bodenarten'. *Forsch. a. d. Geb. Agr. Physik.*, **2**, 164.

SELANDER, S. (1950) 'Floristic phytogeography of southwestern Lule Lappmark. (Swedish Lapland.)' *Acta phytogeogr. Suec.*, **27**.

—— (1955) *Det levande landskapet i Sverige*. Stockholm.

SENEBIER, J. (1782) *Memoires physico-chimiques*. Genève.

SERNANDER, R. (1912) 'Studier öfver lafvarnas biologi I. Nitrofila lafvar'. *Sv. bot. Tidskr.*, **6**, 803.

—— (1917) 'De nordeuropeiska havens växtregioner'. *Ibid.*, **11**, 72.

—— (1918) 'Förna och äfja'. *Geol. fören. förh.*, **40**, 645.

SIGAFOOS, R. S. (1952) 'Frost action as a primary physical factor in tundra plant communities'. *Ecology*, **33**, 480.

SIMONIS, W. (1948) 'CO_2-Assimilation und Xeromorphie von Hochmoorpflanzen in Abhängigkeit vom Wasser- und Stickstoffgehalt des Bodens'. *Biol. Zbl.*, **67**, 77.

SINGH, R. N. (1942) *Ind. J. agric. Sci.*, **12**, 743.

SJÖBECK, M. (1927) *Bondskogar, deras värde och utnyttjande*. Skånska folkminnen.

—— (1931) 'Det äldre kulturlandskapet i Syd-Sverige'. *Sv. Skogsvårdsfören. Tidskr.*, **45**.

—— (1932) Lövängen och dess betydelse för det sydsvenska bylandskapets uppkomst och utveckling'. *Ibid.*, **132**.

—— (1933) 'Den försvinnande ljungheden'. *Svenska turistfören. årsskr.*, **86**.

—— (1933) 'Lövängskulturen i Sydsverige'. *Ymer*, 33.

—— (1936) 'Sydsvensk ängsodling före skiftena'. *Svenska kulturbilder*, **III**, 119.

—— (1946) Utbredningen i Sydsverige av toppbeskuren lind och ask. *Värendsbygden*.

—— (1947) *Allmänningen Kulla fälad*. Hälsingborgs mus. publ., Hälsingborg.

—— (1950) *Blekinge*. Stockholm.

—— (1953) *Studier i Smålands odling, Småland*. Svensk litteratur. Stockholm.

—— (1953) *Dalsland*. Stockholm, 1953.

SJÖRS, H. (1948) 'Myrvegetation i Bergslagen'. *Acta phytogeogr. Suec.*, **21**.

—— (1950) 'On the relation between vegetation and electrolytes in North Swedish mire waters'. *Oikos*, **2**, 241.

—— (1954) 'Slåtterängar i Grangärde Finnmark'. *Acta phytogeogr. Suec.*, **34**.

—— (1956) *Nordisk växtgeografi*. Stockholm.

SLANKIS, V., RUNECKLES, V. C. and KROTKOV, G. (1964) 'Metabolites liberated by roots of white pine (*Pinus strobus* L.) seedlings'. *Pl. Physiol.*, **17**, 301.

SLATYER, R. O. (1967) *Plant–Water Relationships*. London.

SMITH, A. (1932) 'Seasonal subsoil temperature variations'. *J. agric. Res.*, **44**, 421.

SPRENGEL, C. K. (1793) *Das entdecke Geheimnis der Natur im Bau und in der Befruchtung der Blumen*. Berlin.

SPRENGEL, C. K. (1826) *Kastners Arch. ges. Naturlehre* Nürnberg.

STÅLFELT, M. G. (1924) 'Tallens och granens kolsyreassimilation och dess ekologiska betingelser'. *Medd. Stat. skogsförskningsanst.*, 181.

—— (1937) 'Der Gasaustausch der Moose'. *Planta*, **27**, 30.

—— (1944) 'Granens vattenförbrukning och dess inverkan på vattenomsättningen i marken'. *K. Lantbruksakad. tidskr.*, **83**, 3.

—— (1948) 'Soil substances affecting the viscosity of protoplasm'. *Sv. bot. tidskr.*, **42**, 17.

STARKEY, R. L. and WAKSMAN, S. A. (1943) 'Fungi tolerant to extreme acidity and high concentration of copper sulfate'. *J. Bact.*, **45**, 509.

Statistisk årsbok för Sverige (1946 and 1956) (Statistical Yearbook for Sweden). Stockholm.

STEELE, B. (1955) 'Soil-pH and base status as factors in the distribution of calcicoles'. *J. Ecol.*, **43**, 120.

STEEMANN NIELSEN, E. (1944) 'Havets planteverden i ökologisk och produktionsbiologisk belysning'. *Medd. komm. Danmarks fiskeri- og havsundersögelser*, **13**.

—— (1952) 'Production of organic matter in the sea'. *Nature, Lond.*, **169**, 956.

—— (1954) 'On organic production in the ocean'. *Journal du Conseil international pour l'exploration de la mer*, **19**.

—— (1957) 'The autotrophic production of organic matter in the ocean'. *Galathea report*, **1**, 49.

STEINER, M. (1934) 'Zur Ökologie der Salzmarschen der nordöstlichen Vereinigten Staaten von Nordamerika'. *Jb. wiss. Bot.*, **81**, 94.

—— (1937) 'Ökologische Pflanzengeographie'. *Fortschr. d. Bot.*, **7**, 249.

—— (1939) 'Die Zusammensetzung des Zellsaftes bei höheren Pflanzen in ihrer ökologischen Bedeutung'. *Ergebn. d. Biol.*, **17**, 151; **17**, 234.

—— (1948) 'Die Bedeutung autokatalytischer Abbauvorgänge für die Mineralisierung des organischen Phosphors toter Pflanzensubstanz'. *Biol. Zbl.*, **67**, 84.

STEPHENSON, M. (1949) *Bacterial metabolism*. New York.

STERNER, R. (1922). 'The continental element of the flora of South Sweden'. *Geogr. ann.*, Stockholm.

STEWART, W. P. P. (1966) *Nitrogen Fixation in Plants*. London.

STOCKER, O. (1925) 'Standort und Transpiration der Nordsee-Halophyten'. *Z. Bot.*, **17**, 1.

—— (1928) 'Das Halophytenproblem'. *Ergebn. d. Biol.*, **3**, 265.

—— (1929) 'Ungarische Steppenprobleme'. *Die Naturwiss.*, **17**, 189.

STOCKER, O. and HOLDHEIDE, W. (1937) 'Die Assimilation Helgoländer Gezeitenalgen während der Ebbezeit'. *Z. Bot.*, **32**, 1.

STÖCKLI, A. (1950) 'Die Ernährung der Pflanze in ihrer Abhängigkeit von der Kleinlebewelt des Bodens'. *Z. f. Pflanzenern., Düngung, Bodenk.*, **48**, 264.

SUCHTELEN, F. H. VAN (1923) 'Energetik und Mikrobiologie des Bodens'. *Cbl. f. Bakt. Protoz., Parazitenkunde und Inf. Krankh.*, **58**, 413.

SVANBERG, O. (1949) *Mikroelement*. Stockholm.

SVANBERG, O. and NYDAHL, F. (1941) 'Den svenska höskördens kopparhalt'. *K. Lantbruks. akad. tidskr.*, **80**, 457.

SVENDSEN, J. A. (1955) 'Earthworm population studies: a comparison of sampling methods'. *Nature, Lond.*, **175**, 864.

Sveriges offentliga utredningar 1957: 17. (K. Jordbruksdepartementet) Markvård och erosionsskydd. Stockholm.

SVINHUFVUD, V. E. (1937) 'Untersuchungen über die bodenmikrobiologischen Unterschiede der Cajander'schen Waldtypen'. *Acta forest. fenn.*, **44**, 1.

SÖRLIN, A. (1943) 'De mineralogena jordslagen som växtgrund med särskild hänsyn till fruktodlingen'. *Sveriges pomolog. fören. årsskr.*, **17**.

TÄCKHOLM, G. and TÄCKHOLM, V. (1941) *Flora of Egypt*. Cairo.

TAKADA, H. (1957) 'Über Tagesschwankung des osmotischen Wertes in den Blättern von Strandpflanzen in ihrem Zusammenhange mit dem Chloridgehalt'. *J. Inst. Polyt.*, Osaka, **2**, 9.

TAMM, C. O. (1947) 'Markförbättringsförsök på mager hed'. *Medd. Stat. skogsförskningsanst.*, **36**, 1–115.

—— (1951a) 'Removal of plant nutrients from tree crowns by rain'. *Pl. Physiol.*, **4**, 184.

—— (1951b) 'Seasonal variation in composition of birch leaves'. Ibid., **4**, 461.

—— (1953) 'Growth, yield and nutrition in carpets of forest moss'. *Medd. Stat. skogsforskn. inst.*, **43**.

—— (1954) 'Om gödsling av skogsmark. II Den skogliga gödslingsfrågan som pedologiskt problem'. *Sv. Skogsvårdsfören. Tidskr.*, **52**, 317.

—— (1954) 'Some observations on the nutrient turnover in a bog community, dominated by *Eriophorum vaginatum L.*'. *Oikos*, **5**, 189.

TAMM, O. (1920) 'Markstudier i det nordsvenska barrskogsområdet'. *Medd. Stat. skogsförskningsanst.*

—— (1921) 'Om berggrundens inverkan på skogsmarken'. Ibid.

—— (1927) 'Några synpunkter på skogsmarkens fuktighetstillstånd'. *Skogen*, **11**.

—— (1929) 'Die Bodentypen und ihre forstliche Bedeutung'. *Verhandl. d. internat. Kongr. forstl. Versuchsanst.*

—— (1930) 'Om brunjorden i Sverige'. *Sv. skogsvårdsfören. tidskr.*

—— (1931) 'Studier över jordmånen och dess förhållande till markens hydrologi i nordsvenska skogsterränger. *Medd. Stat. skogsförskningsanst.*

—— (1940) *Den nordsvenska skogsmarken.* Stockholm.

—— (1947) 'Skogsproduktionens naturliga förutsättningar'. *K. Lantbruksakad. Tidskr.*, **261**.

—— (1954) 'Till frågan om bestämning av klimatets humuditetsgrad i Sverige'. Ibid., **93**, 105.

—— (1955) 'Fuktighetsproblemet i skogsmarken belyst av svensk forskning under ett halvsekel'. *Medd. Värml. och Örebro läns skogsv. och kol.skol.* Gammelkroppa.

—— (1956) 'Fortsatta studier över klimatets humiditet i Sverige. Beräkning av humiditetsvärden. Ibid., **95**, 401.

—— (1959) 'Studier över Klimatets humiditet i Sverige'. *K. skogshögskolans skrifter*, **32**.

THOMPSON, W. S. (1950) 'Population'. *Scient. Am.*, **182**, 2, p. 11.

THORNTHWAITE, C. W. and MATHER, J. R. 'The role of evapotranspiration in climate'. *Arch. Met. Geophys. u. Bioklim. Ser. B.*, **3**, 16.

THORP, J. (1949) 'Effects of certain animals that live in soils'. *Sci. Monthly*, **68**, 180.

THUNMARK, S. (1942) 'Über rezente Eisenocker und ihre Mikroorganismengemeinschaften'. *Bull. geol. inst. univ. Ups.*, **29**.

—— (1945) 'Zur Soziologie des Süsswasser planktons'. *Folia Limnolog. Scand.*, no. 3.

THURMANN-MOE, P. (1941) 'Om skogens innflydelse på jordens vannförråd'. *Medd. Norg.* Landbrukshögskole.

TIBURTIUS, T. (1761) *Tal om Öster-Götlands Förmoner och Olägenheter*. Hållet för Kongl. Vetenskapsacademien.

TRACEY, M. V. (1955) 'Cellulose and chitinase in soil amoebae. *Nature, Lond.*, **175**, 815.

TROEDSSON, T. (1949/50) 'Bidrag till kännedom om det försämrade marktillståndet till följd av vinderosion på sandjordarna i Skåne'. *Grundförbattring*, **3**, 235.

—— (1952) 'Den geologiska miljöns inverkan på grundvattnets halt av lösta växtnäringsämnen'. *K. Skogshögsk. skr.*, **10**.

—— (1955) 'Vattnet i skogsmarken'. Ibid., **20**.

—— (1956) 'Marktemperaturen i ytsteniga jordarter'. Ibid., **25**.

TUKEY, H. J. JR. (1970) 'The leaching of substances from plants'. *A. Rev. Pl. Physiol.*, **21**, 305–24.

TULL, J. (1731). *Horse Hoeing Husbandry*, London.

UGGLA, E. (1958a) 'Skogsbrandfält i Muddus nationalpark'. *Acta phytogeogr. Suec.*, **41**.

—— (1958b) 'Ecological effects of fire on north Swedish forests'. *Diss.*, Uppsala.

ULRICH, R. (1894) 'Untersuchungen über die Wärmekapacität der Boden-Konstituenten'. *Forschungen a. d. Geb. Agrikultur-Physik*, **17**, 1–31.

UNESTAM, G. (1954) *Söderbykarls socken*. Uppsala.

UPHOF, J. C. Th. (1941) 'Halophytes'. *Bot. rev.*, **7**, 1.

VALLIN, H. (1925) 'Ökologische Studien über Waldund Strandvegetation'. Lunds univ. årsskr. N. F. avd. 2, Bd 21.

VELDKAMP, H. (1952) 'Aerobic decomposition of chitin by microorganisms. *Nature, Lond.*, **169**, 500.

VIRO, P. I. (1953) 'Loss of nutrients and the natural nutrient balance of the soil in Finland'. *Institute forest. fenn.*, **42**, 1.

VIRTANEN, A. I. (1928) 'Über die Einwirkung der Bodenazidität auf das Wachstum und die Zusammensetzung der Leguminosenpflanzen'. *Biochem. Z.*, **193**, 300.

—— (1952) 'Molecular nitrogen fixation and nitrogen cycle in nature'. *Tellus*, **4**, 304.

VIRTANEN, A. I. and LAINE, T. (1939) 'The excretion products of root nodules. The mechanism of N-fixation'. *Biochem. J.*, **33**, 412.

VOGT, W. (1948) *Road to survival*. New York, William Sloane Associates.

VOLK, O. (1931) 'Beiträge zur Ökologie der Sandvegetation der Oberrheinischen Tiefebene'. *Z. Bot.*, **24**, 114.

WAKSMAN, S. A. (1932) *Principles of soil microbiology*. Baltimore.

—— (1938) *Humus*. 2nd ed., London.

—— (1936) *Humus, origin, chemical composition and importance in nature*. London.

—— (1952) *Soil microbiology*. New York.

WAKSMAN, S. A. and IYER, K. R. N. (1932 and 1933) 'Contribution to our knowledge of the chemical nature and origin of humus: I, On the synthesis of the "humus nucleus"'. *Soil. Sci.*, **34**, 43; **36**, 69.

WAKSMAN, S. A. and REUSSER, H. W. (1930) 'Über die chemische Natur und den Ursprung des Humus, etc.'. *Zellulosechemie*, **11**, 209.

WAKSMAN, S. A. and STEVENS, K. R. (1928) 'Chemical nature of organic complexes in peat and methods of analysis'. *Soil Sci.*, **26**, 113.

—— (1930) 'A critical study of the methods for determining the nature and abundance of soil organic matter'. Ibid., **30**, 97.

WAKSMAN, S. A. and TENNEY, F. G. (1927) 'Influence of age of plant upon the rapidity and nature of its decomposition. Rye plants. *Soil Sci.*, **24**, 317, 1927. Baltimore; **24**, 275; **26**, 155–71.

WALDHEIM, S. (1947) 'Kleinmoosgesellschaften und Bodenverhältnisse in Schonen'. *Bot. Not.*, Suppl. **1**, 1.

WALDHEIM, S. and WEIMARCK, H. (1943) 'Skånes myrtyper. *Bot. Not.*

WALFRIDSSON, W. (1956) 'Den nya skogen'. *Skogsägaren*, **8**, 246.

WALLACE, T., HEWITT, E. J. and NICHOLAS, D. J. D. (1945) 'Determination of factors injurious to plants in acid soils'. *Nature, Lond.*, **156**, 778.

WALLÉN, A. (1927) 'Eau tombée, débit et evaporation dans la Suède meridionale'. *Geogr. ann.*

WALLÉN, C. C. (1955) 'Mexicos jordbruk—förutsättningar och problem'. *Ymer*, **75**, 241.

WALLERIUS, J. G. (1761) *Agriculturae Fundamenta Chemica: Åkerbrukets Chemiska Grunder*, Uppsala.

WALTER, H. (1932) 'Die Wasserverhältnisse an verschiedenen Standorten in humiden und ariden Gebieten'. *Beih. bot. Cbl.*, **49**, 495.

—— (1936a) 'Tabellen zur Berechnung des osmotischen Wertes von Pflanzenpressäften, Zuckerlösungen und einigen Salzlösungen'. *Ber. d. d. bot. Ges.*, **54**, 328.

—— (1936b) 'Über den Wasserhaushalt der Mangrovepflanzen. *Ber. schweiz. bot. Ges.*, **46**, 217.

—— (1939) 'Grasland, Savanne und Busch der arideren Teile Afrikas in ihrer ökologischen Bedingtheit'. *Jb. w. Bot.*, **87**, 750.

—— (1936 and 1949) 'Ökologische Pflanzengeographie'. *Fortschr. d. Bot.*, **5**, 245; **12**, 131.

—— (1943) *Die Vegetation Osteuropas*. Berlin.

—— (1947) *Die Grundlagen des Pflanzenlebens*. Stuttgart.

—— (1951) *Grundlagen der Pflanzenverbreitung*. Stuttgart.

—— (1954) *Die Verbuschung, eine Erscheinung der subtropischen Savannengebiete und ihre ökologischen Ursachen. Festschr. f. J. Braun Blanquet.* The Hague.

WALTER, H. and HABER, W. (1957) 'Über die Intensität der Bodenatmung mit Bemerkungen zu den Lundegårdh'schen Werten. *Ber. d. d. bot. Ges.*, **70**, 275.

WALTER, H. and STEINER, M. (1936) 'Die Ökologie der Ost-afrikanischen Mangroven'. *Z. Bot.*, **30**, 65.

WARIS, H. (1939) 'Über den Antagonismus von Wassertoffionen und Metallkationen bei *Micrasterias*'. *Acta bot. fenn.*, **24**, 1.

WARMING, E. (1895) *Plantesamfund. Grundträk af den ökologiske Plantegeografi.* Copenhagen.

—— (1909) *Oecology of plants.* Oxford.

—— (1918) *Lehrbuch der ökologischen Pflanzengeographie.* Berlin.

WATANABE, A., NISKIGAKE, S. and KONISHI, C. (1951) 'Effect of nitrogen-fixing blue-green algae on the growth of rice plants'. *Nature, Lond.*, **168**, 748.

WATSON, D. J. (1971). 'Size, structure and activity of the productive system of crops'. In: *Potential Crop Production*, ed. Wareing, P. F. and Cooper, J. P. Heinemann, London. Pp. 76–88.

WEAVER, J. E. and ALBERTSON, F. W. (1940) 'Deterioration of Mid-western Ranges'. *Ecology*, **21**, 216.

WEAVER, J. E. and ZINK, E. (1946) 'Annual increase of underground material in three range grasses'. *Ecology*, **27**, 119.

WEIMARCK, G. (1952) 'Studier over landskapets förändring inom Lönsboda, Örkeneds socken nordöstra Skåne'. *Lunds Univ. årsskr.* N. F. avd. 2, **48**, no. 10.

WEIMARCK, H. (1939) 'Vegetation och flora i Örkeneds socken'. *Bot. not.*

—— (1950) 'Some phytogeographical aspects of the scanian flora'. Ibid.

WENT, F. W. (1944a) 'Plant growth under controlled conditions. II The thermoperiodicity in growth and fruiting of the tomato'. *Amer. J. Bot.*, **31**, 135.

—— (1944b) 'III. Correlation between various physiological processes and growth in the tomato plant'. Ibid., **31**, 597.

WEST, P. M. and LOCKHEAD, A. G. (1940) 'The rhizosphere in relation to the nutritive requirements of soil bacteria'. *Canad. J. res. Ser. C*, **18**, 129.

WESTLAKE, D. F. (1963) 'Comparisons of plant productivity'. *Biol. Rev.* **38**, 385–425.

WHITE, G. F. *et al.* (1956) *The future of arid lands.* Washington, DC.

WIBECK, E. (1909) 'Bokskogen inom Östbo och Västbo härad av Små-land'. *Sv. Skogsvårdsfören. tidskr.*, **424**.

—— (1910–1911) 'Om ljungbränning för skogskultur. Ibid., 1911, *Medd. Stat. skogsförskningsanst.*, **7–8**, 7.

WIKBERG, E. and FRIES, N. (1952) 'Some new and interesting biochemical mutations obtained in *Ophiostoma* by selective enrichment technique'. *Physiol. Pl.*, **5**, 130.

WIKEN, T. (1942) 'Om metanbakterier och metanjäsning av kloakslam'. *Iva*, 1.

WIKLANDER, L. (1957) 'Om svavlets förekomst i marker'. *Nordisk jordbruksforskning*, **39**, 399.

WIKLANDER, L., HALLGREN, G. and JONSSON, E. 'Studies on gyttja soils. III. Rate of sulfur oxidation'. *K. Lantbrukshögsk. ann.*, **425**.

WILDE, S. A. (1946) *Forest soils and forest growth*. Waltham, Mass.

WILDIERS, E. (1901) 'Nouvelle substance indispensable au developpement de la levure'. *La Cellule*, **18**.

WILLSTÄTTER, R. and STOLL, A. (1918) *Untersuchungen über die Assimilation der Kohlensäure*. Berlin.

WILSON, B. D. and STAKER, E. V. (1932) 'The chemical composition of the muck soils of New York'. *Corn. agric. exp. stat. bull.*, 537.

WINOGRADSKY, S. (1890) *Ann. Inst. Pasteur*, **4**, 213, 257, 760.

WITSCH, H. and FLÜGEL, A. (1953) 'Physiologische Untersuchungen an Pflanzen mit erhöhtem Aneurinhalt'. *Flora*, **140**, 534.

WITTICH, W. (1933) 'Untersuchungen in Nordwestdeutschland über den Einfluss der Holzart auf den biologischen Zustand des Bodens'. *Mitt. Forstwirtsch. Forstwiss.*, **4**, 115.

—— (1952) 'Das heutige Stand unseres Wissens vom Humus and neue Wege zur Lösung des Rohhumus-Problems im Walde'. *Schriftenreihe d. Forstl. Fak. Univ. Göttingen.* **4**.

—— (1957) 'Skogsmarksgödsling i Tyskland'. *Växtnäringsnytt*, **13**, **4**, 2.

WITTING, M. (1947) 'Katjenspestämningand myrvatten'. *Bot. Not.*

—— (1948) 'Preliminärt meddelande om fortsatta katjonsbestämningar i myrvatten sommaren 1948'. *Sv. bot. Tidskr.*, **42**, 116.

—— (1949) 'Kalciumhalten i några nordsvenska myrvatten'. Ibid., **43**, 715.

WOLF, PH. (1956) *Utdikad civilisation*. Gleerup, Lund.

WOLF, F. A. and WOLF, F. T. (1947) *The Fungi*. New York.

WOLLNY, E. (1883) 'Untersuchungen über den Einfluss der Pflanzendecke und der Beschattung auf die physikalischen Eigenschaften des Bodens'. *Forsch. auf d. Geb. d. Agr. Kultur-Chemie*, **6**, 197.

WRIGHT, K. E. (1943) 'Internal precipitation of phosphorous in relation to aluminium toxicity'. *Pl. Physiol.*, **18**, 708.

WRIGHT, J. M. and GROVE, J. F. (1957) 'The production of antibiotics in soil. V. Breakdown of antibiotics in soil'. *Ann. appl. Biol.*, **45**, 36.

ZOHARY, M. and ORSHANSKY, G. (1949) 'Structure and ecology of the vegetation in the Dead Sea region of Palestine'. *Palestine J. Bot.*, **4**, 177.

Index

The late M. G. Stålfelt was one of Europe's best known and respected plant ecologists. He was Professor of Botany in the University of Stockholm from 1941 to 1959 and played a large part in laying the foundations of plant ecology as it is known today. Many years before the current debate about dangers to the environment, he was acutely conscious of the delicate balance of nature and how easily man can upset it.

PLANT ECOLOGY is Stålfelt's *magnum opus* and though originally written for undergraduates, its wealth of references and individuality of approach make it important reading for all plant ecologists. This long overdue translation brings to the attention of an English-speaking readership a large number of classic papers by Swedish and German workers.

The central theme of the book is the balance between the production of vegetation and the crops taken from it by mankind. Starting with a general discussion of the plant environment, it leads through descriptions of the various environmental factors, soil, water, litter etc. to discussions of their interrelationships and interactions with vegetation. Finally, in a discourse on the influence of man on all these things, Stålfelt puts forward a strong plea for urgent conservation measures not only for the plant life of this planet but also for the soil on which it depends.

The translators, Drs. Margaret and Paul Jarvis of the Botany Department, Aberdeen University came into contact with the late Professor Stålfelt during their four years of research work in Sweden. In this translation, whilst preserving the spirit of Stålfelt's work, they have taken the opportunity to update many of the references.